5th International Conference on Competitive Materials and Technological Processes (IC-CMTP5)

IOP Conference Series: Materials Science and Engineering
Volume 613

Miskolc-Lillafured, Hungary
8 - 12 October 2018

ISBN: 978-1-7138-0899-2
ISSN: 1757-8981

Printed from e-media with permission by:

Curran Associates, Inc.
57 Morehouse Lane
Red Hook, NY 12571

proceedings.com

Some format issues inherent in the e-media version may also appear in this print version.

Institute of Physics
Dirac House, Temple Back
Bristol BS1 6BE UK

Phone: 44 1 17 929 7481
Fax: 44 1 17 920 0979

techtracking@iop.org

Additional copies of this publication are available from:

Curran Associates, Inc.
57 Morehouse Lane
Red Hook, NY 12571 USA
Phone: 845-758-0400
Fax: 845-758-2633
Email: curran@proceedings.com
Web: www.proceedings.com

TABLE OF CONTENTS

SESSION 14 - MATERIALS AND NANOMATERIALS FOR MEMBRANES AND CATALYSTS

SESSION 15 - MECHANICAL PROPERTIES AND PROCESSING TECHNOLOGY OF ADVANCED MATERIALS

SESSION 16 - MINERALS FOR ENVIRONMENTAL AND MEDICAL APPLICATIONS

IOP Conf. Series: Materials Science and Engineering **613** (2019) 011001 doi:10.1088/1757-899X/613/1/011001

ic-cmtp5

ic-cmtp5

5TH INTERNATIONAL CONFERENCE ON COMPETITIVE MATERIALS AND TECHNOLOGY PROCESSES

October 8-12, 2018/ Miskolc-Lillafüred, Hungary

Organized by the International Scientific Advisory Board (**ISAB**),

International Organizing Committee (**IOC**) and

Session Chairs

ic-cmtp5

IOP Conf. Series: Materials Science and Engineering **613** (2019) 011001 doi:10.1088/1757-899X/613/1/011001

ic-cmtp5

Abstract

The competitiveness is one of the most important components of our life and it plays key role in efficiency both of organizations and societies. The more scientific supported and prepared organizations develop more competitive materials with better physical, mechanical, chemical and biological properties and the leading companies are applying more competitive equipment and technology processes.

The aims the 5th **International Conference on Competitive Materials and Technology Processes** (**ic-cmtp5**) and the Symposium **is-ism2** are the followings:

- Promote new methods and results of scientific research in the fields of material, chemical, physical, biological, environmental, healts as well as processing and technology sciences;
- Change information between the theoretical and applied sciences and promote their technical and technological implantations.
- Improve the communication between the scientist of different nations, countries and continents.

Among the major fields of interest are traditional and advanced materials with increased physical, chemical, biological, medical, thermal and mechanical properties; including their crystalline and nano-structures, phase transformations as well as methods of their technological processes, tests and measurements. Multidisciplinary applications of material science and technological problems encountered in sectors like metallic and non-metallic materials and composites, including competitive metal-alloys, ceramics, polymers, glasses, thin films, etc. Mashinery, IT tools, aerospace, automotive and marine industry, electronics, energy, construction materials, medicine, biosciences and environmental sciences are of particular interest.

In accordance to the program of the conference **ic-cmtp5**, and its symposiums **is-icbm3** and **is-icm3** we have received more than **450** inquires and registrations frome **40** countries of **5** continents. Finally, from them more than **150** abstracts were accepted for presentation, including the **9** PLENARY and **14** KEYNOTE talks. Scientists and researchers have arrived to Miskolc-Lillafüred (Hungary) from **36** countries of Asia, Europe, Africa, North and South America and Australia. The largest delegations have arrived from **Russian Federation, Japan, China, Hungary** and **Czech Republic**.

Prof. Dr. László A. Gömze
chair, ic-cmtp5

IOP Conf. Series: Materials Science and Engineering **613** (2019) 011001 doi:10.1088/1757-899X/613/1/011001

THE INTERNATIONAL SCIENTIFIC ADVISORY BOARD (ISAB)

Prof. M. E. CONTRERAS-GARCIA, Universidad Michoacana de SNH Michoacán, ***(México)***

Prof. Dr. Emad M.M. EWAIS, Central Metallurgical Institute, ***(Egypt)***

Prof. Dr. Rainer GADOW, University of Stuttgart, ***(Germany)***

Prof. Dr. László A. GÖMZE, University of Miskolc, ***(Hungary)***

Prof. Valerii KULICHIKIN, Russian Academy of Sciences, ***(Russia)***

Prof. Dr. Sergei N. KULKOV, Tomsk State University ***(Russia)***

Prof. Dr. Shigeru NAGASAWA, Nagaoka University of Technology, ***(Japan)***

Prof. Dr. Tomasz SADOWSKI, Lublin University of Technology, ***(Poland)***

Prof. Dr. Tohru SEKINO, The University of Osaka ***(Japan)***

Prof. Dr.Yasunori TAGA, Chubu University, ***(Japan)***

Prof. Dr. Jean Claude TEDENAC, University Montpellier 2, ***(France)***

Prof. Sabu THOMAS, Mahatna Gandhi University Kottayam, ***(India)***

THE INTERNATIONAL ORGANIZING COMMITTEE (IOC)

Prof. Dr. Emad M.M. EWAIS, Central Metallurgical Institute, ***(Egypt)***

Dr. Ludmila N. GÖMZE, Igrex Ltd. ***(Hungary)***

Prof. Dr. Olga KOTOVA, RAS Institute of Geology ***(Russia)***

Dr. Tünde KOVÁCS, University Óbuda, ***(Hungary)***

Dr. Takeshi TOYAMA, Nihon University, ***(Japan)***

Prof. Dr. Sergey N. KULKOV, Tomsk State University ***(Russia)***

Miss Emese KUROVICS, University of Miskolc, ***(Hungary)***

Prof. Limin WU, Fudan University, ***(China)***

Dr. Jean-Marie DREZET, Ecole Polytechnique Federale Lausanne ***(Switzerland)***

Prof. Erwan RAUWEL, Tallin University of Technology Tartu College Tartu, ***(Estonia)***

Prof. Dr. Tomasz SADOWSKI, Lublin University of Technology ***(Poland)***

*Prof. Dr.Yasunori TAGA, Chubu University, **(Japan)***

*Md. Prof. Alexandr L. URAKOV, Izhevsk State Medical Academy **(Russia)***

*Prof. Dr. Tohru SEKINO, The University of Osaka **(Japan)***

*Dr. Daman K. PANESAR, The University of Toronto, **(Canada)***

*Prof. Michel PEREZ, Universite de Lyon - INSA de Lyon, **(France)***

*Dr. Uwe BECK, Federal Institute for Material Research and Testing, **(Germany)***

THE SESSIONS AND SYMPOSIUM

Session 1. Advanced Materials for Bio- and Medical Applications

Session 2. Advanced Materials and Composites for Extreme Applications

Session 3. Advanced Nanomaterials and Competitive Nanocomposites

Session 4. Advanced Optical Materials –Properties and Applications

Session 5. Advanced Processing and Characterization Methods for Nanocomposites

Session 6. Biomaterials Derived Ceramics and Advanced Nanocomposites

Session 7. Competitive Ceramics and Nanoceramics

Session 8. Competitive Light-Weight Metals and Nanostructured Alloys

Session 9. Competitive Nanomaterials for Environment and Health

Session 10. Competitive Polymers and Nanopolymers

Session 11. Glasses, Coatings and Related Competitive Materials

Session 12. Hetero-Modulus and Hybrid Materials– Properties and Applications

Session 13. Interface as a Functional Material – Behavior, Function and Application

Session 14. Materials and Nanomateriasl for Membranes and Catalysts

Session 15. Mechanical Properties and Processing Technology of Advanced Materials

Session 16. Minerals for Environmental and Medical Application

Session 17. Nanomaterials Applied to Biomedical Applications and Nanomedicine

Session 18. Nanomaterials for Electronic and Electrotechnical Applications

Session 19. Novel Synthesis and Processing Technology

IOP Conf. Series: Materials Science and Engineering **613** (2019) 011001 doi:10.1088/1757-899X/613/1/011001

ic-cmtp5

Session 20. Phase Diagram as a Tool of Materials Science

Session 21. Polymer Derived Ceramics and Advanced Nanocomposites

Session 22. Processing, Properties and Application of Ceramics

Session 23. Testing and Characterization of Materials – Methods, Equipment and Errors

Symposium is-ism2. The 2nd International Symposium on Intelligent and Smart Materials

ACKNOWLEDGEMENT

In the name of ic-cmtp5 Conference Board I would like acknowledge and say many thanks to every scientists and colleagues who have supported us in our organizing work. Especially many thanks for their support to societies, scientific journals and organizations like:

cfi – ceramic forum international (Germany)

ECA – European Crystallographic Association

építőanyag – Journal of Silicate Based Composite Materials (Hungary)

GFC – Groupe Francais de la Céramique (France)

ICG – International Commission of Glass

Igrex Ltd. (Hungary)

ME – Miskolci Egyetem / University of Miskolc (Hungary)

refractory WORLDFORUM – Scientific Journal (Germany)

RAS – The Russian Academy of Sciences

SB-RAS Institute of Strength Physics and Materials Science (Russia)

SZTE – Scientific Society of Silicate Materials (Hungary)

TSU – Tomsk State University (Russia)

Many thanks to colleagues of **ISAB**, **IOC** and colleagues who supported and helped in organization work and in successful transaction of the 5th International Conference on Competitive Materials and Technology Processes (ic-cmtp5).

IOP Conf. Series: Materials Science and Engineering **613** (2019) 011001 doi:10.1088/1757-899X/613/1/011001

ic-cmtp5

I would like to say many thank personally to **Prof. Dr. Tohru SEKINO** (The University of Osaka), **Prof. Dr. Sergey N. KULKOV** (Tomsk State University), **Prof. Dr. Tomasz SADOWSKI** (Lublin University of Technology), **Prof. Dr. Olga KOTOVA** (RAS Institute of Geology), **Prof. Mailybi ALDABERGENOV (**Al-Faraby Kazakh National University), **Dr. Tomáš KOVÁŘÍK** (University of West Bohemia)and to **Dr. Takeshi TOYAMA** (Nihon University) for their excellent organization works in Japan, Russia, Asia and in European Union.

Prof. Dr. László A. Gömze
chair, ic-cmtp5

IOP Conf. Series: Materials Science and Engineering **613** (2019) 011002 doi:10.1088/1757-899X/613/1/011002

Peer review statement

All papers published in this volume of *IOP Conference Series: Materials Science and Engineering* have been peer reviewed through processes administered by the proceedings Editors. Reviews were conducted by expert referees to the professional and scientific standards expected of a proceedings journal published by IOP Publishing.

IOP Conf. Series: Materials Science and Engineering **613** (2019) 012001 doi:10.1088/1757-899X/613/1/012001

New adsorbent materials on the base of minerals and industrial waste

Olga Kotova

Laboratory of Mineral Raw Technology, Institute of Geology Komi SC UB RAS

E-mail: kotova@geo.komisc.ru

Abstract. At present, when environmental problems become more pronounced, we need to find special solutions. One of the ways is to use cheap adsorbents that do not need to be restored. An example of developing a class of adsorbent materials is technology for synthesizing nanotubes from refractory ilmenite-leucoxene ore. Another example of the development of a class of zeolite as advanced materials is the technology of their synthesis from ashes of thermal power plants. Synthesis conditions create necessary parameters of adsorption properties to remove heavy metals and radionuclides. "Clean" technologies are formed to replace natural and commercial zeolite. Production of such materials at the national level can be an important source of profit, can create new jobs and increase competitiveness of the industry.

1. Introduction

Sorption and ion exchange technologies are widely used to solve environmental problems (wastes from hazardous hydrometallurgy production, in environmental protection technologies, etc.). The effectiveness of sorbents depends on many factors; the most important are their selectivity and application conditions of sorbents as a consequence of their material composition and structure [1, 2].

The use of sorbents includes various operations from prefiltration with separation of mechanical impurities or oils to supercleaning. Both mineral and synthetic sorbents are used for this purpose. As a rule, mineral sorbents have lower sorption properties, but they are much cheaper than synthetic materials. When using synthetic sorbents, it is possible to obtain more stable results for wastes water treatment, and to increase their service life due to the possibility of their regeneration.

One of the promising directions for creation of sorption materials is the technology of using wastes from the extraction and processing of ores and rocks. The importance of development of this direction lies in increasing stability of the regional development (reducing volume of wastes disposal, additional resources for sorption materials, new jobs for industry).

Production of such materials at the national level can be an important source of profit, can create new jobs and increase competitiveness of the industry. All this stipulates the necessity, firstly, to obtain a deeper knowledge about minerals as carriers of sorption properties, about their actual crystal structure; secondly, to understand the mechanisms of changes of the parameters of the minerals under the influence of natural and anthropogenic impacts; thirdly, to obtain new products competitive in the world market [2–6].

The aim of this work to obtain new knowledge in the field of phase transformations and aggregation of ultrafine mineral substances to create the adsorption materials with desired properties based on natural (ilmenite-leucoxene ore, Pizhemskoe deposit, Russia) and technogenic (fly ash, Pechora coal basin, Russia) materials.

2. Methods and approaches

1. Technology of nanotube synthesis (based on TiO_2) based on the methods and approaches of hydrothermal synthesis. They are adapted for non-magnetic fraction of gravity concentrate of titanium ore from Pizhemskoe deposit [5, 6].
2. Technology of zeolite synthesis (hydrothermal synthesis) from coal fly ash (Pechora coal basin) was described in detail [7]. In short, using a magnetic separator we removed ferriferous phases, which do not participate in the synthesis of zeolites. The suspension (after mixed thoroughly with the solution of sodium hydroxide) was placed in an autoclave.
3. Information on the morphology of raw and processing products was obtained using a scanning electron microscope (TESCAN VEGA 3 LMH) equipped with Oxford Instruments X-Max energy dispersive device. Chemical composition was determined using XRF method (Shimadzu XRF-1800). The mineral composition was determined by XRD (Shimadzu XRD 6000, CuK α radiation).
4. The surface roughness of the samples, shape and dimensions of the electric and magnetic micro- and nano-regions and their topology were determined by ARIS-3500 atomic-force microscope (AFM), magnetic power microscope (MPM) - upgraded AFM

3. Mineralogy – a basis for interactive controlof ore and adsorbent materials

Ores from Russia are predominantly complex, in which useful minerals can be both main and accompanying. The complexity of their mineralogical study is determined by the natural features of the ores:

- a complex textural-structural pattern, a significant amount of fine, metacolloidal units formed by mineral individuals and aggregates of micro- and nanometer-size,
- a polymineral composition, conditioned by simultaneous presence of minerals of different parageneses,
- a variable chemical composition of both the ore and nonmetal minerals caused by wide isomorphous substitutions of atoms of chemical elements in the crystalline structure of minerals,
- phase heterogeneity of mineral grains, conditioned by various causes (decomposition of solid solutions, zonal growth, partial recrystallization, solid-phase secondary transformations, etc.)
- development of metamictization processes of ore minerals containing radioactive elements, with the formation of metamictic (loosing crystal structure) and partially metamictic (with a damaged crystal structure) forms.
- **Initial ore.** Determination of the mineral composition of ore, its texture pattern (mutual orientation of mineral aggregates), structural characteristics (dimensions of useful minerals, nature of their inter-growths with other mineral phases), determination of the nature of distribution of useful elements over ore minerals), determination of properties of minerals) [8].

4. Results and Discussion

As an example we consider ilmenite-leucoxene ore, the Pizhemskoe deposit. Upper Proterozoic and Devonian deposits, which are almost everywhere overlain by Quaternary sediments of various thickness (from 0.2 to 40 m), take part in the geological structure of the Pizhemskoe deposit. A titaniferous productive horizon is determined in the lower part of Devonian section. The titaniferous stratum is considered to be an ancient paleoplacer formed by erosion and redeposition of the weathering crust on underlying Riphean schists [4]. The average content of titanium minerals reaches 200 kg/t. The content of leucoxene in the sandstones reaches 15 %. The titaniferous stratum is overlain by Middle Devonian alluvial quartz sandstones.

In [4] the mineralogical and technological properties of ilmenite-leucoxene ore are considered. According to optical-mineralogical analysis, the heavy fraction is formed by ilmenite, rutile, anatase, zircon, epidote, magnetite, chromespinelide, tourmaline, pyrite, monazite.

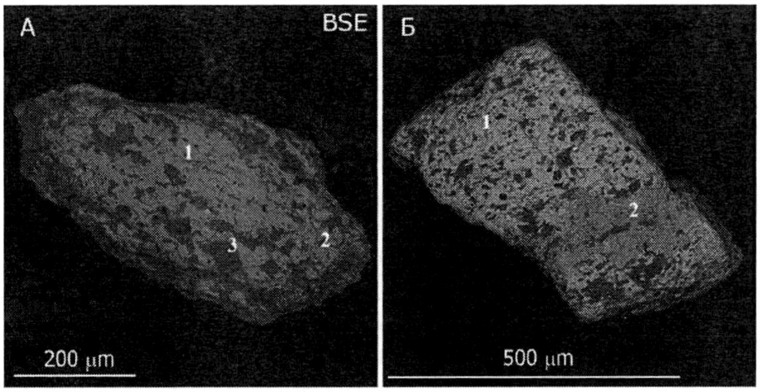

Figure 1. Polymineral aggregate of pseudorutile (1), rutile (2), quartz (3)

Ilmenite and leucoxene, diagnosed in the ilmenite-leucoxene ore, are represented by polymineral aggregates of pseudorutile, rutile, quartz and aluminosilicates. The distribution of titanium and iron in the ilmenite is uniform except of local substitution of ilmenite by rutile. In leucoxene, iron is concentrated within the aggregates and probably associated with relics of pseudorutile. Grains of zircon, xenotime, native gold, copper up to 15 μm in size are found on the surface and in the volume of the aggregates. Quartz and aluminosilicates are represented by mechanical inclusions less than 10 μm in size and constitute about 35 % of the aggregate volume.

The established morphostructural features should be taken into account when choosing methods for processing and separating minerals. The physical parameters of the constituent minerals of the ilmenite-leucoxene ore (density, magnetic susceptibility, etc.) indicate promising physical methods of enrichment. Ore components of mineral sands have a high density (more than 2.85 g/cm^3) and can be quite easily separated from lighter components, such as quartz and aluminosilicates, provided that the intergrowths are well opened. However the established boundary particle size (10 μm) requires, firstly, considerable energy consumption. Secondly, adhesion forces increase and impede separation of particles. The losses of titanium dioxide, associated with ilmenite and leucoxene, are estimated at 20 % of the initial content, the content of titanium dioxide in the collective concentrate does not exceed 62 %. At the same time further processing requires concentrates with a higher content of TiO_2. Thusthe chloride technology uses concentrates from 85 % of titanium dioxide. Therefore the use of combined enrichment methods is increasingly being proposed.

A higher degree of desilification is achieved by applying chemical methods (autoclave leaching and fluoration). The main titanium-containing phase is rutile, which, apparently, does not participate in reactions, retaining its crystalline form. It was established that the degree of desilification during fluoridation is higher (90 %), the fluoration products of the heavy fraction of titaniferous sandstones are more crystalline [7].

The applied technologies do not satisfy modern processing plantsand provide expensive final products. Work is underway to change properties of minerals and (or) to find new methods for their utilization. Alongside with ilmenite, leucoxene and other polymorphs of TiO_2 useful products nowinclude zircon, quartz and others. Active progress is to obtain high-tech materials based on inexpensive raw (leucoxene) including titanium dioxide nanotubes [7, 9].

We obtained structures of nanodisperse structures of titanium minerals using anatase as an example, for them we revealed features of physical-chemical properties resulted from structural reorganization at the nanolevel (surface topography, sorption, electronic properties, including width of forbidden zone, etc.) [9].

Description of practical ideas: The high cost of TiO_2 nanoparticles and their relevance in industry determines the importance of the hydrothermal synthesis method of nanotubes of titanium dioxide from cheap sand (titanium ore, Pizhemskoe deposit).

Leucoxene is a rutile microcrystalline matrix saturated with minute inclusions of quartz, the XRD analysis indicated a high crystallinity of the phases. The synthesized materials are a mixture of two phases: quartz and sodium titanate $Na_2Ti_6O_{13}$, and presented structure ofnanotubes of titanium dioxide. Synthesized TiO_2 nanotubes possess a highly developed surface, which makes the material a good adsorbent and photocatalyst for wastewater purification with organic compounds [6, 9]. Also the changed band gap allows considering the obtained material as raw for production of optical fibers [9]. Other important application of synthetized materials by hydrothermal methods is dialysis of blood. Our laboratory technologies imitate natural “production” - halloysite, found in nature as a hydrated mineral, consisting of rolled in coils aluminum sheets. Halloysite nanotubes are nontoxic and have fast adsorption rates. The nanotubes have a lot of applications: environmentprotection, anticancer therapy, cosmetics, nanotemplate for biocatalyst, tubular container for the corrosion inhibitor, additive in paints, coating material, etc.

The method of zeolite synthesis. When the state of the mineral system changes (P-T-V), phase transformations occur (substance transition from one phase to another). The substance changes its composition and properties. Researches in this area will be useful for the development of theory and experiment of obtaining advanced materials with certain properties. As an example of phase transformations we consider the technology for the synthesis of zeolites from fly ash - wastes of thermal power plants of Pechora coal basin. Synthesis conditions create necessary parameters of adsorption properties for removing heavy metals and radionuclides. We have studied morphology and chemical composition of initial, intermediate and final products.

X-ray diffraction (Fig. 2) showed quartz, mullite, magnetite and hematite in the coal fly ash. The broad "hump" (area of increased background) on the diffractogram in the area 15–35 °2θ indicates the presence of amorphous phase (probably silicate or aluminosilicate glass). The main components of the chemical composition are SiO_2 (57.78 %) and Al_2O_3 (18.25 %), iron oxide content is about 9.0 %, oxides of other elements - 7.42 %, loss on ignition - 7.90 % .

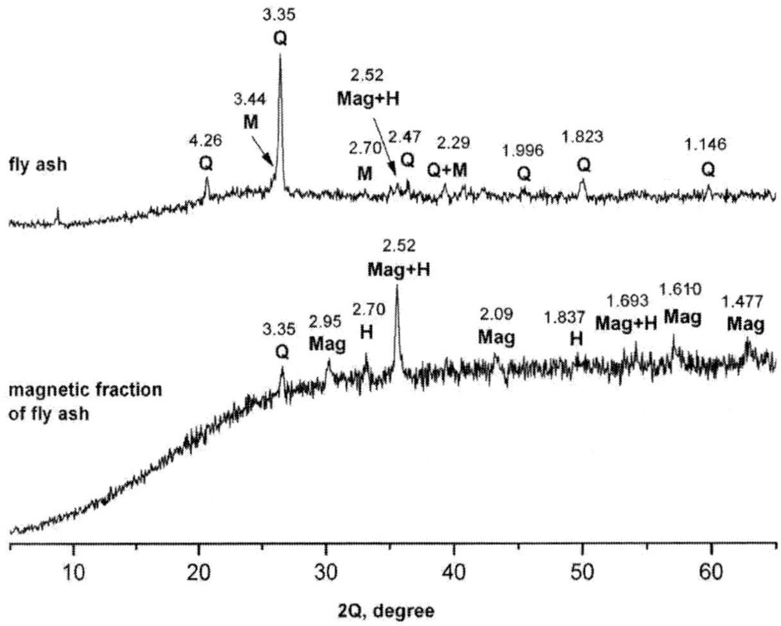

Figure 2. Diffraction patterns of coal fly ash and magnetic fraction of coal fly ash: Q – quartz, Mag – magnetite; M – mullite, H – hematite

The fly ash morphology and composition. The fly ash is represented under the electron microscope by globules (Fig. 3). The globules are divided by the chemical composition to aluminosilicate and iron containing. The aluminosilicate globules composition is predominated by SiO_2 (from 41.82 to 61.27 %) and Al_2O_3 (from 17.03 to 22.8 %); FeO and Fe_2O_3 (up to 8.31 %), MgO (up to 4.83 %), K_2O (up to 3.05 %), TiO_2 (up to 1.04 %) and Na_2O (up to 0.93 %) are also present. Globule size varies from the first to about hundred micrometers; on the surface bubbles and elongated structures are observed.

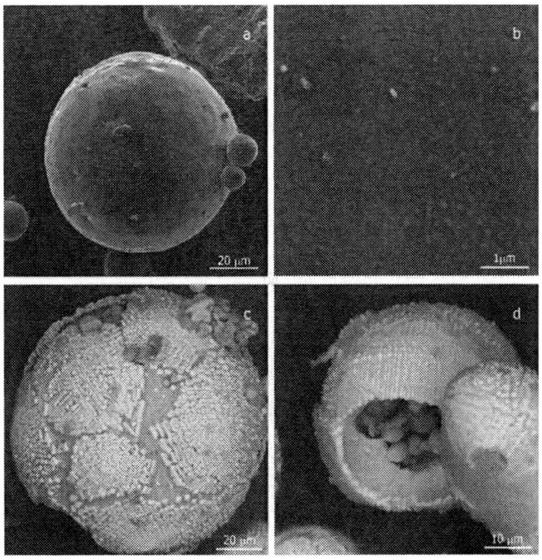

Figure 3. SEM-images of aluminum silicate (a, b) and iron-containing globules (c, d)

On the surface of iron containing globules both flat areas and skeletal forms are observed, which are significantly different from each other by their chemical composition. The skeletal forms have a high content of iron oxides (68.14-74.66 %) and low SiO_2 (1.06-6.22 %), Al_2O_3 (1.33-4.17 %) and CaO (0.48-3.59 %) contents. On the flat areas iron oxides content is greatly reduced (19.29-31.81 %), SiO_2 and Al_2O_3 content increases (27.12-37.86 and 2.06-6.22 %, respectively); CaO presents in amounts of 10.45-25.3 %. Globule size ranges from several to tens micrometers. Globules, which contain smaller globules within, are often observed.

Hydrothermal synthesis. Two sets of experiments were carried out. In the first set the effect of temperature of hydrothermal reaction on zeolite synthesis was studied. The second set of experiments studied the influence of reaction time and concentration of alkali on synthesis process. The synthesis results in powders consisting of the mixture of zeolite and unreacted residue in different proportions, in which output was 70-80 % of the weight of the initial fly ash.

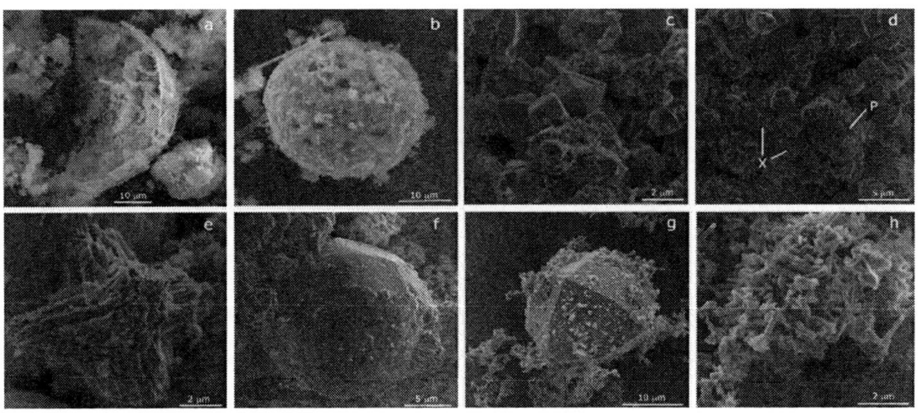

Figure 4. Products of hydrothermal treatment at different temperatures 80 (a, b), 95 (c, d), 140 (e, f) and 180 °C (g, h): a, b - globules exposed to alkaline solution; c - accumulation of zeolite X crystals, d - crystals of zeolites X and P, e – a single zeolite P crystal, f - crystals of analcime and zeolite P, g - a single analcime crystal, h - accumulation of cancrinite crystals.

Effect of reaction temperature on the synthesis of zeolites. At reaction temperatures from 95 to 180 °C several species of zeolites were obtained, different by their morphology and porosity (Fig. 4). It was found that increasing temperature leads to reducing pore size of the zeolites. Thus, at 95 °C zeolites X (a synthetic analogue of natural faujasite, pore size 0.74 nm) formed with an admixture of zeolite P (a synthetic analogue of natural gismondine, pore size 0.31×0.45 and 0.28×0.48 nm). At 140 °C zeolites P formed with an admixture of analcime (pore size 0.26 and 0.42×0.16 nm [10]). At 180 °C, analcime crystallizes with an admixture of cancrinite (a non-zeolitic phase). According to the classification of zeolites by pore size [11, 12], analcime and zeolite P refer to narrow-porous, zeolite X - to wide-porous species.

Effect of reaction time and alkali concentration on zeolite type. The wide porous zeolites X are formed by 4 hours of reaction at a high concentration of alkaline solution (4.5 mol/dm^3). Longer reaction leads to the disappearance of the metastable phases of zeolite X and the occurrence of more thermodynamically stable - zeolite P and then analcime. Zeolite P is crystallized under a wide range of reaction conditions. At the same time, the fields of crystallization of analcime and zeolite P are significantly overlapped, that is, at the same conditions of the hydrothermal reaction, the mixture of zeolites in various quantitative relations is formed. Higher concentrations of alkali results in the increase of the content of narrow porous phases (analcime) compared to zeolite P, and contributes to the formation of non-zeolitic phase - hydrosodalite.

Sorption properties. Synthesized zeolites are characterized by a high sorption activity with respect to long-living natural radionuclides and possess low desorption of uranium and radium. The extraction degree of radium from solutions is 98, thorium - more than 89, uranium – more than 80 %. Thorium is retained in the structure of synthesized sorbents the least strongly. The sorption capacity for barium has values from 113.3 to 157.5 mg/g (1.65-2.29 meq/g), for strontium – 85.9 mg/g (1.94 meq/g), for ammonium 32.2 mg/g (1.78 meq/g). By selectivity of sorption, the above cations can be arranged in a line: Ba^{2+}> Sr^{2+}> NH^{4+} [13].

5. Conclusion

1. We underline an important role of integration of mineralogical and technological research of ores and treatment product, identification of mineralogical features for forecasting of behavior in technological processes.
2. We show that the mechanism of phase transformations in minerals (ores) (example of hydrothermal synthesis of TiO_2 nanotube fromilmenite-leucoxene ore, Pizhemskoe deposit, and example of zeolite synthesis (hydrothermal synthesis) from fly ash (coal of Pechora basin) can form new minerals (new phases), which physical and chemical parameters will satisfy requirements of new adsorbing materials with desired propertiesand reduce environmental risks and expand the list of market products of coal mines - sorbents of various toxicants.

Acknowledgement

The authors express their gratitude to the Center for Collective Use "Geonauka" for their assistance in analytical work. The work was carried out with the partial financial support of UB RAS Programs (project 18-5-5-44), the State task "Scientific basis for effective subsoil use, development and exploration of mineral resource base, development and implementation of innovating technologies and economic zoning of the Timan-Nothern Ural region" No. AAAA- A17-117121270037-4.

Reference

[1] O B Kotova, L N Moskalchuk, D A Shushkov, T G Leontyeva, A A Baklay 2017 *Vestnik of the Institute of Geology, Komi SC UB RAS* **4** 29 https://doi.org/10.19110/2221-1381-2017-4-29-36 (in Russian)
[2] O B Kotova, M Harja, I Cretescu, F Noli, Y Pelovski, D A Shushkov 2017 *Vestnik of the Institute of Geology, Komi SC UB RAS* **5** 49 https://doi.org/10.19110/2221-1381-2017-5-49-53
[3] C N Duduman, J M Gómez de Salazar y Caso de Los Cobos, M Harja, M. Barrena Pérez, C Gómez de Castro, D Lutic, O Kotova, I Cretescu 2018 *Environmental Engineering and Management Journal* **17** (4) 2813
[4] A V Ponaryadov 2017 *Vestnik of the Institute of Geology Komi SC UB RAS* **1** 29 https://doi.org/10.19110/2221-1381-2017-1-29-36 (in Russian)
[5] L A Popescu, R C Ciocinta, D Lutic, G Ciobanu, I Cretescu, O Kotova, M Harja 2018 *Conferința Facultății de Chimie, IasiChem* 76

[6] O Kotova, M Harja, M A T M Broekmans, C Gomez de Castro 2018 *Proceedings of the 10 Edition of Euroinvent European exhibition of creativity and innovation* 550
[7] O B Kotova, A V Ponaryadov, L A Gomze 2016 *Vestnik of the Institute of Geology Komi SC UB RAS* **1** 34 https://doi.org/10.19110/2221-1381-2016-1-34-36
[8] E G Ozhogina, O B Kotova 2018 *XXIX International Mineral Processing Congress, Moscow, Russia* 33
[9] O B Kotova, M Harja, L N Kotov, A V Ponaryadov 2018 *Vestnik of Institute of Geology of Komi SC of UB of RAS* **4** 34. https://doi.org/10.19110/2221-1381-2018-4-34-39
[10] O B Kotova, I N Shabalin, D A Shushkov, L S Kocheva 2016 *Advances in Applied Ceramics: Structural, Functional and Bioceramics* **115**(3) 152 https://dx.doi.org/10.1179/1743676115Y.0000000063
[11] Database of zeolite structures. www.iza-structure.org/database
[12] A Chica 2013 *ISRN Chemical engineering* Art. ID 907425. 19
[13] D A Shushkov, I I Shuktomova, N G Rachkova, M Harja 2018 *Vestnik of Institute of Geology of Komi SC of UB of RAS* **3** 32 https://doi.org/10.19110/2221-1381-2018-3-32-37

IOP Conf. Series: Materials Science and Engineering **613** (2019) 012002 doi:10.1088/1757-899X/613/1/012002

Smart Materials Based on a High and Low Temperatures SME-Alloys

S N Kulkov

Institute of Strength Physics and Materials Science SB RAS, Tomsk, Russia and Tomsk State University 634050, Tomsk, Russia

E-mail: kulkov@ms.tsc.ru

Abstract. The analysis of the nature of low plasticity of composites with strengthening of hard phases are shown that plastic shear constraint in thin interlayer of the binding phase does not provide the small scale of the structural level of deformation. As a result, a material undergoes a brittle fracture. Therefore, one should find another mechanism for relaxation of internal stress concentrations in composites during loading.

The aim of a present paper is to illustrate the concept of how use a shape memory alloys with phase transformation for metal matrix composites. It has been studied shape memory based alloys and deformation and fracture of MMC with phase transformations in matrix.

It have been shown that if plastic deformation is considered as local structural transformation and instability in the material is provided from the very beginning, this will result in a general increase of plasticity due to a decrease in the scale of the structural level of plastic deformation.

Deformation tests of composites near the phase transition temperature show that there are different transformations induced by a highly non-uniform stress state of the binding phase. Under loading the “quasi-amorphous state" is formed in the binding phase after formation of a microcrystalline, highly misoriented structure with the characteristic size of crystallites less than 10 nm. This structure has high plasticity and strong hardening. It governs an efficient transfer of external load to solid particles, inducing dislocation glide even in typically brittle titanium carbide particles during non-uniform loading.

1. Introduction

A considerable interest has been shown in the last decade in developing ceramic-based materials with high mechanical properties [1, 2]. However, the rate of scientific advance requires a non-traditional approach to developing materials of this type [3, 4]. In most cases, material science was developed as a science of describing the properties of existing materials - either present in nature or developed on the basis of the intuition of investigators. Science only described the properties, tried to find a detailed relationship between the macro- and microscopic properties and. in certain cases, managed to predict them. Attempts to design materials have already been made several years ago when investigations were carried out to develop materials with specified properties [5]. However, important results could not be achieved whilst the methodology of material design remained within the limits of conventional considerations.

The aim of a present paper is to illustrate the concept of how use a shape memory alloys with phase transformation for metal matrix composites. It has been studied shape memory based alloys and deformation and fracture of MMC with phase transformations in matrix.

2. Materials and Experimental Procedure

Materials are prepared by standard methods of powder metallurgy – pressing of powders and sintering in vacuum furnace [6, 7]. It was carried out metallography, scanning electron microscopy of section surfaces, transmission electron microscopy and X-ray analysis. Mechanical tests were carried out using Instron-1185 unit with strain rate under loading $5*10^{-3}$ sec^{-1}.

3. Results and Discussion

As indicated by theoretical examination, it may be expected that in a crystal deformed in the vicinity of the point of. a structural phase transition a fragmented structure will appear as a result of formation of rotations of the microregions from the very start of deformation. In fact, if layers of a substance with a structural phase transition are placed between the undeformed carbide particles in a composite of the hard alloy type, deformation of this composite material will be characterized not only by the, uniform compression test state of the interlayer by also by its shear deformation as a result of rotation of the individual carbide grains.

As well-known [8] TiNi alloys near equiatomic concentration has a different states, in Table 1 are presented an initial state and temperatures of phase transformations for different TiNi alloys. On Fig. 1 are shown "stress-strain" curves of studied alloys and one can see that it is possible to obtain a very different its mechanical properties.

Table 1. Phase contents and transition temperatures of TiNi alloys.

Alloy	M_S, K	T_R, K	Initial state (at room temperature)
1	318	≤310	B19′+B2(R)
2	295	302	R+B19′
3	293	≤290	B2(R)+B19′
4	263	283	B2

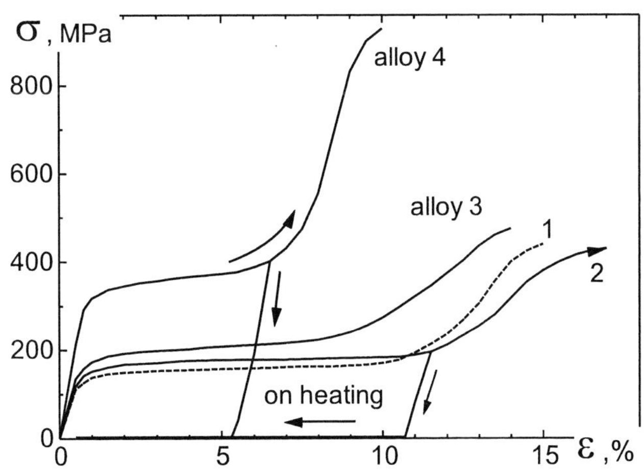

Figure 1. "Stress-strain" dependencies for near equiatomic TiNi alloys

On Figure 2 are shown X-ray pattern for alloy 1 (a) and 4 (b) obtained *in situ* during deformation. As one can see the initial phase content of alloy will stipulate a way of transformations and we can obtain direct B2=>B19 or B2=>R=>B19 phase transform [9].

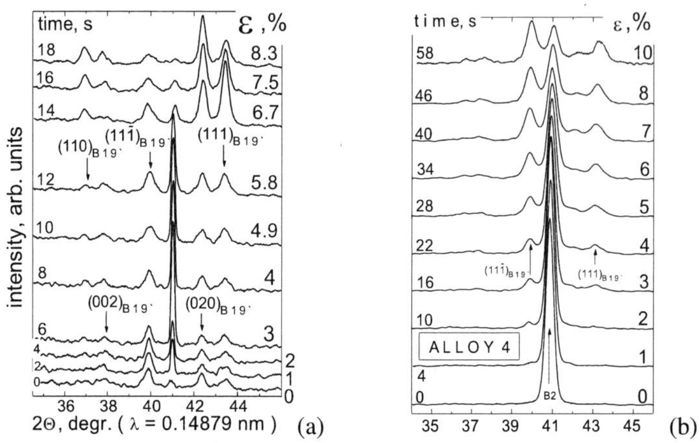

Figure 2. Synchrotron X-ray Pattern (λ=0.14879 nm) for alloy 1 (a) and 4 (b) obtained *in situ* during deformation.

The experiments with composites were carried out with a sintered composite, in which the role of hard undeformed particles was played by titanium carbide particles 5-10 mm in size, situated in an intermetallic compound TiNi characterised by a thermoelastic martensitic transformation. The system was deformed in the vicinity of the critical temperature of the martensitic transformation and at temperatures considerably higher .than this temperature. In cases in which the binding phase was stable, its deformation took place by dislocation slip on main shear planes, mainly (110). The yield limit was inversely proportional to the distance between the carbides and in thin layers of the binder in alloys with a higher content of the hard phase the yield limit was not reached up to failure or was too high to enable sufficient stress relaxation to take place. Beyond yield the limit of these composite cracks appeared at the carbide-carbide and carbide-binder boundary and failure was catastrophically brittle. So, in TiC-TiNi composites we will have different phase states of binder in composite which will stipulate a non-uniform state of TiNi.

The limiting plastic strain of the composites with TiNi in the stable state was low, less than 2%. The limiting strain to failure of alloys with TiNi in the transformation region was 2-3 times higher than that of the alloys with the binder in the stable state, with the strength unchanged.

Deformation of the TiNi system during the loss of shear stability by the lattice is accompanied by large changes of the structural state. Already in the initial undeformed specimens of the- TiC-TiNi composite the structure of binder was highly heterogeneous. The presence of a characteristic contrast in the form of ripple marks on bright field electron microscopic images indicates that the TiNi is in the pre-transition state.

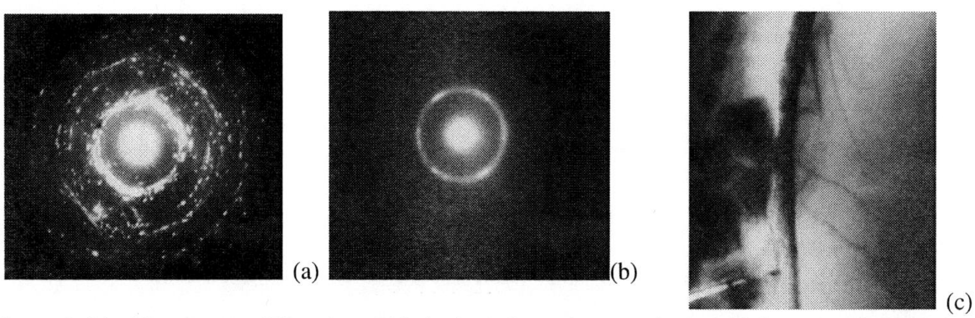

Figure 3. (a) - The electron diffraction of binder in deformed composite in initial stage of deformation. (b) - The quasi-amorphous state of a binder before fracture. (c) - TEM picture of TiC particles after composite deformation. One can see the dislocations.

During loading the examined composites transformations were detected in TiNi. The nature of these transformations greatly differed as a result of a highly heterogeneous stress state of the binder in the composite of this type. The relationships governing the variation of the structure may be described as follows. In loading in the region of elastic behavior of the composite the microstructure of TiNi varied from the dispersed domain Structure to a structure with a banded contrast typical of intermediate shear structures.

Electron diffraction patterns show diffusion scattering and then extra reflections in both commensurate and incommensurate positions with a different incommensurability parameter in different directions of the reciprocal lattice. This indicates that several variants of prematensitic domains with a different domain structure in every domain form. This type of transformation in TiNi is caused by a highly complex stress state formed around solid particles during loading the composite. Under the conditions of steep stress gradients formed in the matrix, the directions of displacement of the atoms in microregions, causing a local loss of stability of the B2-structure, are determined-by the stress conditions existing at the given moment of loading in the given microvolume of the binder [10].

These conditions also determine the orientation of the newly formed martensitic domains. This nature of transformation in titanium nikelide results in a simultaneous reduction of the peak and integral intensity of the line of the B2 phase on x-ray diffraction patterns, but is not accompanied by the growth or appearance of new martensitic peaks. Only widening of the most intensive lines of the monoclinic nickel-titanium to single-diffusion reflection is observed. This is typical of formation of a fine dispersed deformation structure, consisting of misoriented fragments of the B2 phase and martensitic domains. As shown by the transmission electron microscopy results, the degree of misorientation of these fragments and domains exponentially increases with increasing strain. This is also accompanied by rotation of individual grains of the carbides, and the extent of this rotation increases with a reduction of the size of these grains.

Thus, a fine crystalline greatly misoriented structure with a typical crystallite size smaller than 100 A forms during deformation in the binding phase of the TiC-TiNi solid alloy as a result of the B2 => B2+B19' inhomogeneous transition. This structure is characterised by high toughness and hardening capacity and enables an effective transfer of the external load to the hardening agent, causing in this case dislocation slip, even in typically brittle particles of titanium carbides. A subsequent increase of the load results in multiple cracking of plastically deformed solid particles of TiC with the integrity of the material unaffected. In the final analysis, the fracture toughness is high.

It is well-known that high-strength materials are characterised by brittle fracture. There a large number of fracture criteria of the material used to evaluate the strength. Polynomial criteria of strength in stresses, which formally coincide with the Mises-Hill-type plasticity criterion, are used in most cases. Taking this into account, to solve the given problem, we used the model of an elastic ideally plastic solid with the estimate of the onset of plasticity in the first approximation (start of failure) according to Mises criterion.

Under loading, such a matrix undergoes a martensitic-type phase transition which starts in regions of stress raisers and leads to extensive shape changes in these local regions. Shape changes of this type are, in fact, inelastic deformation which causes effective relaxation of high local stresses. On the other hand, the tough behavior of the matrix supports the transfer of load to hard inclusions. Thus, the mesovolume of the material forms a damping structure capable of dissipating the supplied energy and withstanding high loads as a result of a reduction of the level of stress concentration and redistribution of stresses inside the mesovolume. For comparison of mechanical behavior studies were carried out for a material with a damping matrix, capable of phase transformation (Fig.4, curve 2), and with a conventional matrix (curve 1) in which brittle fracture starts after a limiting strain.

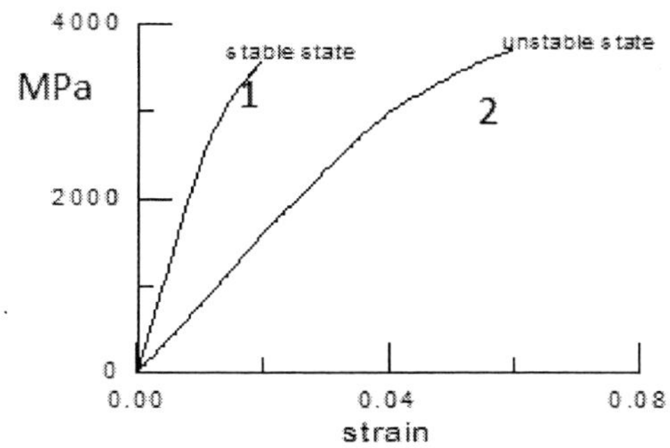

Figure 4. Stress-strain curves for TiC-TiNi composite

Figure 4 shows the pattern of fracture of mesovolumes of the material with different matrices for two consecutive times. Multiaxial loading was applied (combination of tensile, compression and shear loading).

If for the case with a conventional matrix the material may be regarded as completely fractured, then for a structurally unstable matrix, regardless of fractures of inclusions, the material does not lose completely the load-carrying capacity because the matrix is only slightly damaged. The material of the matrix appears to flow around the upper large grain and the extent of fracture of this grain in subsequent stages remains low.

The main work of deformation for the case of the structurally unstable matrix is concentrated in the matrix material. The work of deformation shown in Fig. 3, clearly indicates the characteristic structural elements of the mesovolume and its high value in the central grain is linked with the work of pre-fracture and development of shear deformation in the grain (see Fig.3c).

A very interesting pattern of strain localization (and, consequently, subsequent fracture) due to the heterogeneity of the structure (Fig. 5a) formed in the mesovolume of the ceramic composite during its tensile loading. In the initial stages of deformation in which all stresses are relatively low and elastic, this localization process is not yet completely visible. With increased loading, the strain becomes localized in the direction of the two main tangential stresses.

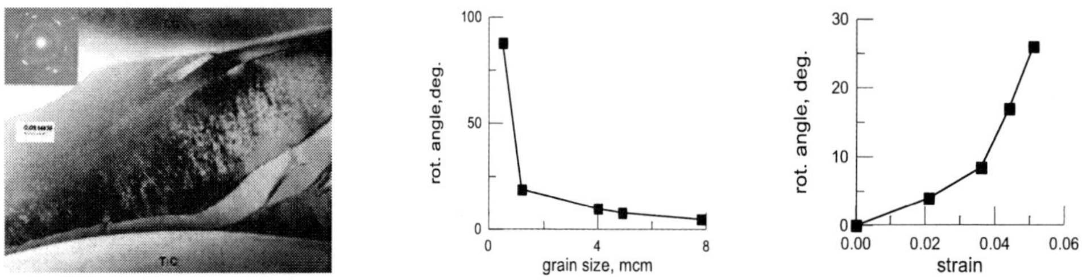

(a) (b) (c)

Figure 5. (a) - TEM image of composite after loading; (b) - The rotation of a carbides vs. its sizes; (c) - The angles between the fragments of binder after deformation

The final stage - design of the material - is the solution of an inverse problem: selection of the optimum physico-mechanical characteristics of the materials of the matrix and solid inclusions and their link with the matrix material, determination of the optimum concentration of hardening particles, *etc.* However, these applied problems can be solved only by taking into account; the specific features of application of the ceramic composite: the materials are to be used for producing dies, cutting tools, or for other purposes.

4. Conclusion

Thus, on the basis of the general concept of the material as a hierarchically organized system of structural levels of strains and fracture we have developed a model of a structural ceramic composite. The representative mesovolume of this material, being a macropaticle (after averaging the parameters over the mesovolume) is in satisfactory agreement with macro-experiments. On the other hand, the special features of deformation and fracture of the mesostructure can be examined in detail by calculations. For example, calculations should show the development of mesovortex structures and strain localization in the meso-volume. The proposed method may be used directly in designing ceramic materials for specific components.

Acknowledgement

This work was carried out with financial support of project 14.584.21.0026 (RFMEFI58417X0026 / 000000007518PRY0002) and these studies were carried out using of equipment of Tomsk regional CKP.

References

[1] Panin V.E. et al. Structural Scale Levels of Plastic Deformation and Fracture of High-Strength Titanium Alloy Welds. Phys. Mesomech., 21, No. 5 (2018) 464.

[2] N.F.Morozov et al. On the Dynamics of the Material with Transformed Microstructure, Phys.

IOP Conf. Series: Materials Science and Engineering **613** (2019) 012002 doi:10.1088/1757-899X/613/1/012002

Mesomech., 21, No. 5 (2018) 379.
[3] Jani J.M., Leary M. A review of shape memory alloy: research, application and opportunities. Mater. Design, 2014, v.56, p.1078-1113.
[4] Otsuka K., Ren X. Physical metallurgy of TiNi based shape memory alloys. Progr Mater. Sci., 2005, v.50, p. 511-678.
[5] Miyasaki S., Otsuka K. and Suzuki Y. Transformation pseudo-elasticity and deformation behaviour in Ti - 50.6 at. % Ni allov. Scripta Met, 15, No-3 (1981) 287.
[6] Kulkov S.N., Poletika T.M., Panin V.E.. Structure, phase composition. and character of fracture of sintered TiC-TiNi composite materials. Powder Metall. Met. Cer.,. 22. No, 7 (1983) 560.
[7]] Kulkov S.N. Formation of micro- and mesostructures in metal matrix composites under mechanical loading, Phys. Mesomech., 9, No. 1-2 (2006),p. 73-80.
[8] Shape memory alloys: modeling and applications / Ed. V.Brailovskii. – Montreal, ETS Publ. 2003. 851p.
[9] Wang X.,Xu B., Yue X. Fracture behaivior of compact tension specimens NiTi alloys. Mater.Sci. Eng. A,2008, v.485(1-2), pp.14-19.
[10] Rahim M. et al. Eggeler impurity levels and fatique lives of pseudoelastic NiTi shape mempry alloys. Acta Mater. 2013, v.61, pp.3667-3686.

Optimisation of plant mediated synthesis of silver nanoparticles by common weed *Plantago major* and their antimicrobial properties

S Küünal[1], M Visnapuu[2,3], O Volubujeva[1], M Soares Rosario[4] and P Rauwel[1,5], E Rauwel[1,5]

[1]Faculty of Engineering, TalTech, Akadeemia tee 15, 12618 Tallinn (Estonia)
[2]Institute of Physics, University of Tartu, W. Ostwaldi Str 1, 50411, Tartu (Estonia)
[3]National Institute of Chemical Physics and Biophysics, 12618, Tallinn (Estonia)
[4]CICECO- Aveiro Institute of Materials, University of Aveiro (Portugal)
[5]Institute of Technology, Estonian Univ. of Life Science, 51074 Tartu (Estonia)

Email: erwan.rauwel@emu.ee

Abstract. Silver nanoparticles synthesized through plant-mediated synthesis have recently gained recognition in the field of biocidal coatings. However, the accurate control of the synthesis using plant extract appears difficult and tends to produce a silver chloride secondary phase. In this study, 2 different methods of synthesis using the same plant extract of *Plantago major* have been investigated to evaluate their influence on the production of AgCl. In both cases the silver nanoparticles have demonstrated efficient biocidal properties against micro-organisms.

1. Introduction

Ag nanoparticles (NPs) have recently spurred a lot of interest due to their biocidal properties and cost-effectiveness [1]. Ag NPs are now investigated for application as antioxidants [2], antimicrobial agents [3, 4] and biocidal coatings [5]. In fact, increased demand of Ag nanoparticles in biomedical fields calls for non-toxic, cost-effective and eco-friendly synthesis procedures [6]. When applications of nanoparticles extend to areas viz., treatment straw bales for green-housing construction [7], economically-friendly approaches need to be implemented, and green routes for nanoparticle production become more important. Plant mediated synthesis methods appear to be the most suitable ones to meet these criteria [8].

Green or biological synthesis of nanoparticles is classified under the bottom-up approach and is a straight-forward and a simple process. Synthesis involves simultaneous reduction-oxidation reactions catalysed by the microbial enzymes or the plant phytochemicals leading to the stabilization of the nanoparticles further supported by the long chain polymers inside the mixture. Another advantage of green synthesis is a direct functionalization of the surface of the nanoparticle by the plant extract through a synergetic effect [9]. Simple procedures indicate also that plant mediated syntheses are reproducible and therefore applicable worldwide whereupon the production of nanomaterials with local flora is realistic.

Our recent investigations have hence focused on the development of methods for the synthesis of Ag NPs. Therefore, an assessment of antimicrobial activity of greenly synthesized nanoparticles was

IOP Conf. Series: Materials Science and Engineering **613** (2019) 012003 doi:10.1088/1757-899X/613/1/012003

conducted. A simple weed Plantago major also known as great plantain was chosen for its well-known properties of wound healing activity, anti-inflammatory, analgesic, antioxidant and weak antibacterial properties. Plantago major is also used as an ingredient in the composition of syrup against bronchitis. Plantago major extract was used as the main ingredient for synthesize of silver nanoparticles using 2 different methods. This paper reports the structural study of the Ag NPs synthesized and the investigation of their biocidal properties.

2. Experimental

2.1. Plant extract preparation

Plantago major plant material was gathered from the humid continental climate in the Baltic region during the growing season. Whole plant without its roots was used. Plants were washed under running tap water. Two different extracts were made. 50g of the plant was dried and 50g was crushed fresh in a mortar. Two liquid media were used for the extraction: ethanol and distilled water. Ethanol extraction was used with the fresh plant material after crushing and distilled water was used with dried plant material (dry mass 6.6g). Ethanol extraction was carried out at room temperature in dark for 24 hours and aqueous extraction under heating at 85°C in dark for 2 hours. Both extracts were then filtered through Whatman No. 1 filter paper

2.2. Synthesis

The synthesis of Ag NPs was carried out under two different conditions already used in reported studies. First sample (Type A) was synthesized in a 50ml Erlenmeyer flask placed in a pressure cooker heated at 85°C for 24hours to accelerate the speed and increase yield of the synthesis as many studies have suggested [10]. Second sample (Type B) was synthesized in a 50ml Erlenmeyer flask at room temperature under the direct exposure of UV light from the sunlight [11].

(a) Type A particles: Aqueous extraction under thermal conditions

The synthesis was carried out in an Erlenmeyer flask containing 50ml of aqueous plant extract mixed with 50ml of aqueous silver nitrate solution (0.035M). Synthesis was carried out in a pressure cooker at 85°C for 2 hours. The temperature increase enables a catalytic effect. The translucent green mixture went cloudy after synthesis and settling began within minutes (Figure 1a). Ag NPs were then gathered and washed by repetitive centrifuging at 3500rpm for 5 minutes. During the last step Ag NPs were dried and crushed into fine powder.

(b) Type B particles: Ethanol extraction with UV radiation

The synthesis was carried out in an Erlenmeyer flask containing 50ml of aqueous Plantago major plant extract that was extracted using ethanol before. 50ml of aqueous silver nitrate solution (0.025M) was then added in the flask. The synthesis was carried at room temperature under direct sunlight, where UV radiation had a catalytic effect to the synthesis. The initial intense dark green mixture became red-brown after synthesis and the produced Ag NPs nanoparticles were clearly settling (Figure 1b). Ag NPs were then gathered and washed by repetitive centrifuging at 3500rpm for 5 minutes. During the last step Ag NPs were dried and crushed into fine powder.

Figure 1. Solution of Ag nanoparticles after synthesis (a) type A synthesized in aqueous medium at 85°C, (b) type B synthesized under sunlight.

2.3. Characterization

X-ray diffraction (XRD) patterns were collected using a Panalytical Empyrean diffractometer with a Cu $K_{\alpha 1}$ radiation source (λ = 0.15406 nm). The particle size was determined using Scherer's formula [12] and quantitative analysis of two phases of Ag@AgCl nanoparticles was also carried out [13]. The shape and size of the synthesized nanoparticles were examined by a high-resolution scanning electron microscope (HR-SEM) Zeiss Merlin with Bruker XFlash 6 EDS detector. The morphology and size distribution of Ag NPs were also studied by transmission electron microscope with a JEOL 2010 LaB_6 filament providing a point to point resolution of 1.94 Å at 200kV acceleration voltage.

2.4. Antimicrobial assays

(a) Antifungal assay on S. Cerevisiae

The methodology of antifungal assays was conducted according to the method described here [14, 15]. Common yeast S. Cerevisiae was chosen as a target organism to evaluate the antifungal effect of the Ag and AgCl nanoparticles synthesized by Plantago major plant extracts.

(b) Antibacterial assay on E. Coli and S. Aureus

The antibacterial testing was carried out by the international standard ISO 20776-1 "Reference method for testing the in-vitro activity of antimicrobial agents against rapidly growing aerobic bacteria involved in infectious diseases" [16]. Target organisms were chosen E. Coli as Gram-negative bacteria and S. Aureus as Gram-Positive bacteria to evaluate the antibacterial effect of the silver and silver chloride nanoparticles synthesised by Plantago major plant extracts.

3. Results and Discussion

3.1. Structural characterizations

The XRD patterns of Ag NPs synthesized with Plantago major highlight the presence of 2 phases for both syntheses (Figure 2a and 2b). The typical XRD pattern of cubic Ag metal nanoparticles is visible on both XRD patterns and the characteristic diffraction peaks at 38.15°, 44.30°, 64.52°, 77.42° correspond to 111, 200, 220 and 311 reflections respectively (ICDD file no. 00-087-0718). A secondary phase that corresponds to the presence of AgCl structure is also visible on the XRD patterns. Typical XRD peaks at 27.88°, 32.26°, 46.25°, 54.85°, 57.50°, 67.51°, 74.45°, 77.40° that correspond to 111, 200, 220, 311, 222, 400, 331, 420 and 422 reflections are clearly identified (ICDD file no. 00-001-1013). AgCl is a well-known secondary phases commonly produced when using plant extracts mediated synthesis [17]. However, this secondary phase is usually neglected in published reports. Here, the XRD pattern shows that synthesis method and conditions highly influence Ag/AgCl secondary phase ratio.

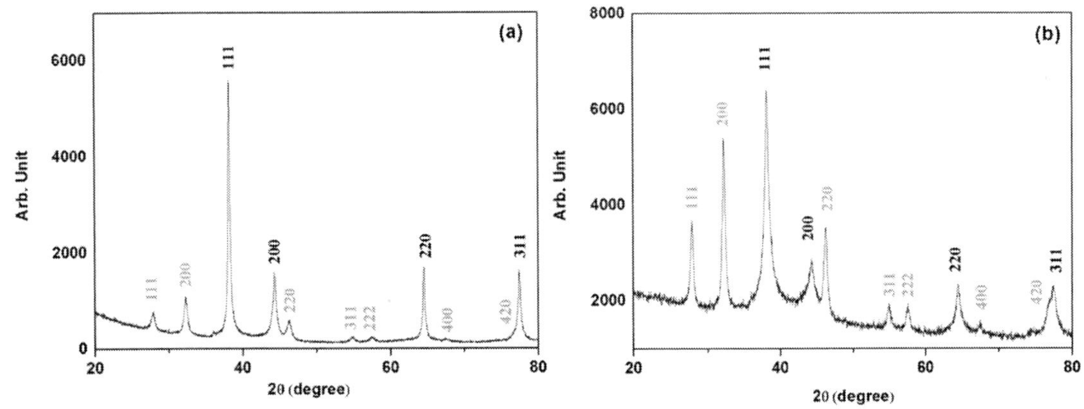

Figure 2. XRD patterns of Ag NPs synthesized (a) in aqueous medium at 85°C (type A), and (b) under sunlight (type B). XRD peaks of Ag NPs and AgCl NPs are indexed in black and green respectively.

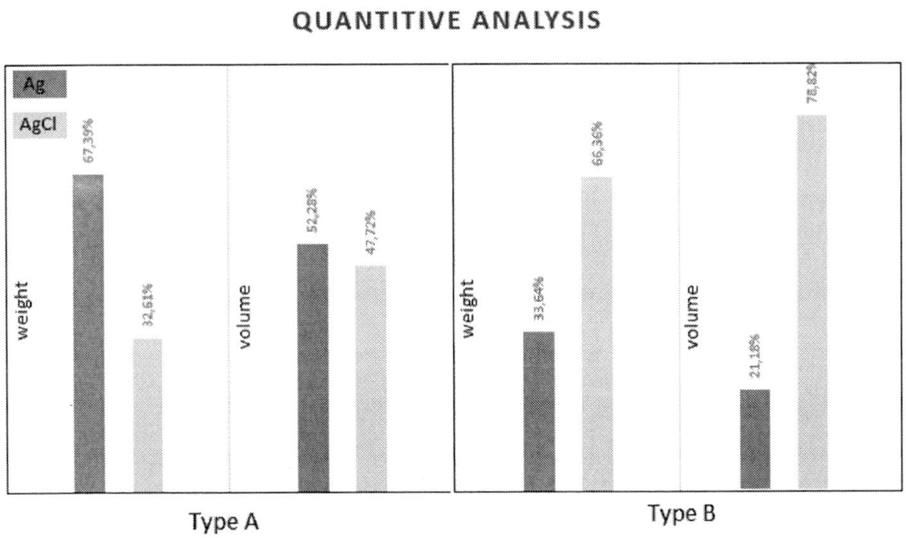

Figure 3. Type A and type B nanoparticles weight and volume percentages based on the XRD data.

XRD data was also used to calculate the weight and volume percentages of the phases present in the nanoparticles [13]. The results are presented in table 1. In the case of aqueous synthesis combined with thermal heating (85°C) (type A), the amount of AgCl secondary phase is lower (figure 3). Ag is the main phase. Form XRD pattern figure 2a and the Debye-Scherrer method, an average nanoparticles size of 25nm and 37nm for Ag NPs and AgCl NPs was calculated respectively. The Ag NPs are smaller than AgCl NPs. In the case of type B (UV sunlight exposure), the amount of AgCl phase is more important and is the dominant phase (figure 3). Debye-Scherrer method was applied to XRD pattern figure 2b and 2 average diameters of 8.5nm and 19nm was calculated for Ag NPs and AgCl NPs respectively. Ag NPs synthesized at room temperature and sunlight exposure are smaller than those synthesized in a pressure cooker at a higher temperature of 85°C. However, the sunlight exposure and biomolecule extracted using ethanol promoted the synthesis of AgCl phase that becomes the primary phase. This study shows that it is possible to promote the synthesis of AgCl under specific synthesis conditions and the proportion of both phases can be controlled.

Figure 4. SEM micrographs of Ag NPs and AgCl NPs synthesized (a) in aqueous medium at 85°C (type A), and (b) under sunlight (type B).

The morphology and size of silver nanoparticles were also studied by SEM. The SEM micrographs in figure 4 give an overview of nanoparticles. The spherical morphology and agglomeration of the particles are clearly visible. Also the size of the Ag NPs appears larger on the SEM micrographs than the size calculated from the XRD patterns. This certainly means that visible nanoparticles correspond to the agglomeration of smaller ones. However, SEM study confirm the XRD results and the Ag NPS synthesized in aqueous medium (Type A) are bigger with a diameter slightly lower than 100 nm (figure 4a) than Ag NPs synthesized in ethanol medium (The type B) that exhibit a diameter ranging from 50 to 70nm (figure 4b). In addition, it is not possible to discriminate between Ag NPs and AgCl NPs.

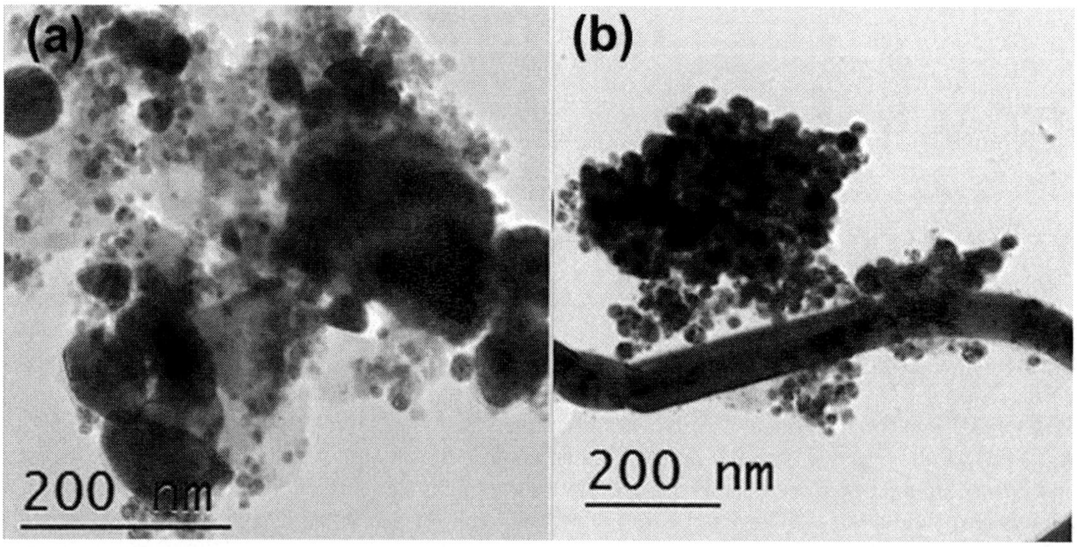

Figure 5. TEM micrographs of Ag NPs and AgCl NPs synthesized (a) in aqueous medium at 85°C (type A), and (b) under sunlight (type B).

IOP Conf. Series: Materials Science and Engineering **613** (2019) 012003 doi:10.1088/1757-899X/613/1/012003

3.2. TEM study

The morphology of Ag nanoparticles was also studied by TEM. Transmission electron micrographs in figure 5 give an overview of the Ag nanoparticles. TEM study confirms the bigger size of Ag NPs synthesized in aqueous medium and thermal heating (type A) (figure 5a). Bigger agglomerates are also visible demonstrating a lower control during the synthesis. TEM micrograph in figure 5b shows a narrower size distribution of the nanoparticles and confirm the average size of 10-20nm.

3.3. Biocidal study

The specific toxicity of silver and silver chloride nanoparticles against microorganisms is already well documented [3, 17]. Bacteria such as *Staphylococcus aureus* and *Escherichia coli* and yeasts such as *Saccharomyces cerevisiae* are very commonly used for testing the antimicrobial properties of metal nanoparticles [18]. Therefore, a similar antimicrobial study was performed. Antibacterial tests were carried out according to the ISO 20776-1 standard showed efficient antibacterial properties (table 1). The minimum inhibitory concentration (MIC) that corresponds to the lowest quantity of Ag NPs to prevent the visible growth of bacteria and the minimum bactericidal concentration (MBC) that corresponds to the lowest quantity of Ag NPs required to kill all bacteria were measured.

Table 1. Antibacterial assay performed on Ag NPs and AgCl NPs synthesized in aqueous medium at 85°C (type A) and under sunlight (type B).

	Type of nanoparticles	MIC ug/mL	MBC ug/mL
E. coli	Type A Ag/AgCl nanoparticles	1.6	3.2
	Type B Ag/AgCl nanoparticles	1.6	3.2
S. aureus	Type A Ag/AgCl nanoparticles	0.8	0.8
	Type B Ag/AgCl nanoparticles	0.8	0.8

This investigation shows that there is no difference between two different samples of nanoparticles (type A and type B) and both samples exhibit strong antibacterial properties whatever the Ag/AgCl ratio. Ag NPs and AgCl NPS proportion does not seems to modify the antibacterial properties and the biocidal properties are high.

Antifungal tests with *S. Cerevisiae* are shown in figure 5. These tests did not show either any difference between the two samples (type A and type B). $AgNO_3$ has been used as a control and by itself shows anti-microbial efficiency. Nevertheless, both plant mediated silver nanoparticles had better biocidal properties against bacteria than yeasts. The minimum bactericidal concentration (MBC) with bacteria is 10 times higher than with *S. Cerevisiae* (Figure 6).

According to the methodology described by Suppi et al., [14]; antifungal properties against *S. Cerevisiae* are observed from a concentration of silver nanoparticles of 30μg/mL.

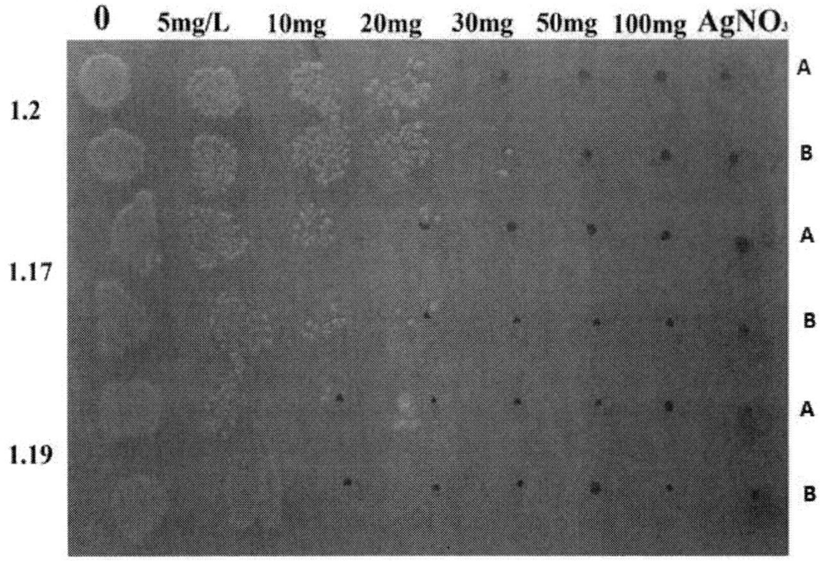

Figure 6. Antifungal tests result with silver nanoparticles synthesized in aqueous medium at 85°C (type A), under sunlight (type B) on yeast *S. Cerevisiae*. Y-axis shows the optical density measured at 600nm by UV-Vis spectrometer, which indicates the number of cells in the tested media. X-axis shows the concentration of silver nanoparticles in the solution.

4. Conclusions

In summary, we have investigated 2 different methods of plant extract mediated-synthesis using Plantago major extract and studied the structural properties of the silver nanoparticles. We observed the presence of AgCl secondary phase and the proportion of AgCl in the sample is strongly dependant on the method of synthesis. It is possible to reduce the amount of AgCl through thermal heating. On the other hand, direct UV sunlight exposure at room temperature promotes the synthesis of AgCl nanoparticles. All the nanoparticles are spherical, but synthesis in aqueous medium under thermal heating produces bigger nanoparticles that tend to agglomerate. The study of the biocidal properties shows that the antibacterial and antifungal properties are not dependent on the Ag NPs and AgCl proportion. In all the cases silver nanoparticles exhibit efficient biocidal properties against bacteria and fungi, with higher toxicity against bacteria. Finally, plantago major extract mediated synthesis appears to be an efficient and cost-effective method for the synthesis of antimicrobial silver nanoparticles.

Acknowledgement

This research has been supported by the European Regional Development Fund project EQUiTANT (F180175TIBT). LMGP, MINATEC is acknowledged for access to their transmission electron microscope.

References

[1] P Rauwel, E Rauwel 2017 *Global journal of nanomedicine* **1** (3) 555562 https://juniperpublishers.com/gjn/pdf/GJN.MS.ID.555562.pdf

[2] C H Ramamurthy, M Padma, I Daisy Mariya Samadanam, R Mareeswaran, A Suyavaran, M Suresh Kumar, K Premkumar, C Thirunavukkarasu 2013 *Colloids Surf. B Biointerfaces* **102** 808 https://doi.org/10.1016/j.colsurfb.2012.09.025

[3] M Rai, A Yadav, A Gade 2009 *Biotechnology Advances* **27** (1) 76 https://doi.org/10.1016/j.biotechadv.2008.09.002

[4] V R Pasupuleti, T N V K V Prasad, R A Shiekh, S K Balam, G Narasimhulu, C S Reddy, I Ab Rahman, S H Gan 2013 *Int. J. of Nanomedicine* **8** 3355 https://doi.org/10.2147/IJN.S49000
[5] A Kumar, P K Vemula, P M Ajayan, G John 2008 *Nature Materials* **7** 236 https://doi.org/10.1038/nmat2099
[6] P Rauwel P Rauwel, S Küünal, S Ferdov, E Rauwel 2015 *Advances in Materials Science and Engineering_Hindawi* 682749 http://dx.doi.org/10.1155/2015/624394
[7] S Küünal, S Kutti, P Rauwel, D Wragg, E Rauwel 2016 *International Nano-Letters* **6** (3) 191 https://link.springer.com/article/10.1007/s40089-016-0186-7
[8] https://www.acs.org/content/acs/en/greenchemistry/principles/12-principles-of-green-chemistry.html
[9] S Küünal, P Rauwel, E. Rauwel 2018 Handbook of nanoparticles and architectural nanostructured materials *Elsevier S&T Books* **14** 411 https://doi.org/10.1016/B978-0-323-51254-1.00014-2
[10] N Vigneshwaran, R P Nachane, R H Balasubramanya, P V Varadarajan 2006 *Carbohydrate Research* **341** (12) 2012 https://doi.org/10.1016/j.carres.2006.04.042
[11] L Rastogi, J Arunachalam 2011 *Materials Chemistry and Physics* **129** (1-2) 558 https://doi.org/10.1016/j.matchemphys.2011.04.068
[12] E Rauwel, A Galeckas, P Rauwel, M Fleissner Sunding, H Fjellvåg 2011 J. Phys. Chem. C **115** (51) 25227 https://pubs.acs.org/doi/abs/10.1021/jp208487v
[13] Y Waseda, E Matsubara, K Shinoda X-Ray Diffraction Crystallography: Introduction, Examples and Solved Problems *Springer London-New-York* ISBN 978-3-642-16635-8 https://www.springer.com/la/book/9783642166341
[14] S Suppi, K Kasemets, A Ivask, K Künnis-Beres, M Sihtmäe, I Kurvet, V Aruoja, A Kahru 2015 *Journal of Hazardous Materials* **286** 75 https://doi.org/10.1016/j.jhazmat.2014.12.027
[15] K Kasemets S Suppi, K Künnis-Beres, A Kahru 2013 *Chem. Res. Toxicol.* **26** (3) 356 https://pubs.acs.org/doi/10.1021/tx300467d
[16] https://www.iso.org/standard/41630.html
[17] N Durán, G Nakazato, A B Seabraet 2016 *Applied Microbiology and Biotechnology* **100** (15) 6555 https://doi.org/10.1007/s0025
[18] J S Kim, E Kuk, K N Yu, J-H Kim, S J Park, H J Lee, S H Kim, Y K Park, Y H Park, C-Y Hwang, Y-K Kim, Y-S Lee, D H Jeong, M-H Cho 2007 *Nanomedicine: Nanotechnology* **3** (1) 95 https://doi.org/10.1016/j.nano.2006.12.001

IOP Conf. Series: Materials Science and Engineering **613** (2019) 012004 doi:10.1088/1757-899X/613/1/012004

Adhesive-free joining and application for flexible devices - a review

Yasunori Taga, Keisuke Nishimura, and Yuri Hisamatsu

Thin Film Research Center, Chubu University Kasugai, Aichi 487-8501, Japan

E-mail: y-taga@i sc .chubu.ac.jp

Abstract. Interface has been recognized for a long time as a part of materials such as surface and grain boundary in bulk in spite of the fact that interface gives decisive influences on the performance and durability of devices in practical use. Interface is also well known to have four principal performances and roles such as mechanical, electrical, magnetic and optical properties. Here, I will at first review the interface behavior with special emphasis on materials science and then introduce typical examples of case study. Mechanical behavior of interface can be attributed to joining of different materials in particularly at low temperature and intrinsic defects introduced in the process of joining caused by difference in thermal expansion. Defects thus introduced is known to result in work hardening of interface. Recently, an epoch-making method of joining to utilize water gas adsorption layer as an adhesive agent. This gas adsorption joining (GAJ) was demonstrated in cycloolefin polymer (COP) film and borosilicate glass system. Both surfaces were first modified by atmospheric pressure plasma, followed by close lamination to reaction distance and heated at about 100°C to result in the formation of a strong covalent bond for joining. Durability thus formed system was found to remain unchanged even after weathering test in 60°C, 95%RH for 2000 hrs. Joining mechanism can be explained in terms of an increase in density of functional molecules at the interface, followed by dehydration. Thickness of joining layer was confirmed by X-ray photoelectron spectroscopy and STEM and EELS to be 1~2nm. Sealing performance of GAJ was examined by measuring Water Vapor Transmission Rate (WVTR) by MOCON system using joined PET films on which high barrier layer was coated. The WVTR of line sealing with 1 mm width and 1-2 nm thick was found to be better than 5 x 10^{-4} g/m^2/day. Practical applications of thus developed GAJ for joining and sealing are now in progress for flexible film devices such as OLED display, solar cells, and healthcare devices. Finally, roles of interface in materials science in future will be critically discussed.

1. Introduction

Great advantage of nanotechnologies brought us highly comfort, convenient, energy-saving and environment-friendly social life, which we are enjoying today [1-3]. In the course, materials science always played a leading role in research and development of device and system. Outstanding examples in materials science will be carbon-nanotube [4], fullerene [5], graphene [6] and mesoporous silica [7], [8]. Basic concept of nanotechnology will be the process innovation of downsizing, which can be expressed simply to manufacture products in light, thin, short and small scale, emphasizing compactness and flexibility. In manufacturing process, law materials of natural resources will become fabricated materials as a result of modification such as refining, distillation and purification. Materials thus modified will be transferred to the most important process of manufacturing including surface treatment and joining for modularization and systemization. Especially, joining with surface treatment is well-known to give a significant impact on the performance and durability of the devices and final products. Nowadays, a huge amount of research and development and accumulation of know-how were devoted to establish the final method of joining with chemical adhesives for all of the joining system by joining system. Newly emerging important nanomaterials [4-8] must be incorporated in the today's device and system by joining using chemical adhesives. In fact, there exist a lot of difficulties in joining of nanomaterials and conventional materials due to the lack of know-how accumulation. Recently, the ultimate bonding method of adhesive-free gas adsorption joining was explored by the present authors [9]. Joining process was well-known to be affected significantly by the conditions of interface such as cleanliness, roughness, mechanical properties such as internal friction and plastic deformability, and chemical properties [10-11]. Particularly, joining by chemical reaction at the interface is attracting much attention from the viewpoint of low temperature joining of different materials for all-solid-state battery and fuel cell. However, interface conditions give a large influence not only on joining but also on a wide range of materials science. Interface has long been considered as a part of bulk materials but as of today, interface became recognized as one of the materials with special properties such as composition and structure. In this review manuscript, we will at first introduce new concept of interface and the classification and then demonstrate GAJ and the application by way of examples.

2. New concept of interface: Function and role

Interface has long been recognized to be very important because of the fatal influence on the performance and durability of device and system made of functional materials [12-15]. Composition and structure of interface are usually different from those of bulk caused by surface treatments in manufacturing process such as physical and chemical and mechanical modifications and annealing. In fact, impurities in the bulk are well known to migrate toward surface and interface by annealing, resulting in the formation of non-stoichiometric interface with large number of defects. Another essential problem of interface is lack of non-destructive method of analytical characterization. Actually, interface is usually buried in deep region from surface and is normally very thin. For example, it is really difficult to evaluate the composition and structure of buried interface layer by analytical method using X-ray beams due to lack of detection sensitivity. Also, in the case of Auger electron spectroscopy and X-ray photoelectron spectroscopy (XPS), Ar ions are usually used for in depth etching from the surface to buried interface in deep region, which will be destructive to alter the original interface properties. On the other hand, interface can be classified into four genres of mechanical, electrical, magnetic and optical properties summarized in Table 1.

Table 1. Function and role of interface

Properties	Function	Role
Mechanical	Joining and adhesion strengthening induced by defects. Cutting and processing	Control of chemical and mechanical re action for adding new functions
Electrical	Ohmic contact. Rectifying contact Barier hight control for electron and hole injection	Design for electric and electronic contact of devices
Magnetic	Magnetic exchange interaction	Giant magnetoresistance effect
Optical	Transmission, absorption, reflection. Birefringence for retardation	Phase retardation plate by obliquely de posited thin films

Mechanical property includes functions of joining, adhesion and strengthening induced by defects, and cutting and processing and modification, and their roles for adding new functions of chemical and mechanical reactions. Among the functions, special attention was focused on joining process with special emphasis of interface chemical reaction. In electrical properties, major function will be electric contact controlling ohmic or rectifying current flow through contact interface and electron and hole injections through at the interface. Main magnetic function will be the so-called exchange interaction between both sides of interface in contact. Typical example of exchange interaction at the interface is Giant magnet-resistance (GMR) effect discovered in 1988 [16] and in 1987 [17]. There exist many functions in optical properties of interface such as transmission, absorption and reflection [18-19]. In addition, growth of retardation film by oblique deposition revealed optical birefringent effect [18] at the interface called Motohiro-Taga interface [20]. Typical example of interface function will be presented in the following sections.

3. Gas adsorption induced adhesive-free joining and application [9]

Looking back the practical joining process as of today, plasma treatment of both surfaces was primarily used to control surface wettability, and then chemical adhesives of 5-10 μm in thickness such as an acrylic resin or a silane coupling agent [21] were coated in solution for adhesion, resulting in strong adhesion by chemical reaction. Challenge for a direct joining of poly (methyl methacrylate) for polymer microchips has been made after plasma treatment without coating chemical adhesives. This type of joining, however, is fundamentally considered to be hydrogen bond at the interface, resulting in insufficient durability. In 2014, Taga and Fukumura [9] developed an epoch-making method of gas adsorption induced adhesive-free joining with enough durability in practical use, where design and control of interface played an important role in interface chemical reaction.

3.1 Overview of GAJ

Process flow chart of GAJ was shown in Figure 1. As mentioned above, joining is so sensitive to the surface conditions such as cleanliness and therefore the surfaces before joining must be at first cleaned ultrasonically using isopropanol to remove organic contamination such as floating oil particles and fingerprint consisting of oleic acid. The surfaces thus cleaned were modified by surface treatments such as plasma exposure, ultraviolet light irradiation and beads blasting.

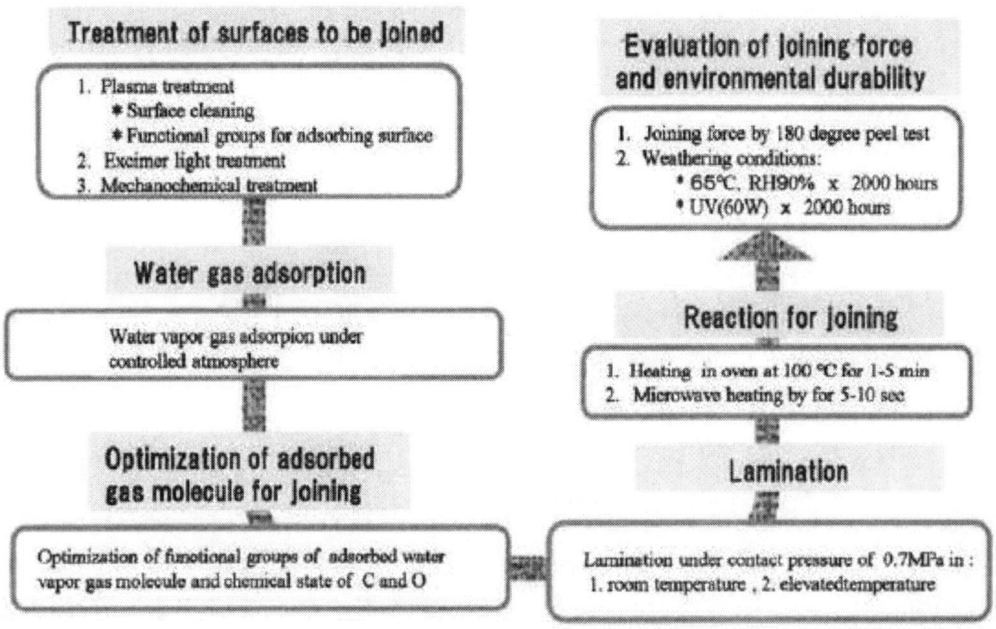

Figure 1. Flow chart of gas adsorption joining process

Typical example of changes surface chemical conditions by air pressure plasma of borosilicate glass and cycloolefin polymer (COP) film, and water contact angle measured using distilled water droplet of 3 μL according to standard method [22] was depicted in Fig.2. As shown by X-ray photoelectron spectroscopy (XPS) analysis using Mg $_{K\alpha}$X-ray source, the spectrum at binding energy of 286 eV~287 eV assigned to surface functional groups of ether or alcohol (C-O, C-OH, C-O-O) and carbonyl (C=O) increased by plasma treatment and the water contact angle drastically decreased, indicating the changes of surface wettability from hydrophobic to hydrophilic. Borosilicate glass sheet of 0.7 mm thick and COP film with 100^m in thickness were directly joined at low temperature by GAJ. A necessary condition for joining of COP and glass is at first to make the contact surfaces clean. After plasma treatment to control surface functional groups on both surfaces, joining was carried out by a roll laminator in surface contact pressure of 0.7 MPa at room temperature, followed by annealing at 100 C for 10 min in an oven. Adhesion strength was evaluated by 180 degree peel test based on ASTM D-903 and durability was examined under the weathering conditions of 60 C and 95 % RH.

Figure 3 showed the results of joining force by water gas adsorption followed by annealing in oven for reaction. Actually, GAJ became possible only after plasma treatment in air atmospheric conditions.

As shown in the figure, joining force after low temperature below 80°C was rather low but steeply increased with temperature to reach high value of over 20 N/25 mm after annealing at 100°C. In addition, joining force can be controllable between 0 to 20N/25 mm depending on annealing temperature. GAJ interface layer is evaluated to be 1-2 nm in thickness by taking inelastic scattering mean free path of photoelectrons into consideration [9]. Durability of this joined couple has been confirmed to remain unchanged under the conditions of the weathering condition of 60°C, 95% RH and Xe lamp irradiation by 60 W/mm^2 for more than 2000 hrs. Reaction mechanism of water GAJ is not yet fully elucidated by data as of today. The reaction mechanism of water GAJ can be explained as follows: Water molecules may adsorb on as received glass and COP in air atmospheric conditions before plasma treatment and no joining is possible at this stage between these glass and COP surfaces. Joining becomes possible after plasma treatment and XPS analyses of both surfaces to be joined revealed an increase in peak at 532 eV of binding energy assigned to OH in O_{1s} spectrum, resulting in possible joining by lamination followed by dehydration to form H-H hydrogen and Si-O-C covalent bonds with strength between 0-20 N/25mm. On the other hand, plasma treatment may also bring alteration of order of orientation and crystallinity of COP surface [23].

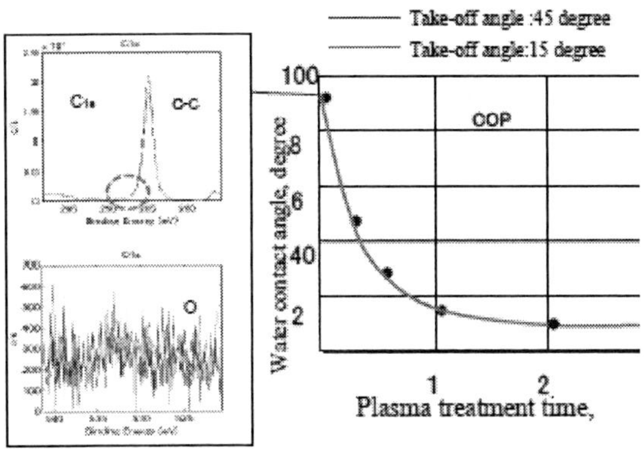

Figure 2. Changes in Cls and Ols spectra, and water contact angle with plasma treatment time

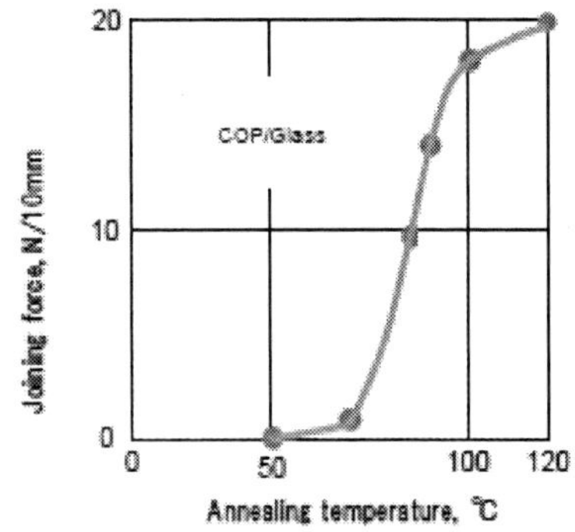

Figure 3. Changes in joining force by GAJ with annealing temperature

3.2 *SFG investigation of COP surface and interface to glass [23]*

Infrared visible SFG spectroscopy has recently been demonstrated to be a highly sensitive surface-specific tool for the examination of polymer surfaces and buried interfaces, as it can probe the orientation and alignment of the surface molecules of thick polymer films. Thus, we investigate here the characterization of the COP surfaces modified by plasma using sum frequency generation (SFG) vibrational spectroscopy [24] to determine how surface pre-treatment improves adhesion between the COP films and borosilicate glass surfaces. The COP/glass buried interfaces and the influence of thermal annealing are also probed directly by SFG measurements. The variation in adhesive force between the borosilicate glass and the COP film with the number of argon/oxygen (Ar-5% O_2) atmospheric pressure plasma scans is shown in Figure 4.

Although an adhesive force of <1 N/ 10 mm was observed in the absence of plasma treatment, this force increased to 18 N/10 mm following the initial plasma scan. Further treatment with the plasma gas resulted in a gradual degrease in the adhesive force to a relatively stable value of 14 N/10 mm. In order to minimize the denaturation of the surface by the leaving time after plasma irradiation, plasma irradiation was performed immediately before SFG measurement (~ 1 minute before). It takes about 30 min for measuring the spectrum. Figure 5 shows the SFG spectra of the COP polymer surfaces both before and after argon/oxygen (i. e., Ar-5% O_2) plasma exposure using the SSP polarization combination, which relates to polarizations of the SFG, visible, and infrared light, respectively.

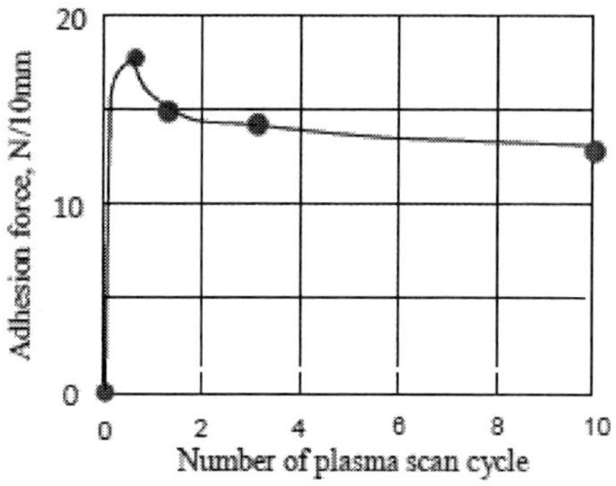

Figure 4. Changes in adhesion force with number of air pressure plasma

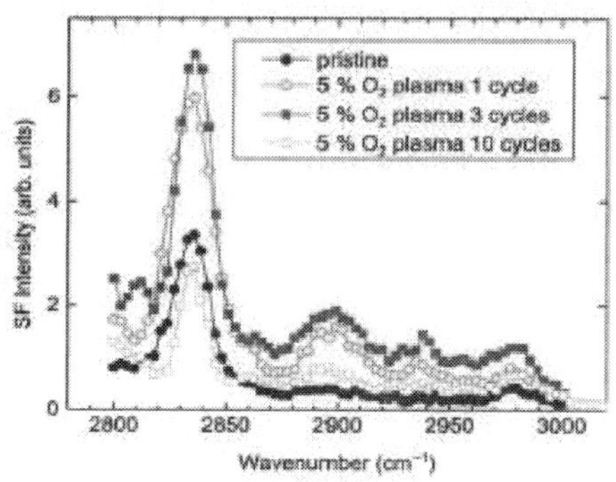

Figure 5. SFG spectra of the COP polymer surfaces both before and after argon/oxygen

In the SFG spectra of the pristine COP surface, symmetric and antisymmetric stretching vibrations corresponding to the CH2 unit of the COP were observed at 2837 and 2907 cm^{-1}, respectively. Upon exposure of the COP surfaces to the argon/ oxygen plasma, an increase in SFG peak intensities was observed. This qualitative increase in the intensities of the antisymmetric/symmetric CH2 stretching signals upon argon/ oxygen plasma treatment therefore indicates that the orientation of the CH2 group becomes slightly tilted the following plasma exposure. Although the observed changes in the SFG intensities are partly caused by changes in the molecular orientation of the surface functional groups following plasma treatment. It should be noted that orientation angle and its distribution will not be quantified, but such a change in the SFG peak intensities can be explained by variation in the orientational order of the functional groups of the surface molecules. As COP is classified as an amorphous polymer, the molecular orientation at the untreated surface should have a relatively broad orientational distribution. Indeed, we note that the formation of oxalic acid on COP surfaces has been previously observed following similar plasma and vaporization treatments, which may influence the COP surface structure, including the degree of the orientational order. More specifically, the increase in SFG signal intensities can be interpreted as an enhancement in the orientational order at the surface COP molecules. Finally, we investigated the COP/glass interfaces in the OH stretching region. Here, it could be expected that if the surface functional groups and surface-adsorbed water molecules reacted to form Si-O-C covalent bonds between the COP and the glass surfaces. However, this hydrogen-bonded OH stretching vibration did not fully disappear in SFG signals by annealing. The improvement in adhesion strength by chemical reaction at the interface is not simply due to the chemical bonding taking place at the glass surface, but an increase in the density of the CH2 groups and an enhancement in the surface molecular orientational order may modify the hardness of the COP at the interface, thereby increasing the adhesion strength.

We could therefore conclude that the increased molecular orientational order and the density of the COP surfaces can be considered a key factor in achieving the strong adhesive force to glass [23].

3.3 *Direct observation of gas adsorption joined interface [25]*

There exists primary interest to observe directly how joined COP and glass without using chemical adhesives. Scanning Transmission Electron Microscope (STEM) equipped with Electron Energy Loss Spectroscope (EELS) was applied for direct observation of SiO_x/COP interface, where Si wafer on which SiOx film of 2 μm thick was used in GAJ process.

Gas adsorption joined specimen of COP/ SiO_x /Si was manipulated for STEM observation by using the Ar ion-milling system. Annular Dark Field (ADF) images of both interfaces of COP/SiO_x and SiO_x /Si was shown in Figure 6 Interfaces at COP/SiO_x and SiO_x/COP can be clearly seen together with mixed amorphous region between SiO_x and COP. As a result, it can be concluded that uniform and tight joining with 1~2 nm thick can be possible by adhesive-free GAJ at low temperature.

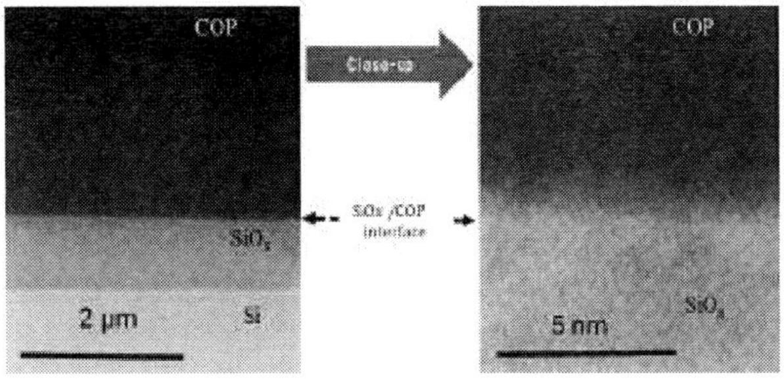

Figure 6. ADF image by STEM-EELS observation of COP/SiO_x/Si interface prepared by GAJ

3.4 *Application for gas adsorption sealing (GAS) [25]*

Recently, electronics and photonics devices on organic film substrate have become popular from the viewpoints of compactness and flexibility, where typical examples can be seen in mobile phone, tablet and displays. In addition, these devices are usually made of organic materials, which are well known to deteriorate easily due to water molecules. In the case of inorganic substrate, however, device sealing was usually carried out by melting of substrate edges by local heating up to higher temperature. Though chemical adhesive is now widely used in manufacturing process for sealing of organic film substrate. In this section, we showed the sealing performance of gas adsorption joined high barrier films for water vapor permeation by monitoring the change in the visible light opaque Mg film to transparent MgO due to oxidation. Figure 7 shows the sample structure and gas adsorption sealing (GAS) processes. Barrier layer formed on 120 μm thick PET film for Water Vapor Transmission Rate (WVTR) measurement. The PET film with barrier layer comprising 3 layers of inorganic/organic/inorganic films, where inorganic films were made of silicon compound was provided by courtesy of Konica-Minolta Corporation [26]. More detailed information of composition and preparation method of the barrier film was not known because of confidential business information. The WVTR value of thus provided barrier layer on PET film was evaluated by Aquatran of MOCON Corporation and confirmed to be less than $5x10^{-5}$ g/m^2/day. As shown in fig.6, Mg film of 10 nm thick was sputter deposited in an area of 20 mm x 20 mm. Metallic mask of thin stainless steel foil for line sealing was set on the Mg film, where line width and length was 1 mm and 120 mm, respectively. Both surfaces were exposed to Ar-O2 air pressure plasma for 30 seconds and then laminated at the pressure of 1 MPa followed by annealing in oven at 80°C, 90°C and 100°C for 5 min for interface chemical reaction. Sample number and sealing conditions were summarized in Table 2. Evaluation of sealing performance was carried out by exposing samples to the conditions of 60°C, 90%RH for 500 hrs. During the course of exposure process, changes in Mg film conditions was checked visually. It was found that metallic Mg films of sample Nos.1 and 2 became transparent gradually from the surrounding of periphery of Mg film with exposure time.

Table 2. Conditions for evaluation of GA sealing

Barrier film	Confirmed limit of Water Vapor Transmission Rate (WVTR) measurement by MOCON : $5X10^{-5}$ g/m^2/day
Sample preparation	3 layer structure of barrier film of inorganic/organic /inorganic with WVTR of $5X10^{-4}$ g/m^2/day No. 1 After heating at 80 °C 1hr No.2 After heating at 90 °C 1 hr No.3 After heating at 100 °C 1 hr
Sealing line structure	Line width : 1 mm Length: 12 cm
Evaluation of WVTR	Visual confirmation of optical transmittance of vacuum deposited metallic Mg film with 100 nm thick

IOP Conf. Series: Materials Science and Engineering **613** (2019) 012004 doi:10.1088/1757-899X/613/1/012004

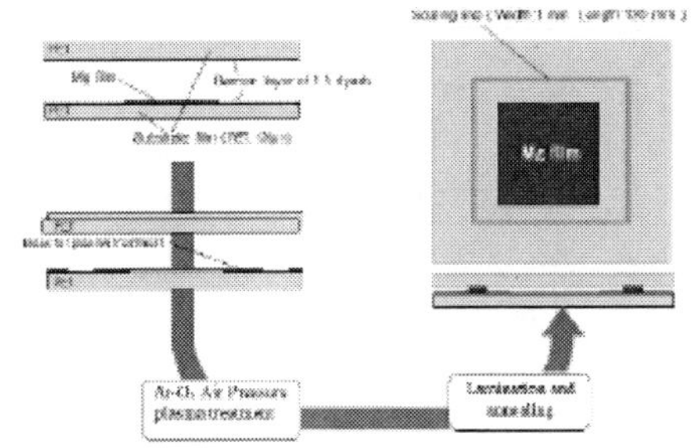

Figure 7. Schematic representation of samples for evaluation of gas adsorption sealing

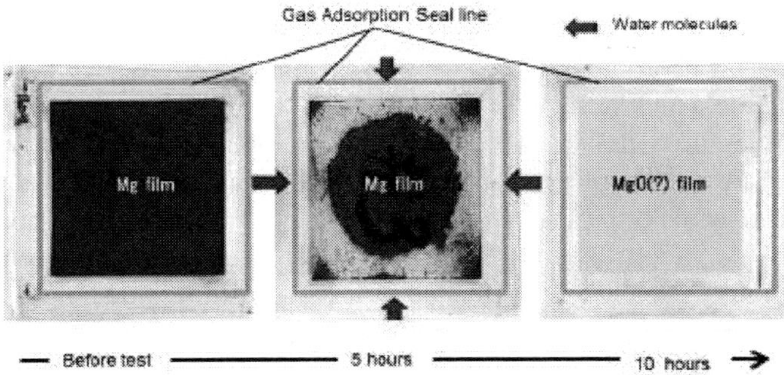

Figure 8. Optical microscope image of sample No. 2 evaluation of GA sealing

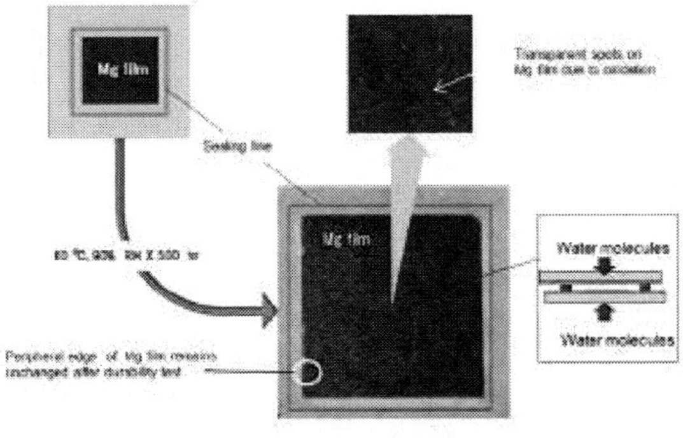

Figure 9. Optical microscope image of sample for evaluation of GA sealingNo.3 for evaluation of GA sealing

Typical example of sample No. 2 was shown in Figure 8, where opaque Mg film to transparent MgO due to oxidation by the water vapor permeated from the directions shown by arrow via sealing line in the periphery. Water vapor permeation will be caused by insufficient chemical reaction at the interface. Mg film of sample No.3, however, showed different performance from Nos.1 and 2. Optical microscope image of No.3 was shown in Figure 9, periphery of Mg film remains unchanged even after exposure to the conditions of 60°C, 90%RH for 500 hrs. However, many transparent spots over the hale area can be seen as shown in the figure. Observation by higher magnification, transparent spots can be seen more clearly. These results showed that permeation of water vapor was caused from the direction perpendicular to the surface of barrier film shown by arrows, indicating that the WVTR value of molecular sealed line of 1 mm width was smaller than the limit of the WVTR value of $5x10^{-5}g/m^2/day$ of barrier layer. This is the first application report of GAJ method for sealing of water permeation at least less than and $5x10^{-5}$ $g/m^2/day$. It can be concluded that no permeation of water vapor through GAS line occurred even exposure for 500 hr, which can be explained in terms of higher degree of chemical

IOP Conf. Series: Materials Science and Engineering **613** (2019) 012004 doi:10.1088/1757-899X/613/1/012004

reaction to form covalent bonds formed by annealing temperature of 100 C.

4. Summary

In this manuscript, we showed composition and structure of the interface is different from the adjacent both layers due to chemical and physical reaction. In addition, the interface thus formed give a fatal influence on the performance and durability of thin film devices. As a result, we can recognize the interface as a functional material.

Interface is normally metastable state and buried in deep region from the surfaces and therefore, especially difficult to investigate. As a result, interface became a key material and new method of research and development of interface analysis is strongly required.

In conclusion, we will be approaching to the advanced stage of materials science by controlling the interface freely.

References

[1] Taga Y 2001 *Thin Solid Films* **C15** 231
[2] Taga Y 2009 *Thin Solid Films* **517** 3167
[3] Asahi R, Morikawa T, Ohwaki T, Aoki K and Taga Y 2001 *Science* **293** 269
[4] Iijima S 1991 *Nature* **354** 56
[5] Kroto H W, Heath R H, Brien S C, Curl R F, and Smally R E 1985 *Nature* **352** 139
[6] Novoselov K S, Geim A K, Morozov S V, and Jiang D 2004 *Science* **306** 666
[7] Wang W Lofgreen J E, Ozin G A 2010 *Small* **6** 2634
[8] Darder E, Aranda P, Ariga K 2010 *Adv. Mater.* **22** 323
[9] Taga Y and Fukumura T 2014 *Appl. Surface Science* **315** 527
[10] Taga Y 2011 IOP Conference Series: Materials Science and Technology **24** 012011 https://doi.org/10.1088/1757-899X/24/1/012011
[11] Taga Y 2011 *Joined structure manufacturing method and joined structure,* PCT W0/2011/010738
[12] Taga Y 1999 *Surface and Coating Technology* **112** 339
[13] Taga Y 1993 *Applied Optics* **32** 5519
[14] Taga Y 1995 J. *Vac. Sci. Technol. A* **13** 990
[15] Taga Y 2001 *Proceeding of European Materials Society* Strasbourg, France
[16] Baibish M N, Broto J M, Fert A, Nguyen Van Dau F, Petroff F, Eitenne P, Greuzet G, Friendrich A and Chazelas J 1988 *Phys. Rev. Lett.* **61** 2472
[17] Binasch G, Grunberg P, Saurebach F, and Zinn W 1989 *Phys. Rev. B* **39** 4828
[18] Taga Y Taga, and Itoh T 1989 *Applied Optics* **28** 2690
[19] Taga Y Taga, and Motohiro T 1990 J. *Crystal Growth* **99** 638
[20] Lalhtakia A and and Messier R 1994 *Optical Engineering* **33** 2529
[21] Arkles B 1977 *Chem. Tech.* **7** 766
[22] Adamson A 1995 Physical chemistry of surface *John Wiley & Sons, Academic Press Inc.* London
[23] Miyamae T, and Taga Y 2017 *J of Adhesion* https://doi.org/10.1018/00218464.2017.1391657
[24] Young M, Rioux R. M, and Somorjai G. A 2005 *J. Catal.* 237 https://doi.org/10.1016/j.jcat.2005.10.025
[25] Taga Y 2018 Int. Conf. on Competitive Materials and Technology Processes Book of Abstract ed. L A Gömze *Published by Igrex Ltd.* (Miskolc, Hungary) ISBN 978-615-00-3114-9
[26] Mori T, Gotou Y, Takemura C, Hirabayashi K 2014 *Konica Minolta Technology Report* **11** 831

IOP Conf. Series: Materials Science and Engineering **613** (2019) 012005 doi:10.1088/1757-899X/613/1/012005

Conventional Brick Clays as a Challenge of Materials Science – New Explanation of Drying Sensitivities

László A. Gömze[1, 2], Ludmila N. Gömze[2], Emese Kurovics[1] and György Benedek[3]

[1]Intsitute of Ceramics and Polymer Engineering, University of Miskolc, Hungary
[2]IGREX Engineering Service Ltd, Igrici, Hungary
[3]CREATON South-East Europe Ltd, Lenti, Hungary

E-mail: femgomze@uni-miskolc.hu, mse205@gmail.com and igrex2009@yandex.ru

Abstract. In this research 13 drilling production, up to 18 m depth of "Lenti" conventional brick clay deposit were examined both on material, mineralogical structures and compositions as well as on technological properties. During the examination the mineralogical and chemical composition of the different layers were determined and tested on physical and technological properties. The authors have found that the grain size distributions and the volumes of X-ray amorphous submicron and nanoparticles of these conventional brick clays do not influence on the dry sensitivity of the extruded and formed green products of ceramics roof tiles. The authors have found also a strong correlation between the amount of X-ray amorphous nanoparticles of the raw materials and the bending strengths of the sintered ceramic rods. The higher the amount (proportion) of submicron and nanoparticles in the used clay minerals, the greater the bending strength of the produced sintered ceramics.

1. Introduction

The conventional brick clays are playing important role in production of ceramic bricks and roof tiles [1-16] and they are widely used also in different segments not only of industry [17-27] but in medicine [28-30] and polymers [31-33]. Generally, the mineralogical composition, grain size structure, specific surface and moisture of the raw materials are strongly influencing the color, quality and the technological properties as well as the chemical and physical parameters of the sintered final ceramic products [34-53]. The technologists and researchers of ceramic brick and tile industries often claim [54-57] that the drying sensitivities of the conventional brick clays are depend on their grain sizes and specific surfaces. This means that as smaller as the average grain sizes, the higher the specific surfaces and stronger the susceptibilities of clays to the drying sensitivities and drying cracks. Thanking to this option it is so important to analyze the layers of mine deposits before their exploitation to produce building ceramics like large borehole bricks or roof tiles.

The aims of this research work are to support the factory management and engineering staffs in the production of ceramic roof tiles with a consistently high quality on competitive prices in an acceptable materials and specific energy consumption.

2. Materials and experiments

To determine the morphological structure, mineralogical composition, extruding, drying and sintering properties of the "Lenti" clay deposit of CREATON South-East Europe Ltd 13 mining holes were drilled in three parallel rows (5-4-4) up to 18 m depths. The drilled materials from each mined hole

IOP Conf. Series: Materials Science and Engineering **613** (2019) 012005 doi:10.1088/1757-899X/613/1/012005

were divided by 3 m as 0-3 m, 4-6 m, 7-9 m, 10-12 m, 13-15 m, 16-18 m and were crashed and mixed on a laboratory pan mill at 90 rpm for 10 minutes. 100 grams of each mixture were taken for the thermo-analytical tests, scanning electron microscopy (SEM and EDAX) and X-ray diffraction. All the remaining clays were extruded on a KEMA-PVP 5/S extruder into cylindrical shape rods with diameter of 33 mm. After extrusions each rod was cut into 7 specimen green body with a length 150 mm before drying, sintering and bending strength tests. The Figure 1 shows the extruded specimens made from mixtures taken from 7-9 m depths of the 2nd and 3rd drillings.

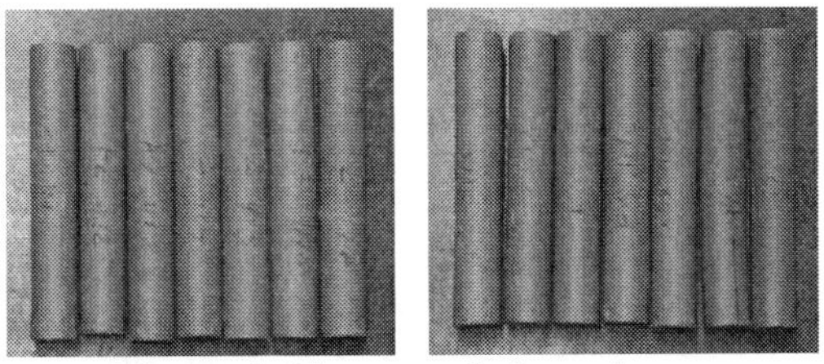

Figure 1. The extruded specimens

Before drying the extruded specimens and creation of Bourry-diagrams the thermo-analytical tests were made for each specimen to determine the **DTA, TG** and **DTG curves** of each mixtures by 3 meters. The thermo-analytical curves of the mixtures were taken from 7-9 m depths of the 2nd and 3rd drillings as shown in Figure 2.

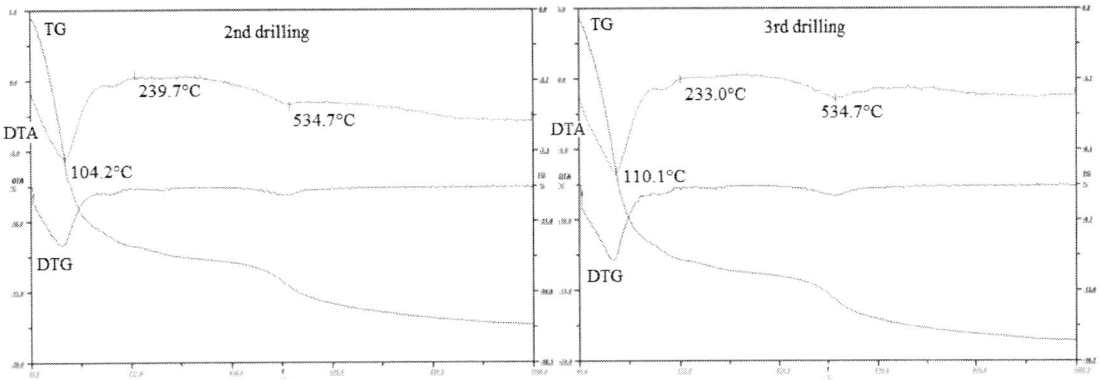

Figure 2. Thermo-analytical curves

These curves are very important to determine the heating rates and optimum temperatures for both drying and sintering (firing) processes. The chemical composition and material structure of clay mixtures taken from 7-9 m depths of the 2nd and 3rd drillings were investigated with scanning electron microscopy and EDAX as shown in Table 1 and Figure 3.

Table 1. The material composition on the surface of the mixtures taken from 7-9 m depths of the 2nd and 3rd drillings

Elemental	O	Na	Mg	Al	Si	K	Ca	Ba	Ti	Fe
2nd drilling (7-9m)	52.57	0.54	0.96	11.96	28.55	1.56	0.42	0.12	0.32	3.00
3rd drilling (7-9m)	56.25	0.38	1.37	12.15	25.01	1.27	0.36	0.02	0.34	2.85

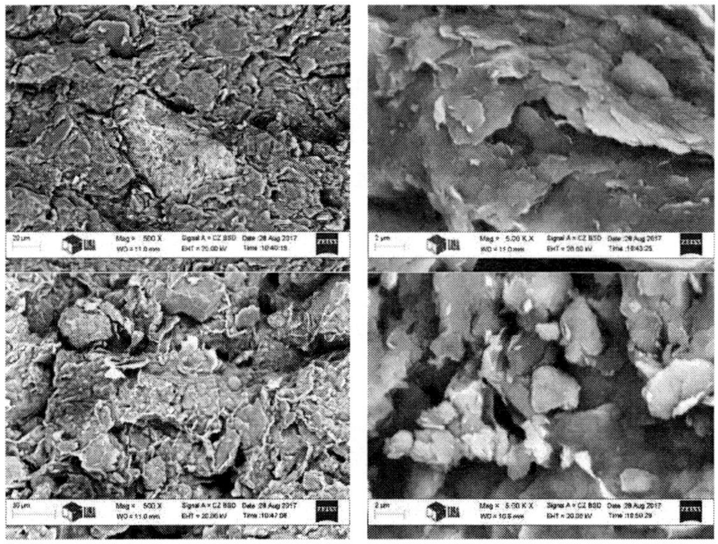

Figure 3. Typical microstructure of the mixtures taken from 7-9 m depths of the 2nd and 3rd drillings

3. Results and discussions

100 g of each mixture were dried in a laboratory chamber-dryer at 75 °C for 96 hours. After drying 30 grams of each mixture were used to determine the mineralogical and chemical compositions using X-ray diffractions. The X-ray diffraction as well as oxide and mineralogical compositions of different layers of the 1st drilling are shown in Figure 4, Table 2 and Table 3.

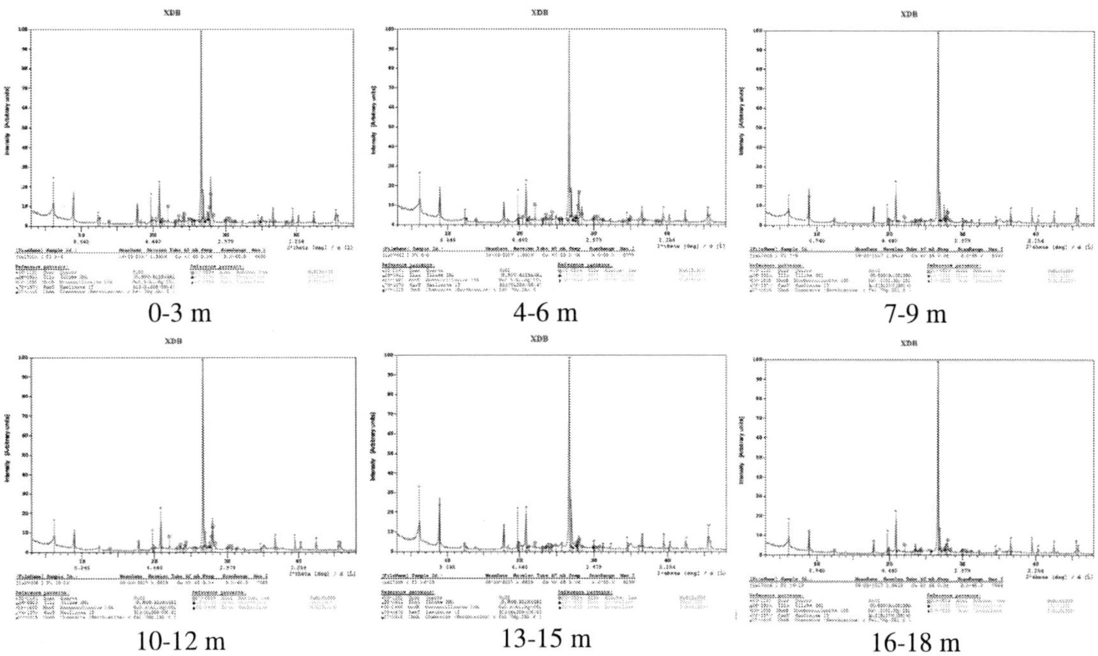

Figure 4. XRD diagrams of F1 drilling samples

IOP Conf. Series: Materials Science and Engineering **613** (2019) 012005 doi:10.1088/1757-899X/613/1/012005

Table 2. The oxide compositions of F1 drilling samples

Depth of drilling		0-3	4-6	7-9	10-12	13-15	16-18
Oxides	Fe_2O_3 %	2.1	1.26	0.84	0.84	0.84	0.84
	K_2O %	2.27	2.27	2.93	1.98	3.33	2.55
	SiO_2 %	54.13	50.99	56.92	55.43	56.72	57.3
	Al_2O_3 %	14.65	13.95	13.97	11.29	16.79	12.7
	MgO %	1.38	1.34	1.11	1.11	1.96	1.32
	Na_2O %	2.19	2.07	1.19	2.14	1.15	1.48
	H_2O %	3.42	3.2	3.09	2.27	4.26	2.87
	LOI %	3.3	3.12	3.04	2.22	4.21	2.82

Table 3. The mineralogical compositions of F1 drilling samples

Depth of drilling			0-3	4-6	7-9	10-12	13-15	16-18
Mineralogical composition	quartz	SiO_2	22	20	30	28	24	30
	illite	$[K(H_3O)^+]Al_2[SiO_3AlO_{10}](OH)_2$	15	15	22	12	28	18
	montmorillonite	$(0,5Ca,Na)_{0,7}(Mg_{0,7}Al_{3,3})[(Si_8O_{20})(OH)_4]$	12	12	10	10	18	12
	kaolinite	$Al_2Si_2O_5(OH)_4$	5	5	3	2	3	2
	chamosite	$(Fe^{2+},Mg)_5Al(AlSi_3O_{10})(OH)_8$	5	3	2	2	2	2
	albite	$NaAlSi_3O_8$	16	15	8	16	6	10
	microcline	$KAlSi_3O_8$	2	2	3	2	2	2
	orthoclase	$KAlSi_3O_8$	3	3	2	3	2	3
	XRD-amorph	-	20	25	20	25	15	21

The investigation of the mineralogical compositions was determined by X-ray diffraction, it can be clearly seen that the total volumes of the clay minerals (illite, montmorillonite and kaolinite) are more than 24 m% in each depth. This means that the mined conventional brick clay "Lenti" is applicable to produce very good quality ceramic roof tiles using plastic forming technology. These clay minerals are phyllosilicates in which the kaolinite is build up from two layers with strong hydrogen-bridge meanwhile the illite has three layers with strong potassium-bridge (Figure 5).

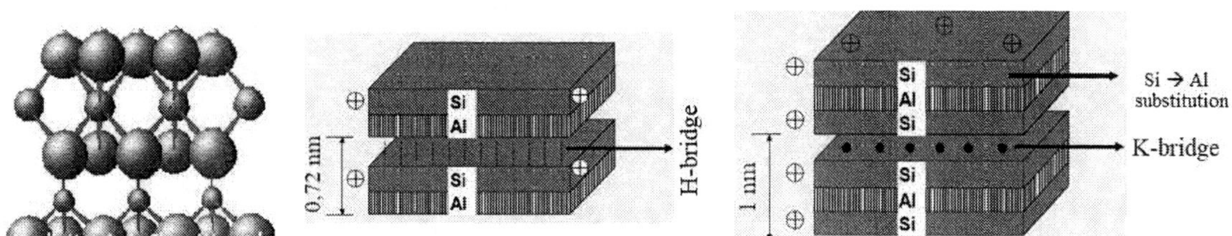

strong hydrogen-bridge (H) strong potassium-bridge (K)

Figure 5. The structure of kaolinite and illite clay minerals (phyllosilicates)

Generally, it can be said that the drying sensitivity of ceramic products made from conventional brick clays are very strongly depend on the amount of clay minerals like kaolinite, illite and the quantities of X-ray amorphous submicron and nanoparticles. The portion of illite is shown in Figure 6 and the amount of X-ray amorphous components is shown in Figure 7 for each mixture.

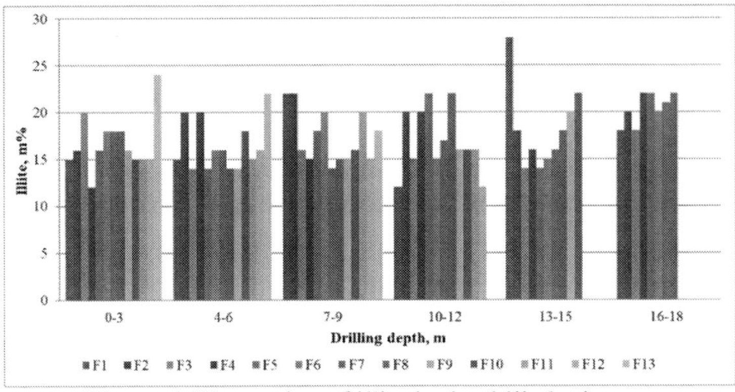

Figure 6. The portion of illite in the drilled mixtures

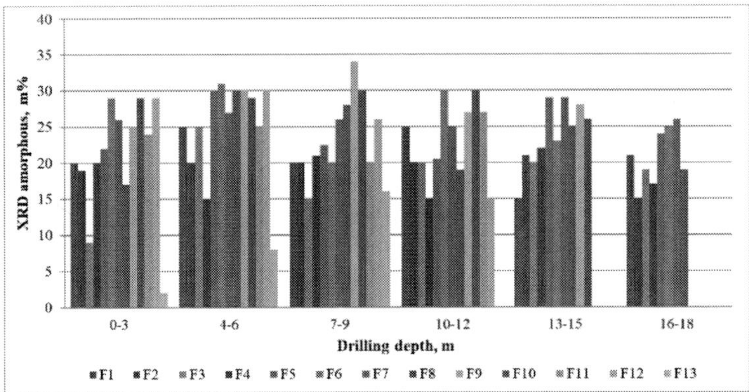

Figure 7. The portion of XRD amorphous components in the drilled mixtures

In most of the mixtures the ratio of the X-ray amorphous particles with more than 15 m% were achieved. Thanking to this relatively large volume of the X-ray amorphous submicron and nanoparticles of the sintered sample, a nanosized porous structure were formed which lead to frost resistant roof tiles ceramic product [58].

After the extrusion the Bourry-diagram of each drilled mixtures were created. For this analysis a cylindrical green body rod with a length of 150 mm were dried in open air at room temperature of about 25-30 °C, then their diameters, lengths and weights were measured each hour in the first 3 hours and after each 2 hours. A typical result of mass (red line) and volume losses (blue line) are presented in Bourry diagrams in Figure 8.

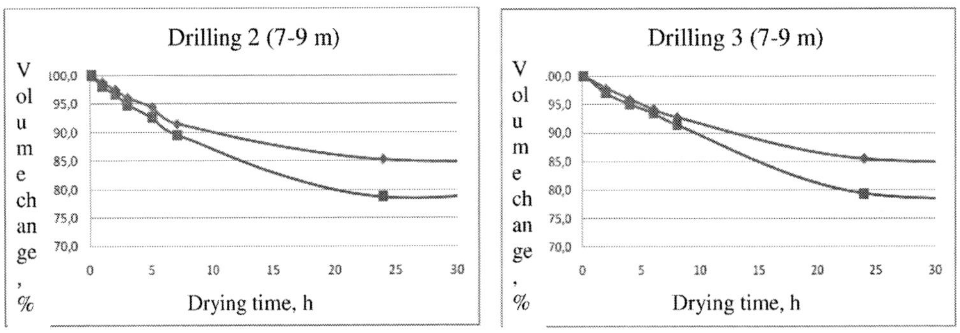

Figure 8. Bourry diagrams (volume and mass losses) during drying of the conventional "Lenti" clay mixtures taken from 7-9 m depths of the 2nd and 3rd drillings

A huge number of literatures were studied which describe the drying properties and drying shrinkage of the conventional brick clays. Many of them [1, 54-57 and 59-62] confirm that, when the amount of the conventional brick clay with a particle size of 2 μm or less are more than 15-20 m%, this will lead to formation of drying cracks due to very fine grain structure in the formed and extruded products. During the drying there were not any cracks or microcracks found not on the surfaces, nor in the cross-section of the extruded ceramic rods, in spite that the "Lenti" clays generally have more than 15 m% up to 33 m% of particles with nanosized. This means that increment of fine particles in the conventional brick clays does not increase the drying sensitivity of the extruded and plastic formed ceramic items. ***The appearances of cracks during drying must be explained with the incorrect forming pressures, residual stresses inside extruded and pressed body and the incorrect heating rate during drying, not with the fine grain size structures.***

After drying, the cylindrical rod specimens were sintered in an electrical chamber kiln with heating rate of 60 °C/hour up to 950 °C and kept at this temperature for 2 hours. The sintered specimens were tested on 3-point bending strength. The bending strength test showed relatively high values which means that the conventional clay minerals from "Lenti" mine can be perfectly used for production of ceramic roof tiles with high quality. The authors could not find a convincing correlation between the bending strength and quartz content but the reasons for the relatively good mechanical bending strength could be attributed to the volumes of XRD amorphous fine particles in the conventional brick clay raw material (Figure 9). So, it can be stated that, the conventional brick clays with higher X-ray amorphous nanoparticles will have good bending strength of the sintered products.

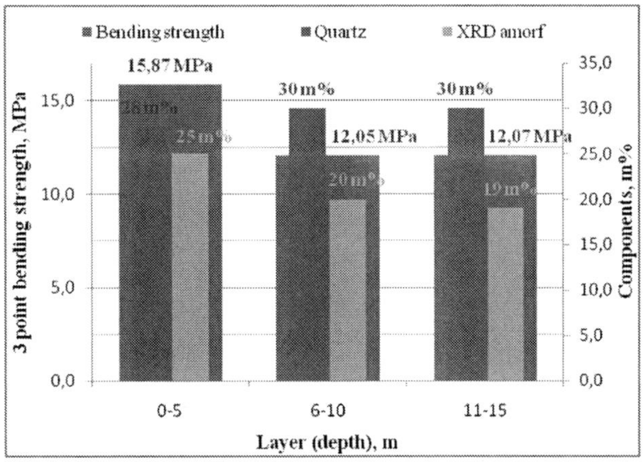

Figure 9. Bending strength, quartz and X-ray amorphous nano particles ratio of specimens made and sintered from the clays of the different layers [14]

4. Conclusions

The most important conclusion of the present research work is that, the "Lenti" conventional brick clay deposit is applicable for producing high quality ceramic roof tiles using relatively not expensive preparation technology (crushing, long rest storage, etc.). The "Lenti" conventional clay is not sensitive not to forming technology, nor to drying. Also, the amount (proportion) of submicron and nanoparticles in the used clay minerals can highly affect the bending strength of the produced ceramics (Figure 9).

Generally, the appearances of cracks during drying must be explained with the incorrect forming pressures, residual stresses inside extruded and pressed body and the incorrect heating rate during drying, not with the fine grain size structures.

IOP Conf. Series: Materials Science and Engineering **613** (2019) 012005 doi:10.1088/1757-899X/613/1/012005

Acknowledgement

The authors acknowledge to the CREATON South-East Europe Ltd and to the IGREX Engineering Service Ltd for their financial and technological support to realize this research work.

"The described study was carried out as part of the EFOP-3.6.1-16-2016-00011 "Younger and Renewing University – Innovative Knowledge City – institutional development of the University of Miskolc aiming at intelligent specialization" project implemented in the framework of the Szechenyi 2020 program. The realization of this project is supported by the European Union, co-financed by the European Social Fund."

References

[1] N A Sarani and A A Kadir 2013 *Advenced Materials Research* **690-693** 919 https://doi.org/10.4028/www.scientific.net/AMR.690-693.919

[2] A A Kadir, H Hinta N A Sarani 2015 *ARPN Journal of Engineering and Applied Science* **10** (15) 6289 ISSN 1819-6608

[3] L Aouba, M Coutand, B Perrin, H Lemercier 2015 *Journal of Building Physics* **38** (6) 531 https://doi.org/10.1177/1744259115571078

[4] J Kalmár, G Kocsis-Buruzs 2016 *Építőanyag-JSBCM* **68** (4) 105 http://dx.doi.org/10.14382/epitoanyag-jsbcm.2016.18

[5] V P Munoz, M P Morales, G V Letelier, M A Mendívil G 2016 *Construction and Building Materials* **125** 241 http://dx.doi.org/10.1016/j.conbuildmat.2016.08.024

[6] L V Korah, P-M Nigay, T Cutard, A Nzihou, S Thomas 2016 *Construction and Building Materials* **125** 654 http://dx.doi.org/10.1016/j.conbuildmat.2016.08.094

[7] I Kocserha, A L Gömze, S Kulkov, E Kalatur, S P Buyakova, R Géber and A Y Buzimov 2017 *J. Phys.: Conf. Ser.* **790** 012013 https://doi.org/10.1088/1742-6596/790/1/012013

[8] M A Tantawy, R S A Mohamed 2017 *Applied Clay Science* **138** 114 http://dx.doi.org/10.1016/j.clay.2017.01.005

[9] A Safeer, I Yassen, M Raz 2017 *Boletin de la Sociedad Espanola de Ceramica y Vidrio* **56** (3) http://dx.doi.org/10.1016/j.bsecv.2017.02.001

[10] P-M Nigay, R Sani, T Cutard, A Nzihou 2017 *Materials Science and Engineering A* 708 http://dx.doi.org/10.1016/j.msea.2017.09.131

[11] M Sveda, R Sokolar, B Janík, Z Stefunková 2017 *Materials Science* **23** (2) http://dx.doi.org/10.5755/j01.ms.23.2.15103

[12] D Dominguez, V P Munoz, V L Munoz 2017 *Materiales de Construcccón* **67** (328) 133 http://dx.doi.org/10.3989/mc.2017.03316

[13] E Kurovics, S N Kulkov, L A Gömze 2018 *Építőanyag-JSBCM* **70** (1) 3 https://doi.org/10.14382/epitoanyag-jsbcm.2018.1

[14] L A Gömze, S N Kulkov, E Kurovics, A S Buyakov, S P Buyakova, A Y Buzimov, R Géber, M V Grigoriev, I Kocserha, A S Kulkov, T Yu Sablina, N L Savchenko, I N Sevostyanova, A Simon 2018 *Építőanyag-JSBCM* **70** (1) 8 https://doi.org/10.14382/epitoanyag-jsbcm.2018.2

[15] L A Gömze, S N Kulkov, E Kurovics, A S Buyakov, A Y Buzimov, M V Grigoriev, B I Kanev, T V Kolmakova, R V Levkov, S A Sitkevich 2018 *Építőanyag-JSBCM* **70** (1) 13 https://doi.org/10.14382/epitoanyag-jsbcm.2018.3

[16] Š Csáki, I Štubňa, V Trnovcová, J Ondruška, L Vozár and P Dobroň 2017 *IOP Conf. Ser.: Mater. Sci. Eng.* **175** 012041 https://doi.org/10.1088/1757-899X/175/1/012041

[17] M Dondi, C Iglesias, E Dominguez, G Guarini, M Raimondo 2008 *Applied Clay Science* **40** 143 https://doi.org/10.1016/j.clay.2007.07.003

[18] I Sperberga, G Sedmale, G Stinkulis, K Zeila and D Ulme 2011 *IOP Conf. Ser.: Mater. Sci. Eng.* **18** 222027 https://doi.org/10.1088/1757-899X/18/22/222027

[19] I Dusenkova, J Malers and L Berzina-Cimdina 2015 *J. Phys.: Conf. Ser.* **602** 012003 https://doi.org/10.1088/1742-6596/602/1/012003

[20] I Sperberga, M Rundans, A Cimmers, L Krage and I Sidraba 2015 *J. Phys.: Conf. Ser.* **602** 012007 https://doi.org/10.1088/1742-6596/602/1/012007
[21] I Calabrese, G Gelardi, M Merli, G Rytwo, L Sciascia and M L Turco Liveri 2013 *IOP Conf. Ser.: Mater. Sci. Eng.* **47** 012040 https://doi.org/10.1088/1757-899X/47/1/012040
[22] G Sedmale, A Korovkins, V Seglins and L Lindina 2013 *IOP Conf. Ser.: Mater. Sci. Eng.* **47** 012056 https://doi.org/10.1088/1757-899X/47/1/012056
[23] M G Khramchenkov, R M Usmanov 2017 *Építőanyag-JSBCM* **69** (4) 110 http://dx.doi.org/10.14382/epitoanyag-jsbcm.2017.19
[24] I Kocserha, A L Gömze, S Kulkov, E Kalatur, S P Buyakova, R Géber and A Y Buzimov 2017 *J. Phys.: Conf. Ser.* **790** 012013 https://doi.org/10.1088/1742-6596/790/1/012013
[25] L Wojcik, P Izak, J Mastalska-Poplawska and M Gajek 2017 *J. Phys.: Conf. Ser.* **790** 012036 https://doi.org/10.1088/1742-6596/790/1/012036
[26] L Wojcik, P Izak, J Mastalska-Poplawska and M Gajek 2017 *J. Phys.: Conf. Ser.* **790** 012036 https://doi.org/10.1088/1742-6596/790/1/012036
[27] Jr. D P Penaloza 2019 *Építőanyag – JSBCM* **71** (1) 5 https://doi.org/10.14382/epitoanyag-jsbcm.2019.2
[28] A Urakov, N Urakova, A Reshetnikov, M Kopylov, A Kasatkin, D Baymurzin and R Gabdrafikov 2017 *IOP Conf. Ser.: Mater. Sci. Eng.* **175** 012012 https://doi.org/10.1088/1757-899X/175/1/012012
[29] A Urakov, N Urakova, A Reshetnikov, A Kasatkin, M Kopylov, D Baimurzin 2016 *Építőanyag – JSBCM* **68** (3) 110 http://dx.doi.org/10.14382/epitoanyag-jsbcm.2016.19
[30] ME Abad-Javier, M Cajero-Juárez, ME Contreras García 2016 *Építőanyag-JSBCM* **68** (4) 124 http://dx.doi.org/10.14382/epitoanyag-jsbcm.2016.22
[31] M G A-A Al-Mosawy, E A J Al-Mulla 2018 *Építőanyag-JSBCM* **70** (4) 116 https://doi.org/10.14382/epitoanyag-jsbcm.2018.21
[32] E Paineau, I Dozov, K Antonova, P Davidson, M Impéror, F Meneau, I Bihannic, C Baravian, A M Philippe, P Levitz 2011 *IOP Conf. Ser.: Mater. Sci. Eng.* **18** 062005 https://doi.org/10.1088/1757-899X/18/6/062005
[33] L H Perng and H -H Huang 2011 *IOP Conf. Ser.: Mater. Sci. Eng.* **18** 062016 https://doi.org/10.1088/1757-899X/18/6/062016
[34] K Tóth, P Bálint, J Bakos 1968 *Építőanyag* **20** (8) 302
[35] P Bálint, K Tóth 1973 *Tonindustrie Ztg.* **97** (5) 122
[36] P Bálint, B Szőke, J Juhász, T Skorecz 1981 *Ceramics International.* **7** (1) 35
[37] L A Gömze, L N Gömze, S N Kulkov, I L Shabalin, I Gotman, F Pedraza, L G Lecomte, T Mayorova, E Kurovics, A Hamza, A 2015 *Építőanyag-JSBCM* **67** (4) 143 http://dx.doi.org/10.14382/epitoanyag-jsbcm.2015.24
[38] M V Grigoriev, N L Savchenko, T Yu Sablina, E Kurovics, I N Sevostyanova, S P Buyakova, L A Gömze, S N Kulkov 2018 *Építőanyag-JSBCM* **70** (1) 18 https://doi.org/10.14382/epitoanyag-jsbcm.2018.4
[39] A Y Buzimov, W Eckl, L A Gömze, I Kocserha, E Kurovics, A S Kulkov, S N Kulkov 2018 *Építőanyag-JSBCM* **70** (1) 23 https://doi.org/10.14382/epitoanyag-jsbcm.2018.5
[40] A S Buyakov, R V Levkov, S P Buyakova, E Kurovics, L A Gömze, S N Kulkov 2018 *Építőanyag-JSBCM* **70** (1) 27https://doi.org/10.14382/epitoanyag-jsbcm.2018.6
[41] M V Vasić, L Pezo, J Zdravković, S Stanković, Z Radojević 2018 *Ceramics International* **44** (2) 1269 https://doi.org/10.1016/j.ceramint.2017.10.191
[42] N Phonphuak and P Chindaprasirt 2018 *Chiang Mai J. Sci.* **45** (4) 1855 http://www.thaiscience.info/Journals/Article/CMJS/10990152.pdf
[43] Gagandeep, Punita Manocha, Rohit Datyal, Amritpal Singh 2018 *International Journal of Management. Technology and Engineering* **8** (9) 1856 ISSN NO: 2249-7455
[44] Anuwat Srisuwan, Nonthaphong Phonphuak and Chiawchan Saengthong 2018 *Suranaree J. Sci. Technol.* **25** (1) 49

[45] P-M Nigay, A Nzihou and T Cutard 2018 *Journal of Materials in Civil Engineering* **30** (4) https://doi.org/10.1061/(ASCE)MT.1943-5533.0002221

[46] David Rieger, Michal Pola, Petr Franče, Jaroslav Kadlec and Tomáš Kovářík 2018 *J. Phys.: Conf. Ser.* **1045** 012037 https://doi.org/10.1088/1742-6596/1045/1/012037

[47] M Martín-Morales, D Eliche-Quesada, M López-Alonso, J Martín-Pascual, L Pérez-Villarejo, D P Ruiz-Padillo, M Zamorano 2018 *Tecnología. Ciencia y Educación* **11** ISSN: 2444-250X / ISSN-e: 2444-288

[48] O Kotova 2013 *IOP Conf. Ser.: Mater. Sci. Eng.* **47** 012037 https://doi.org/10.1088/1757-899X/47/1/012037

[49] O B Kotova, A V Ponaryadov, L A Gömze 2016 *Vestnik IG Komi SC UB RAS* **1** 34 https://doi.org/10.19110/2221-1381-2016-1-34-36

[50] M Rundans, I Sperberga and G Sedmale 2016 *IOP Conf. Ser.: Mater. Sci. Eng.* **123** 012042 https://doi.org/10.1088/1757-899X/123/1/012042

[51] A M Szőke, M Muntean, M Sándor and L Brotea 2016 *IOP Conf. Ser.: Mater. Sci. Eng.* **123** 012043 https://doi.org/10.1088/1757-899X/123/1/012043

[52] O Yu Vaulina, E A Darenskaia, Y V Myachin, I E Vasilyeva and S N Kulkov 2017 *IOP Conf. Ser.: Mater. Sci. Eng.* **175** 012038 https://doi.org/10.1088/1757-899X/175/1/012038

[53] Š Csáki, V Trnovcová, J Ondruška, I Štubňa, P Dobroň, T Václavů, M Záleská and P Bačík 2017 *AIP Conference Proceedings* **1866** (1) https://doi.org/10.1063/1.4994489

[54] G Sopronyi1978 *Építőanyag* **30** (6) 219

[55] K Vértesffy, S Verdes 1985 *Építőanyag* **37** (7) 193

[56] K Junge, A Tretan, E Specht 2007 Energy expenditure for drying of green bricks in chamber dryers; *Zi-ANNUAL* 2007 25 by Müller W. and Fisher A.; Bauverlag BV GmbH; Gütersloh

[57] K Junge, A Tretan, E Specht Drying of green bricks – Material-related principles. kinetics and energetics; *Zi-ANNUAL* 28 Edited by Fischer A, Bauverlag BV GmbH. Gütersloh

[58] L A Gömze, L N Gömze 2008 *Építőanyag-JSBCM* **60** (4) 102 http://dx.doi.org/10.14382/epitoanyag-jsbcm.2008.16

[59] P Aungatichart, S Wada 2009 *Applied Clay Science* **43** (2) 182 http://dx.doi.org/10.1016/j.clay.2008.08.001

[60] I Aksu, E Bazilevskaya, Z T Karpyn 2015 Geo. Res. J. **7** 1 http://dx.doi.org/10.1016/j.grj.2015.02.003

[61] Christopher 2017 Lenthall Destroying the structure of halloysite clay through oven drying, 6th International Young Geotechnical Engineers' Conference (*iYGEC6*)

[62] Haluk Celik 2017 Technological characterization and comparison of two ceramic clays used for manufacturing of traditional ceramic products in Turkey, *Madencilik* **56** (4) 137

IOP Conf. Series: Materials Science and Engineering **613** (2019) 012006 doi:10.1088/1757-899X/613/1/012006

Texture and structure formation of low-carbon low-alloy pipe steel after TMCP and heat treatment

Sergey Danilov[1*], Iana Semkina[1], Igor Pyshmintsev[2] and Mikhail Lobanov[1]

[1]Ural Federal University, Yekaterinburg (Russia)
[2]RosNITI JSC, Chelyabinsk (Russia)

E-mail: s.v.danilov@bk.ru

Abstract. The structure and texture of low-carbon low-alloy pipe steel after Thermo-Mechanical Controlled Processing (TMCP) and subsequent isothermal quenching were studied by the method of Electron Backscatter Diffraction (EBSD) analysis. The texture, obtained after TMCP, was formed mainly by two strong scattered orientations from {112}<110> and two weaker scattered orientations from {110}<223> and one weak orientation (001)[110]. Complex multicomponent textures were observed after heat treatment. The regions with "ferrite" and martensitic structures were identified and analyzed using Oxford Instruments HKL software. Orientations, constituting textures for both selected areas types, are associated with the basic orientations of the initial deformed austenite grains, formed during the hot deformation by orientation relationships (OR), transitory between Kurdjumov-Sachs OR and Nishiyami-Wasserman OR.

1. Introduction

Low-carbon low-alloy steels with ferritic, ferritic-bainitic and bainitic structure are used for the production of pipes applied for the transportation of gas and oil products. An important feature of these steels is a high level of strength properties and a high level of crack resistance [1, 2].

Thermo-Mechanical Controlled Processing (TMCP) is used in the manufacturing of main pipeline sheets. It allows to decrease a metal intensity (pipe wall thickness) and increase the reliability of main pipelines under construction [3]. The transition from ferritic-pearlitic microstructures to structures with a predominance of intermediate transformation products is realized through the application of the accelerated cooling [2, 4].

A crystallographic texture is an additional resource that allows to attach products an improved complex of physicomechanical properties in certain directions. The emergence and development of texture occurs at the stage of manufacturing the product with the direction of deformation and thermal effects on the material [5]. Accounting patterns of texture formation allows you to optimize the production of materials due to the rational choice of temperature-time and deformation intervals of technological operations [6].

In [7–9], it is shown, that a significant role in the processes of pipe steel sheet destruction is played not by the integral texture of the product, but by one of its weak components – (001)[110]. The presence of sufficiently extended regions with a corresponding orientation along the entire length exceeding the critical crack size is important for the development of a crack.

This paper is devoted to the study of the features of the structural-textural state formation in low-carbon low-alloy sheet steel, after heat treatment and TMCP, as a result of which a ferrite-martensitic mixture is formed.

IOP Conf. Series: Materials Science and Engineering **613** (2019) 012006 doi:10.1088/1757-899X/613/1/012006

2. Research Methodology

As materials for research, we used samples of low-carbon low-alloyed tube steel 06G2MB with predominantly bainitic structure after TMCP, characterized within the strength class (X80) by similar levels of mechanical properties ($\sigma_{0.2}$ in the direction of deformation ~ 575 – 585 H/mm^2). Samples were heat treated in laboratory muffle furnaces and tank furnaces. The process of austenitization was carried out at 1000 °C. Some samples were water quenched. Another part of the samples was transferred to the tank furnace at a temperature of 700 °C, where they were aged for 10 and 20 minutes and then water quenched. Metallographic thin sections on the RD-ND plane were prepared on the samples (where RD is the rolling direction at TMCP, ND is the normal direction to the rolling plane).

The electron microscopic study of the structure was carried out on a ZEISS CrossBeam AURIGA scanning microscope with an accelerating voltage of 20 kV. EBSD HKL Inca prefix with the Oxford Instruments analysis system was used to determine the orientation of individual grains and analyze the local texture. Scanning step – 0.1 microns. The error in determining the orientation of the crystal lattice is no more than ± 1° (on average ± 0.6°). Low-angle boundaries between crystallites were built on orientation maps with misorientations of 2 – 7° (the thickness of the boundaries in the figures is 1 pixel) and 8 – 15 ° (thickness is 2 pixels); high-angle boundaries were drawn with misorientations ≥ 15° (3 pixels thick). The texture was studied using the construction of orientation distribution functions (ODF). The analysis of the coincident site lattice (CSL) between the individual grains was carried out by constructing them on orientation maps, taking into account the Brandon standard ± $\Delta\Theta$ criterion incorporated in the software. It constitutes a specific value for each boundary: $\Delta\Theta = 15°/(\Sigma n)^{1/2}$, where Σn is the number of coinciding nodes when three-dimensional crystal lattices are superimposed.

3. Results and Discussion

The initial samples after TMCP were characterized by a predominantly bainitic structure (grain size 5 – 30 μm) formed within the limits formed during isothermal hot rolling of deformed austenite grains (ND size 10–30 μm) [7]. The texture of the initial samples consisted mainly of the same scattered components: two strongly pronounced orientations of {112}<110>, two weaker orientations of {110}<223> and weak orientation (001)[110] [7].

A package structure characteristic of martensite consisting of alternating plates with a thickness of 0.5 to 2 μm and a length of up to 20 μm was observed in the samples in the case of water quenching [10]. About 2 – 3 crystallographic orientations of crystallites were observed within the limits of one martensitic package. The spectra of the intercrystalline boundaries corresponded to a martensitic structure [11, 12]. Only CSL boundaries were fixed in the spectrum of special boundaries: Σ3, Σ11, Σ25b, Σ33c, Σ41c. In [12], it was shown that the appearance of this spectrum is the result of a shear phase transformation in accordance with the orientation relations (OS) intermediate between the Kurdumov-Sachs (K-S) OS and the Nishiyama-Wasserman (N-W) OS.

A ferritic-martensitic mixture was fixed in the structure in the samples after quenching from a temperature of 700 °C in the case of isothermal exposures during 10 and 20 minutes (Figure 1, a–c). In this case (Figure 1, a, b), martensite differed from the "direct" quenched martensite in greater dispersion and a higher level of microstresses, expressed by a high density of low-angle boundaries. Obviously, this is due to the increased amount of carbon in this martensite compared with martensite, formed from austenite, the chemical composition of which corresponded to the concentration of elements in steel. Ferritic grains in ferritic-martensitic structure (Figure 1, a, c) had a size slightly smaller compared to the pure ferritic structure obtained in the work [10] after cooling with furnace to room temperature.

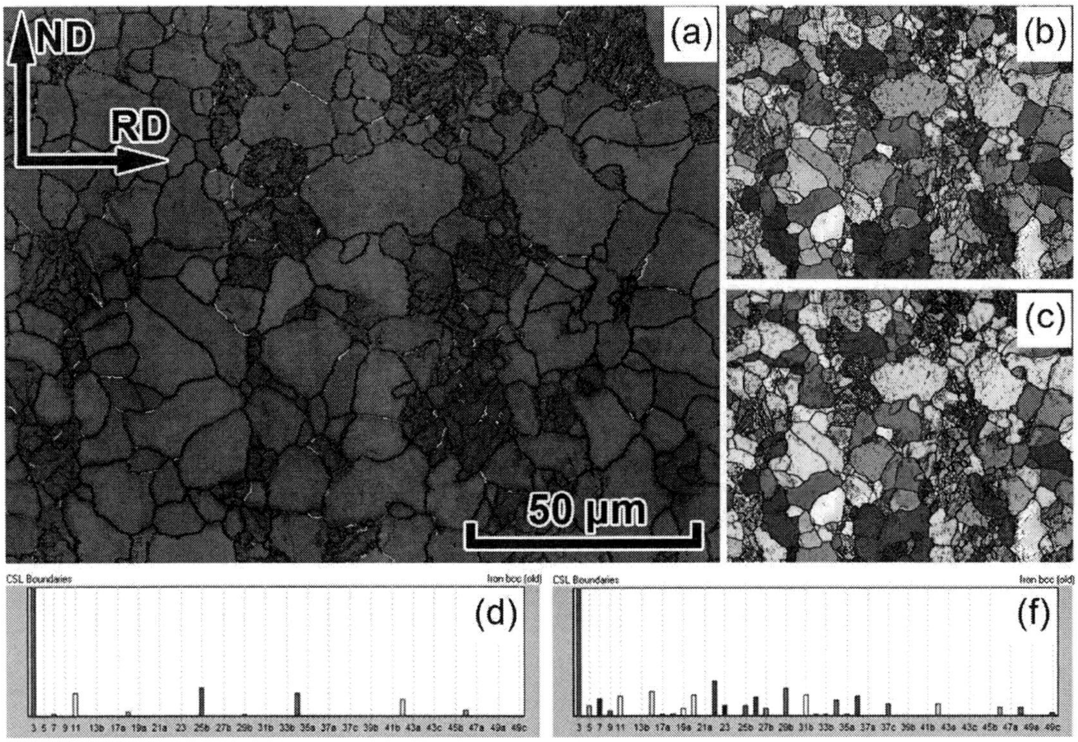

Figure 1. The microstructure of 06G2MB steel after isothermal quenching with 20 min exposure at the temperature of 700 °C obtained by EBSD: a - orientation map with CSL boundaries; b - orientation map, colored martensite; c - orientation map, colored ferrite; d - spectra of CSL boundaries in martensite; f - spectra of CSL boundaries in ferrite

Martensitic and ferritic structural components were separated and individually analyzed almost "manually" using Oxford Instruments HKL software (Figure 1, b, c). Martensitic regions differed from ferritic in size, more elongated shape and increased number of low-angle boundaries. During processing of orientation maps using software tools, areas revealed as a result of boundaries detection were identified and summarized by the presence of several characteristic features. The resulting combined regions (Subsets in software) were collections of martensitic or ferritic texture components for which ODFs were built. Areas that were in doubt (not part of any of the sets) were discarded.

Interestingly, the distribution of intergranular and special boundaries of misorientation angles defined separately for martensitic and ferritic structures constituting substantially repeated spectra corresponding in the martensite and hardening ferrite [10] obtained during the isothermal exposure (Figure 1, d, f).

It is important to emphasize that crystallographic textures obtained in samples as a result of shear and diffusion transformation demonstrate significant similarity, while repeating the bainite texture that existed before heat treatment (Figure 2). All components of these textures can be obtained in accordance with the intermediate between the OS K-S and N-W from the main orientations of rolled austenite (that is, a material with FCC lattice): {112}<111>, {110}<111>, { 110}<112>, {110}<001> [13]. Obviously, the severity of individual texture components in different samples is explained by the relatively small EBSD survey areas containing a small amount of crystallites.

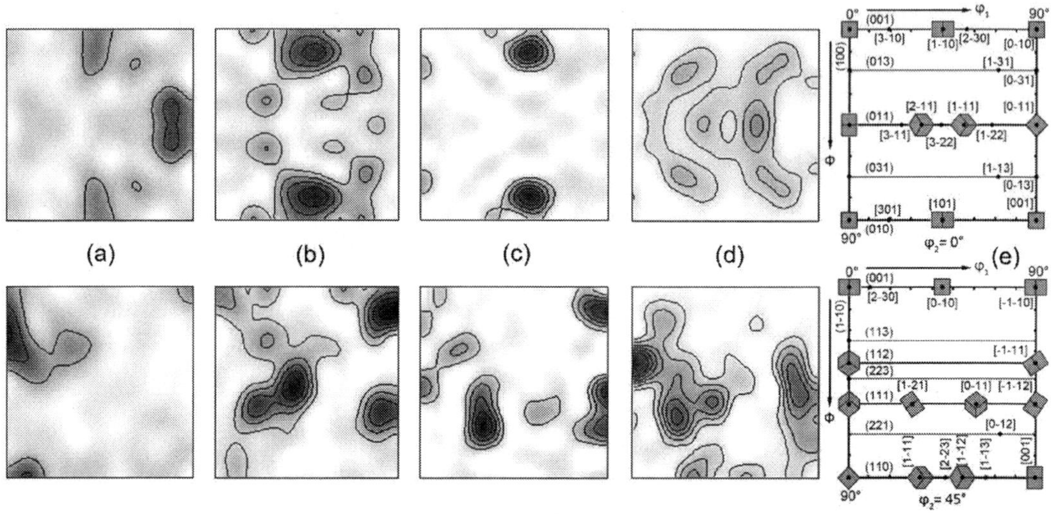

Figure 2. 06G2MB steel texture in the form of ODF after TMCP and various heat treatments: a, b, c, d – section of the Euler angles of space at $\varphi_2 = 0°$, $\varphi_2 = 45°$; e – standard grids for cross sections of ODF $\varphi_2 = 0°$ and $\varphi_2 = 45°$ with the application of ideal orientations in the form of elementary crystallographic cells (view from TD); a – martensite; b – martensite in ferritic-martensitic structure; c – ferrite; d – ferrite in ferritic-martensitic structure

4. Conclusion

It was established that in samples of low-carbon low-alloy pipe steel with the structure formed as a result of TMSP, with their further heat treatment, a pronounced textural heredity is observed. During heat treatments of steel, including double phase recrystallization of $\alpha_{Bainite} \rightarrow \gamma \rightarrow \alpha'$or α, formed by both the shear and diffusion mechanism, the structure is characterized by the same crystallographic texture as the original bainite.

Acknowledgment
This article was prepared with the financial support of Competitiveness Enhancement Program – CEP 3.1.1.1-18

References

[1] K Hulka, P Peters, F Heisterkamp 1997 *Steel.* **10** 62 (In Russ)
[2] A B Arabey 2010 *Steel Transl.* **40** 601 http://dx.doi.org/10.3103/S0967091210070016
[3] D J Stalheim 2013 *Metallurgist.* **55** 53 (In Russ)
[4] E Shigeru, N Naoki 2015 *JFE Technical Report.* **20** 1
[5] M Shtremel 1997 *Strength alloys. Part II. Deformation* (Moscow: Publisher MISIS) 527
[6] A S Belyaevskikh, M L Lobanov, G M Rusakov, A A Redikul'tsev 2015 *Steel Transl.* **45** 982 http://dx.doi.org/10.3103/S0967091215120037
[7] I Y Pyshmintsev, A O Struin, A M Gervasyev, M L Lobanov, G M Rusakov, S V Danilov, A B Arabey 2016 *Metallurgist.* **60** 405 http://dx.doi.org/10.1007/s11015-016-0306-7
[8] I Pyshmintsev, A Gervasyev, R H Petrov, V C Olalla, L Kestens 2012 *Mater. Sci. Forum.* **702–703** 770 http://dx.doi.org/10.4028/www.scientific.net/MSF.702-703.770
[9] M A Mohtadi-Bonab, M Eskandari, J A Szpunar 2014 *Mater. Sci. Eng., A.* **620** 97 http://dx.doi.org/10.4028/10.1016/j.msea.2014.10.009
[10] M L Lobanov, M D Borodina, S V Danilov, I Y Pyshmintsev, A O Struin 2017 *Steel Transl.* **47** 710 http://dx.doi.org/10.3103/S0967091217110055
[11] M Lobanov, S Danilov, A Struin, M Borodina, I Pyshmintsev 2016 *Bulletin of the South Ural State University. Ser. Metallurgy.* **16** 46 (In Russ) http://dx.doi.org/10.14529/met160207
[12] G Rusakov, M Lobanov, A Redikul'cev, A Belyaevskikh 2014 *Phys. Metals Metallogr.* **115** 775 http://dx.doi.org/10.1134/S0031918X14080134
[13] M Hölscher, D Raabe, K Lücke 1994 *Acta metall. mater.* **42** 879 http://dx.doi.org/10.1016/0956-7151(94)90283-6

IOP Conf. Series: Materials Science and Engineering **613** (2019) 012007 doi:10.1088/1757-899X/613/1/012007

An "in situ" study of study of quenched Ti-Al system samples during heating

Sergey Demakov, Iana Semkina, Irina Kirillova

Ural Federal University, 620002 Yekaterinburg, Russia

Iana.semkina@urfu.ru

Abstract. The study of the Ti-Al system samples containing up to 10 wt.% Al water quenched from different temperatures in β-phase region, was carried out in situ during heating on X-ray diffractometer. XRD analysis of quenched samples showed the presence of α'-martensite only. After quenching the increase of Al content in alloys displayed the decrease of the lattice parameters «a» and «c» and the rise of «c/a» ratio, since the lattice parameter «c» dropped slightly in contrast with the lattice parameter «a». Furthermore, Thermo-XRD (T-XRD) analysis represented the anisotropy of the coefficient of thermal expansion (CTE). Besides, T-XRD analysis indicated, that «c/a» ratio of samples Ti-0.5 wt.% Al with the least amount of Al revealed the fall of «c/a» ratio and at the same time samples contained 7 and 8 wt.% Al remained practically the same during heating. Whereas, samples Ti-10 wt.% Al with the largest quantity of Al showed the rise of c/a ratio during heating.

1. Introduction

The objective of this work was the study of the «c/a» parameter changes during the heating compared with room temperature values. Regularities of parameters «c/a» change allow more accurately determine the HCP lattice deformation behavior of α-, pseudo-α- and (α+β)-titanium alloys of different composition with changing deformation temperature, since it directly depends on the alloy composition and, accordingly, on the parameter «c/a» [1, 2].

2. Experiment and Research Methodology

The research material were samples of plates with a thickness of 3 mm made of titanium alloys with different aluminum content. The specimens were water quenched and annealed at 300 ° (1 hour). The charge and obtained compositions of the studied alloys are shown in table 1.

Table 1 Chemical composition and quenching temperatures of studied specimens

№	*Al content, weight %*		*Quenching temperature, °C*	*Exposure time, h*
	charge composition	*chemical analysis*		
1	0	0,25	930	0,5
2	7	5,91	1100	0,5
3	8	7,12	1100	0,5
4	10	9,85	1100	0,5

Cutting of samples was carried out on a wire erosion machine ECOCUT. Hot pressing was carried out on a CitersPress-20 press by Struers, PolyFast resin was used. Micrographs were prepared on a LaboPol-5 grinding and polishing machine LaboForce-1 with the sequential use of SiC sanding paper for removing scale and leveling the surface of the sample. Electropolishing of thin sections was carried out in a solution of 20% HClO4 + 80% CH3COOH (glacial acetic acid), at a temperature of 4–8 ° C, current density I = 1–2 A / cm2. The samples were etched in an aqueous solution of nitric and hydrofluoric acids (1 h. HNO3, 1 h. HF, 8 h. H2O).

The study of quenched and annealed samples was carried out by X-ray diffraction analysis (XRD) and Thermo-XRD (TXRD). Both analysis were performed by a BRUCKER D8 X-ray diffractometer equipped electric furnace with a vertical θ-θ-goniometer in the angular range 2θ = 30 ... 120 ° in Kα Cu radiation (λ = 0.154178 nm) in the temperature range 30 ... 290 ° C (temperature step of 65 °). Diffractograms were also obtained after cooling at 30 °C.

The diffraction patterns were processed using TOPAS software.

3. Results and Discussions

Figure 1 illustrates the coincidence of the position of the experimentally obtained (blue) and reference diffractograms for the α-phase (red). The crystal lattice parameters «a» and «c» were calculated using the TOPAS software. Table 2 represents the results of the calculation of the periods «a», «c», the «c/a» ratio as well as the measurement errors.

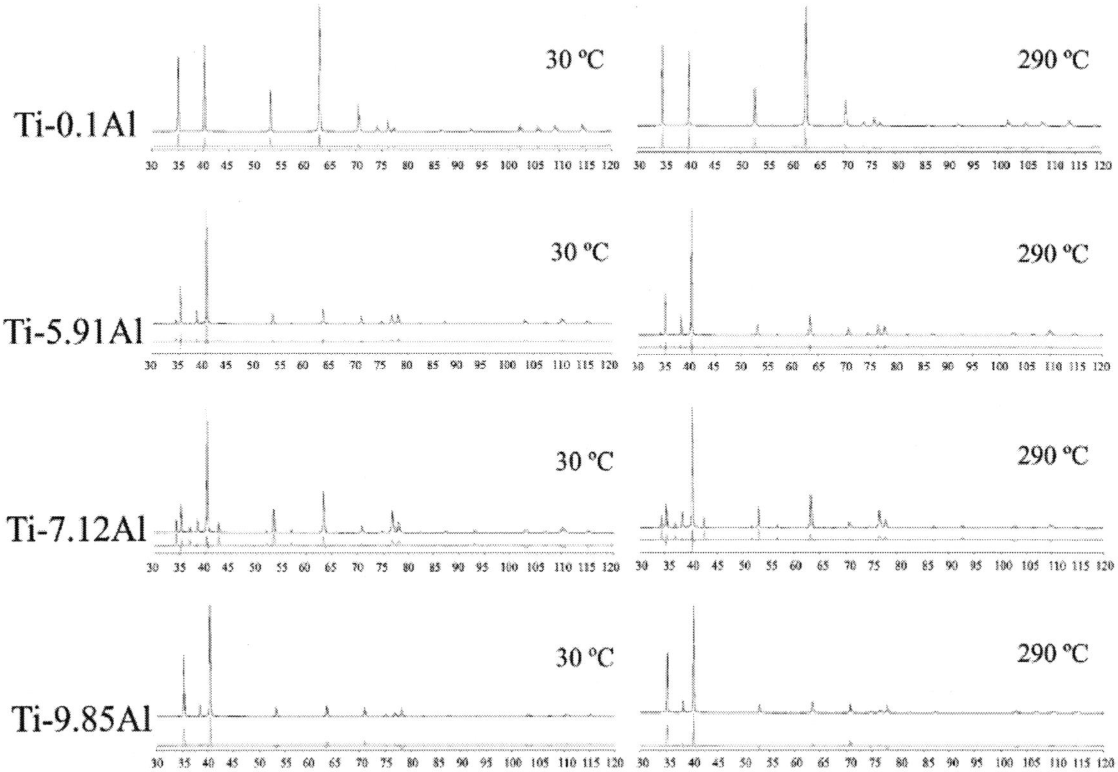

Figure 1. The diffraction patterns taken at the initial and maximal temperatures in the angular range 2θ = 30 ... 120 ° and the second line is the difference between theoretically calculated by the Rietveld method and experimental diffractograms

Table 2 Orthorhombic martensite lattice parameters of WQ and annealed specimens

№	*Al content*		*a, nm·10*	*Δa, nm·10*	*c, nm·10*	*Δc, nm·10*	*c/a*	*Δ(c/a)*
	weight. %	*at. %*						
1	0,25	0,44	2,9501	0,0001	4,6845	0,0002	1,5879	0,0002
2	5,91	10,03	2,932	0,0002	4,6723	0,0004	1,5935	0,0002
3	7,12	11,98	2,929	0,0001	4,6712	0,0002	1,5948	0,0002
4	9,85	16,25	2,9215	0,0002	4,670	0,0003	1,5984	0,0001

A graphic illustration of the obtained values of the periods of the crystal lattice is presented in Figure 2. The resulting graphs clearly show an increase in the periods «a» and «c» depending on temperature, and the pattern obtained is well described by a linear function (coefficient of determination R^2 in all cases > 0.97). The calculation of the periods on the diffraction patterns obtained at 30 °C before and after cooling showed the same values within the error. A similar experiment for technically pure titanium [3] also revealed the existence of a «c/a» parameter decrease in the temperature range from 25 to 480 °C. About 6% aluminum alloying leads to temperature stabilization of this parameter, and this ratio begins to increase as the temperature rises to 290 °C at higher concentrations of aluminum.

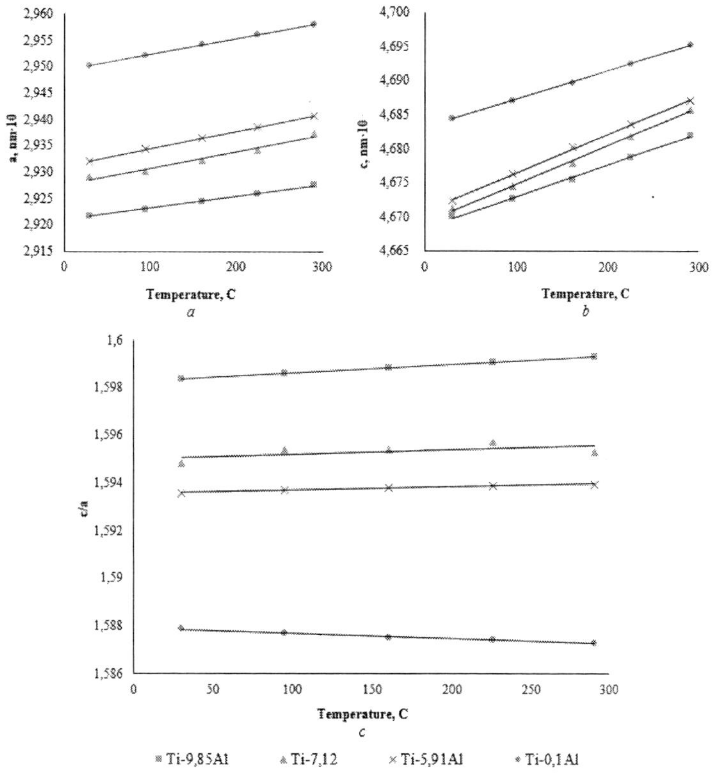

Figure 2. The graphs of the parameters «a» *(a)*, «c» *(b)* and «c/a» *(c)* on the aluminum content in the alloys studied during heating in the temperature range of 30 ... 290 ° C

IOP Conf. Series: Materials Science and Engineering **613** (2019) 012007 doi:10.1088/1757-899X/613/1/012007

4. Conclusions

The Ti-alloys of the Ti-Al system, the ratio c/a does not remain constant at room and T=290 °C temperatures. During heating from 30 °C to 290 °C of an alloy with a minimum Al content (Ti-0.1Al), the initial value of the ratio decreases to «c/a» = 1.5873. In case of Ti-5.91Al sample, this ratio remained almost the same in the process of heating. In Ti-7.12 Al and Ti-9.85Al alloys the ratio of lattice parameters «c/a» increased.

As a result of the study, data available in the literature on the behavior of α'-HCP martensite lattice alloying with an aluminum and under temperature exposure were confirmed. Also, new regularities were revealed in the change of the ratio of the lattice parameters «c/a» alloying and temperature changed.

Acknowledgement

This article was prepared with the financial support of Competitiveness Enhancement Program – CEP 3.1.1.1-18

References

[1] Zwikker Ulrich 1974 Titan und Titanlegirungen *Springer-Verlag* 717 https://doi.org/10.1007/978-3-642-80587-5

[2] S L Demakov, O A Elkina, A G Illarionov et al. 2008 *Phys. Metals Metallogr.* **105** 602 https://doi.org/10.1134/S0031918X08060112

[3] M Jafari, M Vaezzadeh and S Noroozizadeh 2010 *Metall and Mat Trans A* **41** 3287 https://doi.org/10.1007/s11661-010-0393-1

IOP Conf. Series: Materials Science and Engineering **613** (2019) 012008 doi:10.1088/1757-899X/613/1/012008

Engineered composite polymer sheets with enhanced thermal conductivity

Oguzhan Der[1], Stuart Edwardson[2], Marco Marengo[3] and Volfango Bertola

[1] Laboratory of Technical Physics, School of Engineering, University of Liverpool, Liverpool L69 3GH, United Kingdom
[2]Centre for Laser Manufacturing, School of Engineering, University of Liverpool, Liverpool LG9 3GH, United Kingdom
[3]School of Computing, Engineering and Mathematics, University of Brighton, Brighton BN2 4GJ, United Kingdom

E-mail: Volfango.Bertola@liverpool.ac.uk

Abstract. An innovative type of engineered composite polymer sheet with enhanced thermal conductivity is described and tested. The concept is to sandwich a closed-loop serpentine channel, cut out in a polypropylene sheet and containing a self-propelled liquid-vapour mixture, between two sheets of the same material bonded by selective laser welding. Polymer materials are widely used to replace metals in different applications, however sometimes this is not possible because of their poor thermal conductivity. Thin, flexible and low-weight polymer sheets can be advantageous in various aerospace, aircraft and portable electronic applications where the device weight and its mechanical flexibility are crucial. The transient and steady-state thermal response of a prototype engineered polymer sheet was characterised for different heat power levels and spatial orientations. The equivalent thermal conductance, calculated from the surface temperatures at opposite ends of the sheet, increases four to six times in comparison with a composite polypropylene sheet without working fluid.

1. Introduction

The use of polymeric materials to replace metallic parts is the obvious choice to address weight and cost constraints in a large number of devices and applications, including space, aircraft and portable electronics applications. Whilst polymeric materials offer excellent features of mechanical flexibility, resistance to fatigue, low weight and low cost in comparison with metallic materials, they exhibit poor heat transfer performances due to their low thermal conductivity, which for commercial polymers is often less than 1-2 W/mK.

Recently, there were several attempts to increase the thermal conductivity of polymers by means of high-thermal conductivity additives and fillers, such as minerals, fibres and metals [1]. The fillers size often reaches the nanometric scale to obtain a direct and efficient interaction between the composite components. The final properties of such nanocomposites depend on the properties of their components and on the shape, size, distribution and orientation of the dispersed phase. Fillers can be added in-situ during the synthesis process or ex-situ by mechanically dispersing the filler into a molten polymer using a turbomixer or a twin-screw extruder. Commonly used fillers include particles [2, 3], fibres [4], metal powders or particles [5, 6], and carbon nanotubes [7-9].

IOP Conf. Series: Materials Science and Engineering **613** (2019) 012008 doi:10.1088/1757-899X/613/1/012008

In the present work, it is proposed to enhance the thermal conductivity of polymer sheets by embedding a self-driven liquid-vapour mixture, which transfers heat from an evaporator to a condenser region of the material according to the well-known working principle of pulsating (or oscillating) heat pipes (PHP) [10]. The pulsating heat pipe is a thermally driven heat transfer device, which usually consists of a long tube in a serpentine configuration, which is firstly evacuated and then partially filled with a working fluid. Due to capillarity, the working fluid initially resides as an alternation of liquid plugs and vapour slugs. When heat is applied to the evaporator zone, the fluid inside the tube is activated: liquid plugs receive sensible heat while vapour slugs expand due to evaporation and push the heated fluid towards the condenser zone. There, vapour slugs undergo condensation, releasing heat to a cold source. This promotes a pulsating/circulating slug flow.

A simple and cost-effective manufacturing process to embed the heat transfer fluid into composite polymer sheets is described. Using this process, a number of prototype composite polypropylene sheets were fabricated and tested. Results indicate the equivalent thermal conductivity of the engineered composite material shows a four- to six-fold increase with respect to the conductivity of the composite polypropylene sheet without working fluid.

2. Design and manufacturing

Prototype composite sheets consisted of one black polypropylene layer (0.7 mm thick) sandwiched between two transparent polypropylene layers (0.4 mm thick), all with a length of 250 mm and a width of 98 mm. The material was chosen based on: (i) mechanical properties (elastic modules, yield stress, resistance to fatigue); (ii) compatibility with a range of organic heat transfer fluids (ethanol, acetone, refrigerant fluids); (iii) suitability to laser cutting and welding process. The maximum continuous service temperature is 130°C.

A serpentine shaped channel was cut-out in the central black layer, and featured either three or five turns, as illustrated in Figure 1. The channel had a width of 9 mm, which was determined so that the hydraulic diameter, $D_H = 1.3$ mm, satisfies the design criterion given in Eq. (1), which ensures surface forces prevail on gravity [11]:

$$0.7\sqrt{\frac{\sigma}{g(\rho_L-\rho_G)}} \leq D_H \leq 1.8\sqrt{\frac{\sigma}{g(\rho_L-\rho_G)}} \tag{1}$$

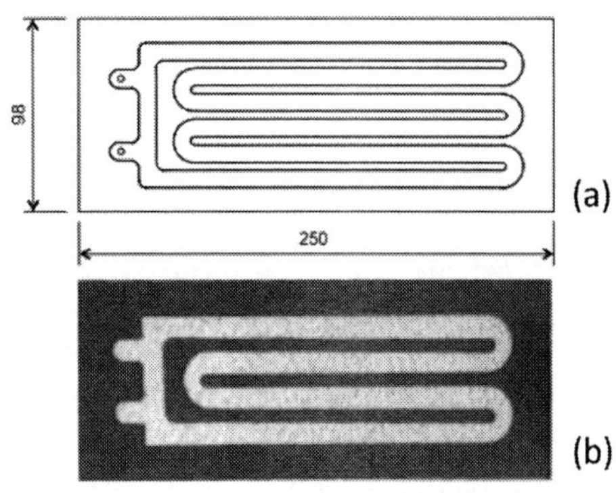

Figure 1. Technical drawing of the five-turns channel (a) and top view of the three-turns channel cut-out in the polypropylene sheet (b).

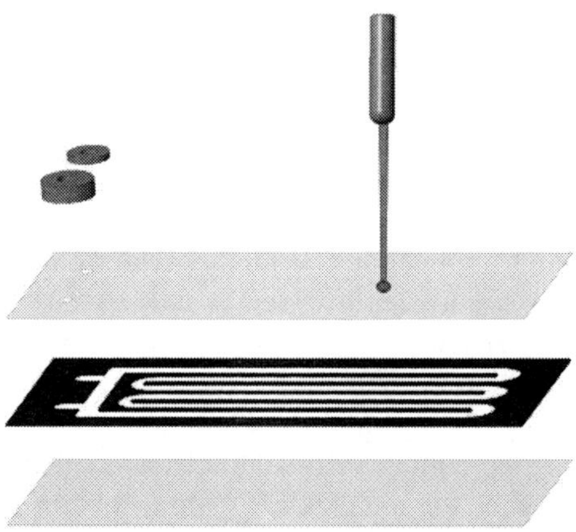

Figure 2. Schematic of the PHP assembly and laser welding process.

IOP Conf. Series: Materials Science and Engineering **613** (2019) 012008 doi:10.1088/1757-899X/613/1/012008

According to Eq. (1) the hydraulic diameter depends on the fluid properties; in particular, for ethanol (ρ = 789 kg/m^3; σ = 22 mN/m) Eq. (1) yields 1.18 mm $\leq D_H \leq$ 3.04 mm, while for the refrigerant fluid FC-72 (ρ = 1680 kg/m^3; σ = 10 mN/m) the criterion becomes 0.54 mm $\leq D_H \leq$ 1.4 mm. The five-turns channel (Figure 1a) had a total volume of 8.4 ml, while the three-turns channel (Figure 1b) had a total volume of 5 ml.

The transparent polypropylene layers were bonded on the two sides of the channel by selective laser welding [12, 13], as shown schematically in Figure 2, to create a strong and seamless joint between adjacent layers; while the external sheets are transparent to the laser wavelength, the central black layer absorbs the same wavelength, to enable polymer bonding exactly at the interface, without affecting the rest of the material. Two polypropylene rings were glued (Loctite All Plastic) in correspondence of two holes in one of the transparent sheets, to fit a pressure transducer and a micro-metering valve used for introducing the heat transfer fluids (FC-72 for the five-turns channel and ethanol for the three-turns channel, respectively).

3. Experimental setup and procedure

A schematic view of the experimental setup is displayed in Figure 3. The polymer sheet sample was mounted on a vertical support, with the heated zone (evaporator) at the bottom and the cooled zone (condenser) at the top. The heat supply was provided to both sides of the evaporator by two ceramic heaters (100 W each); two copper plates were used to distribute heat uniformly, and a thin layer of heat sink paste was applied on surfaces to minimise the contact resistance. The heaters were connected to a regulated DC power supply (Circuit Specialists CSI 120001X). Two fan-assisted heat sinks (Malico) were used to remove heat from the condenser section.

The polymer sheet was connected to a pressure transducer (GEMS 3500, 0-160kPa) and to a micro-metering valve (Upchurch Scientific); the pressure transducer DC output was sampled at 1 Hz by a data acquisition system (LabJack U6). Eight surface K-thermocouples (Omega Engineering) were placed on the heat pipe surface, four in the evaporator and four in the condenser zone. The temperature distribution in the adiabatic region was monitored by a FLIR One infra-red camera (FLIR Systems Inc.).

To introduce the working fluids (ethanol and refrigerant FC-72), composite sheets were vacuumed to a pressure of 0.2±0.5 kPa (abs) using a two-stage vacuum pump (Bacoeng) connected to the micro-metering valve; after disconnecting the pump, the valve was connected to an external reservoir

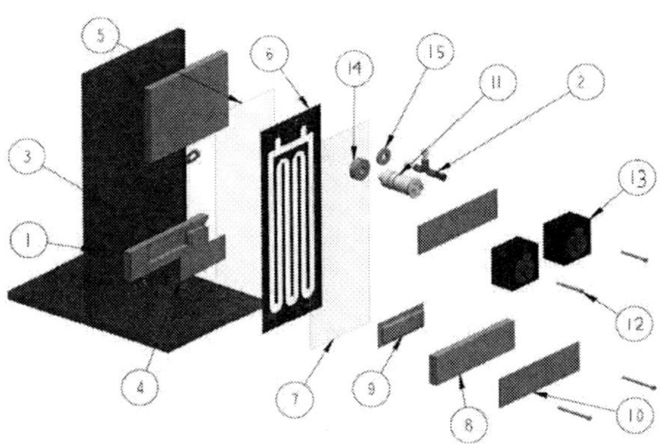

Figure 3. Experimental setup assembly: (1) support frame; (2) valve; (3,8) insulation; (4,9) ceramic heaters; (5,6,7) polypropylene sheets; (10) holder; (11) pressure transducer; (13) cooling fans.

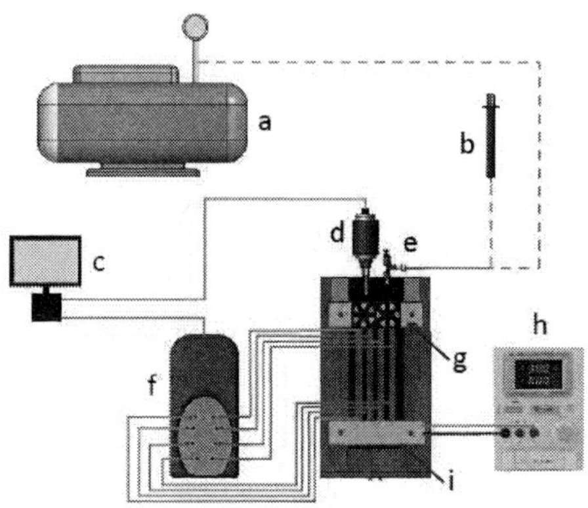

Figure 4. Schematic layout of the experimental setup: (a) vacuum pump; (b) syringe; (c) data logger and PC; (d) pressure transducer; (e) micro-metering valve; (f) thermocouples unit; (g) cooling fans; (h) power supply; (i) ceramic heaters.

IOP Conf. Series: Materials Science and Engineering **613** (2019) 012008 doi:10.1088/1757-899X/613/1/012008

containing the heat transfer fluid and gently re-opened until the desired amount (3 ml for ethanol and 4 ml for FC-72, corresponding to 60% and 50% of the total hollow volumes, respectively) flowed into the composite sheet.

Experiments were conducted by applying to the evaporator section an ascending/descending stepped heating power ramp ranging approximately between 2 W and 30 W, and measuring the temperatures on the composite polymer sheet surface. For each power step, the heat supply was kept constant until a pseudo steady-state regime was attained. Tests were interrupted earlier in case any point of the material reached a temperature of 110°C.

The equivalent thermal conductance of the engineered composite polypropylene sheet in the longitudinal direction was calculated as:

$$K_{eq} = \frac{\dot{Q}}{T_{ev} - T_{cond}} \tag{2}$$

where T_{ev} and T_{cond} are the averages of the four thermocouples temperatures measurements of the evaporator and condenser sections, respectively, and $\dot{Q}$ is the heating power supply.

4. Results

An example of the temperature distribution in the adiabatic region when the composite polymer sheet is heated in the evaporator zone, obtained from FLIR measurements, is shown in Figure 5. Unlike materials that are homogeneous in the direction of the heat flux, where one can identify a temperature gradient with isothermal lines orthogonal to the heat flux, the engineered composite sheet displays alternating hot and cold stripes, corresponding to the ascending boiling fluid and the descending condensate, respectively.

Temperatures measured in the evaporator and in the condenser zones of the composite polymer sheet during the ascending/descending heating power supply ramp are displayed in Figure 6. Due to the relatively small thermal conductivity of polypropylene, and to the intrinsically unstable gas-vapour flow, the composite sheet exhibits significant thermal inertia, and reaches a pseudo-steady state in about 40 minutes after each step change in the heating power supply.

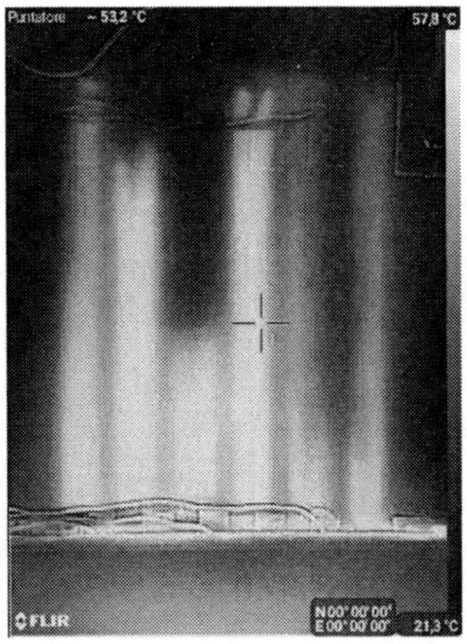

Figure 5. FLIR images of the adiabatic region during the heating ramp.

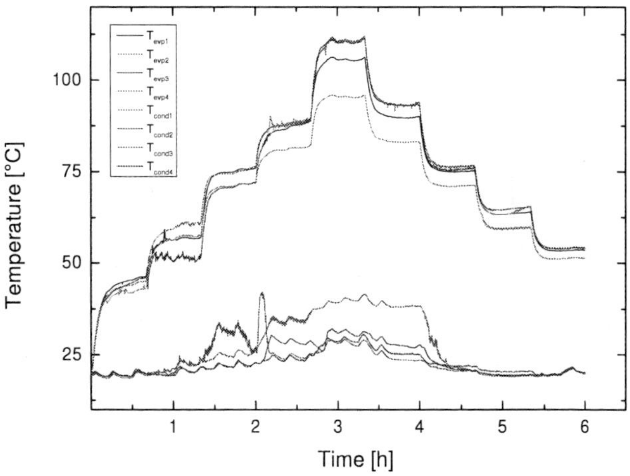

Figure 6. Temperatures measured in the evaporator and in the condenser zones of the composite sheet during the ascending /descending power supply ramp.

The overall thermal performance of the composite polymer sheet is shown in Figures 7 and 8, which display the equivalent thermal conductance of the material between a hot end and a cold end, calculated according to Eq. (2), as a function of the heating power supply for the three-turn-channel sheet filled with ethanol and the five-turn-channel sheet filled with FC-72, respectively. To quantify the heat transfer enhancement due to the fluid circulation, these values are compared with the equivalent thermal conductance of the vacuumed composite sheets without heat transfer fluid.

The highest value of the equivalent thermal conductance is observed the five-turn-channel sheet filled with FC-72, at the maximum heating power supply of 31.3 W, where it attains a maximum value of 0.74 W/°C; this means the equivalent thermal conductance of the PHS increases of 585% with respect to the composite polypropylene sheet without working fluid (0.126 W/°C). The corresponding increase of thermal conductivity observed in the three-turn-channel sheet filled with ethanol is about 250% higher in comparison with the composite sheet without working fluid (0.067 W/°C), i.e. 0.16 W/°C at the maximum heating power supply of 14.7 W. The different performance is obviously due to the different number of turns of the channel, and the different fluid properties, which make evaporation of FC-72 (heat capacity: 1.1 kJ/kgK; latent heat; 88 kJ/kg) easier compared with ethanol (heat capacity: 2.5 kJ/kgK; latent heat; 846 kJ/kg).

In both cases, the maximum performance is limited by the temperature of the composite sheet in the evaporator region, which is limited by the maximum continuous service temperature of the material used.

5. Conclusions

A novel concept of engineered composite polymer sheet was designed and manufactured using three polypropylene sheets, bonded together by selective laser welding, where the central sheet contains a serpentine channel filled with a heat transfer fluid. The thermal response was evaluated for two different geometries of the channel and two different fluids, for different values of the heat input at the evaporator.

Preliminary results indicate a 250% increase of the equivalent thermal conductance for the three-turn-channel sheet filled with ethanol and a 585% increase for the five-turn-channel sheet filled with FC-72, respectively. The proposed technology represents a promising route to produce composite polymeric materials with enhanced thermal characteristics.

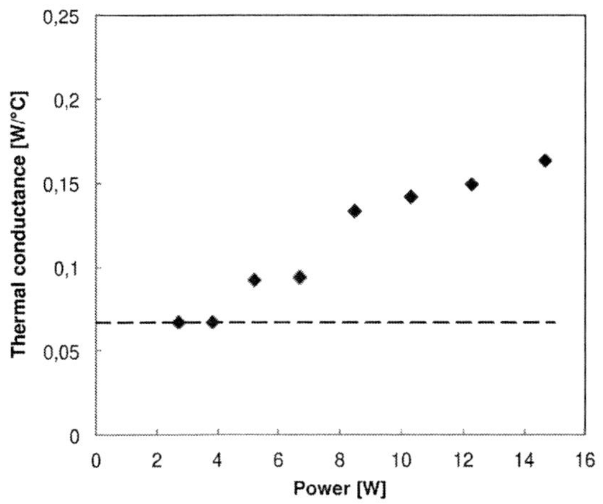

Figure 7. Equivalent thermal conductance of three-turn sheet filled with ethanol, as a function of the heat rate supplied at the evaporator. The dashed line indicates the thermal conductance of the composite sheet without fluid.

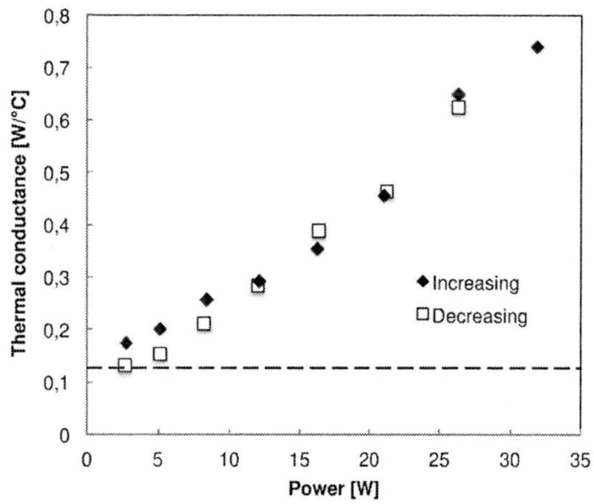

Figure 8. Equivalent thermal conductance of five-turn sheet filled with FC-72, as a function of the heat rate supplied at the evaporator. The dashed line indicates the thermal conductance of the composite sheet without fluid.

Acknowledgments
O. Der gratefully acknowledges a YLSY doctoral studentship from the Republic of Turkey, Ministry of National Education.

References
[1] Mazumadar S K 2002 Composite manufacturing: materials, product and process engineering *Boca Raton, Fla, CRC Press* London https://www.taylorfrancis.com/books/9781420041989
[2] Kumlutas D and Tavman I H 2006 *Journal of Thermoplastic Composite Materials* **19** (4) 441 https://doi.org/10.1177/0892705706062203
[3] Krupa I, Boudenne A and Ibos L 2007 *European Polymer Journal* **43** (6) 2443 https://doi.org/10.1016/j.eurpolymj.2007.03.032
[4] Tavman I H, Akinci H 2000 *International Communications in Heat and Mass Transfer* **27** (2) 253 https://doi.org/10.1016/S0735-1933(00)00106-8
[5] Maiti S N, Ghosh K 1994 *Journal of Applied Polymer Science* **52** (8) 1091 https://doi.org/10.1002/app.1994.070520810
[6] Chae D W, Hwang S S, Hong S M, Hong S P, Cho B G, Kim B C 2007 Influence of High Contents of Silver Nanoparticles on the Physical Properties of Poly(Vinylidene Fluoride) *Taylor & Francis* Great Britain **464** 233 https://tandfonline.com/doi/abs/10.1080/15421400601031140
[7] Biercuk M J, Llaguno M C, Radosavljevic M, Hyun J K, Johnson A T, Fischer J E 2002 *Applied Physics Letters* **80** 2767 https://doi.org/10.1063/1.1469696
[8] Haggenmueller R, Guthy C, Lukes J R, Fischer J E, Winey K I 2007 *Macromolecules* **40** (7) 2417 https://pubs.acs.org/doi/abs/10.1021/ma0615046
[9] Winey K I, Kashiwagi T, Mu M 2007 *MRS Materials Research Society* **32** 348 https://doi.org/10.1557/mrs2007.234
[10] Ma H 2015 Oscillating Heat Piper *Springer New York* https://link.springer.com/book/10.1007/978-1-4939-2504-9
[11] Paudel S B, Michna G J 2014 *ASME Proceedings* ICNMM2014-22016 V001T05A006 https://doi.org/10.1115/ICNMM2014-22016
[12] Acherjee B Kuar A, Mitra S, Misra D 2015 *International Journal of Advanced Manufacturing Technology* **78** (5-8) 853 http://doi.org/10.1007/s00170-014-6693-7
[13] Lui H, Jiang H, Chen G, Guo D, Yan Z, Li P, Wang X 2016 *International Journal of Advanced Manufacturing Technology* **86** (1-4) 809 http://doi.org/10.1007/s00170-015-8242-4

IOP Conf. Series: Materials Science and Engineering **613** (2019) 012009 doi:10.1088/1757-899X/613/1/012009

The Influence of Cr doping on the Structural and Magnetic Properties of $HoMnO_3$ Multiferroic Ceramics

J. F. M. Ibrahim[1,2,3], A. Mergen[1], Umut Parlak[1,4], Emese Kurovics[2]

[1]Department of Metallurgical and Materials Engineering, Marmara University, Turkey.
[2]Department of Materials Sciences and Engineering, University of Miskolc, Hungary.
[3]Department of Materials Technology, University of Bahri, Sudan.
[4]Peter Grünberg Institute, Research Centre Jülich, Jülich, Germany.

jamalfadoul@gmail.com, fememese@uni-miskolc.hu, umut.parlak@marmara.edu.tr.

Abstract. In this research single-phase Cr-doped $HoMnO_3$ ceramics were synthesized via solid state reaction method. The structural and magnetic characteristics of the prepared samples were examined using various characterization techniques, including x-ray diffraction (XRD), scanning electron microscopy (SEM), energy dispersive spectroscopy (EDX) and vibrating sample magnetometer (VSM). All samples exhibit single phase that emphasized by SEM and XRD. Magnetic properties of the prepared samples at low temperature (10K) showed the presence of different magnetic characteristics. The change in magnetic behavior could probably occur due to slight decrease in the unit cell volume that leads to a structural distortion in the lattice. Cr-doped $HoMnO_3$ could be a good candidate with a better ferromagnetic property for practical multiferroic applications.

1. Introduction

Nowadays the research, development and applications of ceramic materials are the most fast growing field of material science [1-16]. Multiferroic materials which show coexistence of (anti) ferromagnetic and ferroelectric orders have achieved a great deal of interest for many years [17-19]. In principle, these materials could enable single device components to perform multiple functions [20]. Due to magnetoelectric effect in these materials many expected technical applications are recommended such as, future multiple-state memory technologies, traditional magnetic data storage tools and high-density low-power data storage devices [21]. Furthermore, they could be applied in various ways for sensors, transducers as well as fiber optical communications [22]. Although these materials are highly interesting, different challenges have emerged in dealing with multiferroicity, which hinder the practical applications of these materials, the first dilemma is that multiferroic materials are rare and normally show small coupling of multiferroic characteristics or exist only at low temperatures [23]. Also, synthesizing these kind of materials normally produces a mixture of the main phase along with other impurity phases [24]. To address these obstacles, intensive work has been done to discover new single phase multiferroic materials, as well as to improve the multiferroic properties through doping processes.

Single-phase rare-earth manganites are one of the well-known multiferroic materials, it is of huge interest in the context as different cations can be chosen in order to combine magnetic and polar properties. The multiferroic phenomenon normally occurs in perovskite-structure oxide [25]. In these materials multiferroic properties can be enhanced through A-site doping which is a partial replacement in the rare-earth site, or B-site doping which is a partial replacement in manganese site.

Multiferroicbehavior has been reported in hexagonal and orthorhombic *RE*MnO_3 [26, 27], *RE*Mn_2O_5 [28, 29] and some other compounds lately. Strong enhancement of magnetoelectric coupling has been reported in Dy^{3+} doped $HoMnO_3$, Fe-substituted Mn in $YMnO_3$ and $ErMnO_3$ respectively [30-32]. Moreover, substantial changes in the electric and magnetic properties have been observed in Yttrium doped $GdMnO_3$ [33]. Since $HoMnO_3$ (HMO) compounds in either the orthorhombic or hexagonal structure are multiferroic and exhibiting both antiferromagnetic and ferroelectric order below antiferromagnetic transition temperature, Cr-doped HMO samples may become interesting due to formation of ferromagnetism of Mn moments and structural distortion in the lattice, which can enhance the dipole ordering. Cr-doped HMO samples could be the promising candidate for multiferroic applications performing good ferroelectric and ferromagnetic properties. As far as is concerned, the effect of Cr doping in $HoMnO_3$ will be performed first time in the literature. This research work aims at substituting Cr into Mn site in the $HoMnO_3$ to produce single phase perovskite multiferroic ceramics.

2. Experimental methods

2.1. Synthesis of $HoMn_{1-x}Cr_xO_3$

Single-phase $HoMn_{1-x}Cr_xO_3$ (x= 0.2 and 0.8) bulk ceramic materials were prepared by conventional solid state synthesis technique. A high purity Ho_2O_3 (99.9%; Alfa Aesar), Mn_2O_3 (98%; Alfa Aesar) and Cr_2O_3 (99.9%; Alfa Aesar) powders were used as reactants. Stoichiometric proportions of these powders were milled to obtain a homogeneous mixture of green powder of fine particle size. The green powder was then mixed in ethanol in a plastic container using zirconium balls for 6 hours. Two different compositions were prepared; $HoMn_{1-x}Cr_xO_3$ where x= 0.2 and 0.8. After drying the liquid mixture in an oven at 95°C for 24 h, the powder mixture was calcined at 600°C for 10 h, then ground in a mortar to refine particles and sintered at different temperatures between 1100-1350°C for 10 h with a heating and cooling rate of 5°C/min to obtain the phases development. Part from the powders were used to produce dense cylindrical pellets with radius about 5 mm and thickness 2 mm, using pressing device with pressure of 10 MPa. Heat treatment was carried out in a high temperature programmable furnace at 1400°C for 24 h with a heating and cooling rate of 2°C/min.

2.2. Characterization

he structural properties of the sintered ceramic materials were investigated using a Bruker D2 PHASER powder x-ray diffractometer, operated in the Bragg-Brentano geometry with Cu Kα radiation at 1.54 Å in the angular range of $10° \leq 2\theta \leq 70°$ with a scanning rate of 1°/min and a step size of 0.01016°. FullProf/Rietveld software was used for computer based examination. Microstructures of the sintered samples were studied using scanning electron microscopy (SEM, JEOL5910 LV) with different magnification values using secondary and back-scattered electrons and surface analyses of samples were investigated by energy dispersive X-ray spectrometer (EDX). Furthermore, the magnetic measurements of Cr doped $HoMnO_3$ samples were made at low temperature (10 K) using the vibrating sample magnetometer (VSM).

3. Results and discussion

3.1. Structures

Figure 3.1 shows the XRD pattern of B-site doped $HoMn_{1-x}Cr_xO_3$ (x= 0.2, 0.8) ceramics carried out at ambient temperature. The unit cell parameters were obtained using FullProf/Rietveld software. The observed and standard diffraction peaks from Inorganic Crystal Structure Database showed good matching and no additional peaks were seen. Rietveld refinements on all the samples confirmed the synthesizing of pure single-phase perovskite structure crystallized in an orthorhombic structure (Figures. 1, 2 and 3).

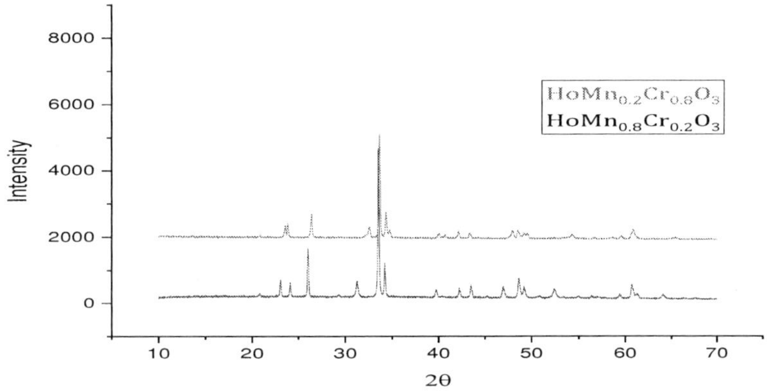

Figure 1. Room temperature XRD patterns of $HoMn_{1-x}Cr_xO_3$ ceramics compound (x=0.2 and 0.8) sintered at 1350°C.

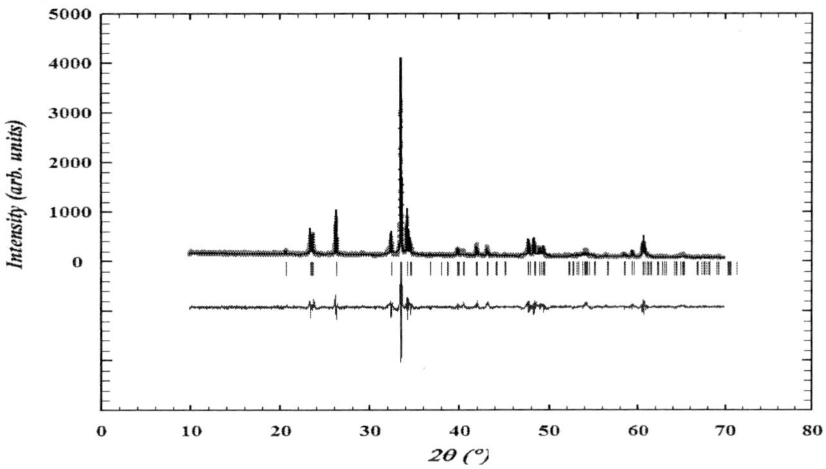

Figure 2. XRD pattern for $HoMn_{1-x}Cr_xO_3$ ceramics sample at (x= 0.2) Rietveld fit agreement factors are Rp= 56.8, Rwp = 37.0 Chi2=3.48.

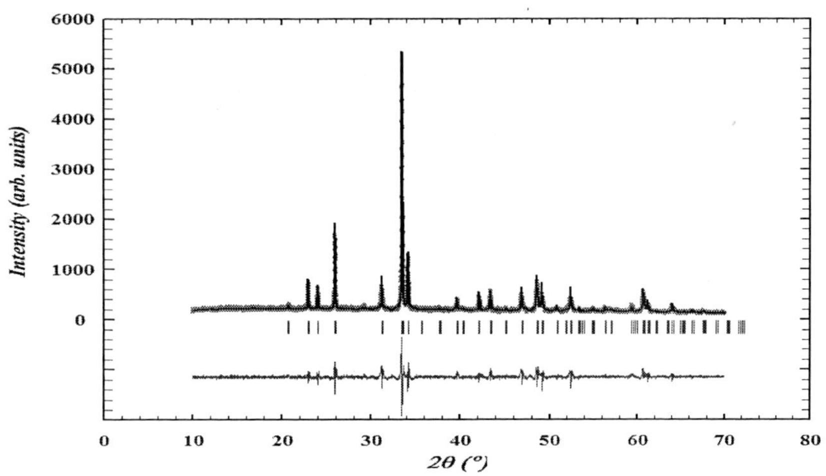

Figure 3. XRD pattern for $HoMn_{1-x}Cr_xO_3$ ceramics sample at (x= 0.8) Rietveld fit agreement factors are Rp= 73.4, Rwp = 48.0, Chi2=4.32.

The most important notice to be made here is that a close investigation of the peaks shows a presence of a shift in the peak positions to the higher angles with increasing the amount of doping elements. This feature agrees with Vegard's law [34], as the replacement of large radius Mn_{q+} ion (rMn= 1.79 Å) by smaller Cr_{q+} (rCr=1.18 Å) result on decrease in the unit cell volume that may cause a slight distortion in the lattice and a modification of the crystal structure. This latter might be a crucial factor controlling the magnetic behavior of the material. FullProf/Rietveld software was used to evaluate the unit cell parameters (Table 1). With increasing the amount of Cr the units cell volume is decreasing due to the smaller ionic radius of Cr (1.18 Å) compared to Mn (1.79 Å).

Table 1. Lattice parameters resulted from refinement procedure of powder XRD pattern of $HoMn_{1-x}Cr_xO_3$ ceramics compound (x=0.2, 0.8).

Composition	a (Å)	b (Å)	c (Å)	Volume (Å)^3	Space group
$HoMnO_3$	5.2700	5.8400	7.360	226.52	Pnma
$HoMn_{0.8}Cr_{0.2}O_3$	5.2720	5.7950	7.380	225.45	Pnma
$HoMn_{0.2}Cr_{0.8}O_3$	5.2277	5.5080	7.508	215.92	Pnma

3.2. Scanning electron microscopy (SEM) and EDS investigation of $HoMn_{1-x}Cr_xO_3$

Microstructure investigations of Cr-doped $HoMnO_3$ are carried out by SEM and shown in Fig. 4(a), (b), (c) and (d). From the secondary and backscattered SEM photographs of the two samples ($HoMn_{0.8}Cr_{0.2}O_3$and $HoMn_{0.2}Cr_{0.8}O_3$), it is clearly visible that there is no any phase contrast, which is a good indication for the formation of the single phase. The grain sizes of the prepared samples were between 2-4 μm and 1-3 μm respectively, showing that the grains sizes were decreasing when the doping concentration were increased.

The results of EDS analysis taken from grains of $HoMn_{0.8}Cr_{0.2}O_3$ and $HoMn_{0.8}Cr_{0.2}O_3$ showed that the theoretical weight percentages were near to the calculated values. Au and Pd were due to coating made on the samples.

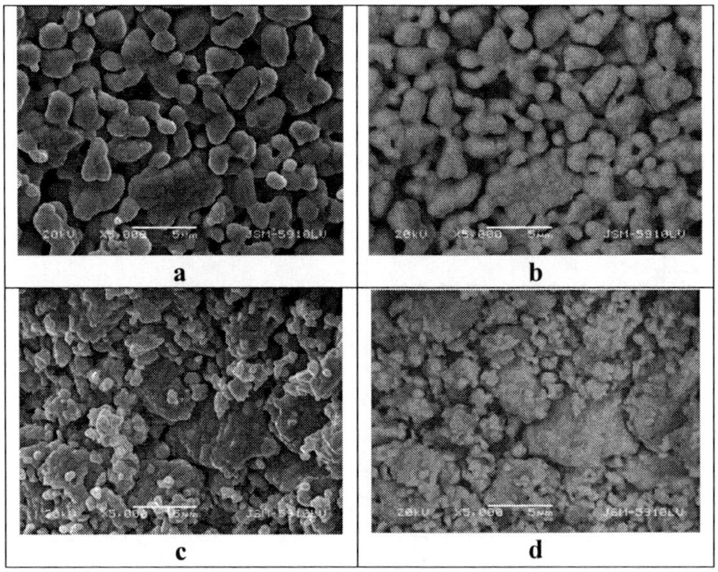

Figure 4. SEM micrographs of Cr doped $HoMnO_3$ sample at ×5.000
a) SEI of $HoMn_{0.8}Cr_{0.2}O_3$, **b)** BEI of $HoMn_{0.8}Cr_{0.2}O_3$,
c) SEI of $HoMn_{0.2}Cr_{0.8}O_3$ and **d)** BEI of $HoMn_{0.2}Cr_{0.8}O_3$
samples ($HoMn_{1-x}Cr_xO_3$ ceramic where x=0.2 and x=0.8).

IOP Conf. Series: Materials Science and Engineering **613** (2019) 012009 doi:10.1088/1757-899X/613/1/012009

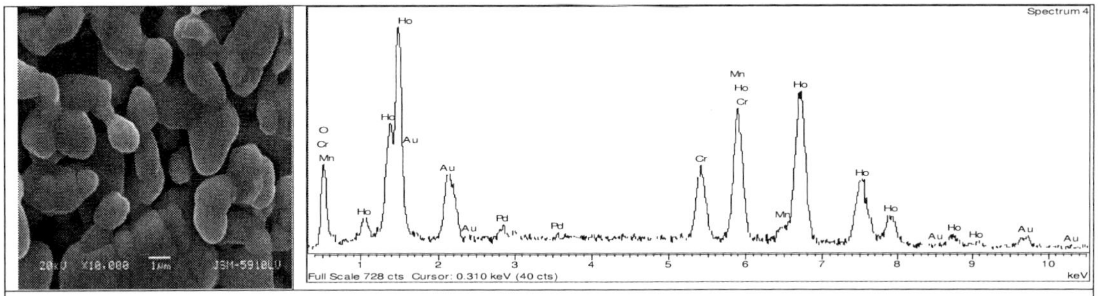

Elements	O	Cr	Mn	Ho	Total
Wt (%)	13.11	6.95	13.51	66.43	100.00
Calculated (Wt %)	17.95	3.89	16.44	61.70	100.00

a

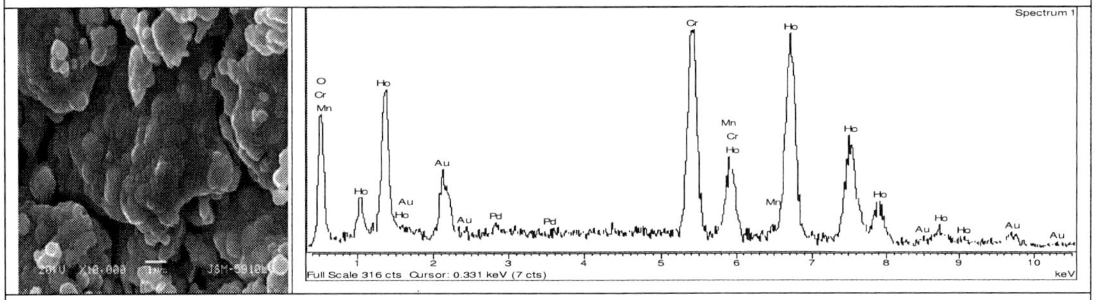

Elements	O	Cr	Mn	Ho	Total
Wt (%)	13.87	16.48	3.54	66.10	100.00
Calculated (Wt %)	18.07	15.66	4.138	62.11	100.00

b

Figure 5. EDS analysis Cr doped $HoMnO_3$ samples **a)** EDS analysis of $HoMn_{0.8}Cr_{0.2}O_3$ and b) EDS analysis of $HoMn_{0.2}Cr_{0.8}O_3$ sample (Tables show the theoretical weight fractions of the elements and practical values from EDS results).

3.3. Magnetic properties

The magnetic characteristics of $HoMn_{1-x}Cr_xO_3$ (x= 0.2, 0.8) multiferroic ceramics were studied at low temperature (10 K) to reveals the influence of Cr doping in the magnetic behaviors. Magnetization of $HoMn_{0.8}Cr_{0.2}O_3$ and $HoMn_{0.8}Cr_{0.2}O_3$ at 10 K as a function of magnetic field (M–H) is shown in Fig. 5.The enhancement in the magnetization of $HoMn_{0.2}Cr_{0.8}O_3$ multiferroic ceramic might be attribute to the change in canting angle [35] or suppression of the complicated spiral spin structure [36] along with coupling interaction between the Mn^{3+} and Cr^{3+} ions in the sample [37].

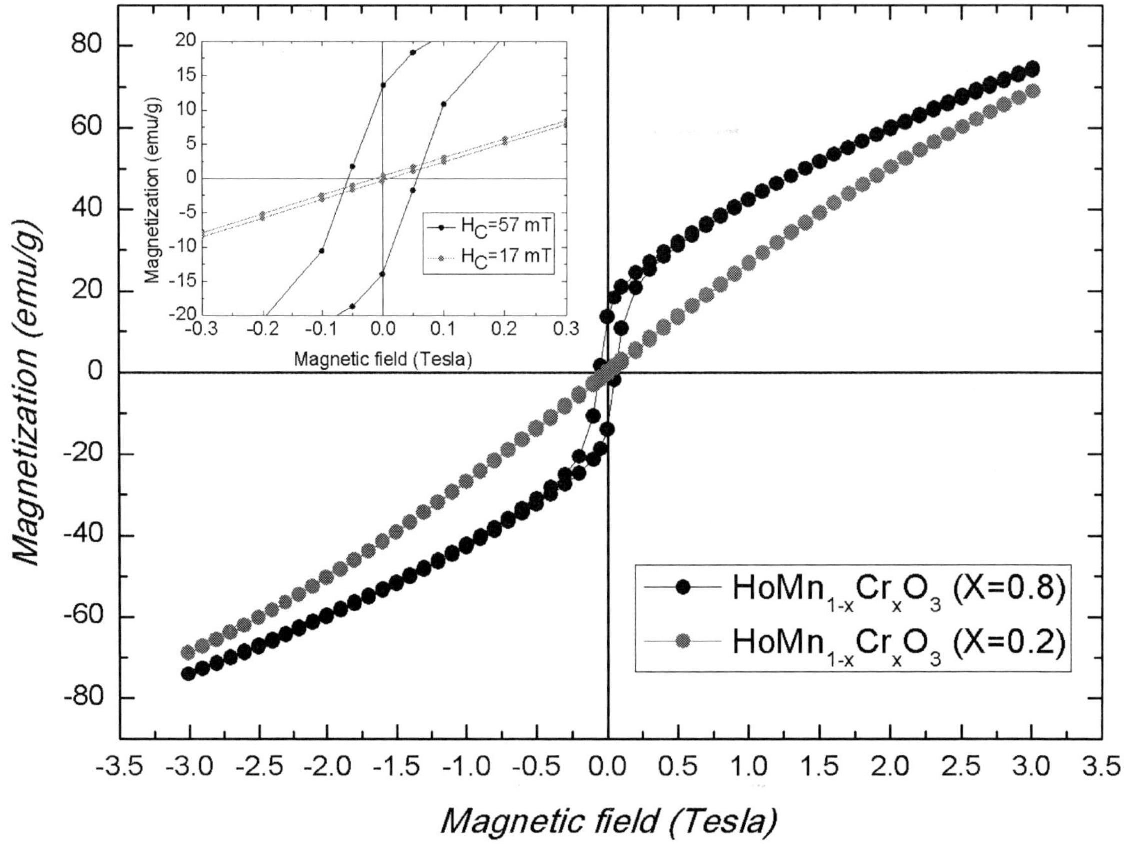

Figure 6. The magnetic hysteresis loops of $HoMn_{1-x}Cr_xO_3$ ceramics compound (0.2 and 0.8) sintered at 1350°C

4. Conclusion

The work in this research verifies the influence of B-site doping on the structural and magnetic properties of bulk HMO multiferroic ceramics. Through tuning the reaction conditions such as, sintering temperature and the chemical synthesis procedure, a variety of single phase orthorhombic $HoMn_{1-x}Cr_xO_3$ bulk samples with different morphologies, structure and compositions were successfully prepared via conventional solid state reaction route at ambient conditions, all samples showed distorted orthorhombic perovskite structures with space group Pnma. The produced doped samples exhibit significant change of the magnetic characteristics probably due to the structural distortion.

5. Acknowledgments

We would like to express our massive thank to Prof. Ayhan MERGEN, for all the help and academic guidance he has contributed to this work, who unfortunately passed away, may his soul rest in peace. We also like to thank Turkish government for their financial support in this research.

"The described study was finalized as part of the EFOP-3.6.1-16-2016-00011 "Younger and Renewing University – Innovative Knowledge City – institutional development of the University of Miskolc aiming at intelligent specialization" project implemented in the framework of the Szechenyi 2020 program. The realization of this project is supported by the European Union, co-financed by the European Social Fund."

References

[1] L A Gömze, L N Gömze 2008 *Építőanyag–JSBCM* **60** (4) 102 http://dx.doi.org/10.14382/epitoanyag-jsbcm.2008.16
[2] L A Gömze, L N Gömze 2009 *Építőanyag–JSBCM* **61** (2) 38 http://dx.doi.org/10.14382/epitoanyag-jsbcm.2009.7
[3] L A Gömze, L N Gömze 2010 *Építőanyag–JSBCM* **62** (4) 98 http://dx.doi.org/10.14382/epitoanyag-jsbcm.2010.18
[4] L A Gömze 2010 *Mater. Sci. Forum* **659** 19 http://dx.doi.org/10.4028/www.scientific.net/MSF.659.19
[5] L A Gömze, L N Gömze 2011 *IOP Conf. Ser.: Mater. Sci. Eng.* **18** 082001 http://dx.doi.org/10.1088/1757-899X/18/8/082001
[6] L A Gömze, L N Gömze 2013 *IOP Conf. Ser.: Mater. Sci. Eng.* **47** 012033 http://dx.doi.org/10.1088/1757-899X/47/1/012033
[7] N Savchenko, L A Gömze, S N Kulkov 2014 *AIP Conference Proceedings* **1623** (1) 547 http://dx.doi.org/10.1063/1.4899003
[8] S N Kulkov, S R Buyakova, L A Gömze 2014 *Építőanyag–JSBCM* **66** (1) 2 http://dx.doi.org/10.14382/epitoanyag-jsbcm.2014.1
[9] A S Apkaryan, S N. Kulkov and L A Gömze 2014 *Építőanyag–JSBCM* **66** (2) 38 http://dx.doi.org/10.14382/epitoanyag-jsbcm.2014.8
[10] N L Savchenko, I N Sevastyanova, T Yu Sablina, L Molnár, R Géber, L A Gömze, S N Kulkov, L N Gömze 2014 *Építőanyag–JSBCM* **66** (2) 44 http://dx.doi.org/10.14382/epitoanyag-jsbcm.2014.9
[11] L A Gömze, L N Gömze, S N Kulkov, I L Shabalin, I Gotman, F Pedraza, L G Lecomte, T Mayorova, E Kurovics, A Hamza 2015 *Építőanyag–JSBCM* **67** (4) 149 http://dx.doi.org/10.14382/epitoanyag-jsbcm.2015.24
[12] L A Gömze 2016 Applied Materials Science I. Compilation of Selected Scientific Papers, *Published by IGREX Ltd,* Hungary ISBN 978-963-12-6600-9
[13] L A Gömze, L N Gömze 2017 *IOP Conf. Ser.: Mater. Sci. Eng.* **175** 012001 http://dx.doi.org/10.1088/1757-899X/175/1/012001
[14] E Kurovics, A Shmakova, B Kanev, L A Gömze 2017 *IOP Conf. Ser.: Mater. Sci. Eng.* **175** 012013 http://dx.doi.org/10.1088/1757-899X/175/1/012013
[15] L A Gömze 2018 Applied Materials Science II. Compilation of Selected Scientific Papers, *Published by IGREX Ltd,* Hungary ISBN 978-615-00-3118-7
[16] A Y Buzimov, S N Kulkov, L A Gömze, R Géber, I Kocserha 2018 *Inorg. Mater. Appl. Res.* **9** 910. https://doi.org/10.1134/S2075113318050040
[17] W Eerenstein, N D Mathur, and J F Scott 2006 *Nature* **442** (7104) 759 https://doi.org/10.1038/nature05023
[18] T Kimura, T Goto, H Shintani, K Ishizaka, T H Arima, and Y Tokura, 2003 *Nature* **430** (6962) 55 https://doi.org/10.1038/nature02018
[19] T Lottermoser, T Lonkai, U Amann, D Hohlwein, J Ihringer, and M Fiebig, 2004 *Nature* **430** (6962) 55 https://doi.org/10.1038/nature02728
[20] D P Norton 2009 Synthesis and Characterization of $BaFeO_3$,(Ba, Bi)FeO_3, and Related Epitaxial Thin Films and Nanostructures https://apps.dtic.mil/dtic/tr/fulltext/u2/a510215.pdf
[21] L Gerhard, T K Yamada, T Balashov, A F Takács, R J H Wesselink, M Däne, M Fechner, S Ostanin, A Ernst, I Mertig, and W Wulfhekel, 2010 *Nature Nanotechnology* **5** (11) 792 https://doi.org/10.1038/nnano.2010.214
[22] Y N Venevtsev, V V Gagulin and I D Zhitomirsky 1987 *Ferroelectrics* **73** (1) 221

[23] J W Kim, K Nenkov, L Schultz, K Dörr 2009 *Journal of Magnetism and Magnetic Materials* **321** (11) 1727 https://doi.org/10.1016/j.jmmm.2009.02.007
[24] A Ghazala 2011 Multiferroic properties of Mn doped $La_{0.8}Bi_{0.2}FeO_3$ system and effect of SHI irradiation http://hdl.handle.net/10603/12935
[25] A K Kundu, R Ranjith, B Kundys, N Nguyen, V Caignaert, V Pralong, W Prellier, and B Raveau, 2008 *Applied Physics Letter* **93** (5) 052906 https://doi.org/10.1063/1.2961022
[26] Z J Huang, Y Cao, Y Y Sun, Y Y Xue, C W Chu, C.W. 1997 *Journal of Physical Review B* **56** (5) 2623 https://doi.org/10.1103/PhysRevB.56.2623
[27] B Lorenz, Y Q Wang, Y Y Sun, C W Chu 2004 *Journal of Physical Review B* **70** (21) 212412 https://doi.org/10.1103/PhysRevB.70.212412
[28] D Higashiyama, S Miyasaka, N Kida, T Arima, Y Tokura 2004 *Journal of Physical Review B* **70** (17) 174405 https://doi.org/10.1103/PhysRevB.70.174405
[29] S Kobayashi, T Osawa, H Kimura, Y Noda, I Kagomiya, K Kohn 2004 *Journal of the Physical Society of Japan* **73** (6) 1593 https://doi.org/10.1143/JPSJ.73.1593
[30] J Magesh, P Murugavel, R V K Mangalam, K Singh, C Simon and W Prellier 2012 *Applied Physics Letters* **101** (2) 022902 https://doi.org/10.1063/1.4733367
[31] B B Van Aken, T T Palstra, A Filippetti, N A Spaldin 2004 *Journal of Nature materials,* **3** (3) 164 https://doi.org/10.1038/nmat1080
[32] P Liu, Z X Cheng, X L Wang, Y Du, Z W Yu, S X Dou, H Kimura 2012 *Journal of Nanoscience and Nanotechnology* **12** (2) 1238 http://works.bepress.com/sxdou/449/
[33] N V Andreev, T A Sviridova, N Y Tabachkova, V I Chichkov, Y M Mukovskii 2012 *Journal of Russian Microelectronics* **41** (8) 511 https://doi.org/10.1134/S1063739712080033
[34] L Vegard 1921 *Journal of Zeitschrift für Physik,* **5** (1) 17
[35] J A Wang, H Scholl, S B Zheng, D Ogale, D G Viehland, N A Schlom, Spaldin et al. 2005 *Science* **307** (5713) 1203 https://doi.org/10.1126/science.1103959
[36] P Uniyal, and K L Yadav 2009 *Journal of Physics: Condensed Matter* **21** (40) 405901. https://doi.org/10.1088/0953-8984/21/40/405901
[37] L Bingcheng, C Chang-le and J Ke-xin 2011 *Solid State Communications* **151** (9) 712 https://doi.org/10.1016/j.ssc.2011.02.014

IOP Conf. Series: Materials Science and Engineering **613** (2019) 012010 doi:10.1088/1757-899X/613/1/012010

Computer simulation of damage to metal-ceramic composites under shock loading

V V Karakulov[1], I Yu Smolin[1,2] and S N Kulkov[1,2]

[1] National Research Tomsk State University, 634050 Tomsk, Russia
[2] Institute of Strength Physics and Materials Science SB RAS, 634055 Tomsk, Russia

E-mail: valery@ftf.tsu.ru

Abstract. In this work the results of studying of the mechanical behaviour of metal-ceramic composites with an aluminium matrix on mesoscale level are submitted. Computer simulation of mechanical reaction of representative volume of composite material, considered as ensemble of the interacting structural elements (ceramic particles and metal matrix), is used for studying processes of the nucleation and growth of damages on mesoscale level of metal-ceramic composites under the loading by shock pulses. The sizes of representative volume give the possibilities to enter effective mechanical parameters of a composite material. The mechanical behaviour of aluminium matrix is described by the model of the damaged elastic-plastic medium. The model of the damaged brittle solid is used for ceramics. The problem is solved in 2D statement with application of finite-difference method. The results of computer simulation show that the local tensile stresses can appear in unloading wave and can cause damages in ceramic particles. In the region of composite where two opposing unloading waves interacts (spall zone) the calculation predicts the existence of cracked ceramic particles, local damages in metal matrix and existence of mesoscale and macro-scale cracks.

1. Introduction

Metal-ceramic composite materials are widely used now in various fields of industry and are often used under extreme operating conditions [1, 2]. In this connection, it is necessary to adequately numerically evaluate the elastic and strength properties and predict the mechanical behaviour of these materials under intensive energy impacts.

In the simulations of mechanical behaviour, the composites are often taken as homogeneous or quasi-homogeneous, but these materials represent a complex of components with different physical and mechanical properties. The components interconnected on inner contact surfaces form the structure of composites. The structure and its evolution during deformation can have a significant influence on the mechanical properties and the mechanical behaviour of composite materials.

The results of experimental investigations show that the actual strength of the metal-ceramic composites under shock loading is lower than the theoretical strength estimates. The results of investigations of microstructure in the composites after loading show the fractured ceramic particles and cracks between particles and matrix. This specificity of the mechanical behaviour of composites can be caused by processes on mesoscale level.

The present work aims at the numerical simulation of the mechanical behaviour of stochastic metal-ceramic composites on mesoscopic scale level under loading by shock pulses and study processes of the nucleation and growth of damages on mesoscale level of composites with different structure parameters.

IOP Conf. Series: Materials Science and Engineering **613** (2019) 012010 doi:10.1088/1757-899X/613/1/012010

2. Simulation of damage to the metal-ceramic composite under loading by the shock pulse

In this paper, the loading by a shock pulse of a plate of stochastic metal-ceramic composite material consisting of a metal matrix and reinforcing ceramic inclusions is considered. The numerical simulation is performed on a rectangular fragment of the plane section of the plate along the direction of the shock pulse.

The physical-mathematical model of the two-phase condensed heterogeneous medium with an explicit description of its structure is used to describe the mechanical behaviour of the composite on mesoscopic scale level under the considered loading conditions.

The used model represents the heterogeneous medium as a complex of interconnected structural elements – matrix and inclusions. Inclusions have different shapes and are randomly distributed in the matrix. Within interfaces of each structural element, the medium is taken as homogeneous and isotropic while in transition through the interface mechanical properties of the medium change abruptly. The Johnson–Holmquist model of the damaged elastic-brittle medium is used for the mechanical behaviour of ceramic inclusions and the Johnson–Cook model of the elastic-viscoplastic medium for the metal matrix.

The fragments of simulated areas of the two-phase heterogeneous medium with different concentrations of inclusions are shown in figure 1.

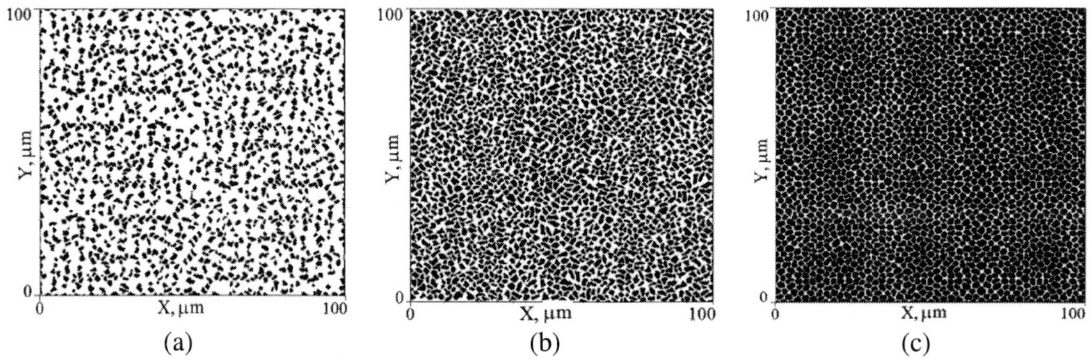

Figure 1. Simulated areas of the two-phase heterogeneous medium with the model structure of the composite composed of the matrix (light region) and arbitrary-shaped inclusions (dark regions). The characteristic size of inclusions is 5 μm, the volume concentration is: (a) 25%, (b) 50%, (c) 75%.

Dimensions of the simulated area and the number of structural elements are chosen in such a way as to determine the effective values of parameters of the mechanical state of the medium. Effective mechanical properties of the heterogeneous medium loaded by a plane shock wave are determined by volume averaging of local values of the state parameters in thin flat layers perpendicular to the shock front direction. The physical-mathematical model and the method for determining effective parameters are described in [3, 4].

The model of the behaviour of the composite material under shock loading and method for determining effective parameters of the mechanical state are adopted here for metal-ceramic composites Al-B_4C, Al-SiC and Al-Al_2O_3 with different concentration of inclusions.

3. Results of the computer simulation of damage to metal-ceramic composites

Propagating in a composite material the shock pulse interacts with the inner boundaries of structural elements. These processes cause the distribution of local values of the parameters of the mechanical state on mesoscale level.

The results of computer simulation show that there is a strong variation of stresses and strains at the mesoscale level in metal-ceramic composites under shock pulse loading. The distribution of stresses depends on mesoscale structure of composites, but has no essential dependence on the shock pulse amplitude. The figure 2 shows the distribution of pressure and strains in the composite Al-75%B_4C.

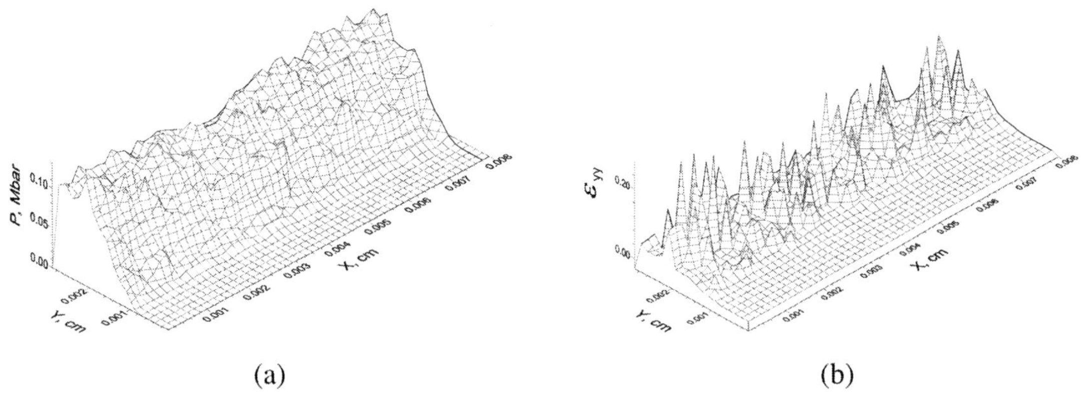

Figure 2. The distributions of local values of pressure (a) and strain (b) in the shock pulse at the mesoscale in the metal-ceramic composite Al-75%B_4C.

The results of calculations forecast the existence of the local tensile stresses in unloading wave. The strength of ceramics is very different in compression and in tension. For this reason, some local damages appear in ceramic particles after passing of unloading wave. The figure 3 shows that local tensile stresses may appear in unloading wave in the composite Al-75% B_4C and the figure 4 shows that this may be the reason of the appearance of damages in ceramic particles.

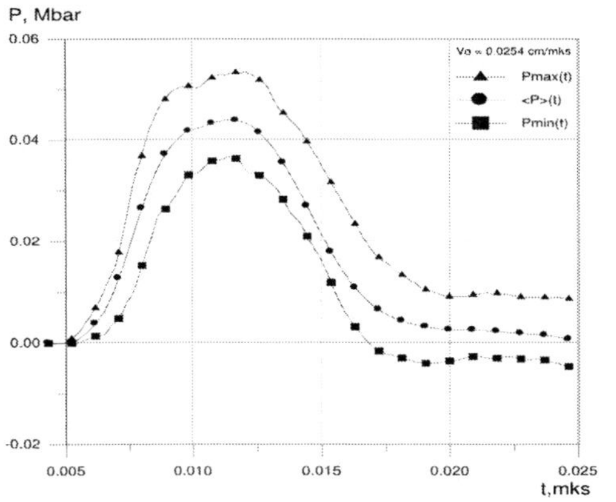

Figure 3. The structure of shock pulse in metal-ceramic composite Al-75%B_4C: P_{max} is maximum pressure, <P> is effective pressure, P_{min} is minimum pressure.

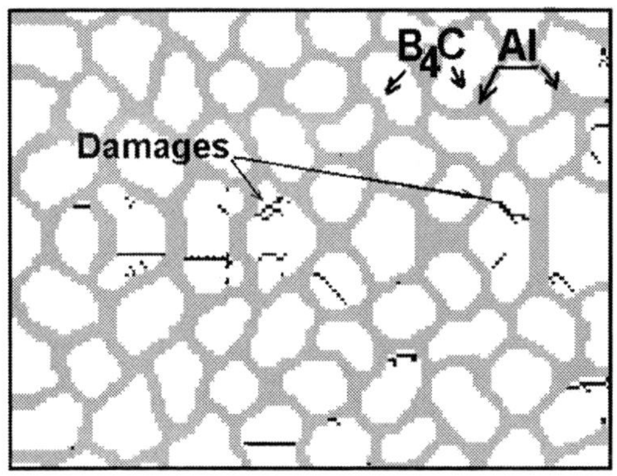

Figure 4. Damages in ceramic particles after the passage of a shock pulse due to local tensile stresses in the unloading wave.

There are more significant damages of the metal-ceramic composite in the region where two opposing unloading waves interact (spall zone). The calculation predicts the existence of cracked ceramic particles, local damages in metal matrix and the existence of mesoscale and macro-scale cracks. The spall zone in metal-ceramic composite has larger dimensions than in metals and ceramics. There is the effect of bridging meso-cracks in composite materials under loading by shock pulses. The efficiency of strengthening of metal-ceramic composites depends on not only the concentration of ceramics but on meso-structure of composites.

The figures 5 and 6 show the formation of the spall zone in composite Al-75% B_4C under tensile stresses. The volume of the damaged composite material catastrophically increases in the spall zone. There are three kinds of damages in spall zone: (i) cracks in ceramic particles; (ii) cracks between ceramic particles and matrix; (iii) damages in aluminium matrix.

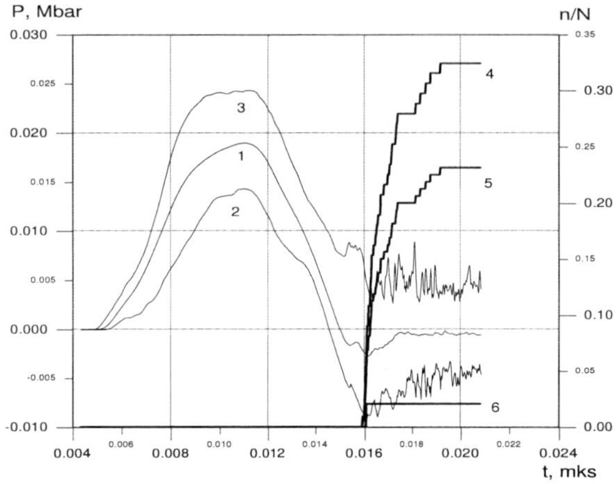

Figure 5. The structure of shock pulse under spallation in metal-ceramic composite Al-75%B_4C, (1) effective pressure <P>, (2) minimum pressure P_{min}, (3) maximum pressure P_{max}, (4) the specific number of damaged cells, (5) the specific number of damaged ceramic mesh cells, (6) the specific number of damaged aluminium matrix mesh cells.

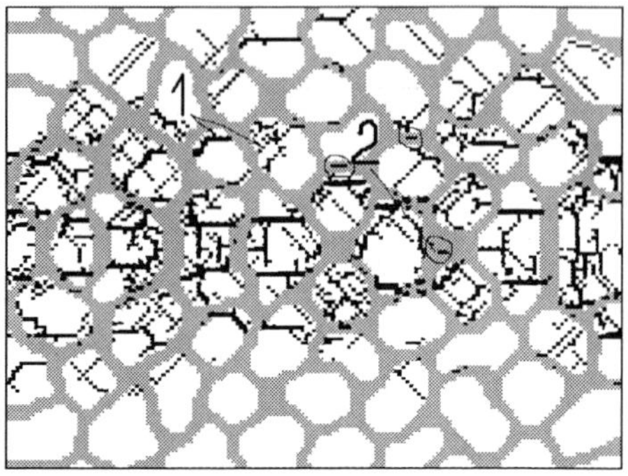

Figure 6. The structure of spall zone in metal-ceramic composite Al-75%B_4C: (1) damage of ceramic particles, (2) damage of aluminium matrix.

Cracks between ceramic particles and matrix and damages in aluminium matrix may appear under tensile stresses when two opposing unloading waves interact. Cracks in ceramic particles may appear under local tensile stresses in unloading wave.

4. Conclusion

The simulation results show:

- there is a strong variation of local values of stresses at the mesoscale level in metal-ceramic composites under shock pulse loading and local tensile stresses may appear in unloading wave and may be the reason of the appearance of damages in ceramic inclusions;
- the spall zone in composites has larger dimensions than in metals and ceramics, there are cracks in ceramic particles, cracks between particles and matrix, damages in the matrix in spall zone;
- the efficiency of strengthening of metal-ceramic composites depends on not only the concentration of ceramics but on meso-structure of composites.

Acknowledgements

The work was carried out with the support of the Tomsk State University Competitiveness Improvement Program and partly within the framework of the Fundamental Research Program of the State Academies of Sciences for 2013–2020, line of research III.23.

References

[1] Gömze L A and Gömze L N 2013 *IOP Conf. Ser.: Mater. Sci. Eng.* **47** 012033 http://dx.doi.org/10.1088/1757-899X/47/1/012033
[2] Kulkov S N, Buyakova S and Gomze L A 2017 *J. Phys.: Conf. Ser.* **790** 012015 https://doi.org/10.1088/1742-6596/790/1/012015
[3] Karakulov V V, Smolin I Yu and Skripnyak V A 2016 *AIP Conf. Proc.* **1783** 020081 http://dx.doi.org/10.1063/1.4966374
[4] Karakulov V V, Smolin I Yu and Kulkov S N 2017 *IOP Conf. Ser.: Mater. Sci. Eng.* **175** 012029 http://dx.doi.org/10.1088/1757-899X/175/1/012029

IOP Conf. Series: Materials Science and Engineering **613** (2019) 012011 doi:10.1088/1757-899X/613/1/012011

Elastic, Electronic, and Optical properties of NaSnX (X=Sb, Bi, As): First principle calculations

Husnu Koc[1], Amirullah M Mamedov[2,3], Ekmel Ozbay[2]

[1]Siirt University, Faculty of Science and Letters, Department of Physics, Siirt, Turkey
[2]Nanotechnology Research Center, Bilkent University, Bilkent, Ankara, Turkey
[3]Baku State University, International Scientific Center, Baku, Azerbaijan

E-mail: mamedov@bilkent.edu.tr

Abstract. In the present work, the structural, mechanical, electronic and optical properties of NaSnX (X=Sb, B, As) compounds have been investigated by means of first principles calculations. The generalized gradient approximation has been used for modeling the exchange-correlation effects. It has been observed that the calculated lattice parameters are in good agreement with the experimental and theoretical lattice parameters. Bulk modulus, shear modulus, Young's modulus Poisson's ratio, sound velocities and the Debye temperatures using the calculated elastic constants for NaSnSb, NaSnBi and NaSnAs compounds have been obtained. The electronic band structure and the projected density of states corresponding to the electronic band structure have calculated.and interpreted. The obtained electronic band structure for NaSnSb and NaSnBi compounds are metallic in nature, and the NaSnAs compound is also a narrow semiconductor. Based on the obtained electronic structures, we further calculated the frequency-dependent dielectric function, the energy-loss function, optical conductivity, and reflection along the x- and z- axes.

1. Introduction

The topological insulators (TI) have attracted a lot of attention since their theoretical discovery. The topological materials exhibit interesting paradoxes. The bulk and surface properties of these materials show different physical properties. The surface properties of these materials are different and they exhibit good conductive or superconducting properties. The finding of 2 and 3 dimensional TI materials and the use of these materials in different application areas have been the main target of the researchers [1-3]. The HgTe/CdTe quantum well was theoretically estimated as the first TI, and this estimate was confirmed by transport experiments. Later, the two-dimensional (2D) InAs / GaSb quantum well was theoretically discovered. Other 2D TI compounds such as the bismuth double layers and monolayer Bi_4Br_4, $ZrTe_5$ and $HfTe_5$ were discovered theoretically using first principle methods. Three-dimensional (3D) TIs were initially estimated at $Bi_{1-x}Sb_x$ and then Bi_2Se_3, Bi_2Te_3 and Sb_2Te_3 compounds with larger band gaps from this material were estimated [3-14].

We have determined that there is not much work in our literature research about these compounds that exhibit interesting physical properties. Dai et al. [3], using the DFT-GGA approach, calculated the bulk and surface electronic band structure of NaSnX (X = Sb, Bi, As) compounds and tried to explain the topological properties of these compounds. Jia et al. [2] studied NaXBi (X = Sn, Pb) compounds both experimentally and theoretically. In theoretical studies, they investigated the presence of the topological phase and Rashba effect by calculating the electronic band structures of the NaXBi (X = Sn, Pb) compounds using the DFT approach, and examined the dynamic and thermal properties of these compounds by phonon spectra and DFT calculations. Our aim in this study is to fill this emptiness by examining the mechanical and optical properties of NaSnX (X = Sb, Bi, As) compounds which have not been studied theoretically as far as we know.

2. Method of calculation

In all of our calculations that were performed using the ab-initio total-energy and molecular-dynamics program VASP (Vienna ab-initio simulation program) [15-18] that was developed within the density functional theory (DFT) [19], the exchange-correlation energy function is treated within the a spin polarized GGA (generalized gradient approximation) by the density functional of Perdew et al. [20]. The potentials used for the GGA calculations take into account the $2p^6 3s^1$ valence electrons of each Na-, $4d^{10} 5s^2 5p^2$ valence electrons of each Sn-, $4d^{10} 5s^2 5p^3$ valence electrons of each Sb-, $5d^{10} 6s^2 6p^3$ valence electrons of each Bi-, and $3d^{10} 4s^2 4p^3$ valence electrons of each As-atoms. When including a plane-wave basis up to a kinetic-energy cutoff equal to 12.16 Ha for NaSnSb/NaSnAs and 15.31 Ha for NaSnBi, the properties investigated in this work are well converged. The Brillouin-zone integration was performed using special k points sampled within the Monkhorst-Pack scheme [21]. We found that a mesh of 8x8x3 k points for NaSnSb, 12x12x4 k points for NaSnBi and 9x9x3 k point for NaSnAs was required to describe the structural, mechanical, electronic, and optical properties. This k-point mesh guarantees a violation of charge neutrality less than 0.008e. Such a low value is a good indicator for an adequate convergence of the calculations.

3. Result and discussion

3.1 Structural properties

NaSnX compounds crystallize in a hexagonal structure with the space group $P6_3mc$ (No = 186). The compounds have optimized using atomic positions and lattice parameters given in Ref.(3). The lattice parameters obtained as a result of the optimization are given in Table 1. The calculated lattice parameters are in good agreement with the experimental and theoretical lattice parameters given in Table 1.

Table 1. The calculated equilibrium lattice parameters (a and c) together with the theoretical and experimental values for NaSnX (X=Sb, Bi, As) compounds in hexagonal structure

Lattice	Material	Reference	a (Å)	c (Å)	V_0 (Å^3)
	NaSnSb	Present	4.3608	11.8135	194.56
Hexagonal		Experimental[3]	4.3610	11.8360	194.94
P6_3mc	NaSnBi	Present	4.4983	11.6911	204.87
(No:186)		Experimental [3]	4.5000	11.6950	205.10
	NaSnAs	Present	4.0684	11.7676	168.68
		Theory [3]	4.0010	11.7290	162.60

3.2. Elastic Properties

The elastic constant of a crystal is a measure of the crystalline response to the external strain applied. We used the strain-stress method [22] to calculate the elastic constants of NaSnX compounds. The obtained elastic constants are given in Table 2. The calculated elastic constants provide the mechanical stability criteria [$C_{11}>|C_{12}|$, $C_{44}>0$, $(C_{11}+2C_{12})C_{33}>2C_{13}^2$] [23,24]. The C_{11} and C_{33} elastic constants in hexagonal crystals show resistance to linear compression in the a- and c-directions. Considering the elastic constants C_{11} and C_{33} calculated in NaSnSb and NaSnBi compounds are $C_{11}> C_{33}$ and in NaSnAs compound is $C_{33}> C_{11}$. Therefore, NaSnSb and NaSnBi compounds are more highly compressible in the c- direction, whereas NaSnAs is more highly compressible in the a- direction.

Table 2. The calculated elastic constants (in GPa) for NaSnX (X=Sb, Bi, As) compounds

Material	Reference	C_{11}	C_{12}	C_{13}	C_{33}	C_{44}	C_{66}
NaSnSb	Present	56.17	12.68	23.45	40.69	21.75	22.85
NaSnBi	Present	52.67	16.58	23.27	38.27	18.04	20.11
NaSnAs	Present	30.97	3.79	11.78	32.62	13.59	10.90

Using the calculated elastic constants, the bulk (B) and shear G) moduli are obtained from the Voigt (V) -Reuss (R) -Hill (H) approximation [25-27]. Young's modulus, Poisson's ratio, sound velocities, and Debye's temperature are also calculated from obtained bulk and shear moduli. The results are given in Tables 3 and 4. The value of the Poisson's ratio is always 0.1 for covalent materials and 0.25 for ionic materials [28-30]. For this reason, the ionic character from the obtained Poisson's ratio (υ = 0.25 for NaSnSb, υ = 0.27 for NaSnBi and υ = 0.22 for NaSnAs) for all three compounds is dominant in the atomic bonding. We cannot say that these materials are stiff when we consider the value of Young's modulus (46.30 GPa, 41.16 GPa, and 27.72 GPa for NaSnSb, NaSnBi, and NaSnAs compounds, respectively), which is a measure of stiffness. if the B/G ratio is less (high) than 1.75, a material is brittle (ductile) [31,32]. Therefore, NaSnSb and NaSnAs compounds are brittle while NaSnBi compound is ductile. The calculated Debye temperature value (210.7 K) for NaSnSb is higher than the calculated Debye temperature value (172.4 K) and (176.9 K) for NaSnBi and NaSnAs. Generally, the Debye temperature is small for soft materials and large for hard materials [33-35]. We can say that all three compounds are not hard materials, but the NaSnSb compound is harder than the NaSnBi and NaSnAs compounds.

Table 3. The calculated isotropic bulk modulus (B, in GPa), shear modulus (G, in GPa), Young's modulus (E, in GPa), Poisson's ratio for NaSnX (X=Sb, Bi, As) compounds

Material	Reference	B_V	B_R	B_H	G_V	G_R	G_H	E	υ	B/G
NaSnSb	Present	30.24	30.16	30.20	19.72	17.48	18.60	46.30	0.25	1.62
NaSnBi	Present	29.98	29.73	29.86	17.02	15.38	16.20	41.16	0.27	1.84
NaSnAs	Present	16.58	16.19	16.39	11.56	11.19	11.38	27.72	0.22	1.44

Table 4. The calculated sound velocities ($\upsilon_t, \upsilon_l, \upsilon_m$) and the Debye temperatures for NaSnX (X=Sb, Bi, As) compounds

Material	Reference	v_t (m/s)	v_l (m/s)	v_m (m/s)	θ_D (K)
NaSnSb	Present	2034	3497	2256	210.7
NaSnBi	Present	1688	3008	1879	172.4
NaSnAs	Present	1634	2720	1807	176.9

3.3. Electronic properties

The electronic band structures of NaSnX (X = Sb, Bi, As) compounds along the high symmetry points in the k space have calculated. The results are given in Fig. 1. As can be observed from the band structures, the NaSnSb and NaSnBi compounds are metallic but the NaSnAs compound is a narrow semiconductor with Eg = 0.04 eV. In the electronic band structure of the NaSnAs compound, the maximum of the valence band is between the Γ-K points but close to the Γ point while the minimum of the conduction band is between the Γ-M points but close to the Γ point. The results obtained for NaSnSb and NaSnBi compounds in comparison with the theoretical values [2,3] are consistent with the theoretical results, but the obtained Eg value for the narrow semiconductor NaSnAs compound is smaller than the theoretical value.

The projected density of states corresponding to the electronic band structure of all three compounds have also calculated and given in Fig. 2. In the NaSnSb compound, It is understood that the bands between -25 and -24 eV are occupied by the Na p states, and the bands between -23 and -21 eV are occupied by the Sn d states. The bands between -11 and -9 eV are occupied by the Sb s and Sn s states (but the Sb s states are more dominant than the Sn s states) while the bands between -8 and -5 eV are occupied by the Sn s and Sb s states (but the Sn s states are more dominant than the Sb s states). The bands just above and below the fermi level are occupied by all three states. In the NaSnBi compound, the lowest valence bands are occupied by Na p states. the bands between -23 and -21 eV are occupied by the Sn d and Bi d states. The bands between -11 and -9 eV are occupied by the Sb s and Sn s states (but the Bi s states are more dominant than the Sn s states) while the bands between -8 and -5 eV are occupied by the Sn s and Bi s states (but the Sn s states are more dominant than the Sb s states). The bands just above and below the fermi level are occupied by all three states. In the NaSnAs compound, the lowest valence bands are occupied by Na p states. The bands between -23 and -21 eV are occupied by the Sn d. The bands between -11 and -9 eV are occupied by the As s and Sn s states (but the As s states are more dominant than the Sn s states) while the bands between -8 and -5 eV are occupied by the Sn s and As s states (but the Sn s states are more dominant than the Sb s states). The bands just above and below the Fermi level are occupied by all three states.

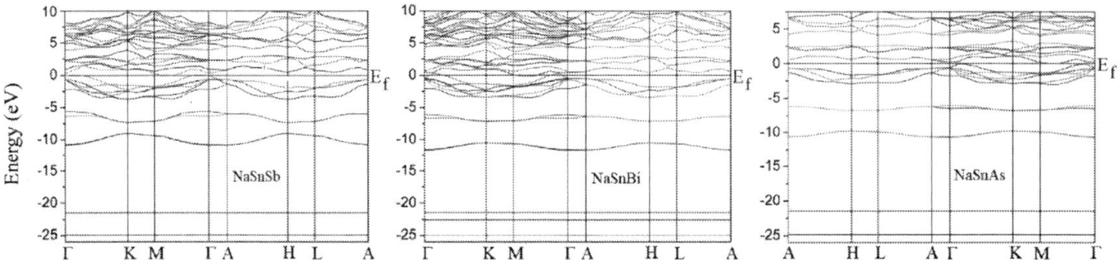

Figure 1. Energy band structures for NaSnSb, NaSnBi, and NaSnAs compounds

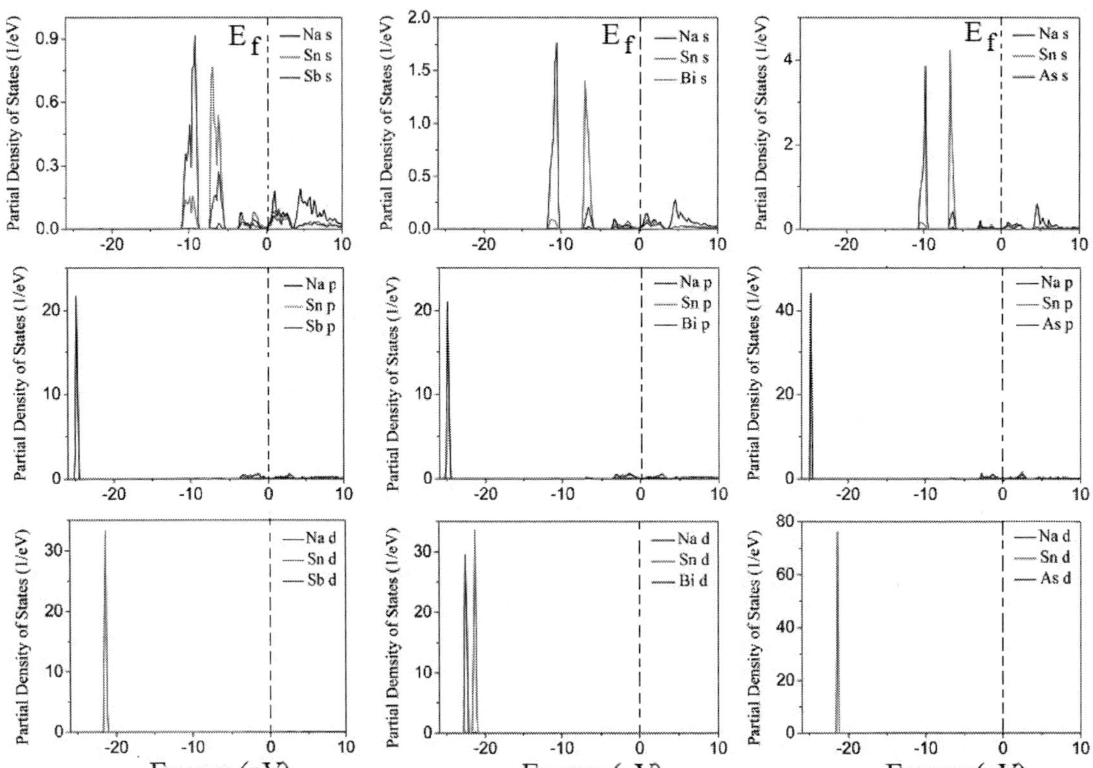

Figure 2. The projected density of states for NaSnSb, NaSnBi, and NaSnAs compounds

3.4. Optical properties

The optical spectra of NaSnX, which result from the band structure in Fig.1, are plotted in Fig. 3. More precisely, the real (ε_1) and imaginary parts (ε_2) of the dielectric function (ε) [36,37] are displayed for the x- and z-axis. Above ~2 eV they show more or less the same line-shape as the corresponding spectrum. The pronounced reflection peaks between $\hbar\omega = 2.0$ eV and 10 eV in Fig.4 still appear with almost the same intensity. However, the spectra for small photon energies $\hbar\omega < 1.5$ eV are totally different. The spectral weights are redistributed to guarantee the oscillator strength sum rule. An additional pronounced peak appears at $\hbar\omega \approx 10.5$ eV, whereas the imaginary parts approach a small positive value for ω → 0. Both observations are consequences of the modified band structure. The additional peak is due to van Hove singularities in the joint density of states at the highest optical transition energies near H and L points. The low-energy behavior is instead dominated by the non-conical 3D Dirac cones (Table 5). For intermediate energies the structure of the energy loss spectra in Fig. 3 is due to the details of the band structure. However, the high-energy behavior of the loss functions is also influenced by the collective plasmonic excitations of the valence electron with approximate plasma frequencies ω and corresponding valence electron densities. According to the band structure in Figs. 1 and 2, the valence electrons can be grouped into such due to Na p, s and Sn d electrons. Since only the valence s and p electrons can be nearly treated as free electrons in the formula for the plasma frequency, two peaks in the loss spectrum near 10.6 eV (only p electrons) and 13. eV (s and p electrons) can be explained. Because of the stronger binding, the right effective mass of the X p-d electrons is larger than the free-electron mass. Consequently, two other structures near 12.5 and 15.1 eV may be related to collective excitations where the delectrons partially contribute. The pronounced peak around $\hbar\omega$ = 10.3 eV can be explained by contributions from the lowest Na s and X p bands into higher-lying conduction bands with p character.

Table 5. Some of the principal features and singularities of the linear optical responses for NaSnSb, NaSnBi and NaSnAs

Material	ε_1(eV)	$d\varepsilon_1/dE<0$	$d\varepsilon_2/dE>0$	ε_2(eV)		L (eV)	
NaSnSb	ε_1^x	3.02	12.71	$\varepsilon_{2,max}^x$	2.42	L_X	13.32
	ε_1^z	4.84	12.71	$\varepsilon_{2,max}^z$	4.24	L_Z	13.31
NaSnBi	ε_1^x	2.76	11.06	$\varepsilon_{2,max}^x$	2.07	L_X	11.06
	ε_1^z	4.15	11.75	$\varepsilon_{2,max}^z$	3.46	L_Z	13.82
NaSnAs	ε_1^x	2.82	11.87	$\varepsilon_{2,max}^x$	1.77	L_X	13.34
	ε_1^z	3.67	12.85	$\varepsilon_{2,max}^z$	3.06	L_Z	13.34

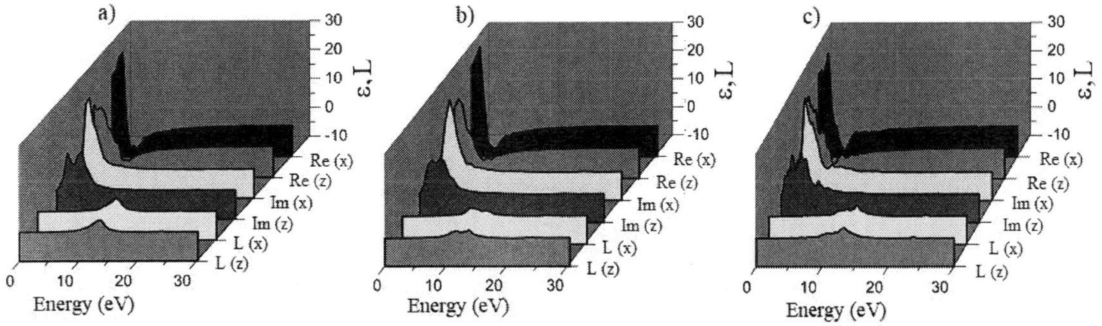

Figure 3. The real (ε_1) and imaginary (ε_2) parts of the dielectric function (ε) and energy-loss function (L) along the x- and z- axes for a) NaSnSb, b) NaSnBi and c) NaSnAs

The real(σ1) and imaginary σ(2) parts of optical conductivity (σ= σ1+ iσ2) are shown in Fig. 4 for the whole investigated spectral range. In inspecting the spectra of the real part of σ first, we recognize a metallic-like contribution merging into a Drude-like resonance well below 2 eV. All other peaks above 2 eV dominate σ1 in the investigated compounds. We have proposed to ascribe all of these absorption peaks to generic electronic interband transitions. Their resonance frequencies closely coincide indeed with the direct transition along the Γ-K-L directions in the reciprocal space, as predicted by band structure calculations. The main panel of Fig. 4 emphasizes σ1 in the spectral range of relevance for the following discussion. Again, we can appreciate the narrow metallic contribution and the two of the dominant absorptions pointed out here in above. These data, supported by band structure calculations, indicate that the degeneracy at the Dirac points in NaSnX is lifted by additionality breaking the time-reversal symmetry with a canted magnetic order of the X-atoms, resulting in the formation of Weyl nodes.

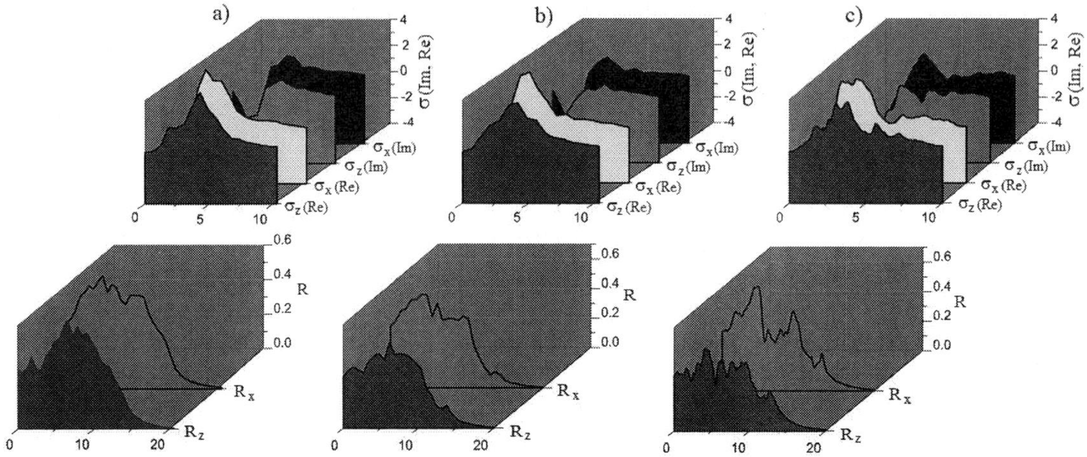

Figure 4. The real (σ1) and imaginary σ(2) parts of optical conductivity (σ= σ1+ iσ2) and reflection (R) along the x- and z- axes for a) NaSnSb, b) NaSnBi and c) NaSnAs

4. Conclusion

We have performed the structural, mechanical, electronic, and optical properties of the NaSnSb, NaSnBi and NaSnAs compounds using the density functional theory. The lattice parameters obtained as a result of the optimization are in good agreement with the experimental and theoretical lattice parameters. Considering the calculated elastic constants of C_{11} and C_{33}, the NaSnSb-NaSnBi and NaSnAs compounds have a higher compressible in the c- and a-directions, respectively. From the calculated Young's modulus and Debye temperature values, it is concluded that these compounds are soft materials. The ionic character from obtained Poisson's ratio for all three compounds is dominant in the atomic bonding. The electronic band structures and the projected density of states corresponding to these band structure have been calculated. The NaSnSb and NaSnBi compounds are metallic in nature, but the NaSnAs compound is a narrow semiconductor. The real and imaginary parts of the dielectric function, energy-loss function, optical conductivity, and reflection for these compounds have been calculated and interpreted.

5. References

[1] Hasan M Z and J. E. Moore J E 2011 *Annu. Rev. Condens. Matter Phys.* **2** 55 https://doi.org/10.1146/annurev-conmatphys-062910-140432

[2] Jia Y Z, Ji W X, Zhang C W, Li P and Wang P J 2018 *RSC Adv.* **8** 27995 https://doi.org/10.1039/C8RA05458D

[3] Dai X, Le C C, Wu X X, Qin S S, Lin Z P and Hu J P 2016 *Chin. Phys. Lett.* **12** 12730 http://cpl.iphy.ac.cn/10.1088/0256-307X/33/12/127301

[4] Bernevig B A, Hughes T L and Zhang S C 2006 *Science* **314** 1757 https://doi.org/10.1126/science.1133734

[5] Moore J E 2010 *Nature* **464** 194 https://www.nature.com/articles/nature08916

[6] Konig M, Wiedmann S, Brüne C, Roth A, Buhmann H, Molenkamp L W, Qi X L and Zhang S C 2007 *Science* **318** 766 https://doi.org/10.1126/science.1148047

[7] Hughes T L, Qi X L, Wang K and Zhang S C 2008 Phys. Rev. Lett. 100 236601 https://doi.org/10.1103/PhysRevLett.100.236601

[8] Knez I, Du R R and Sullivan G 2011 Phys. *Rev. Lett.* **107** 136603 https://doi.org/10.1103/PhysRevLett.107.136603

[9] Spanton E M, Nowack K C, Du L J, Sullivan G, Du R R and Moler K A 2014 *Phys. Rev. Lett.* **113** 026804 https://doi.org/10.1103/PhysRevLett.113.026804

[10] Murakami S 2006 *Phys. Rev. Lett.* **97** 236805 https://doi.org/10.1103/PhysRevLett.97.236805
[11] Zhou J J, Feng W, Liu C C, Guan S and Yao Y 2014 *Nano Lett.* **14** 4767 https://doi.org/10.1021/nl501907g
[12] Weng H, Dai X and Fang Z 2014 *Phys. Rev.* X 4 011002 https://doi.org/10.1103/PhysRevX.4.011002
[13] Teo J C Y, Fu L and Kane C L 2008 *Phys. Rev. B* **78** 045426 https://doi.org/10.1103/PhysRevB.78.045426
[14] Zhang H J, Liu C X, Qi X L, Dai X, Fang Z and Zhang S C 2009 *Nat. Phys.* **5** 438 https://www.nature.com/articles/nphys1270
[15] Kresse G and Hafner J 1993 *Phys. Rev. B* **47** 558 https://doi.org/10.1103/PhysRevB.47.558
[16] Kresse G and Furthmüller J 1996 *Comput. Mater. Sci.* **6** 15 https://doi.org/10.1016/0927-0256(96)00008-0
[17] Kresse G and Joubert D 1999 *Phys. Rev. B* **59** 1758 https://doi.org/10.1103/PhysRevB.59.1758
[18] Kresse G and Furthmüller J 1996 *Phys. Rev. B* **54** 11169 https://doi.org/10.1103/PhysRevB.54.11169
[19] Hohenberg P and Kohn W 1964 *Phys. Rev.* **136** A1133 https://doi.org/10.1103/PhysRev.136.B864
[20] Perdew J P and Burke S 1996 *Phys. Rev. Lett.* **77** 3865 https://doi.org/10.1103/PhysRevLett.77.3865
[21] Monkhorst H J and Pack J D 1976 *Phys. Rev. B* **13** 5188 https://doi.org/10.1103/PhysRevB.13.5188
[22] Le Page Y and Saxe P 2001 *Phys. Rev. B* **63** 174103 https://doi.org/10.1103/PhysRevB.63.174103
[23] Wu Z J, Zhao E J, Xiang H P, Hao X F, Liu X J, and Meng J 2007 *Phys. Rev. B* **76** 054115 https://doi.org/10.1103/PhysRevB.76.054115
[24] Watt J P and Peselnick L 1980 *J. Appl. Phys.* **51** 1525 https://doi.org/10.1063/1.327804
[25] Voight W *1928 Lehrbook der kristallphysik* (Leipsig: Teubner) p 962

[26] Reuss A 1929 *Z. angew. Math. Mech.* **9** 49 https://doi.org/10.1002/zamm.19290090104
[27] Hill R 1952 *Proc. Phys. Soc. London Sect A* **65** 349 https://doi.org/10.1088/0370-1298/65/5/307
[28] Bannikov V V, Shein I R and Ivanovskii A L 2007 *Phys. Stat. Sol. (RRL)* **3** 89 https://doi.org/10.1002/pssr.200600116
[29] Koc H, Yildirim A, Tetik E and Deligoz E 2012 *Computational Materials Science* **62** 235 https://doi.org/10.1016/j.commatsci.2012.05.052
[30] Koc H, Mamedov A M, Deligoz E and Ozisik H 2012 *Solid State Science* **14** 1211 https://doi.org/10.1016/j.solidstatesciences.2012.06.003
[31] Shein I R and Ivanovskii A L 2008 *J. Phys. Considens. Matter.* **20** 415218 https://doi.org/10.1088/0953-8984/20/41/415218
[32] Pogh F 1954 *Philos. Mag.* **45** 823 https://doi.org/10.1080/14786440808520496
[33] Johnston I, Keeler G, Rollins R and Spicklemire S 1996 *Solids state physics simulations, the consortium for upper level physics software* (New York: Wiley)
[34] Anderson O L 1963 *J Phys Chem Solids* **24** 909 https://doi.org/10.1016/0022-3697(63)90067-2
[35] Schreiber E, Anderson O L and Soga N 1973 *Elastic constants and their measurements* (McGraw-Hill, New York)
[36] Philipp H R and Ehrenreich H 1963 *Phys. Rev.* **129** 1550 https://doi.org/10.1103/PhysRev.129.1550
[37] Marton L 1956 *Rev. Mod. Phys.***28** 172 https://doi.org/10.1103/RevModPhys.28.172

IOP Conf. Series: Materials Science and Engineering **613** (2019) 012012 doi:10.1088/1757-899X/613/1/012012

Electrochemical study of the corrosion inhibition effect of 3-cyanophenyl compound on 420 martensitic stainless steel in strong acid media

Roland T. Loto[1], Bryan U. Ayozie[1] and Cleophas A. Loto[1]

[1]Department of Mechanical Engineering, Covenant University, Ogun state, Nigeria

Email: tolu.loto@gmail.com

Abstract. The corrosion inhibition performance of 3-cyanophenyl organic compound on heat treated 420 martensitic stainless steel in 6 M H_2SO_4 was evaluated through potentiodynamic polarization test, ATR-FTIR spectroscopy and adsorption isotherm studies. Results obtained proves the compound to be effective with highest inhibition performance value of 99.4%. The inhibition property was determined to be mixed type inhibition with significant anodic inhibition tendencies. Adsorption of the compound onto the steel was through chemisorption mechanism in agreement with the Langmuir and Freundlich isotherms models at correlation coefficients of 1 and 0.7570. The transmittance of detected functional groups of 3-cyanophenyl diminished at distensible wavelengths because of excellent surface coverage and adsorption of the compound onto the steel

1. Introduction

Stainless steels are metallic alloys with high corrosion resistance in comparison to carbon steels due to varying weight content of chromium, nickel etc. alloyed with the iron substrate metal. 420 martensitic steel, an alteration of 410 steel, has slightly greater volumetric composition of carbon which enhances the hardness, strength and wear resistance of the steel. They basically derive their corrosion resistance from the presence of a passive protective film formed on their surfaces. Destruction of the film leads to localized corrosion deterioration and failures. Acids are extensively applied in numerous industrial technological processes leading to corrosion. Corrosion is a major problem in the petrochemical industries resulting in high maintenance cost [1-2]. Research on 420 steel concludes the overwhelming effect quenching on their corrosion resistance by as much as 60% [3]. Further increase in the versatility and application of 420 steels requires the use of chemical compounds (inhibitors) capable of inhibiting its corrosion in corrosive environments. Most corrosion inhibitors act through adsorption, retarding the cathodic, anodic or both electrochemical corrosion reactions. Most promising inhibitors are organic compounds that contain heteroatoms which chemically adsorbed onto metallic surfaces [4, 5]. Research on organic derivatives of 3-cyanophenyl has proven their corrosion inhibition effectiveness however only in low molar concentrations of acid media [6-8]. This research evaluates the corrosion inhibiting action of 3-cyanophenyl on 420 martensitic stainless in 6M H_2SO_4 acid solution.

2. Material and methods

420 stainless steel (420SS) procured from the open market has the following percentage content of alloying elements after analysis; shown in Table 1. The steel samples are of cylindrical dimension with an unconcealed area of 0.79 cm^2 after embedding in resin mounts. 420SS samples were cut and sectioned before being abraded with silicon carbide papers (80, 320, 600, 800 and 1000 grits). They

IOP Conf. Series: Materials Science and Engineering **613** (2019) 012012 doi:10.1088/1757-899X/613/1/012012

were washed with distilled water and acetone for electrochemical tests. 420SS samples were heat heated in a muffle furnace to 1000°C before quenching in distilled water. The furnace temperature was controlled at an accuracy ± 10°C coupled with thermocouple (K-Type). 3-cyanophenyl (3CL) shown in Fig. 1 was procured from BOC Sciences, USA is a achromatic organic liquid with a sweet almond odour. It has a density of 103.12 g/mol and a chemical formula of C_7H_5N. 3CL was prepared in quantitative concentrations of 9.25 x 10^3, 1.85 x 10^2, 2.77 x 10^2, 3.70 x 10^2 and 4.62 x 10^2 in 400mL of 6 M H_2SO_4 acid prepared from standard grade (98% purity) with distilled water. Electrochemical evaluation through polarization was performed out at 34°C with a triple electrode cell system, consisting of 200 mL of the electrolyte and 420SS electrodes connected to Digi-Ivy 2311 potentiostat. Graphical plots were produced at a scan rate of 0.0015 V/s from -1.5 V and +1.5 V. The acid inhibitor solution, before and after polarization tests were subjected to infrared ray beams using Bruker Alpha FTIR spectrometer from 375 to 7500 cm^{-1} wavelength and 0.9 cm^{-1} resolution. Graphical plots of ATF-FTIR absorption including of spectra peaks were evaluated and correlated with theoretical ATF-FTIR absorption Table for identification of functional groups involved in the corrosion inhibition of 420SS.

Table 1. Percentage content of alloying elements in 420SS

Element	C	Mn	P	S	Si	Cr	Ni	Mb	N	Fe
% Composition	0.28	0.8	0.04	0.02	0.8	14	0.4	-	-	79.15

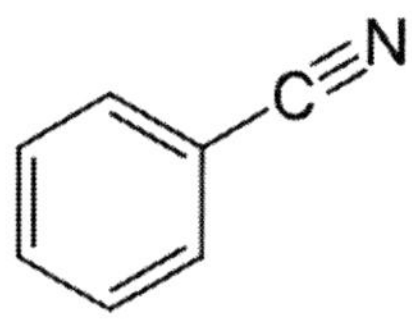

Figure 1. Molecular structure of benzonitrile

3. Results and discussion

3.1 Potentiodynamic polarization studies

The results of the polarization test are given in Table 2. 420SS at 0% 3CL concentration exhibited a corrosion rate value of 3.30 mm/y, analogous to corrosion current density of 3.04 x 10^4 Acm^2. This value is significantly higher than the corrosion rates of 420SS at 0.13% - 0.63% 3CL due to the activity of SO_4^{2-} ions in the aqueous media. The strong ionization potential of 6M H_2SO_4 in H_2O at 0% 3CL causes the release two protons which aggressively reacts with and deteriorates 420SS surface. This also shows 3CL alters the mechanism of electrochemical reactions responsible for 420SS corrosion through adsorption, surface coverage, and suppression of the reduction reactions. Addition of 3CL (0.13% - 0.63% 3CL) to the acid solution significantly altered the corrosion resistance properties of 420SS as shown from the changes in corrosion potential values from -0.306 V at 0% 3CL to -0.274 V at 0.13% 3CL due to anodic inhibition through surface coverage and precipitation of 3CL on reactive sites on the steel surface. The visible change in anodic Tafel slope values after 0% 3CL is due to changes in the electrode substrate, rate controlling step and influence of potential controlled conditions. This is consistent with the earlier observation of the corrosion potential values. The maximum change in corrosion potential with respect to the value at 0% 3CL is less than 85 mV as a result 3CL is a mixed type inhibitor with dominant anodic inhibiting properties.

Table 2. Results of 420SS potentiodynamic polarization behaviour in 6 M H_2SO_4 solution at 0% - 0.63% 3CL concentration

Sample	3CL Conc. (%)	3CL Conc. (M)	420SS Corrosion Rate (mm/y)	3CL Inhibition Efficiency (%)	Corrosion Current (A)	Corrosion Current Density (A/cm²)	Corrosion Potential (V)	Polarization Resistance, R_p (Ω)	Cathodic Tafel Slope, B_c (V/dec)	Anodic Tafel Slope, B_a (V/dec)
A	0	0	3.30	0	2.40E-04	3.04E-04	-0.306	107.10	-10.50	8.06
B	0.13	9.25E-03	0.19	99.40	1.35E-05	1.71E-05	-0.274	1906.00	-9.23	20.97
C	0.25	1.85E-02	0.19	99.39	1.37E-05	1.74E-05	-0.267	1874.00	-7.58	23.03
D	0.38	2.77E-02	0.21	99.32	1.52E-05	1.92E-05	-0.282	1693.00	-7.25	17.36
E	0.50	3.70E-02	0.33	98.93	2.40E-05	3.03E-05	-0.285	1073.00	-7.77	16.27
F	0.63	4.62E-02	0.22	99.29	1.59E-05	2.02E-05	-0.348	1612.00	-9.68	9.28

3.2 Mechanism of Inhibition

3CL consist of a 5-membered 1, 3-thiazole ring fused to a benzene ring. The presence of heteroatoms of N and S, and π-electrons on the thiazole rings of 3CL strongly influences its adsorption and mode of interaction with 420SS surface, and the resulting formation of metallic complexes. The lone pair of electrons in the heteroatoms within 3CL molecules facilitates transfer of electrons from 3CL to the metal. The S atom enables the formation of dπ-dπ bond resulting from the overlapping of 3d-electrons, while the double bonds in 3CL molecule reverse release of metal d-electron to the π-orbital. The wide variation in corrosion rate values of 420SS before and after 3CL inhibition shows that 420SS strongly oxidizes in 6M H_2SO_4 solution releasing metallic cations. This behaviour results in vacant'd' orbital of the iron substrate metal leading to the formation of co-ordination bond between Fe–N-S, hence adsorption. Electrostatic attraction occurs between 3CL inhibitor and 420SS; the positively charged 420SS surface electrostatically attracts SO_4^{2-} ions, which preadsorbed onto steel surface. The excess negative charge draws cationic 3CL molecules forming a protective $(FeSO_4^{2-}3CL\text{-}H^+)_{ad}$ layer. 3CL cations are also attracted to cathodic sites on 420SS surface, competing with H^- ions for reduction, hence stifling the hydrogen evolution reaction.

3.3 ATF-FTIR Spectroscopy analysis

Functional groups within 3CL molecules i.e. atoms and bonds responsible for electrostatic attraction, molecular adsorption and invariably corrosion inhibition of 420SS surface were identified by ATF-FTIR spectroscopy and correlated with the ATR-FTIR Theoretical Table [9, 10]. Fig. 2 shows the spectra plots of the 6 M H_2SO_4 /3CL solutions before and after 420SS corrosion. The transmittance of the spectra plot of 6 M H_2SO_4/3CL before corrosion decreased significantly after the corrosion test by comparison with the transmittance of the spectra plot of 6 M H_2SO_4/3CL after corrosion at wavelengths between of 526 - 673 cm^{-1}, 826 - 1030 cm^{-1}, 1360 - 1433 cm^{-1}, 1529 - 1597 cm^{-1}, 1776 - 2822 cm^{-1} and 3004 - 3489 cm^{-1} due to corrosion inhibition resulting from adsorption of specific 3CL functional groups. The identified moieties are alkynes, aromatics, primary, secondary amines, carboxylic acids, alkenes, alkanes, nitro compounds, aldehydes, alkynes (terminal), amides and alcohols, phenols which all consists of bonds such as –C(triple bond)C–H: C–H bend, C–H "oop", N–H wag, O–H bend, =C–H bend, C–H rock, C–C stretch (in–ring), N–O asymmetric stretch, N–H bend, H–C=O: C–H stretch, =C–H stretch, C–H stretch, –C(triple bond)C–H: C–H stretch, N–H stretch, O–H stretch and H–bonded. The decrease in transmittance at the specific wavelength range confirmed the surface coverage effect and adsorption of 3CL compound on 420SS resulting on MCS surface.

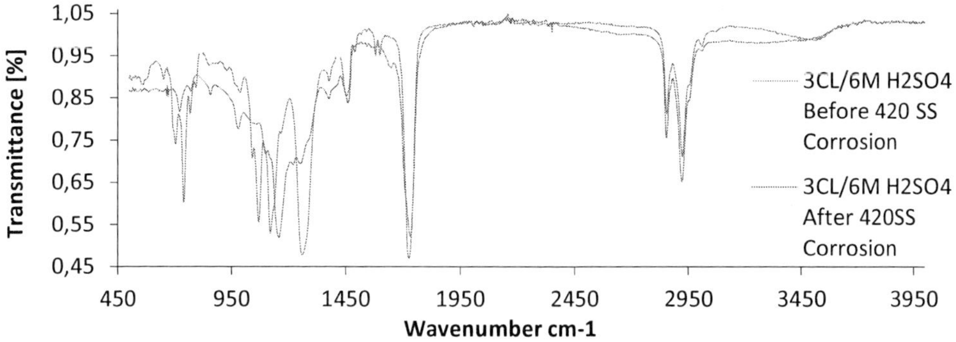

Figure 2. ATF-FTIR spectra of 3CL/6M H_2SO_4 before and after 420SS corrosion inhibition

3.4 Adsorption Isotherm

3CL adsorption on 420SS surface was further determined and studied through adsorption models. Studies show 3CL agrees with the Langmuir and Freundlich isotherm at correlation coefficients of 1 and 0.7570 respectively. Langmuir isotherm plot of $\frac{C_{\mathrm{BEZT}}}{\theta}$ versus $C_{3\mathrm{CL}}$ [Fig. 3(a)], calculated from equation 1 assumes that; (i) The molecular interactions at the metal- solution interface is constant, (ii) Gibbs free energy results (equation 3) is non-dependent on the extent of molecular protection and (iii) The effect of interaction of adsorbates on the value of Gibbs free energy is negligible [11]. The Frumkin plot of Log $\frac{\theta}{1-\theta}$ versus Θ [Fig. 3(b)], determined from equation 2 proposes complete adsorbate covering at peak 3CL concentrations for a non-disparate alloy surface coupled with negligible effect of lateral interaction.

$$\theta = \left[\frac{K_{\mathrm{ads}} C_{3\mathrm{CL}}}{1 + K_{\mathrm{ads}} C_{3\mathrm{CL}}}\right] \quad (1)$$

$$\mathrm{Log}\,[C_{3\mathrm{CL}} * (\theta/_{1-\theta})] = 2.303 \log K_{\mathrm{ads}} + 2\alpha\theta \quad (2)$$

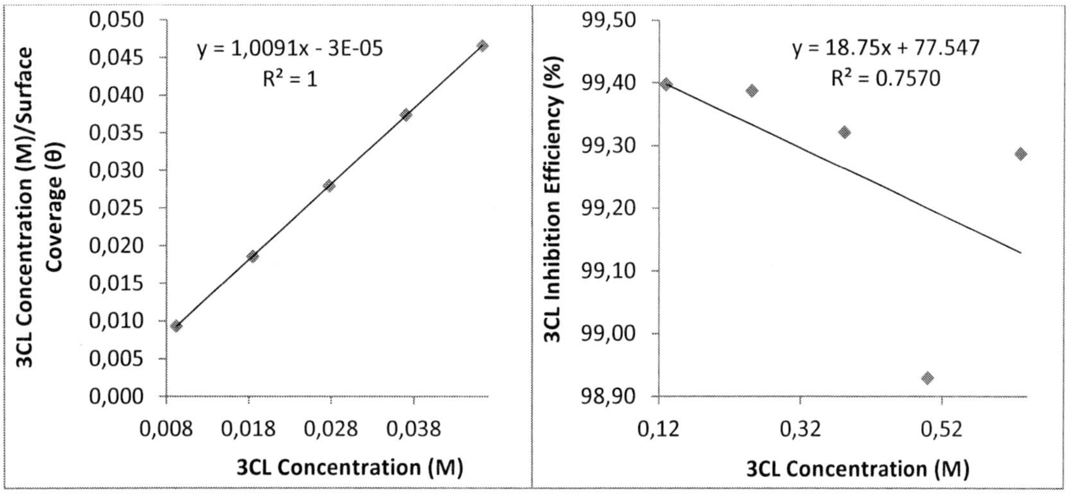

(a) (b)

Figure 3. Graphical plot of (a) $\frac{C}{\theta}$ versus 3CL concentration, (b) $\frac{\theta}{1-\theta}$ versus 3CL concentration in 6M H_2SO_4

3.5 Thermodynamics of the corrosion inhibition

Data for Gibbs free energy ($\Delta G^{o}_{\mathrm{ads}}$) for the molecular interaction are shown in Table 3 and were calculated from the mathematical relationship below.

$$\Delta G_{\mathrm{ads}} = -\,2.303RT \log\,[55.5K_{\mathrm{ads}}] \quad (3)$$

55.5 is a constant for standard water concentration in the acid media, R represents the universal gas constant, T stands for the absolute temperature and K_{ads} is the equilibrium constant of adsorption. K_{ads} is related to surface coverage (θ) from the Langmuir equation (equation 1). The surface properties and non-homogeneous nature of 420SS strongly influences the ΔG^o_{ads} values of 3CL adsorption with respect to the extent of 3CL coverage value. Negative values of ΔG^o_{ads} signifies the spontaneous behaviour of the adsorption species on the metal surface. The highest ΔG^o_{ads} value obtained for 3CL adsorption on 420SS is -42.22 KJmol^{-1} at 2.5% 3CL, while the lowest value in both acids are -38.89 KJmol^{-1} at 1.5% 3CL and -41.53 KJmol^{-1} at 6.5% 3CL. This observation is consistent with chemisorption adsorption mechanism [12, 13].

Table 3. Result for Gibbs free energy (ΔG^o_{ads}), surface coverage (θ) and equilibrium constant of adsorption (K_{ads}) for 3CL adsorption on 420SS in 6 M H_2SO_4

420SS Samples	3CL Concentration (M)	Surface Coverage (θ)	Equilibrium Constant of adsorption (K_{ads})	Gibbs Free Energy, ΔG_{ads} (Kjmol^{-1})
A	0	0	0	0
B	9.25E-03	0.994	17838.8	-34.21
C	1.85E-02	0.994	8768.9	-32.45
D	2.77E-02	0.993	5276.3	-31.19
E	3.70E-02	0.989	2498.3	-29.34
F	4.62E-02	0.993	3013.8	-29.80

4. Conclusion

3-cyanophenyl compound effectively suppressed the degradation and surface deterioration of 420 martensitic steel in 6M H_2SO_4 solution. The compound acted through surface coverage and chemisorption adsorption molecular interaction with respect to the Langmuir and Freundlich isotherm model. The inhibition property of 3-cyanophenyl was calculated to be mixed type with dominant anodic inhibition effect.

Acknowledgement

The author is grateful to Covenant University Ota, Ogun State, Nigeria for the sponsorship and provision of research facilities for this project.

References

[1] El-Etre, A Y 1998 *Corros. Sci.* **40** (11) 1845. https://doi.org/10.1016/S0010-938X(98)00082-1
[2] Parikh K S and Joshi K J 2004 *Trans. SAEST.* **39** (1-2) 29.
[3] Loto R T, Aiguwurhuo O and Evana U 2016 *Rev. Téc. Ing. Univ. Zulia.* **39** (7) 35. https://doi.org/10.21311/001.39.7.04
[4] Loto C A, Joseph O O, Loto R T and Popoola A P I 2014 *Int. J. Electrochem. Sci.* **9** 1221.
[5] Loto C A, Loto R T 2013 *Int. J. Electrochem. Sci.* **8** 12434.
[6] Zaiz T and Lanez T 2012 *J. of Fund. & Appl. Sci.* **4** (2). https://doi.org/10.4314/jfas.v4i2.8.
[7] Saha S K, Dutta A, Ghosh P, Sukulc D, Banerjee P 2016 *Phys. Chem. Chem. Phys.* **27** (18) 17898. https://doi.org/10.1039/C6CP01993E
[8] Loto R T 2012 *Sci. Rep.* **7** 17555. https://doi.org/10.1038/s41598-017-17867-0
[9] Table of Characteristic IR Absorptions. https://orgchem.colorado.edu/Spectroscopy/specttutor/irchart.pdf
[10] George S 2014 *Infrared and Raman Characteristic Group Frequencies: Tables and Charts.* John Wiley & Sons, New York.
[11] Ashish K S, Quraishi M A 2011 *Corros. Sci.* **53** (4) 1288. https://doi.org/10.1016/j.corsci.2011.01.002
[12] Bouklah M, Hammouti B, Lagrene M and Bentiss F 2006 *Corros. Sci.* **48** (9) 2831. https://doi.org/10.1016/j.corsci.2005.08.019
[13] Loto R T 2016 *Cogent Eng.* **3** 1242107. https://doi.org/10.1080/23311916.2016.1242107

IOP Conf. Series: Materials Science and Engineering **613** (2019) 012013 doi:10.1088/1757-899X/613/1/012013

Effect of the combined admixture of 4-methyl-norvalin and 2-Methoxy-4-formylphenol on the corrosion inhibition of low carbon steel in artificial seawater solution

Roland T. Loto[1], Cleophas A. Loto[1] and Ogaga Osamaige[1]

[1]Department of Mechanical Engineering, Covenant University, Ogun state, Nigeria

Email: tolu.loto@gmail.com

Abstract. The synergistic effect of 4-methyl-norvalin and 2-methoxy-4-formylphenol on the corrosion inhibition of low carbon steel in artificial seawater (3.5 wt.% NaCl solution) was studied with potentiodynamic polarization technique, open circuit potential measurement and optical microscopic analysis. Results obtained from the electrochemical test showed the organic derivatives performed effectively with optimal inhibition performance of 89%. Mixed type inhibition behavior was observed. Anodic corrosion potential shift occurred because of effective surface coverage of the inhibitor molecules. The adsorption mode was determined to occur through physisorption reaction from thermodynamic calculations in agreement with the Langmuir, Frumkin and Freundlich isotherms at correlation coefficients beyond 0.7. Corroded morphology of steel occurred without with inhibitor compared to the steel from solution at highest inhibitor concentration.

1. Introduction

The corrosion process results in deterioration of the surface properties and metallic substrate of carbon steels. Despite the weak resistance of carbon steel to corrosion, it has extensive industrial application across most industries. Marine environments have abundance of chloride anions hence the serious debilitating effect of corrosion encountered metal alloys in such environments. Corrosion prevention by chemical compounds known as inhibitors has been proven to be one of the most feasible, reliable and cost-effective methods [1]. Previous individual research on the corrosion inhibition effect of 4-methyl-norvalin and 2-methoxy-4-formylphenol compounds have been done in acid solution with above average results [2-6]. The purpose of this study is to assess the adsorption and corrosion inhibition performance of 4-methyl-norvalin and 2-methoxy-4-formylphenol admixture on low carbon steel in artificial seawater solution.

2. Experimental methods

Low carbon steel (LCS) with cylindrical configuration (diameter of 1.2 cm^2) has nominal (wt. %) composition shown in Table 1. The steel was machined and underwent metallographic preparation before washing with deionized H_2O and methyl-ketone. 4-methyl-norvalin with molecular formular $C_6H_{13}NO_2$ and molar mass of 131.17 g/mol, and 2-Methoxy-4-formylphenol with a molecular formular $C_8H_8O_3$ and molar mass of 152.15 g/mol were evaluated in the admixed form for corrosion inhibition. The compounds (4MMP) were prepared in combined molar concentrations of 1.76 x 10^2, 3.53 x 10^2, 5.29 x 10^2, 7.06 x 10^2, 8.82 x 10^2, 1.06 x 10^1 per 200 mL of 3.5% NaCl solution. Potentiodynamic polarization analysis was performed at 35 °C with a triple electrode system Graphs were produced at a scan rate of 0.0015 V/s between potentials of -1.25 V and +1.5 V. Optical images of corroded and 4MMP inhibited LCS morphology were studied after electrochemical analysis with

IOP Conf. Series: Materials Science and Engineering **613** (2019) 012013 doi:10.1088/1757-899X/613/1/012013

Omax trinocular metallurgical microscope. Open circuit potential measurement (OCP) was performed at 0.05 V/s step potential for 3000s to study the active-passive behavior of LCS.

Table 1. Composition (wt. %) of Low carbon steel

Element Symbol	C	Si	Mn	P	S	Cu	Ni	Al	Fe
% Composition	0.401	0.169	0.440	0.005	0.012	0.080	0.008	0.025	98.86

3. Result and discussion

3.1 Potentiodynamic polarization

The potentiodynamic polarization plots indicating the anodic-cathodic behavior of LCS in 4MMP /3.5%NaCl solution is shown in Fig. 1., Table 2 depicts the data obtained. LCS at 0% and 0.5% 4MMP severely deteriorated in the presence of Cl^- ions attaining corrosion rate values of 2 mm/y and 1.99 mm/y. 4MMP at 0.5% was unable to inhibit the corrosion of LCS due to insufficient protonated molecules to oppose the deterioration effect of Cl^- ions. At 1% - 2% 4MMP, the corrosion rate decreased to values between 0.96 mm/y and 0.75 mm/y corresponding to inhibition efficiencies of 51.84%, 55.45% and 62.73% respectively due to the availability of enough 4MMP molecules to inhibition the electrochemical action of chlorides. While changes in the cathodic Tafel slope a marginal a significant decrease in anodic Tafel slope was observed beyond 0.5% 4MMP due to decrease in anodic reactions on the steel. Observation of the polarization plot shows the cathodic branches of the plot with respect to 4MMP concentration are quite similar indicating the cathodic reaction mechanism is under activation control. The anodic branch of the polarization plot at 0% 4MMP and 0.5% 4MMP were linear due to oxidation reaction on LCS surface, however beyond 0.5% 4MMP short passivation behavior was observed due to inhibiting action of 4MMP compound. It shows the inhibition mechanism is through surface coverage. At 2.5% 4MMP and 2% 4MMP, the inhibition efficiency has attained values of 80.85% and 89% respectively signifying effective inhibition of LCS corrosion. The maximum change in corrosion potential of LCS between 0% 4MMP and 1% 4MMP in the reduction reaction direction is 81 mV thus it is a mixed type inhibitor [7].

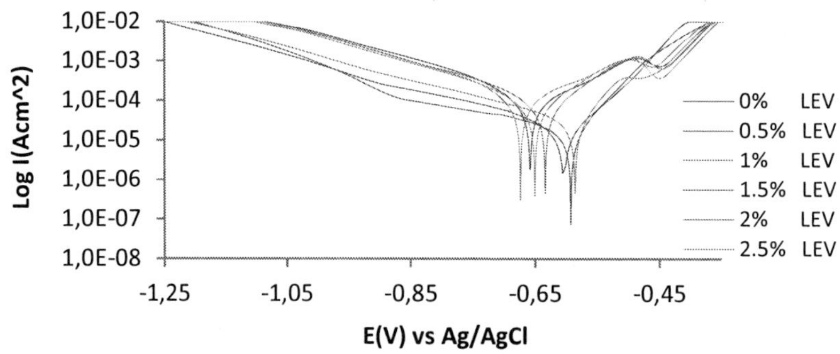

Figure 1. LCS corrosion polarization plots from 3.5% NaCl/0% - 2.5% 4MMP

Table 2. Results of potentiodynamic polarization test of LCS corrosion in 3.5% NaCl/0% - 2.5% 4MMP

Sample	4MMP Conc. (%)	4MMP Conc. (M)	LCS C_R (mm/y)	4MMP Y, (%)	C_I (A)	C_{ID} (A/cm²)	C_P (V)	R_p (Ω)	B_c (V/dec)	B_a (V/dec)
A	0	0	2.00	0	1.95E-04	1.73E-04	-0.593	116.00	-4.167	16.140
B	0.5	1.76E-02	1.99	0.73	1.93E-04	1.71E-04	-0.604	108.00	-6.002	13.630
C	1	3.53E-02	0.96	51.84	9.38E-05	8.30E-05	-0.674	273.80	-6.573	5.744
D	1.5	5.29E-02	0.89	55.45	8.68E-05	7.68E-05	-0.659	296.00	-7.401	7.001
E	2	7.06E-02	0.75	62.73	7.26E-05	6.42E-05	-0.634	353.90	-6.639	6.681
F	2.5	8.82E-02	0.38	80.85	3.73E-05	3.30E-05	-0.586	688.90	-4.579	6.306
G	3	1.06E-01	0.22	89.00	2.14E-05	1.90E-05	-0.651	759.70	-6.560	6.010

3.2 Inhibition Mechanism of 4MMP

4MMP consisting of 4-methyl-norvalin and 2-Methoxy-4-formylphenol are compounds consisting of organic molecules which adsorbs on the metallic surface during corrosion inhibition reactions inhibiting the diffusion and electrochemical action of corrosive species. These compounds form precipitates with metal cations on the ionized metal surface [8]. Observation of the inhibition reaction mechanism in the presence of 4MMP shows passivation resulting from film formation (surface coverage) of protonated 4MMP molecules occurred on LCS surface as evident on the anodic branch of the polarization plots.

3.3 Open circuit potential measurement (OCP)

The OCP plot of LCS in 3.5% NaCl/0%, 0.5% and 3% 4MMP versus exposure time is shown in Fig. 2. The OCP value of LCS (3.5% NaCl/0% 4MMP) at 26.45 s is the most electronegative (-0.680 V) due to surface oxidation reactions. The value later progressed sharply at short time interval to -0.503 V at 200 s due to short instantaneous passivation. However, the oxide formed are porous in nature, as a result the OCP immediately decreased to -0.551 V at 500 s before remaining marginally stable till 750.01 s (-0.554 V. Beyond this point the OCP values of LCS at 3.5% NaCl/0% 4MMP continue to decrease till 3200 s. The presence of 4MMP at 0.5% concentration caused an instantaneous increase in OCP due to marginal passivation of LCS 250 s at -0.559 V, beyond which the OCP value remained thermodynamically stable and at higher OCP values than LCS at 0% 4MMP. At 3% 4MMP the OCP value of LCS starting at -0.605 V (0 s) increased to -0.460 V at 300 s due to the corrosion inhibiting action of 4MMP. A progressive but stable marginal decrease later occurred till the end of the OCP test.

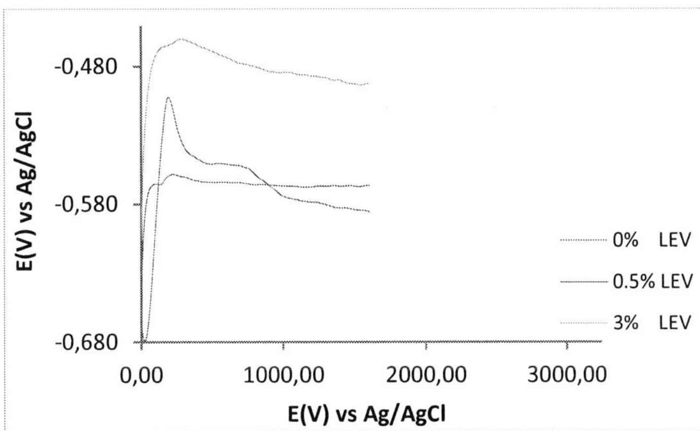

Figure 2. Variation of LCS OCP values versus exposure time in 3.5% NaCl/4MMP compound

3.4 Adsorption isotherm studies

Langmuir, Frumkin and Freundlich isotherm models produced significantly high correlation coefficients amongst other isotherms evaluated to further understand the nature of 4MMP adsorption and corrosion inhibition of LCS [9]. Fig. 3 shows the graphical illustration of $\frac{C_{4MMP}}{\theta}$ vs C_{4MMP} at correlation coefficient of 0.8024 with respect to Langmuir equation below.

$$\theta = \left[\frac{K_{4MMP}C_{4MMP}}{1+K_{4MMP}C_{4MMP}}\right] \qquad (1)$$

θ is the sum of 4MMP adsorbed per unit gram on LCS surface at constant temperature. C_{4MMP} is 4MMP concentration and K_{4MMP} is the equilibrium constant of adsorption. Fig. 4(a) shows the Frumkin isotherm graph of graph of $\log\left[\theta/(1-\theta)c\right]$ against θ with correlation coefficient of 0.9813 according to equation 2

$$\theta/_{1-\theta} = K_{\mathrm{TC}}ce^{2\alpha\theta}, \tag{2}$$

α is the lateral synergism parameter determined from the gradient of the graphical illustration of the Frumkin isotherm and K is the adsorption-desorption constant. Fig. 4(b) shows the Freundlich isotherm graph for 4MMP adhesion and assimilation with correlation coefficient of 0.7570 according to equation 3 and 4.

$$\theta = \mathrm{KC}^{\mathrm{n}} \tag{3}$$

$$\log\theta = \mathrm{n}\log\mathrm{C} + \log K_{\mathrm{ads}} \tag{4}$$

n is a constant subject to the properties of the inhibitor molecules involved in the adsorption reaction. K_{ads} is the adsorption–desorption equilibrium constant showing the adsorption strength.

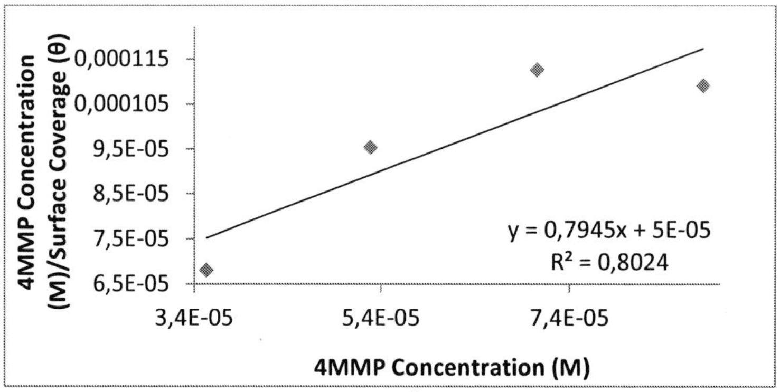

Figure 3. Langmuir isotherm plot of $\frac{C_{4\mathrm{MMP}}}{\theta}$ versus 4MMP concentration

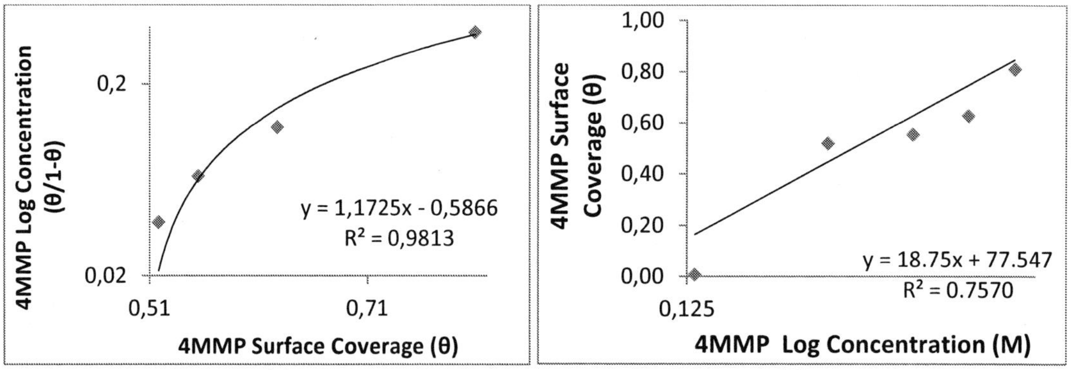

(a) (b)

Figure 4. (a) Graph of Frumkin isotherm, $\log\left[\theta/_{(1-\theta)c}\right]$ versus θ and (b) Graph of Freundlich isotherm, log 4MMP concentration vs log surface coverage.

3.5 Thermodynamics of inhibitor adsorption

Gibbs free energy data values are shown in Table 3. The highest $\Delta G^{o}_{\mathrm{ads}}$ value obtained is -20.70 KJmol^{-1} while the lowest significant $\Delta G^{o}_{\mathrm{ads}}$ value obtained is -18.42 KJmol^{-1}. The $\Delta G^{o}_{\mathrm{ads}}$ values shows the mechanism of adsorption is through physical attraction which shows that surface coverage of the inhibiting compound hinders the movement of the corrosive anions [10].

IOP Conf. Series: Materials Science and Engineering **613** (2019) 012013 doi:10.1088/1757-899X/613/1/012013

Table 3. Gibbs free energy (ΔG^{o}_{ads}), surface coverage (θ) and equilibrium constant of adsorption (K_{ads}) data for 4MMP adsorption on LCS

Sample	4MMP Concentration (*M*)	Surface Coverage (*θ*)	Equilibrium Constant of adsorption (*K*)	Gibbs Free Energy, ΔG ($Kjmol^{-1}$)
A	0	0	0	0
B	1.76E-02	0.007	0.42	-7.79
C	3.53E-02	0.518	30.49	-18.42
D	5.29E-02	0.555	23.53	-17.78
E	7.06E-02	0.627	23.84	-17.81
F	8.82E-02	0.809	47.87	-19.54
G	1.06E-01	0.890	76.33	-20.70

3.6 Morphological studies

Fig. 5(a) to 5(b) shows the optical images of LCS before corrosion, and after corrosion in 3.5% NaCl/0%, 0.5% and 3% 4MMP solution. The image in Fig. 6(b) differs from Fig 6(a) due to the destructive effect of Cl^- anions. Corrosion pits coupled with general surface deterioration are visible with corrosion products of iron oxides. Addition of 4MMP at 0.5% concentration [Fig. 6(a)] did not offer any significant protection. General corrosion seems to have been inhibited however the corrosion pits on the surface appears larger. The LCS morphology at 3% 4MMP [Fig. 6(b)] shows a remarkable improvement in comparison the earlier observed surfaces.

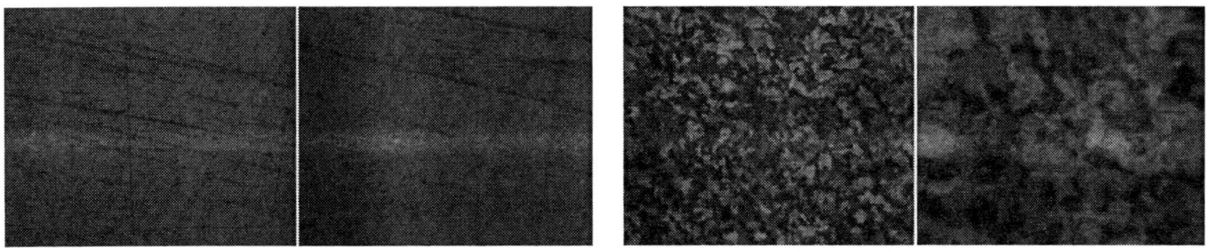

(a) (b)

Figure 5. Optical images of LCS at mag. x40 and x100 (a) LCS morphology before corrosion, (b) LCS morphology after corrosion in 3.5% NaCl

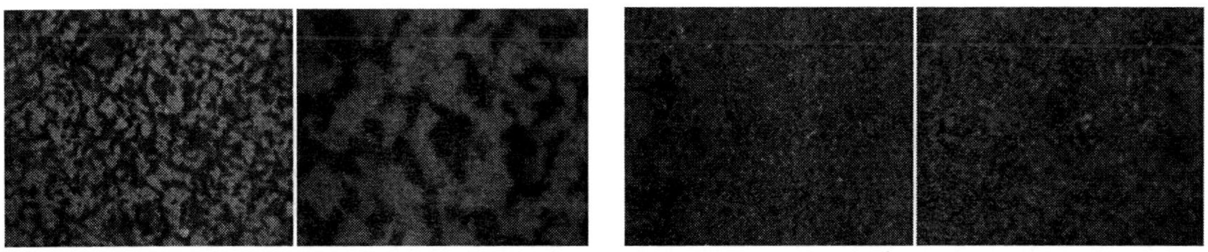

(a) (b)

Figure 6. Optical images of LCS at mag. x40 and x100 (a) LCS morphology after corrosion in 3.5% NaCl/0.5% 4MMP, (b) LCS morphology after corrosion in 3.5% NaCl/3% 4MMP

4. Conclusion

4-methyl-norvalin and 2-Methoxy-4-formylphenol synergistically suppressed the electrochemical action of chlorides on the carbon steel surface in artificial seawater. Inhibition efficiency performance of the compounds was well above 80%. The mechanism of inhibition was determined to be through surface coverage. This caused a large anodic shift in Corrosion potential shifted anodically in the presence of the compound. Physisorption adsorption dominated the mechanism of inhibitor-metal interaction.

Acknowledgement
The author acknowledges Covenant University Ota, Ogun State, Nigeria for the sponsorship and provision of research facilities for this project.

References

[1] Laamari R, Benzakour J, Berrekhis F, Abouelfida A, Derja A and Villemin D 2011 *Arabian J. of Chem.* **4** (3) 271 https://doi.org/10.1016/j.arabjc.2010.06.046
[2] Ashassi-Sorkhabi H, Majidi M R and Seyyedi K 2004 *Appl. Surf. Sci.* **225** (1-4) 176 https://doi.org/10.1016/j.apsusc.2003.10.007
[3] Loto R T 2017 *J. Mater. Res. Technol.* http://doi.org/10.1016/j.jmrt.2017.09.005
[4] Singh P, Bhrara K and Singh G 2008 *Appl. Surf. Sci.* **254** (18) 5927 https://doi.org/10.1016/j.apsusc.2008.03.154
[5] Samy M S, Ismail A, Mohamed M E, Soliman E A and Moshira Y E 2015 *J. of Mol. Liq.* **203** 20 https://doi.org/10.1016/j.molliq.2014.12.033
[6] Loto R T 2016 *Cogent Eng.* **3** 1242107 http://doi.org/10.1080/23311916.2016.1242107
[7] Loto R T, Loto C A, Popoola A P I and Fedotova T I 2016 *J. Cent. South Univ.* **23** 258 https://doi.org/10.1007/s11771-016-3069-1
[8] Abdul Amir H K, Abu Bakar M, Leiqaa A H, Ahmed A A, Ng H S and Ahmed Y M 2014 *Materials MDPI.* **7** 4335 http://doi.org/10.3390/ma7064335
[9] Bulut E, Ozacar M and Sengil I A 2008 *Microporous Mesoporous Mater.* **115** (3) 234 https://doi.org/10.1016/j.micromeso.2008.01.039
[10] Loto R T, Loto C A, Joseph O O and Olanrewaju G 2016 *Results Phys.* **6** 305 https://doi.org/10.1016/j.rinp.2016.05.013

Incommensurate Phase Transition and Electronic Properties of $BaMnF_4$

Selami Palaz[1], Sevket Simsek[2], Husnu Koc[3], Rena Babayeva[4], Amirullah M. Mamedov[*5,6], Ekmel Ozbay[5]

1Harran University, Faculty of Sciences, Department of Physics, Urfa, Turkey
[2] Hakkari University, Faculty of Engineering, Department of Material Science and Engineering, 3000, Hakkari, Turkey
[3] Siirt University, Faculty of Sciences, Department of Physics, Siirt, Turkey
[4]Department of Physics and Chemistry, Azerbaijan State University of Economics, Baku, Azerbaijan
[5] Bilkent University, Nanotechnology Research Center, 06800, Ankara, Turkey
[6] Baku State University, International Scientific Center, Baku, Azerbaijan

*E-mail: mamedov@bikent.edu.tr

Abstract. We present the ab initio study the electronic, mechanical and structural properties of $BaMnF_4$. We duscuss the trends in the electronic and mechanical properties of $BaMnF_4$ under pressure up to 80 GPa. $BaMnF_4$ belongs to the family of $BaMF_4$-type fluorides (M = Mn, Fe, Co, Ni, Mg, Zn) which share the same orthorhombic structure. The main focus of this study is to elaborate the changes brought about in the electronic and the structural properties by applied pressure. The calculated lattice parameters have been in agreement with the available experimental and theoretical value. Band gap of $BaMnF_4$ in our calculation is about 2.0 eV, separating the empty upper-Hubbard t_{2g} bands and occupied lower-Hubbard e_g bands. The total and partial DOS corresponding to the electronic band structure are calculated. Comparative analysis of the results of these calculations shows that the band-gap energy of $BaMnF_4$ decreases with increasing pressure and has a minima value at a critical pressure (appr. 65 GPa), after which it increases again. Some fundamental physical parameters such as elastic constants, bulk modulus, Poisson's ratio, sound velocities and Debye temperature were calculated and interpreted, too.
Keywords: *$BaMnF_4$, multiferroic, incommensurate phase transition, band structure*

1. Introduction

One of the very promising approaches to create novel materials is to combine in one material different physical properties to achieve rich functionality. The attempts to combine in one system both the ferromagnetic and ferroelectric properties started in 1960's [1], Materials combining these different "ferroic" [2] properties were later on called "multiferroics" [3].

Barium manganese fluoride $BaMnF_4$ one of the most favorite compound of the barium fluorides $BaMF_4$, where M is Mn, Fe, Co, or Ni. This series of multiferroic compounds presents numerous interesting physical properties and has been the subject of experimental or theoretical studies [3]. It also presents interesting structural properties and, in particular, a structural transition at 250 K from a high-temperature (HT) phase to an incommensurate modulated phase which has already been widely investigated.

IOP Conf. Series: Materials Science and Engineering **613** (2019) 012014 doi:10.1088/1757-899X/613/1/012014

We hope that our work will inspire further research in this interesting class of multiferroic materials, with the goal of exploring a broader range of non-oxide-based materials as candidates for magnetoelectric device applications.

Here we present a comprehensive computational study of the structural, electronic, mechanical, and optical properties of $BaMnF_4$ using first principles electronic structure calculations. The goal is to elucidate the origin of ferroelectricity in these materials. We have determined that there is not much work in our literature research about these compounds that exhibit interesting physical properties. Spaldin et al. [3], using the DFT-GGA(+U) approach, calculated the density of states for $BaMF_4$ compounds and tried to explain the origin of ferroelectricity in these compounds. Our aim in this study is to fill this emptiness by examining the mechanical and optical properties of BaMnF4 multiferroic compound under pressure which have not been studied theoretically as far as we know.

This paper is organized as follows. We first describe the methods we use in our calculations, together with some technical details. We then present the results of our calculations for the structural, electronic, and mechanical properties of $BaMnF_4$. Finally, we analyze the mechanism underlying the polar structural distortions in this compound. We end with a discussion and summary.

2. Method of calculation

In all of our calculations that were performed using the ab-initio total-energy and molecular-dynamics program VASP (Vienna ab-initio simulation program) [4-7] that was developed within the density functional theory (DFT) [8], the exchange-correlation energy function is treated within a spin polarized GGA (generalized gradient approximation) by the density functional of Perdew et al. [9]. The potentials used for the GGA calculations take into account the $5s^2 5p^6 6s^2$ electrons for Ba-atom, the $3p^6 4s^2 3d^5$ electrons for Mn-atom and the $2s^2 2p^5$ electrons for F-atom as valence states. To get good convergence, the kinetic energy cutoff for the total energy calculation is found to be 554 eV for BaMnF4 compound. The Brillouin-zone integration was performed using special k points sampled within the Monkhorst-Pack scheme [10]. We found that the mesh of 5x5x4 and 9x9x9 k points was required to describe well of structural, electronic, and elastic properties. This k-point mesh guarantees a violation of charge neutrality less than 0.008e. Such a low value is a good indicator for an adequate convergence of the calculations.

3. Results and discussion

3.1. Structural properties

$BaMnF_4$ compound have the orthorhombic symmetry with space group Cmc2_1 (No:36). There are four molecules with 24 atoms in the unit cell of $BaMnF_4$ compound. As a first step of our calculations, we obtained equilibrium lattice parameters and atomic potions of $BaMnF_4$ using experimental data given in Ref. [11]. These obtained results were used in all subsequent calculations. The obtained results are given in Table 1. We see that the results of our structural optimizations are in very good agreement with the experiment data.

Table 1. The calculated and experimental lattice parameters for $BaMnF_4$ compound.

Material	Reference	a (Å)	b (Å)	c (Å)
	Present	4.285	15.355	6.086
	Exp. [11]	4.221	15.107	5.998
	Exp. [12]	4.22	15.08	5.99
	Exp. [13]	4.221	15.098	5.984
BaMnF4	Exp. [14]	4.211	14.853	5.924
	Exp. [15]	4.23	15.1	6.0
	Exp. [16]	4.215	15.096	5.966
	Exp. [17]	4.221	15.098	5.984
	Exp. [18]	4.220	15.100	6.00

IOP Conf. Series: Materials Science and Engineering **613** (2019) 012014 doi:10.1088/1757-899X/613/1/012014

3.2. Elastic properties

Elastic constants are vital in determining physical properties of solids, such as their mechanical stability, stiffness, the brittle and ductile behaviours, Debye temperature, thermal expansion coefficient and bonding nature between atoms. The elastic constants of orthorhombic crystals are characterized with nine independent elastic constants, that is, C_{11}, C_{22}, C_{33}, C_{44}, C_{55}, C_{66}, C_{12}, C_{13} and C_{23} [19,20]. The elastic constants of $BaMnF_4$ compound were calculated using the strain–stress method [21] as implemented in the VASP [6,7]. Mechanical stability criteria for orthorhombic crystals [22],

$C_{11}>0$, $C_{22}>0$, $C_{33}>0$, $C_{44}>0$, $C_{55}>0$, $C_{66}>0$, $[C_{11}+C_{22}+C_{33}+2(C_{12}+C_{13}+C_{23})]>0$, $(C_{11}+C_{22}-2C_{12})>0$, $(C_{11}+C_{33}-2C_{13})>0$, $(C_{22}+C_{33}-2C_{23})>0$,

The bulk modules (B), shear modules (G), Young's modulus (E) and Poisson ratios (ν) for orthorhombic crystals can be calculated using the Voigt-Reuss-Hill (VRH) method [23]. The transverse, longitudinal, average elastic wave velocities (V_t, V_l and V_m) and Debye temperatures (Θ_D) of $BaMnF_4$ compund have been calculated using the common relations given in Ref. [24,25].

Table 2. The Poisson's ratio (ν), density (ρ in g/cm^3), transverse, longitudinal, average elastic wave velocities (V_t, V_l and V_m, in m/s) and Debye temperature (Θ_D in K) for $BaMnF_4$ compound.

Pressure	ν	B/G	ρ	V_t	V_l	V_m	Θ_D(K)
0	0.316	2.387	4.449	2525	4872	2827	329.36
5	0.314	2.359	4.874	2040	3921	2283	274.28
30	0.324	2.518	6.374	3300	6477	3697	485.63
40	0.325	2.527	6.718	3451	6781	3867	516.95
60	0.324	2.517	7.275	3763	7385	4217	578.76
70	0.309	2.295	7.510	3791	7222	4240	588.18
80	0.326	2.543	7.728	3964	7806	4443	622.22

3.3. Electronic properties

The electronic structure plays an important role in determining the multiferroic properties of the $BaMnF_4$. The band structure of along the principal symmetry directions have been calculated by using the equilibrium lattice constants as shown in Table 1.

The majority spin (spin-up) (a) and minority spin (spin-down) (b) electronic band structures and partial and total DOS (c) of the $BaMnF_4$ compound were calculated using GGA along the high symmetry directions in the first Brillouin zone (BZ) and, shown in Fig. 1. It is seen that for $BaMnF_4$, the majority spin states cross the Fermi level and thus have the semiconductor characters, while the minority spin states open the band gaps around the Fermi level and thus have the insulating nature with indirect transiton between valence and conducting bands (Γ – S).

IOP Conf. Series: Materials Science and Engineering **613** (2019) 012014 doi:10.1088/1757-899X/613/1/012014

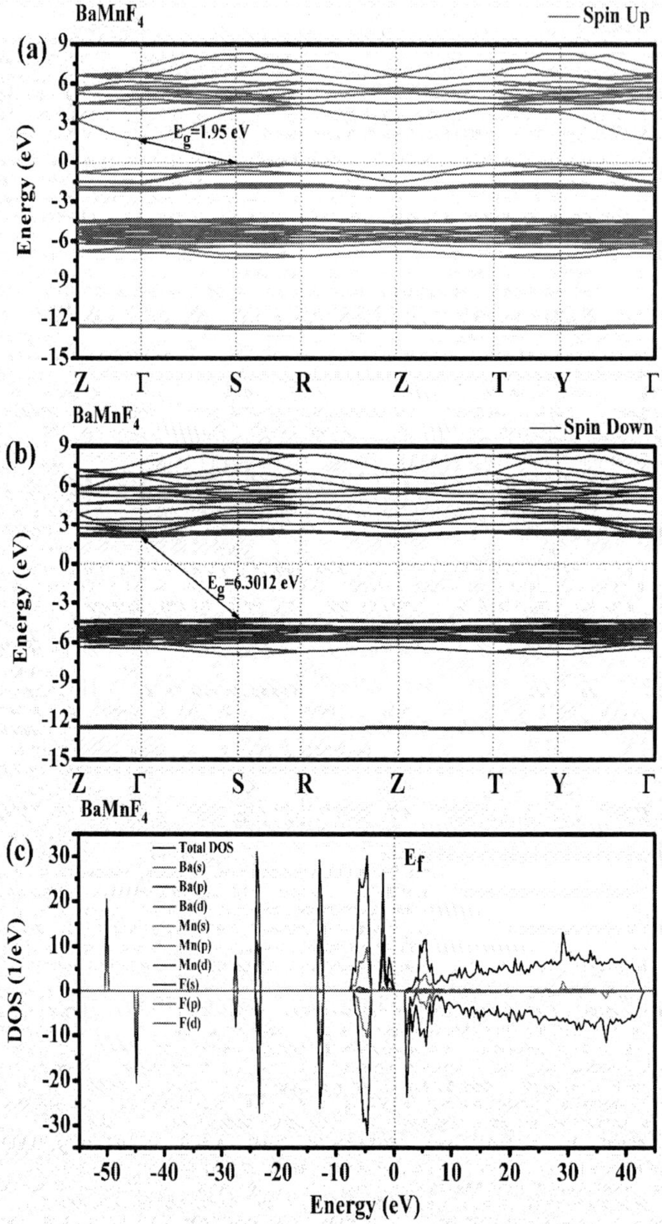

Figure 1. The calculated electronic band structure and total density of states for the majority spin (spin-up) and minority spin (spin-down) of the BaMnF4 compound.

The calculated total densities of states as well as the partial densities of the F p and Mn d states for $BaMnF_4$ is shown in Fig. 1. In the $BaMnF_4$ compound the gap is larger and is produced mainly by the strong exchange splitting between the two spin channels. In $BaMnF_4$ the forbidden gap is between occupied and unoccupied d states of the transition metal ion.

These results indicate that the use of the GGA is inadequate for the $BaMF_4$ systems, whereas the GGA method with an appropriate U value leads to a good description of the electronic structure. Since the overall features of the transition-metal d states in the present fluoride are very similar to the oxide cases. It can be seen from Fig. 1 that transition-metal cation are in a high-spin configuration, where the local majority spin states are fully occupied, and the minority spin states are filled. In the case of the GGA, the transition-metal d states are energetically well separated from the fluorine p states, leading to only negligible hybridization between the two sets of states. The use of GGA lowers the energy of the filled transition-metal d states, leading to energetic overlap of these states with the fluorine p levels and a certain degree of hybridization. The evolution of the electronic structure of $BaMnF_4$ under compression was studied by employing DFT, too. Comparative analysis of the results of these calculations shows that the band-gap energy of $BaMnF_4$ decreases with increasing pressure and has a minima value at a critical pressure (appr. 65 GPa), after which it increases again.

4. Conclusion

We have performed the structural, mechanical, and electronic properties of the $BaMnF_4$ multiferroic under pressure using density functional theory within the GGA approximation. The lattice parameters obtained as a result of geometric optimization are in good agreement with the experimental values. In the calculation of the calculated electronic band structure, $BaMnF_4$ compound is found as indirect band structured semiconductors in nature. The calculated elastic constants provide the mechanical stability. In addition, the calculated mechanical properties show that these compounds are ionic in character, less compressible, nearly rigid, low anisotropy, and dynamically stable.

References

[1] Roginskaya Y, Venevtsev Y, and Zhdanov G 1965 *Sov. Phys. JETP* **21** (5) 817
[2] Izyumskaya N, Alilov Y and Morkoc H 2009 *Critical Rev. Sol. State and Mat. Sci.* **34** (3-4) 89
[3] Ferroic and Multiferroics 2012 *ed. Virk H and Kleemann W* TTP Ltd.
[4] Kresse G and Hafner J 1993 *Phys. Rev. B.* **47** 558 https://doi.org/10.1103/PhysRevB.47.558
[5] Kresse G and Furthmuller J 1996 *Comput. Mater. Sci.* **6** 15 https://doi.org/10.1016/0927-0256(96)00008-0
[6] Kresse G and Joubert D 1999 *Phys. Rev. B.* **59** 1758 https://doi.org/10.1103/PhysRevB.59.1758
[7] Kresse G and Furthmuller J 1996 *Phys. Rev. B.* **54** 11169 https://doi.org/10.1103/PhysRevB.54.11169
[8] Hohenberg P and Kohn W 1964 *Phys. Rev.* **136** A1133 https://doi.org/10.1103/PhysRev.136.B864
[9] Perdew J P, Burke S and Ernzerhof M 1996 *Phys. Rev. Lett.* **77** 3865 https://doi.org/10.1103/PhysRevLett.77.3865
[10] Monkhorst H J and Pack J D 1976 *Physical Review B.* **13** 5188 https://doi.org/10.1103/PhysRevB.13.5188
[11] Cos D E, Shapiro S M, Cowley R A, Eibschütz M and Guggenheim H J 1972 *Physical Review, Serie 3. B Condensed Matter* **19** 5754
[12] Cousseins J C and Samouel M 1967 *Sciences Chimiques* **265** 1121
[13] Keve E T, Abrahamas S C and Bernstein J L 1969 *Journal of Chemical Physics* **51** 4928 https://doi.org/10.1063/1.1671885
[14] Posse J M, Friese K and Grzechnik A 2011 *Journal of Physics: Condensed Matter* **23** 1
[15] Petrov S V and Ippolitov E G 1971 *Inorganic Materials* **7** 769
[16] Yoshimura M and Hidaka M 2005 *Journal of Physical Society of Japan* **74** 1181 https://doi.org/10.1143/JPSJ.74.1181
[17] Venevtsev Y N and Gagulin V V 1995 *Inorganic Materials* **31** 797
[18] Posse J M, Grzechnik A and Friese K 2009 *Acta Crystallographica* **65** 576 https://doi.org/10.1107/S0108768109027177
[19] Deligoz E and Ozisik H 2015 *Philosophical Magazine* **95** 2294 https://doi.org/10.1080/14786435.2015.1056854
[20] Bhardwaj P and Singh S 2016 *Materials Science-Poland* **34** (4) 715
[21] Le Page Y and Saxe P 2011 *Phys. Rev. B.* **63** 174103 https://doi.org/10.1103/PhysRevB.63.174103

[22] Wu Z-J, Zhao E-J, Xiang H-P, Hao X-F, Liu X-J and Meng J 2007 *Physical Review B.* **76** 054115 https://doi.org/10.1103/PhysRevB.76.059904
[23] Chen Q and Sundman B 2001 *Acta mater.* **49** 947 https://doi.org/10.1016/S1359-6454(01)00002-7
[24] Koc H, Mamedov A M, Deligoz E and Ozisik H 2012 *Solid State Sciences* **14** 1211 https://doi.org/10.1016/j.solidstatesciences.2012.06.003

IOP Conf. Series: Materials Science and Engineering **613** (2019) 012015 doi:10.1088/1757-899X/613/1/012015

Comparative investigations of the selective Co etching from hardmetal support layer of diamond coating

Peter Rugóczky[1,2], Gábor Muránszky[2] and János Lakatos[2]

[1]Fraisa Hungária Kft, Sárospatak (Hungary)
[2] Institute of Chemistry, University of Miskolc (Hungary)

E-mail: prugoczky@gmail.com, gabor.muranszky@gmail.com, mtasotak@uni-miskolc.hu

Abstract. Selective cobalt etching from WC–Co composite usually carry out prior diamond deposition was investigated. Diluted Caro's acid as primarily utilized etching reagent was used. The cobalt and tungsten dissolution from hardmetals (HM) were determined as a function of time and a comparison were done in case of different grade hardmetals. For comparison beside the hardmetals the pure metallic cobalt and binderless, plasma sprayed tungsten carbide dissolution also has been tested. Hot filament chemical vapor deposition (HFCVD) technic was used to investigate the diamond nucleation on the differently pre-treated samples. The investigation highlighted for that: a) the selective cobalt etching can due to the different rate of dissolution of the two components, however the dissolution of the WC was not zero, b) the dissolution rate governed by the composite substance structure by the pore size which has formed during the selective dissolution process, therefore the dissolution rate of both components was found to be influenced by the grain size of WC and the cobalt content of the hardmetal. c) The removed cobalt quantity from the hardmetal surface has strong effect on the diamond nucleation during HFCVD method. Based on this research a better understanding of the mechanism of the pre-treatment and a chance for optimizing the selective cobalt etching from hardmetals prior diamond deposition will be achieved.

1. Introduction

Diamond is an attractive material in a wide range of technical applications due to its exceptional optical, chemical, and mechanical properties. This generates a huge needs of synthetic diamond production now days. The methods available for producing synthetic diamonds luckily let the producers to tailor the produced formations to the demands. One of the typical cases of this is the production of diamond on hardmetals to get wear resistant drilling and cutting tools.

The hardmetal which in this case works as a support of the deposited diamond is a metal matrix composite: consisting of different size tungsten carbide particles sintered by cobalt.

The hot filament diamond deposition method beside of the instrumental conditions as the gas composition, the gas flow, the temperature of the filament and the substrate sensitive to the chemical nature and the elemental composition of the substrate. Hwang et al. [1] has demonstrated that in the same CVD condition different carbon forms can be produced if the metallic substrate is different, e.g. silicon, iron or palladium. Diamond could form only on the silicon, graphite soot will deposit on the iron while nanotubes on the palladium. This means that the chemical nature of the substrate determines the structure of the deposited carbon form during CV deposition. Based on the literature it can state the preferred carbon forms is graphite if the substrate is Fe, Co, Rh, Ir, Ni, Pd or Pt, despite in case of Ti, Nb, Ta, Cr, Mo, W, Cu, Ag, Au or Si the diamond deposition is possible and preferred [1]. Not only the pure metals but metallic compounds and composites as well can be used as support. Researches demonstrated the possibility of diamond deposition onto carbides and nitrides as TiC, Ti SiC , WC, SiC, Si_3N_4, CrN, VC also [2-6].

The above described dependence of diamond formation from the chemical nature of the substrate surface showing that the well adherent diamond film deposition onto hardmetal can form if the support composition and chemical nature are suitable. This is does not prevail in case of the WC-Co type hardmetals since the cobalt content of the hardmetals has a detrimental effect on the diamond deposition. There are ways which generally used to eliminate the detrimental effect of cobalt. The first one is to cover the hardmetal surface with a substance which adherent to the hardmetal and promote the sp^3 bonded diamond formation. The second one is to convert the cobalt content of the hardmetal surface formatting stable intermetallic Co compounds. The third one is to remove the cobalt from the surface of the hardmetal [7].

The removal of cobalt can base on evaporation using high energy dens heat treatment like LASER ablation of the surface. The cobalt can transform to different compounds during heat treatment which later removed by chemical process. E.g. heat treatment at 1000 C° in a CH_4/H_2 atmosphere followed by Caro's acid Co removal; or surface treatment at 900 C° using microwave plasma leaded to $CoWO_4$ formation on the surface which removable in alkaline solution prior to Co etching in HCl solution [8-10].

The cobalt removing is possible using strong chemicals like alkaline media and different acids. One of the sophisticated two-steps method based on the selective WC etching in aquatic solution of $K_3[Fe(CN)_6]$ and KOH , called Murakami's reagent; followed by a second step which is the dissolution of the Co from the upper surface layer of hardmetal with an acid solutions containing different combination of H_2SO_4, H_2O_2, HNO_3 and HCl. The so-called diluted Caro's acid is the most frequently used for Co etching, which is the aquatic solution of the H_2SO_4 and H_2O_2 mixture [10-15].

The two-step Co removal pre-treatment method prior the diamond deposition is facilitating the CVD diamond formation but has also unfavourable effect cause a weakening of the outermost layer of the hardmetal leaves behind a porous WC skeleton.

The presented research focuses on: (i) determination of the cobalt dissolution rate as a function of the WC grain size and the Co content of the hardmetals; (ii) demonstrate that the Co dissolution mechanism is different in case of pure metallic Co and the Co content of the hardmetal; (iii) finding the optimum, a minimum of the removed cobalt quantity which enough to let the diamond nucleation during CVD process and the removal of cobalt does not cause large change in the hardmetal surface strength which must hold and fix the diamond film during the tool application.

2. Materials and experiments

The Co dissolution experiment were carried out on three types of different grade of hardmetals. The used samples were ø 14 mm and 5 mm thick. The two flat sides of the samples were polished using metallurgical grinding unit. Then was selectively etched the WC of the sample surface in Murakami's reagent containing 100 g $K_3[Fe(CN)_6]$, 100 g KOH, and 1000 ml H_2O for 25 minutes. The second step of the treatment was the cobalt solution in so called Caro's acid aqua solution containing of $C_{H_2SO_4} =$ 0.29 mol/dm^3 and $C_{H_2O_2}$= 2 mol/dm^3.

Further, as reference pure cobalt metal and plasma spray made WC samples were also etched in Caro' acid. The cobalt samples dimension were 9 x 11 x 2 mm and the WC samples were 12 x 11 x 5 mm. The table 1. summarize the tested materials.

Each of the dissolution tests were carried out on different sets which runs different length of time. The time varied between 20 and 180 seconds in case of hardmetals, between 60 and 420 seconds in the case of the metallic Co; between 60 and 540 seconds in the case of WC.

The volume of the Caro's acid was 100 ml at each dissolution experiment. The Co concentrations in the separated solutions were determined by Atomic Absorption Spectrophotometry (AAS) why in the W case Inductively Coupled Plasma Optical Emission Spectroscopy (ICP-OES) was used. The dissolution rate of the investigated materials was determined by using linear fit onto the measured points.

The integrity of the modified outermost layer of the treated hardmetal samples were investigated using standard Rockwell-C indentation adhesion test applying 1471 N pressing force. The test pieces were the same as which were used to the Co and W dissolution experiment.

The diamond nucleation was investigated using the ultrafine hardmetal type in form of Ø 6 mm rods. The samples were treated in the same way than the samples used for the dissolution rate investigation

IOP Conf. Series: Materials Science and Engineering **613** (2019) 012015 doi:10.1088/1757-899X/613/1/012015

and the dissolved cobalt was calculated according that measurement data. The diamond deposition of the samples was performed in the HFCVD diamond coating unit of the Fraisa Hungária Kft using the company's standard deposition parameter for diamond coating, but the deposition was stopped after one hour. Then to determining the diamond nucleation density the individual diamond crystals were counted on the SEM made pictures where the magnification was 10'000x.

Table 1. Characteristic data of tested materials

No.	Commercial name	WC grades, (typical grain size)	Co binder, m/m %
HM1	**CTF12E** (Ceratizit GmbH)	***Fine*** (0,8-1,3 μm)	6,0
HM2	**CTS12D** (Ceratizit GmbH)	***Submicron*** (0,5-0,8 μm)	6,0
HM3	**TSF22** (Ceratizit GmbH)	***Ultrafine*** (0,2-0,5 μm)	8,2
Co	Co metal	-	100
WC	WC (plasma sprayed)	(1-2 μm)	0

3. Results and discussions

3.1. Dissolution rate of Co and W from hardmetal and reference materials

The dissolution rate of the cobalt and tungsten are presented in Fig. 1. The dissolution rate of Co is higher in the case of the **Fine** hardmetal (HM1) than the **Submicron** hardmetal (HM2). The most probably cause of the higher solubility of HM1 is the higher average grain size of WC in the HM1. The two HM has the same initial Co content. The higher WC grain size mean higher space between the grains filled with Co. Since the HM1 and HM2 hardmetal have the same of WC to Co ratio and the higher grain size of WC mean lower specific surface area. Hence, in HM1 the same volume of Co cover lower surface of WC particles than the average Co film between the WC grain must be thicker compared to HM2 (in so far as the shape distribution of WC particles have the same in the two HM). During the Co dissolution larger pores has formed which more favourable for the transport processes.

The **Ultrafine** HM (HM3) has the highest Co content and this shows the highest dissolution rate. Our finding is that the higher WC grain size and the higher initial Co content of HM result higher dissolution rate of the cobalt. However, this picture can be more complex since it is known that the Co content of HM during the HM sintering process solute some W and that can modify the solubility of Co. Further investigation needed to clarify the effect of the dissolved W in Co at the contacting layer.

Determine the tungsten concentration in the etching solutions it shows that the tungsten solubility from the hardmetal in Caro's acid does not zero. Even the binderless plasma sprayed WC has shown some W solubility. Based on this it can state the selective Co etching in Caro' acid can due to the large difference of the solubility of the two components. If someone can improve this difference the selective Co removal will be more pronounced.

3.2. Cobalt dissolution mechanism from composite materials

According to the presented data of the Fig. 1. the cobalt dissolution rate shows great difference in case of HM compared to pure Co. It can be explained by supposing different dissolution mechanism take palace in the compared two cases. Concerning the dissolution of a piece of pure metal or its powder it is obvious the surface will govern the rate of dissolution. The rate is goeverned how fast the solvent and the dissolved ions can move towards the surface from the solution phase. But in case of HM composites the solution can access to the cobalt in the developed pores formed during the dissolution as it showed on Fig. 2. The rate difference of the transport in pores and free space can create this low solubility for the composites compared to the pure metals.

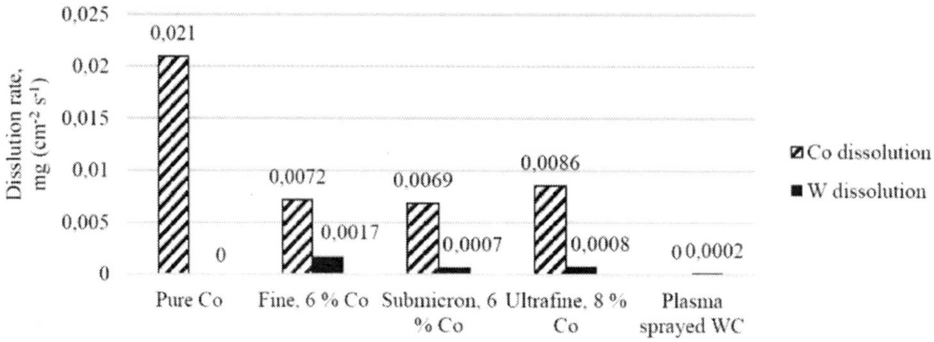

Figure 1. Cobalt and tungsten dissolution rate comparison from different materials in Caro's acid.The ratio of liquid to solid: in case of HM, pure Co and plasma spray made WC were 130, 500, 150 respectively, temperature 20 °C, static systems.

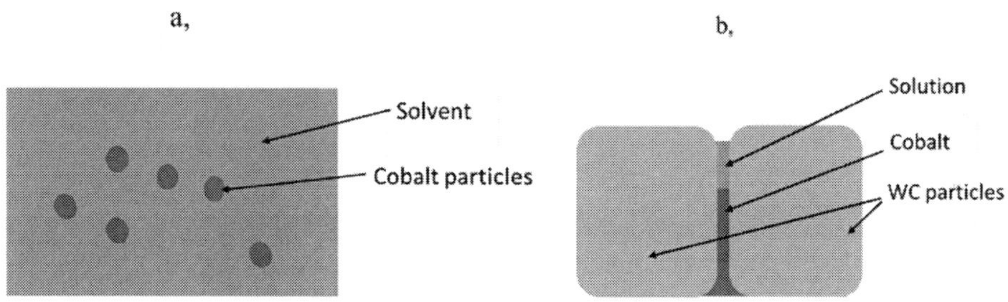

Figure 2. Different mechanism of the cobalt dissolution. In the left side the cobalt powder dissolution in solvent. The right-side cobalt dissolution between tow WC particle, in that case the accessibility of the solvent to the Co is limited by the WC particles.

3.3. Diamond nucleation enhancement during CVD process
The diamond nucleation were tested on hardmetal probes etched differently which means increasing amounts of Co was removed from the boundary layers where the diamond can formed and deposited. The Fig. 3. shows this in the case of two samples. On this SEM pictures it can be seen the coverage is different with diamond which equal the different manner of the diamond nucleation. The Fig. 4. presents the found nucleation density of diamond as a function of etched Co from the sample surface. These data prove that better quality diamond film can form if more Co removed from the boundary layer.

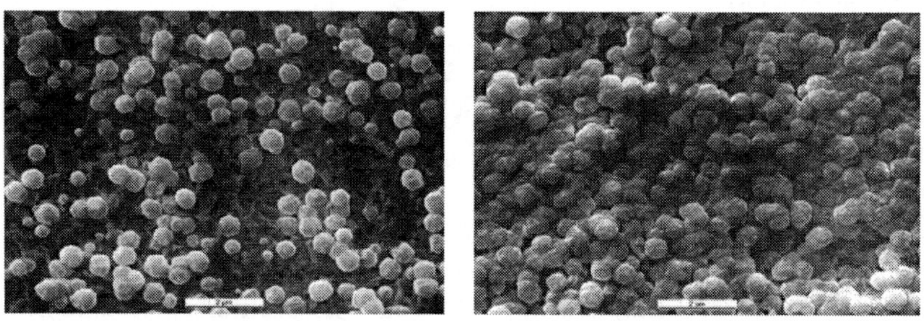

Figure 3. SEM pictures of the early stage of CV diamond deposition onto ultrafine hardmetal samples on the left-side poor diamond nucleation density cause of the higher Co content of the surface, on the right-side higher diamond nucleation density cause of the more removed cobalt from the surface. (Left-side sample: 0,33 mg cm^{-2} dissolved Co, 2,4 μm^{-2} diamond nucleation density; the right-side sample: 0,57 mg cm^{-2} dissolved Co, 3,9 μm^{-2} diamond nucleation density.)

3.4. Integrity of the modified outermost layer of hardmetal
The Rockwell indentation test usually used to determine the adhesion of coatings [16]. We found that test is suitable to characterise the modified Co-depleted or Co etched layer of HM prior diamond

deposition. The Fig. 5. shows the outermost Co-depleted layer of hardmetal which is weakened because the Co was removed from it during the pre-treatment. In the section 3.3. and Fig. 4. it was shown that larger amount of Co removal helps the diamond nucleation, but the Fig. 5. shows another considerable fact must relate to Co removal, this is the strength of the boundary layer which function is to support the deposited diamond.

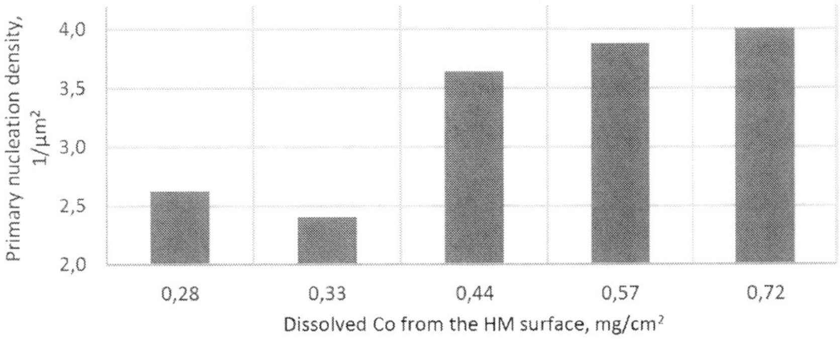

Figure 4. Nucleation density of diamond during HFCVD process onto differently Co etched samples.

Using the Rockwell indentation test we have found that the Co-depleted outermost layer of the HM exceeded a certain removed cobalt become instable and around the imprint of the indenter delaminate. In other words, the modified layer after too much etched Co behave like a coating having bad adhesion. The Fig. 6. shows that delamination effect. [17]

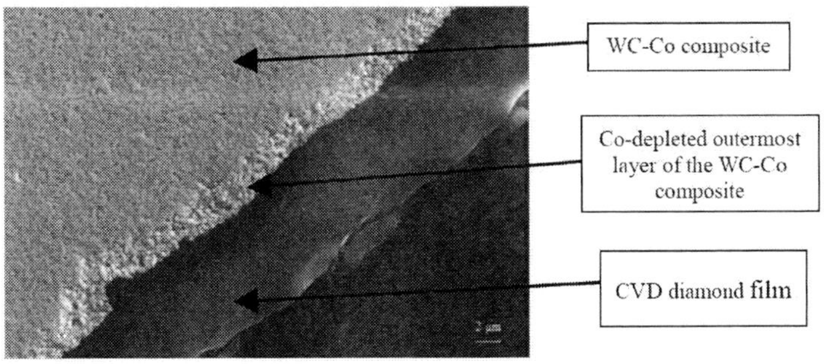

Figure 5. SEM picture of CVD diamond coated HM3 type hardmetal. The Co-depleted layer of the hardmetal is well visible between the bulk hardmetal and the diamond film. The polished sample was etched during 10 s in 3 % of Nital solution.

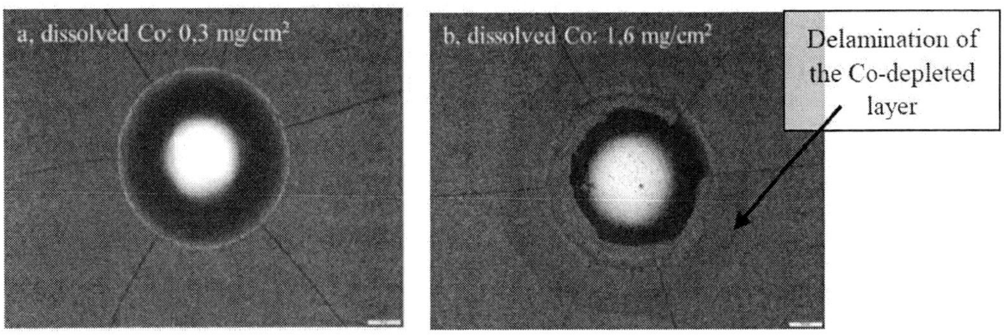

Figure 6. Light microscope pictures of the Rockwell imprints on two steps treated ultrafine hardmetal samples. The sample (b) around the imprint it can be seen delamination of the outermost Co-depleted layer as a cause of the too much dissolved cobalt; the sample (a) don't show the delamination phenomena due to the less removed cobalt than the sample (b). Magnification: 200x. [17]

4. Conclusions

The presented research highlighted the following aspects of the CV deposition of diamond onto hardmetal substrate:

In point of the engineering of a proper diamond coating onto the hardmetal support need to find the balance between removing cobalt from the surface to enhance diamond nucleation but left enough Co in the modified hardmetal layer which able to fix the WC particles and strongly hold the diamond film.

The Co dissolution rate in case of hardmetal is affected by the grain size and the Co content of hardmetals, but to establish correlation between the dissolution rate and the grain size of hardmetal are needed furthermore investigation which considers the effect of the dissolved tungsten in the WC sintering metallic Co.

The Co dissolution mechanism in the case of composite materials was found different from that of the bulk or the powder-like sample.

The selective Co etching from hardmetal has based on that the Co dissolution rate is much higher than the W dissolution rate from WC, however the W dissolution rate is not zero.

References

[1] N. M. Hwang 2016 "Non-Classical Crystallization of Thin Films and Nanostructures in CVD and PVD Processes" *Springer Nature* https://doi.org/10.1007/978-94-017-7616-5

[2] S. Konoplyuk, T. Abe, T. Takagi, T. Uchimoto 2007 *Diamond and Related Materials* **16** (3) 609 https://doi.org/10.1016/j.diamond.2006.11.044

[3] T .Abe, T .Takagi, Z .M .Sun, T .Uchimoto, J .Makino, H .Hashimoto 2004 *Diamond and Related Materials* **13** (4) 819 https://doi.org/10.1016/j.diamond.2003.10.044

[4] K. Mallika, R. Komanduri 2001 *Thin Solid Films* **396** 145 https://doi.org/10.1016/S0040-6090(01)01262-7

[5] M. Fischer, M. Chandran, R. Akhvlediani, A. Hoffman 2016 *Diamond and Related Materials,* **70** 167-172. https://doi.org/10.1016/j.diamond.2016.10.021

[6] D.D. Damm, A. Contin, M.C.C.B. Gomes, V.J. Trava-Airoldi, D.M. Barquete, E.J. Corat 2018 *Diamond and Related Materials* **89** 218 https://doi.org/10.1016/j.diamond.2018.09.013

[7] R. Haubner, W. Kalss 2010 *Int. Journal of Refractory Metals & Hard Materials* **28** 475 https://doi.org/10.1016/j.ijrmhm.2010.03.004

[8] M. Barletta, G. Rubino, A. Gisario 2011 *Wear* **271** 2016 https://doi.org/10.1016/j.wear.2011.01.042

[9] Z.M. Zhang, X.C. He, H.S. Shen, F.H. Sun, M. Chen, Y.Z. Wan 2000 *Diamond and Related Materials* **9** 1749 https://doi.org/10.1016/S0925-9635(00)00328-9

[10] R. Polini, F. Bravi, F. Casadei, P. D'Antonio, E. Traversa 2002 *Diamond and Related Materials,* **11** 726 https://doi.org/10.1016/S0925-9635(02)00020-1

[11] T. Yang, Q. Wei, Y. Qi, Y. Wang, Y. Xie, J. Luo, Z. Yu 2015 *Journal of Alloys and Compounds* **639** 659 https://doi.org/10.1016/j.jallcom.2015.03.134

[12] S. Kim, B. Seo, S.-Ho Son 2014 *Hydrometallurgy* **143** 28 https://doi.org/10.1016/j.hydromet.2014.01.004

[13] H. Gomez, D. Durham, X. Xiao, M. Lukitsch, P. Lu, K. Chou, A. Sachdev, A. Kumar 2012 *Journal of Materials Processing Technology* **212** 523 https://doi.org/10.1016/j.jmatprotec.2011.10.020

[14] F.A. Almeida, J.M. Carrapichano, A.J.S. Fernandes, J. Sacramento, R.F. Silva, F.J. Oliveira 2011 *International Journal of Refractory Metals and Hard Materials* **29** 618 https://doi.org/10.1016/j.ijrmhm.2011.04.008

[15] H.Sein, W. Ahmed, M. Jackson, R. Woodwards, R. Polini 2004 *Thin Solid Films* **447-448** 455 https://doi.org/10.1016/j.tsf.2003.08.044

[16] N. Vidakis, A. Antoniadis, N. Bilalisb, 2003 *Journal of Materials Processing Technology* **143-144** 481 https://doi.org/10.1016/S0924-0136(03)00300-5

[17] P. Rugoczky, G. Lassú, J. Lakatos 2018 *International Journal of Advances in Science Engineering and Technology* **6** 47 http://www.iraj.in/journal/journal_file/journal_pdf/6-480-153493770247-52.pdf

Formation of metastable zirconium oxides using pulsed laser deposition of ZrO based target

Tomáš Křenek[1], Petr Mikysek[2,3], Michal Pola[1], Lukáš Vála[1], Estela Melaré[5], Věra Jandová[4], Veronika Vavruňková[1], David Rieger[1]

[1] University of West Bohemia, Research Centre New Technologies 30614 Plzeň, Czech Republic
[2] The Czech Academy of Sciences, Institute of Geology, Rozvojová 269, 165 00 Prague 6, Czech Republic
[3] Masaryk University, Faculty of Science, Institute of Geological Sciences, Kotlářská 2, 611 37 Brno, Czech Republic
[4] Institute of Chemical Process Fundamentals, Czech Academy of Sciences, Rozvojova 135, 160 00 Prague 6, Czech Republic
[5] Federal University of São Carlos, Department of Chemistry, Interdisciplinary Laboratory of Electrochemistry & Ceramics (LIEC), Rodovia Washington Luis, Km 235, SP-310, Monjolinho, 13565905 - São Carlos, SP, Brasil

E-mail: tkrenek@ntc.zcu.cz

Abstract
There is a growing interest about the possibilities for preparation of various zirconium oxides from scientific as well as application point of view. Laser ablation of solid target consisting of sintered grains of metallic hexagonal Zr_3O and monoclinic ZrO_2 results in evaporation of Zr, O species (ions, atoms) and subsequent deposition of Zr-O film. Ta and Cu have been used as substrates. The films were analyzed by SEM-EDX, Raman and FTIR spectroscopy and X-ray diffraction. SEM analysis revealed μm/sub-μm sized roundshape and ring-like objects on continuous coat. Using Raman spectroscopy broad peaks which can suggest partial laser induced amorphization of Zr-O deposits were detected. FTIR spectroscopy shows bands which are assignable to Zr-O vibration of nanostructured zirconia. X-ray diffraction analysis provides more explicit assignment of deposited phases where the deposit on Ta exhibits presence of monoclinic ZrO_2, oxygen deficient rhombohedric Zr_3O and interestingly indicates presence of high-pressure orthorhombic ZrO_2 phase. The coat on Cu consists of monoclinic ZrO_2, rhombohedric Zr_3O, and metastable tetragonal ZrO_2 phase.

1. Introduction

There is a growing interest about the properties of zirconium oxides in a wide range of applications. Zr and ZrO_2 have been proved as useful materials for orthopedic applications because ZrO_2 possesses high hardness, wear and corrosion resistance, and good biocompatibility [1–4]. Though ZrO_2 is mostly considered as bioinert material also its bioactivity is possible to enhanced [5]. ZrO_2 belongs among wide band gap (with band gap of ~5 eV) n-type semiconductors which represent suitable photocatalysts for the decomposition of water pollution [6-11]. Among these wide band gap semiconductors, especially ZrO_2 nanoparticles have been attracted considerable attention because of their various potential applications [6], and also play a significant role in photocatalysis research due to their high surface to volume ratio [12]. Nanostructured zirconia represents promising material also because of its unique optical and electrical properties as well as other potential applications in transparent optical devices, oxygen sensors, fuel cells, electrochemical capacitor electrodes, and advanced ceramics [13-17]. The applications of zirconia strongly depend on both crystal structure and phase transformations [15-17].

Pure Zr possesses two crystalline modifications: hcp crystal structure (α-Zr) which transforms to the bcc crystal structure (β-Zr) above 889°C [18]. Both hcp and bcc polymorphs of Zr possess significantly high oxygen solubility. The β-Zr is able to dissolve 10.5 at% oxygen at high temperatures, whereas the solubility of α-Zr achieves values as high as 28.6–35 at% of oxygen. Exceedance of the solubility limit of oxygen leads to the formation of ZrO_2 which crystallize in three polymorphs according the temperature: the monoclinic α-ZrO_2 phase is stable below 1000°C, the intermediate tetragonal β-ZrO_2 phase is stable at 1200°C and is transformed to a cubic γ -ZrO_2 phase above 2285 °C [19]. Moreover to γ -, β-, and α-ZrO_2, at least two orthorhombic phases are stabilized at high pressure [20]. Contrary to well described ZrO_2 there is not fully known stable bulk form of ZrO.

Here we report on pulsed laser deposition of the Zr-O target consisting of monoclinic ZrO_2 and hexagonal Zr_3O sintered grains. The laser ablative deposition represents challenging method for preparation of smooth and nanostructured films as well as nanosized particles from elemental, inorganic and organic bulk materials [eg. 21, 22]. Thermodynamically non-equilibrium conditions of laser ablation [21] are challenging for deposition of metastable Zr-O phases.

2. Experimental

A 3rd harmocnic of pulsed Nd:YAG laser (model Q SMART 850, base wavelength: 1064 nm, used wavelength: 355 nm, energy per pulse: 180 ± 5 mJ, pulse duration: 10 ns, repetition rate: 10 Hz) was used for ablation of a ZrO target. The irradiation was focused by lens (f = 15 cm) on the spot area of 0.02 cm^2. Laser irradiation experiments were carried out in a simple tubular Pyrex reactor (70 mL in volume) continuously pumped to high vacuum by using a turbomolecular source (base pressure 5×10^{-2} Pa). The reactor was furnished with borosilicate glass windows. The duration of irradiation was 6 minutes. The target of ZrO pellet with diameter 14 mm and height 5 mm was positioned vertically in the center of the reactor and the substrate was situated perpendicularly above the target. The tantalum and copper foils were used as a substrate. For each sample deposition, first the vacuum chamber is opened and the clean substrate is placed inside of the Pyrex chamber. Then, the reactor is closed and the pressure is lowered. After the irradiation the pressure is increased up to atmospheric, the chamber is opened, the sample with the resulting coats is taken for examination and a new sample is placed.
Raman spectra were obtained using a DXR Raman microscope with Diode-pumped solid state laser emitting at 532 nm using high resolution gratings working in the range of 50 – 1800 cm^{-1} and spectral resolution 2 cm^{-1} FWHM.

Infrared absorbance spectra were performed on the Nicolet 380 FTIR spectrometer in the range of 650–4000 cm^{-1} using the ATR (attenuated total reflection) accessory with trapezoidal silicon crystal with a beveled edge of 45 degree. ATR technique offers a high sensitivity of absorbance measurements due to special data acquisition by multi-passing the investigated thin films by IR radiation.

X-ray diffraction investigation was carried out with a Bruker D8 Discover diffractometer equipped with a silicon-strip linear LynxEye detector and a focusing germanium primary monochromator of Johansson type providing Cu$K\alpha_1$ radiation (λ = 1.54056 Å). Data for mineral identification were collected in the 2θ range of 20–125° with a step size of 0.016° and a counting time of 8 second at each step, and detector angular opening of 3.757°. The phase identification was performed with Diffrac.Eva software v4.2.2 and ICDD PDF-2 database (Bruker AXS GmbH, Karlsruhe, Germany; 2011–2016). ZrO target represents commercially available product: Zirconium monooxide ZrO tablets (umicore).

3. Results and discussion

The UV laser irradiation of ZrO target in vacuum leads to ablation and creation of visible bluish luminescence zone suggesting ionization. The laser ablation results in the deposition of the coat which exhibit similar interference appearance on both Ta as well as Cu substrate (Fig. 1). Ta has been selected as an inert substrate and Cu as a substrate with reactive potential.

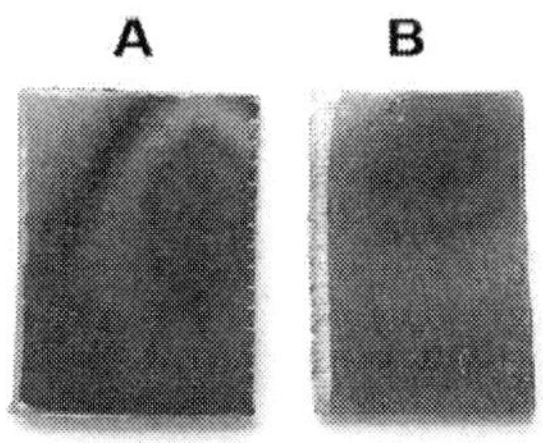

Fig. 1 The coated Ta (a), Cu (b) substrates.

The SEM images of the deposit on Ta (Fig.2 A) reveal μm/sub-μm sized roundshape and ringlike objects on continuous coat. The elemental mapping (Fig. 2 B) shows homogenous distribution of Zr with higher concentration in μm-sized particles. The deposit morphology on Cu substrate (Fig. 2 C) exhibits spherical particles with the size spanning from tens of nm up to units of μm. Elemental mapping (Fig. 2 D) depicts large shapeless agglomerates with high Zr concentration whereas the continuous surrounding seems to possess lower Zr content compared to the coat on Ta. The presence of larger shapeless/ringlike objects is in accordance with vaporized and plasma-produced clusters, and liquid droplets expelled from the target surface and quenched upon deposition. The roundshaped smaller particles indicate the rapid cooling of the solidifying droplets, which may occur in a metastable/amorphous state. SEM-EDX analysis provides the ratio of presented elements (Zr, O, Ta/Cu) where is detected excess of oxygen in Zr:O ratio which can indicate small oxidation of the substrates.

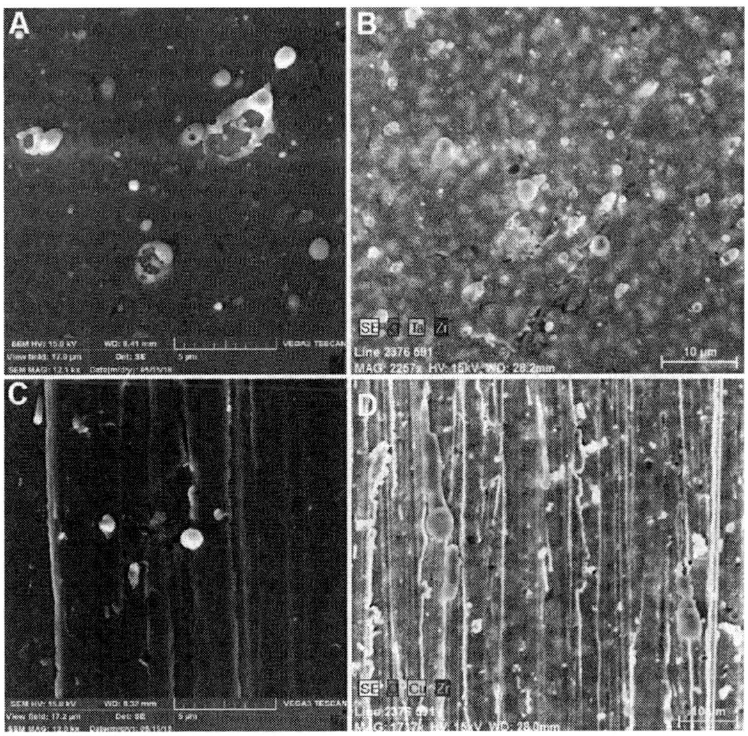

Fig. 2 SEM images of coats deposited on Ta (A) and Cu (C);
elemental mapping on Ta (B) and Cu (D)

Raman spectroscopy analysis revealed the presence of two phases in original target (Fig. 3 A). The target consists of sintered grains where one kind of the grains does not possess any vibrational peaks and thus can belongs to metallic Zr or solid solution of Zr and O and the second one whose peaks agrees with monoclinic ZrO_2 [23, 24]. The peak position slightly differs compared to literature [23, 24] which can be caused by stress in the target which was prepared obviously by sintering under the press. The Raman spectroscopy of Zr-O coat on Ta and Cu (Fig. 3 B) shows different features compared to original target which suggest that some phase transformation was carried out. The deposit on Ta reveals the broad peaks centered at ~ 167 cm^{-1}, 241 cm^{-1}, 620 cm^{-1}. The most significant broad peaks of the coat on Cu are positioned at ~ 290 cm^{-1}, 620 cm^{-1}. The peak at 620 cm^{-1} is similar for both Ta and Cu substrate whereas to position of other peaks varies. The broad peaks occur in the same regions like the sharp peaks of crystalline monoclinic ZrO_2 in original target. From the broad shape of the peaks we can tentatively conclude that partial laser induced amorphization of Zr-O phase took place.

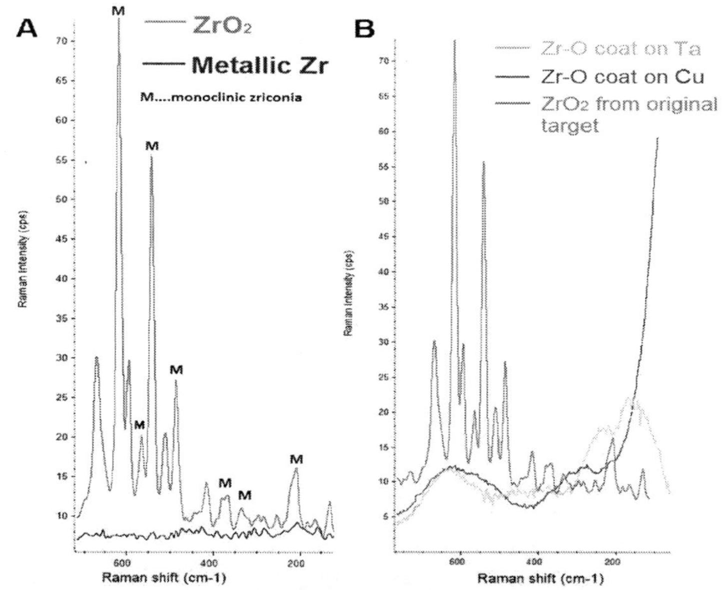

Fig.3 Raman spectroscopy of the original target (A) and of the coats on Ta and Cu substrates (B)

FTIR spectra exhibit very similar features for the deposits on Ta and Cu (Fig. 4). It revealed presence of bands at 630, 654 and 842 cm^{-1} which is assigned to strong stretching vibrations of the Zr-O group [25]. The vibration band around 1652 cm^{-1} is attributed to Zr-OH vibration [25] and the absorbance band at around 3300 cm^{-1} agrees with the adsorbed water which is consistent with the published FTIR spectra of nano-ZrO_2 [26].

Fig. 4 Typical FTIR spectrum for Zr-O coat on Ta and Cu

XRD of the ablated pellet exhibits in agreement with Raman spectroscopy presence of two phases: monoclinic ZrO_2 and hexagonal Zr_3O. Oxygen deficient hexagonal Zr_3O represents solid solution of O in hcp α-Zr. The weight fraction of present phase corresponds with approximately 1/3 of monoclinic ZrO_2 and 2/3 of hexagonal Zr_3O. XRD analysis of the deposit on Ta (Fig. 5) shows intensive peaks corresponding with cubic body centered Ta of the substrate. In the detail of the XRD spectrum less intensive diffraction peaks of the polycrystalline coat are visible. The peaks are clearly assignable to monoclinic ZrO_2 and to oxygen deficient rhombohedric Zr_3O. Moreover XRD analysis suggests interestingly also the presence of orthorhombic ZrO_2 phase which is stabilized only under the high pressure [20]. The brief crystallographic description of detected phases summarizes Tab.1.

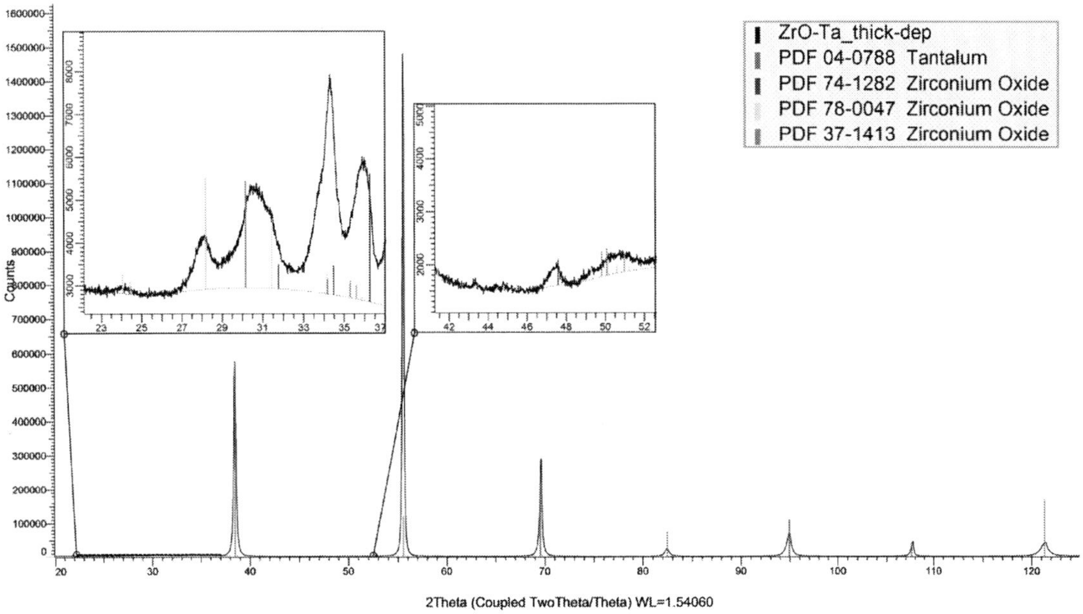

Fig. 5 XRD of Zr-O coat on Ta

Tab. 1 List of phases of the coat on Ta assigned using XRD

Pattern #	Compound Name	Formula	System	Space Group
PDF 04-0788	Tantalum	Ta	Cubic	Im3m (229)
PDF 74-1282	Zirconium Oxide	Zr3O	Rhombo.H.axes	R-3c (167)
PDF 78-0047	Zirconium Oxide	ZrO2	Monoclinic	P21/c (14)
PDF 37-1413	Zirconium Oxide	ZrO2	Orthorhombic	P212121 (19)

XRD of Zr-O coat on Cu (Fig. 6) revealed the most intensive peaks belonging to cubic Cu of the substrate. Less intensive peaks of the coat are in good agreement with monoclinic ZrO_2 and to oxygen deficient rhombohedric Zr_3O as in the case of Ta substrate. However in coat on Cu XRD indicates the peaks corresponding with tetragonal ZrO_2 phase which is considered as unstable phase under the room temperature. Contrary to the coat on Ta there is no evidence of high-pressure orthorhombic ZrO_2 polymorph. Summarization of the phases detected in the coat on Cu is given in Tab. 2.

In case of both deposits the assignment of monoclinic ZrO_2 and rhombohedric Zr_3O is unambiguous however the clear confirmation of high-pressure orthorhombic and unstable tetragonal structure should be investigated and clearly confirmed by next analytical methods such as back-scattered electrons and/or electron diffraction.

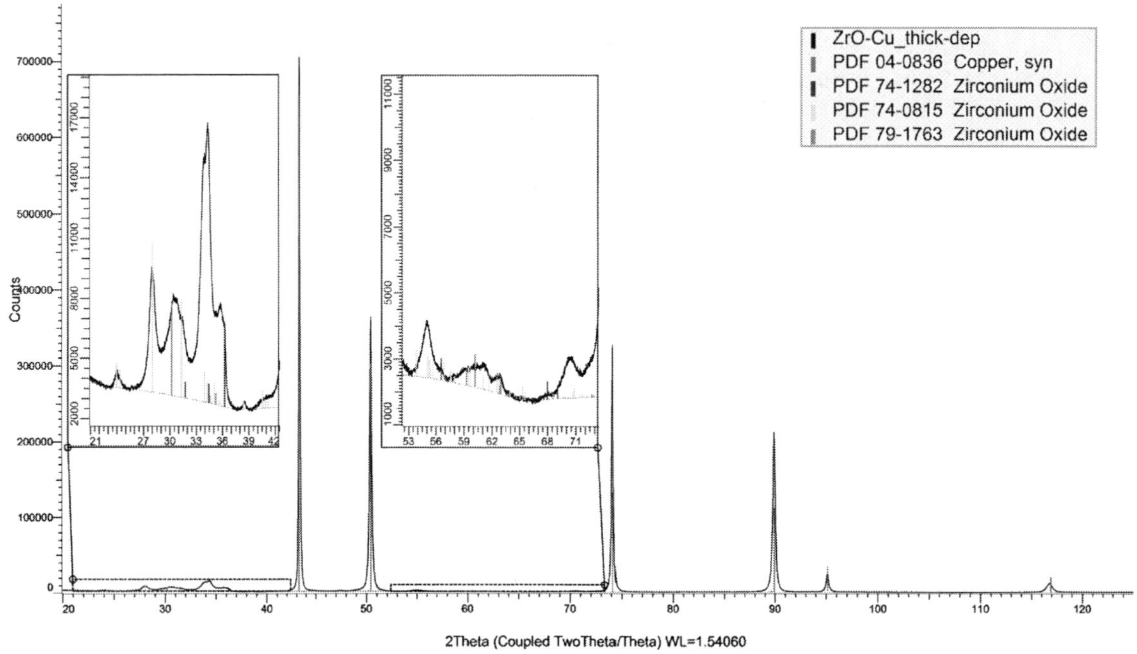

Fig.6 XRD of Zr-O coat on Cu

Tab. 2 List of phases of the coat on Cu assigned using XRD

Pattern #	Compound Name	Formula	System	Space Group
PDF 04-0836	Copper, syn	Cu	Cubic	Fm-3m (225)
PDF 74-1282	Zirconium Oxide	Zr3O	Rhombo.H.axes	R-3c (167)
PDF 74-0815	Zirconium Oxide	ZrO2	Monoclinic	P21/c (14)
PDF 79-1763	Zirconium Oxide	ZrO2	Tetragonal	P42/nmc (137)

These complementary analyses are in line with the deposition of monoclinic ZrO_2 and rhombohedric Zr_3O in case of both Ta as well as Cu coats. Moreover XRD indicates high-pressure orthorhombic ZrO_2 phase on Ta and metastable tetragonal ZrO_2 on Cu. The laser induced formation of high-pressure polymorph has been previously published for eg. orthorhombic Fe_3O_4 nanograins [27] or high-pressure Ga nanostructures [28].

These results are consistent with rapid evaporation of both hexagonal Zr_3O and monoclinic ZrO_2 from an irradiated original target which results in intermixing events of atomic/ionic species of Zr and O in the gas phase and subsequent rapid deposition of the multi-phase structure containing also metastable zirconium oxides.

4. Conclusion

Laser ablation of the target based on sintered grains of Zr and ZrO_2 allows evaporation/ionization of present species and following deposition of Zr-O coats. The deposits on Ta and Cu exhibits roundshape and ringlike particles whose sizes span from nm up to units of μm. Raman and FTIR spectroscopy suggested formation of nanostructured partially amorphous Zr-O films. X-ray diffraction reveals formation of monoclinic ZrO_2 and rhombohedric Zr_3O on both Ta and Cu substrate. Moreover XRD suggests formation of high-pressure ortorhombic ZrO_2 phase on Ta and metastable tetragonal ZrO_2 on Cu. These results show that intermixing events in the gas phase together with highly non-equilibrium conditions of laser ablation are favorable for the deposition of various metastable Zr-O phases.

Acknowledgements
The result was developed within the CENTEM project, reg. no. CZ.1.05/2.1.00/03.0088, cofounded by the ERDF as part of the Ministry of Education, Youth and SportsOPRDI programme and, in the follow up sustainability stage, supported through CENTEM PLUS (LO1402) by financial means from the Ministry of Education, Youth and Sports under the National Sustainability Programme I. The study was also conducted under institutional support from the Institute of Geology of the Czech Academy of Sciences, v.v.i. (RVO67985831). Computational resources were provided by MetaCentrum (LM2010005) and CERIT-SC (CZ.1.05/3.2.00/08.0144) infrastructures.

This work was supported also by the project n. 201 "Advanced porous biomaterials functionalized with stem cells for enhancing of osseointegration of implants: **MATEGRA**" realised within the frame of the Program IN-TERREG V-A: Cross-border cooperation between the Czech Republic and the Federal State of Germany Bavaria, Aim European Cross-border cooperation 2014-2020. The realisation is supported by financial means of the European Regional Development Fund (85% of the costs) and the state budget of the Czech Republic (5%).

References

[1] Spector M, Ries M D, Bourne R B, Sauer W S, Long M, Hunter G 2001 *J. Bone Joint Surg.* **83** 80
[2] Ferraris M, Verne E, Appendino P, Moisescu C, Krajewski A, Ravaglioli A, Piancastelli A 2000 *Biomaterials* **21** 765
[3] Suzuki T, Fujibayashi S, Nakagawa Y, Noda I, Nakamura T 2006 *Biomaterials* **27** 996
[4] Uchida M, Kim H M, Kokubo T, Nawa M, Asano T, Tanaka K, Nakamura T 2002 *Jour. Biomed. Mater. Res.* **60** 277
[5] Zhang L, Han Y 2011 *Materials Science and Engineering C* **31** 1104
[6] Gröttrup J, Schütt F, Smazna D, Lupana O, Adelung R, Mishra Y K 2017 *Ceram. Int.* **43** 14915 http://dx.doi.org/10.1016/j.ceramint.2017.08.008
[7] Mishra Y K, Adelung R 2018 *Mater. Today* http://dx.doi.org/10.1016/j.mattod.2017.11.003
[8] Gröttrup J, Paulowicz I, Schuchardt A, Kaidas V, Kaps S, Lupan O, Rainer Adelung R, Mishra Y K 2016 *Ceram. Int.* **42** 8664
[9] Babu B, Mallikarjuna K, Reddy Ch V, Park J 2016 *Mater. Lett.* **176** 265
[10] Reddy Ch V, Babu B, Shim J 2017 *Mater. Sci. Eng. B.* **22** 131
[11] Davar F, Loghman-Estarki M R 2014 *Ceram. Int.* **40** 8427
[12] Renuka L, Anantharaju K S, Sharma S C, Nagabhushana H, Vidya Y S, Nagaswarupa H P, Prashantha S C 2017 *J. Alloy. Compd.* **695** 382
[13] Jamali H, Shoja-Razavi R, Mozafarinia R, Ahmadi-Pidani R, Loghman-Estarki M R 2012 *Curr. Nanosci.* **8** 402
[14] Loghman-Estarki M R,et al. 2014 *Ceram. Int.* **40** 3721
[15] Salavati-Niasari M, Dadkhah M 2009 *Polyhedron* **28** 3005
[16] Loghman-Estarki M R,et al. 2014 *Ceram. Int.* **40** 1405
[17] Davar F, Hassankhani A, Loghman-Estarki M R 2013 *Ceram. Int.* **39** 2933
[18] Abriata J P, Garc´es J, Versaci R 1986 *Bull. Alloy Phase Diagrams* **7** 116
[19] Atkinson A 1985 *Rev. Mod. Phys.* **57** 437
[20] Bouvier P, Djurado E, Lucazeau G, Le Bihan T 2000 *Phys. Rev. B* **62** 8731
[21] Chrisey D B, Hubler G K 1994 Pulsed Laser Deposition of Thin Films *Wiley* New York
[22] Geohegan D, Stuke M 1995 *MRS Symp. Proc.* **55**
[23] Kurpaska L et al. 2015 *Journal of Nuclear Materials* **466** 460
[24] Keramidas V G, White W B 1974 *Journal of The American Ceramic Society* **57** (1)
[25] Kolvari E, Koukabi N, Hosseini M M, Vahidian M, Ghobadi E 2016 *RSC Adv.* **6** 7419
[26] Chakravarty R, Shukla R, Ram R, Tyagi A, Dash A, Venkatesh M 2010 *Chromatin* **72** 875
[27] Pola J, Gondal M A, Urbanová M, Pokorná D, Masoudi H M, Bakardjieva S, Bastl Z, Šubrt J, Siddiqui M N 2012 *Journal of Photochemistry and Photobiology A: Chemistry* **243** 33
[28] Pokorna D, Urbanova M, Bakardjieva S, Šubrt J, Pola J 2010 *Journal of Photochemistry and Photobiology A: Chemistry* **215** 164

IOP Conf. Series: Materials Science and Engineering **613** (2019) 012017 doi:10.1088/1757-899X/613/1/012017

Synthesis and Characterization of Carbon Nanosheets from Stinging Nettle (Urtica Dioica)

K Nuilek[1], **A Simon**[1] **and P Baumli**[2]

[1]Institute of Ceramics and Polymer Engineering, University of Miskolc, 3515-Miskolc Egyetemvaros, Hungary
[2]Institute of Nanotechnology, University of Miskolc, 3515-Miskolc Egyetemvaros, Hungary

E-mail: k.nuilek@gmail.com

Abstract. Extensive research in nanotechnology has allowed us to improve and revolutionize technology and industry. This had led to an interest in the synthesis of other forms of carbon nanomaterial, such as graphene, fibre, carbon nanotubes and particles. Nowadays, a variety of carbon materials is available offering the flexibility to modify carbon properties. Therefore, the manufacture of carbon materials in different shapes is a likely way to take advantage of the morphology and relate it to the properties of the carbon nanomaterial. Stinging nettle is a new material to study for the production of carbon nanostructured materials. In this research, carbon nanosheets were successfully prepared from stinging nettle (nettle stem and leaf) using potassium hydroxide (KOH), sulphuric acid (H_2SO_4) and thermal treatment. This method is simple for producing carbon nanosheets from natural materials. Treating the nettle stem with KOH (solid) produces carbon nanosheets with a thickness of ~159.2– 252.3 nm, while KOH (solute) treatment led to thinner nanosheets (~70.43–111.2 nm). SEM micrographs of the nettle leaf with KOH (solute) activation show a smooth surface and clearly reveal the formation of carbon nanosheets. However, in the samples activated by KOH (solute) a greater number of micropores were found on the wall. The microstructure and morphology of the samples were investigated by scanning electron microscopy (SEM), the chemical composition was analysed by energy-dispersive X-ray spectrometry (EDS).

1. Introduction

Carbon is one of the most widely used materials in various industries, such as automotives, electronics, etc. [1-7]. It can be found in various forms such as graphene, carbon nanotubes (CNT), multi-walled carbon nanotubes (MWCNT), carbon nanofibre, carbon nanosheets (CNS), etc. Carbon nanosheets are a new kind of two-dimensional material. A nanosheet is a two-dimensional nanostructure with thickness ranging from 1 to 100 nm. Carbon nanostructured materials have become important due to their good physical and mechanical properties and high performance in composite materials. Extensive research in nanotechnology has allowed us to improve and revolutionize technology and industry. Carbon nanostructures can be prepared by microwave plasma irradiation, pyrolysis in a tubular reactor, chemical vapour deposition, microwave plasma-enhanced chemical vapour deposition hydrothermal carbonization and mechanical exfoliation through probe sonication [8-15].

The production of nanostructured materials from waste materials such as pineapple leaves, water hyacinth, peanut shell, corncob, rice husk and sunflower piths has appeared in recent years [16-21]. Carbon nanomaterials derived from natural waste materials reveal diversity and high performance.

Exfoliated graphite, an important industrial raw material for the production of flexible graphite sheets, is usually prepared by rapid heating of residue compounds of natural graphite flakes with acid, through formation of intercalation compounds [22]. In a previous study [23], carbon nanostructured material was produced from activated and exfoliated peanut shell. There are different methods for preparing carbon materials. Therefore, the search is on for cheap, abundant natural waste materials to protect the environment in the production of carbon materials [24-28].

This paper focuses on the effect of chemical activation with potassium hydroxide (solid and solute) and the sulphuric exfoliation on the formation of carbon nanosheets from stinging nettle (Urtica Dioica). This low-cost process (chemical activation/exfoliation and carbonization) is used to produce carbon nanostructured material from agro waste in simple steps, adding value to the products.

2. Materials and experiments

2.1 Materials

In the present work, nettle stem and nettle leaf were used as carbonaceous material precursors. Chemicals were used as follows: hydrochloric acid (HCl) as cleaner, potassium hydroxide (KOH) as activator and sulphuric acid (H_2SO_4) as exfoliator.

2.2 Experiments

2.2.1 Sample preparation and pre-carbonization process

The nettles (nettle stem and nettle leaf) were washed several times and dried at 80 °C for 24 hours, then treated by HCl 0.5 M for 24 hours to remove any metallic oxide or organic compounds. Finally, they were washed by distilled water and dried at 80 °C for 24 hours. The pre-carbonization process was carried out at 450 °C for 2 hours in a stainless-steel tube furnace under argon atmosphere.

2.2.2 Chemical activation and characterization process (a) Chemical activation by KOH (solid) method: The pre-carbonized samples (nettle stem and nettle leaf) were mixed with KOH (solid) at a weight ratio of 1:1 and milled for 1 hour by mortar. (b) Chemical activation by KOH (solute) method: First, the pre-carbonized samples were milled for 1 hour. Secondly, the pre-carbonized sample and KOH (solid) with a weight ratio of 1: 1 were put into a beaker, filled with water and stirred into aqueous KOH (solute) for 1 hour then dried at 80 °C for 24 hours. For both methods, after being activated by KOH, samples were heat treated at 800 °C for 1 hour in a stainless-steel tube furnace under argon atmosphere. Then KOH was extracted from the mixture by ethyl acetate [29].

2.2.3 Exfoliation process Finally, to separate the layered structure of carbon, the carbonized samples were treated with 10% H_2SO_4 solution, stirred for 1 hour and then washed with distilled water and dried at 80 °C for 24 hours. The schematic illustration of the preparation of carbon nanosheet from nettle is shown in Figure 1.

2.3 *Characterization* The microstructure and morphology of the samples were investigated by scanning electron microscopy (SEM, HITACHI S-4800 and ZEISS EVO-MA10), the chemical composition was analysed by energy-dispersive X-ray spectrometry (EDS, BRUKER AXS) and the crystal structure and the phase purity were examined using X-ray diffraction (XRD, BRUKER D8 ADVANCE CoKα X-ray source).

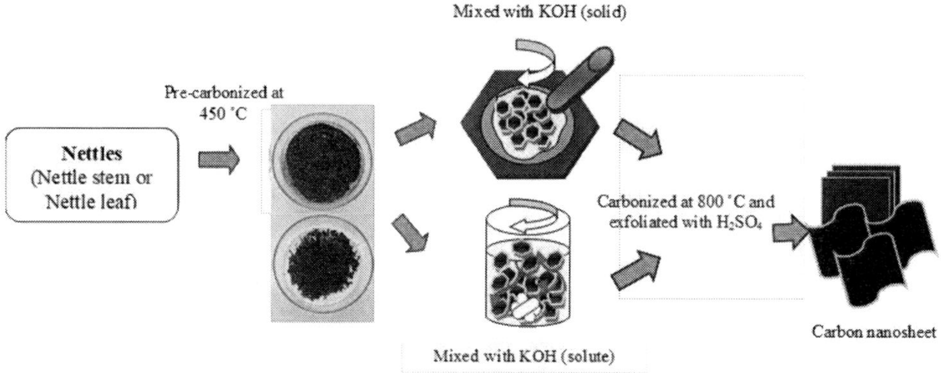

Figure 1. Schematic illustration of the preparation of carbon nanosheet

3. Results and discussion

The purpose of this work was to investigate the effect of chemical treatment (activation by using difference state of KOH and exfoliating with H_2SO_4 acid) and to study the microstructure of the carbon nanosheets produced from nettle stem or nettle leaf with a focus on the effect of chemical treatment by using different states of KOH activation and exfoliating process by H_2SO_4 acid.

The microstructure of the dried nettle stem was examined (Figure 2a). The cell wall of a fibre is not a homogeneous layer and composed mainly from cellulose, lignin and hemicellulose [30]. The chemical composition investigated by EDS showed a high weight percentage of carbon (50.8 wt.%) and low weight percentage of other elements, as shown in Figure 2b. The bristly surface of a dried nettle leaf is shown in Figure 3(a). Nettle leaf contains veins covered by fine stinging hairs and a higher amount of silica, potassium and calcium (Figure 3b) than nettle stem.

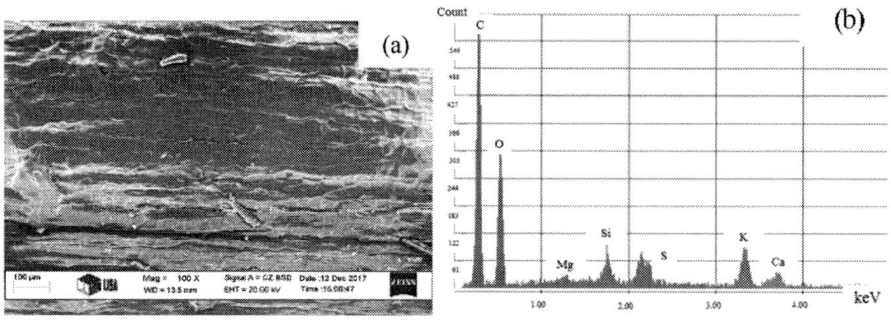

Figure 2. (a) SEM micrograph and (b) EDS spectrum of dried nettle stem

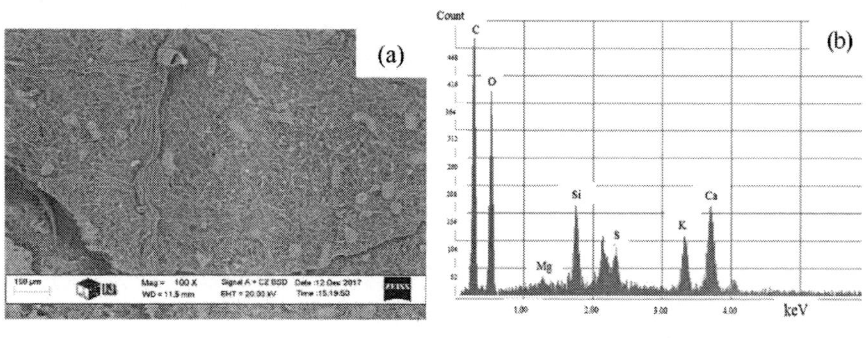

Figure 3. (a) SEM micrograph and (b) EDS spectrum of dried nettle leaf

Figure 4 represents the microstructure of nettles activated with potassium hydroxide. It is difficult to find the carbon nanosheet after this step, as agglomerates of small particles were found in both nettle stem and nettle leaf. Only a slight difference is observed between the KOH (solid) and KOH (solute) activator.

Figure 4. SEM micrographs of the samples after activating by potassium hydroxide

When comparing the microstructure of the stinging nettle before and after exfoliation (Figures 4-6), it can be concluded that the thickness of the layers decreases during the activation process, as melted potassium-hydroxide penetrates into the pores of the carbonized materials and rips apart the carbon layers. Thus, the exfoliator acid can intercalate between the carbon layers and break bonds between them [31]. The potassium hydroxide and the sulphuric acid can form compounds and move between the carbon layers to increase their distance. As a result, carbon layers are easily separated from each other and clear surfaces are found after acid exfoliation.

Carbon nanosheets found in nettle stem after activation by KOH (solid) and exfoliation with H_2SO_4, have thicknesses varying between 159-252 nm (Figure 5a). The SEM micrograph (Figure 5b) of the nettle stem activated with KOH aqueous solution and exfoliated by H_2SO_4 shows a smooth surface and clearly reveals the formation of carbon nanosheets with a thickness of 70 to 111 nm. Analysis shows that fully treated carbon nanosheets from nettle stem mainly contain carbon (88.83 wt.%) (Figure 7).

IOP Conf. Series: Materials Science and Engineering **613** (2019) 012017 doi:10.1088/1757-899X/613/1/012017

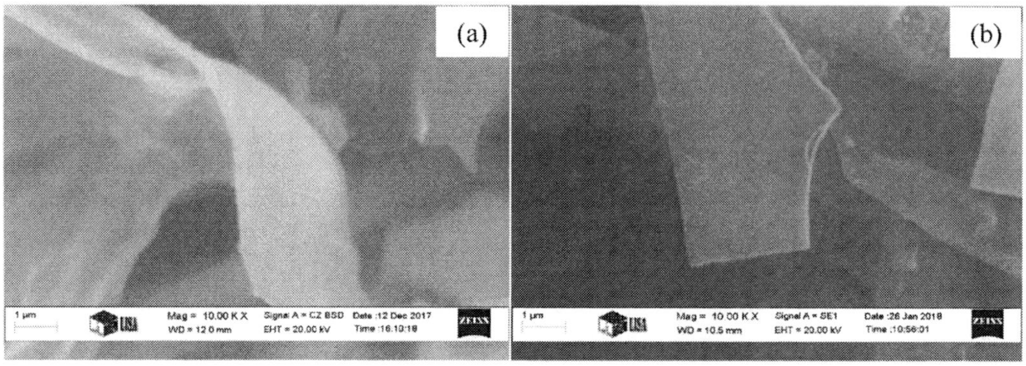

Figure 5. SEM micrographs of nettle stem activated with (a) KOH (solid) (b) KOH (solute) and heat treated at 800°C for 1 h after exfoliation processing by H_2SO_4

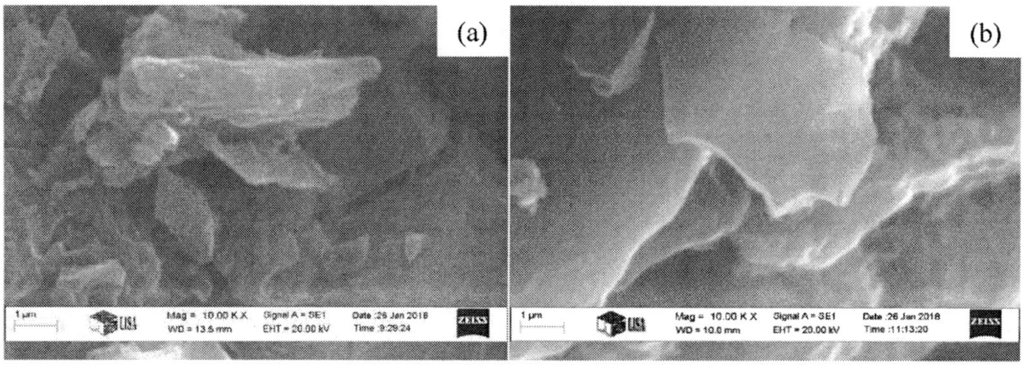

Figure 6. SEM micrographs of nettle leaf activated with (a) KOH (solid) (b) KOH (solute) and heat treated at 800°C for 1 h after exfoliation processing by H_2SO_4

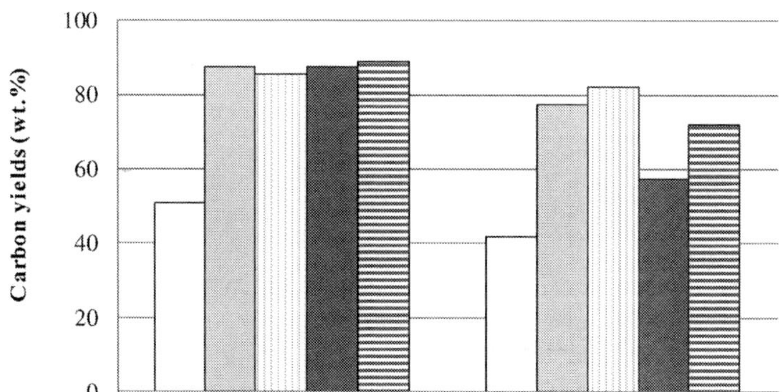

	Nettle stem	Nettle leaf
□Dried sample	50.8	41.86
▨Activated with KOH(solid)	87.35	77.34
□ Activated with KOH(solid) and exfoliated by sulfuric acid	85.54	82.14
■Actiivated with KOH(solute)	87.56	57.4
⊟ Activated with KOH(solute) and exfoliated by sulfuric acid	88.83	72.01

Figure 7. Comparison of carbon yields of the samples

Carbon nanosheets can be also found in nettle leaf (Figure 6) but are not as sharply formed compared with nettle stem (see Figure 5). The SEM micrographs of the nettle leaf activated by KOH (solute) activation show a smooth surface and clearly reveal the formation of carbon nanosheets with 41-183 nm thickness, whereas the formation of carbon nanosheets in KOH (solid) activated samples reveal numerous multilayers and agglomerated small plates. The carbon yield of nettle leaf is less than that of nettle stem due to its original structure and chemical composition. The sulfuric exfoliation affects the weight percentage of carbon yields (Figure 7).

The microstructure of nanosheets activated with the KOH (solute) and exfoliated by H_2SO_4 shows clear formation of carbon nanosheet structures with flat surfaces. Activation with KOH (solute) also provides better dispersion of carbon nanosheets.

As for the carbon yield in nettle stem samples, an increase of 35-38 wt.% (68-75%) was achieved with a slight difference between the methods. Nettle leaf originally contains a lower amount of carbon (~42 wt.%) but the activation and exfoliation increase it by 30-40 wt.% (72-96%), meaning that with this method a significant enhancement is reached for both materials and a higher level of impurities can be removed.

4. Conclusion

Carbon nanosheets were successfully synthesized from stinging nettle. The SEM micrographs of the nettle stem with KOH (solid and solute) activation show a smooth surface and clearly reveal the formation of carbon nanosheets, whereas the formation of carbon nanosheets of nettle leaf also show carbon nanosheet formation but not as good as nettle stem.

The chemical (KOH and H_2SO_4) and thermal activation affects the formation of carbon and the separation of carbon layers to form nanosheets of graphite. This method is a simple and easy way for producing carbon nanosheets from natural materials and makes it possible to improve the purity of materials and to remove contaminants (metallic oxides and organic compounds) effectively.

Acknowledgements

The authors would like to thank Dr. Maria Sveda, Dr. Anna Sycheva, Dr. Daniel Koncz-Horvath, Mr. Arpad Kovacs and Mr. Tibor Ferenczi for examining and characterizing (SEM, EDS and BET) samples in this research work. The described article was carried out as part of EFOP-3.6.1-16-2016-00011 "Younger and Renewing University-Innovative Knowledge City-institutional development of the University of Miskolc aiming at intelligent specialisation" project implemented in the framework of the Szechenyi 2020 program. The realization of this project is supported by the European Union, co-financed by the European Social Fund.

References

[1] Futaba D N, Hata K, Yamada T, Hiraoka T, Hayamizu Y, Kakudate Y, Tanaike O, Hatori H, Yumura M and Iijima S 2006 *Nature Materials* **5**(12) 987 https://doi.org/10.1038/nmat1782

[2] Popov V N 2004 *Materials Science and Engineering: R: Reports* **43**(3) https://doi.org/10.1016/j.mser.2003.10.001

[3] Frackowiak E 2007 *Physical Chemistry Chemical Physics* **9**(15) 1774 https://doi.org/10.1039/b618139m

[4] Strauss V, Marsh K, Kowal M D, El-Kady M and Kaner R.B. 2018 *Advanced Materials* **30**(8) 1704449 https://doi.org/10.1002/adma.201704449

[5] Arici E, Kaplan B Y, Mert A M , Gursel S A and Kinayyigit S 2019 *International Journal of Hydrogen Energy* https://doi.org/10.1016/j.ijhydene.2018.11.210

[6] Hoppe B, Hindricks K D, Warwas D P, Schulze H A, Mohmeyer A, Pinkvos T J, Zailskas S , Krey M R, Belke C, König S, et al. 2018 *CrystEngComm* **20**(41) 6458 https://doi.org/10.1039/C8CE01264D

[7] Zhou J, Guo Y, Liang C, Yang J, Wang J and Nuli Y 2018 *Electrochimica Acta* **273** 127 https://doi.org/10.1177/0263617418807856

[8] Wang Z, Shoji MBaba, K, Ito T and Ogata H 2014 *Carbon* **67** 326 https://doi.org/10.1016/j.carbon.2013.10.002
[9] Wang Z, Ogata H, Morimoto S, Ortiz-Medina J, Fujishige M, Takeuchi K, Muramatsu H, Hayashi T, Terrones M, Hashimoto Y, et al. 2015 *Carbon* **94** 479 https://doi.org/10.1016/j.carbon.2015.07.037
[10] Bernd M, Bragança S, Heck N and Filho L 2017 *Journal of Materials Research and Technology* **6**(2) 171 https://doi.org/10.1016/j.jmrt.2016.11.003
[11] Ma H, Pan L and Nakayama Y 2011 *Carbon* **49**(3) 854 https://doi.org/10.1016/j.carbon.2010.10.029
[12] Wang Z, Shoji M and Ogata H 2011 *Applied Surface Science* **257**(21) 9082 https://doi.org/10.1016/j.apsusc.2011.05.104
[13] Wang Z, Ogata H, Morimoto S, Fujishige M, Takeuchi K, Hashimoto Y and Endo M 2014 *Carbon* **68** 360 https://doi.org/10.1016/j.carbon.2015.07.037
[14] Wang S, Wang C and Ji X 2017 *RSC Advances* **7**(82) 52252 https://doi.org/10.1039/C7RA07489A
[15] Purkait T, Singh G, Singh M, Kumar D and Dey R S 2017 *Scientific Reports* **7**(1) 15239 https://doi.org/10.1038/s41598-017-15463-w
[16] Sodtipinta J, Amornsakchai T, Pakawatpanurut P 2017 *Advances in Natural Sciences: Nanoscience and Nanotechnology* **8**(3) 035017 https://doi.org/10.1088/2043-6254/aa7233
[17] Wu K, Gao B, Su J, Peng X, Zhang X, Fu J, Peng S and Chu P K 2016 *RSC Advances* **6**(36) 29996 https://doi.org/10.1039/C5RA25098F
[18] Wu H, Chen R, Du H, Zhang J, Shi L, Qin Y, Yue L and Wang J 2018 *Adsorption Science & Technology* 0263617418807856 https://doi.org/10.1177/0263617418807856
[19] Romanos J, Beckner M, Rash T, Firlej L, Kuchta B, Yu P, Suppes G, Wexler C and Pfeifer P 2012 *Nanotechnology* **23**(1) 015401 https://doi.org/10.1088/0957-4484/23/1/015401
[20] Cheah W, Ooi C and Yeoh F 2016 *Open Material Sciences* **3**(1) 27 https://doi.org/10.1515/mesbi-2016-0004
[21] Baysal M, Bilge K, Yılmaz B, Papila M and Yurum Y 2018 *Journal of Environmental Chemical Engineering* **6**(2) 1702 https://dx.doi.org/10.1016/j.jece.2018.02.020
[22] Inagaki M, Tashiro R, Washino Y and Toyoda M 2004 *Journal of Physics and Chemistry of Solids* **65**(2-3) 133 https://doi.org/10.1016/j.jpcs.2003.10.007
[23] Nuilek K, Simon A and Baumli P 2018 *Resolution and Discovery* **3**(2) 29 https://doi.org/10.1556/2051.2018.00060
[24] Kumar R, Singh R and Singh D 2016 *Renewable and Sustainable Energy Reviews* **58** 976 https://doi.org/10.1016/j.rser.2015.12.120
[25] Wang Z, Ogata H, Melvin G, Obata M, Morimoto S, Ortiz-Medina J, Cruz-Silva R, Fujishige M, Takeuchi K, Muramatsu H, et al. 2017 *Carbon* **121** 426 https://doi.org/10.1016/j.carbon.2017.06.003
[26] Wang B, Gao B and Fang 2017 *Critical Reviews in Environmental Science and Technology* **47**(22) 2158 https://doi.org/10.1080/10643389.2017.1418580
[27] Cheah W, Ooi C and Yeoh F 2016 *Open Material Sciences* **3**(1) https://doi.org/10.1515/mesbi-2016-0004
[28] Yallappa S, Deepthi D, Yashaswini S, Hamsanandini R, Chandraprasad M, Kumar S A and Hegdea G 2017 *Nano-Structures & Nano-Objects* **12** 84 https://doi.org/10.1016/j.nanoso.2017.09.009
[29] Gabor T, Karman F H, Sytchev J, Kalman E and G. Kaptay 2009 *Carbon* **47**(4) 1195 https://doi.org/10.1016/j.carbon.2009.01.006
[30] Smole M, Hribernik S, Stana K and Kreze T 2013 *In Advances in Agrophysical Research.* InTech. https://doi.org/10.5772/52372
[31] Singh P, Bahadur J and Pal K 2017 *Graphene* **6** 61 https://doi.org/10.4236/graphene.2017.63005

Band Structure of Phononic Crystal Consist of Hollow Aluminum Cylinders in Different Media; Finite Element Analysis

Zafer Ozer[1], Mehtap Demir[2], Amirullah M. Mamedov [3, 4], Ekmel Ozbay[4]

[1]Mersin Vocational High School Electronic and Automation Department, Mersin University (Turkey)
[2]Faculty of Engineering, Department of Metallurgy and Material Engineering, Adiyaman University (Turkey)
[3]Nanotechnology Research Center (NANOTAM), Bilkent University (Turkey)
[4]International Scientific Center, Baku State University, Baku (Azerbaijan)

E-mail: zaferozer@mersin.edu.tr

Abstract. In this study, we investigated band structure of a 2D Phononic Crystal (PnC) which was consist of hollow aluminum cylinders in different media. To validate the finite element predictions, we made a square lattice PnC which has lattice constant a=14mm, outer radius r_1=5mm and inner radius r_2=4mm of cylinders. In our experimental measurement, by using a signal generator software, we sent 100 Hz-20kHz sine signal for 5 seconds duration from speaker to the PnC where placed 1m away from. With the help of a microphone, we recorded the sound in front of the PnC (P_{in}) and backside the PnC (P_{out}). Then we were converted our recording into frequency domain by standard FFT algorithm, we calculated the Transmission Loss value according to the formula $TL = 20*log_{10} (P_{out} / P_{in})$ and obtained the experimental Transmission Loss. After experimental validation, we conducted band structure analysis of different combination of the 2D PnC by finite element method.

1. Introduction

Phononic Crystals are periodic structures made of two elastic materials with different mechanical properties. The basic property of PnC is that mechanical (either elastic or acoustic) waves, having frequencies within a specific range, are not able to propagate within the periodic structure. This range of forbidden frequencies is called a phononic band gap. This is analogous to electrons in a crystal, where classical waves propagate in a structure with periodically modulated material parameters [1]. In recent years, many researchers have studied PnC. This is important for applications, such as acoustic filter, noise control, and subwavelength imaging [2, 3]. To acquire wide acoustic band gaps at some desired frequency ranges, several factors are mainly considered, including scatter shapes, filing fraction, shear modulus ratio, mass density ratio, Poisson's ratio, and lattice structures [4].

In this study, we experimentally demonstrate the existence of complete band gaps in 2D phononic crystal consisting of a square array parallel aluminum cylinders in different media. The Finite Element Method (FEM) is used to calculate the band structure of phononic crystal. Experimental measured data are obtained and compared with the range of frequencies of complete band gaps.

2. Materials and experiments

In the PnC, the propagation of acoustic Bloch waves can be expressed by the acoustic wave equation as in Eq. (1).

IOP Conf. Series: Materials Science and Engineering **613** (2019) 012018 doi:10.1088/1757-899X/613/1/012018

$$\nabla.(\frac{\mathbf{1}}{\boldsymbol{\rho}}\nabla p) = -\frac{1}{\rho c^2}\left[\omega(\tilde{k})\right]^2.p \tag{1}$$

where, $c=c(\tilde{r})$ is longitudinal velocity of acoustic waves inside the PnC, $\rho=\rho(\tilde{r})$ is density of the material and $p = p_k(\tilde{r})$ is the spatial part of $p_k(\tilde{r}, t)$ an acoustic Bloch wave within the PnC with wavevector $\tilde{k}$.

In figure 1a the basic vectors of the square lattice are $a_1 = a\vec{x}$, $a_2 = a\vec{y}$ and reciprocal lattice vectors in figure 1(b) are $b_1 = \frac{2\pi}{a}\vec{x}$, $b_2 = \frac{2\pi}{a}\vec{y}$. In reciprocal space high symmetry points are; Γ=(0,0), $X=\left(\frac{\pi}{a},0\right)$, $M=\left(\frac{\pi}{a},\frac{\pi}{a}\right)$.

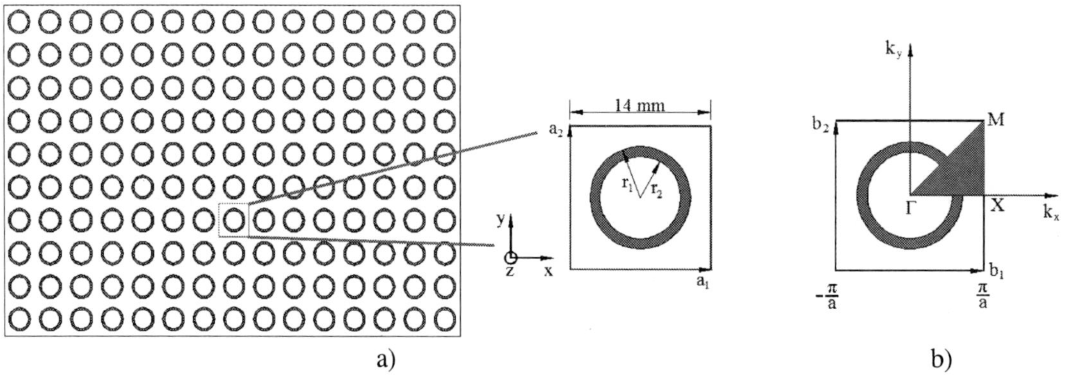

Figure 1. a) Phononic crystal in rectangular lattice in the direct space and unit cell b) high symmetry points (*Γ-X-M*) irreducible Brillouin Zone in reciprocal space

To obtain the band structure of PnC we applied Floquet periodic conditions edges of the unit cell in figure 1(a). The obtained acoustic band structure was plotted along the *M–Γ–X–M* path for the two dimensional (2D) PnC as seen in fig. 2.

For validating the attenuation of acoustic waves as seen in band structure of unit cell, we conclude a measurement setup by finite array of PnC as seen in fig 1(a) and we designed the finite array as seen in fig. 2. In finite array FEM analysis, we applied Symmetry Boundary Conditions upper and lower edges, we assumed Perfectly Matched Layer left and right ends, and we applied 1 Pa pressure to the back pressure field in *x* direction as seen in fig. 2. In fig.2 the points *A* and *B* are P_{in} and P_{out} observation points in following *TL* formula.

Figure 2. Boundary conditions of finite array FEM model

The experimental setup consists of 150 aluminum cylinders with an internal diameter of 8 mm an external diameter of 10 mm and length of 29.5 mm in the form of square lattice with a lattice constant a = 14 mm were placed in the holes on the fiber substrate in a 10x15 array.

In our experimental study, we placed a speaker 1 m away from the PnC and we sent a sound signal from 100 Hz to 20000 Hz on to the PnC for 5 seconds. With the microphone we placed behind the PnC, we recorded the sound signal to the computer via an external sound card and converted it to the frequency domain with the standard FFT algorithm. Then we obtained the experimental transmission loss according to the formula $TL = 20 * \log_{10} (P_{out} / P_{in})$ where P_{in} the data obtained from the microphone that we placed on the front side of the PnC and P_{out} the data we obtained from the microphone which we placed behind the PnC.

Furthermore, we performed parametric scanning to examine the effect of the thickness of the aluminum cylinder on the band structure by FEM. The band structure of PnC is calculated, with the geometrical parameters r_1 = 5 mm, r_2 = 4, 3.5, 3, 2.5 mm where r_1 and r_2 are the outer and inner radius of the hollow cylinder of large size, respectively.

3. Results and discussions

Obtained acoustic band structure in Brillouin zone of the square lattice was plotted along the M–Γ–X–M path for the two dimensional (2D) PnC as seen in fig. 3, the narrow full band observed in the dispersion curves between 13.2-13.8 kHz and a partial band between 8-14 kHz in Γ-X direction.

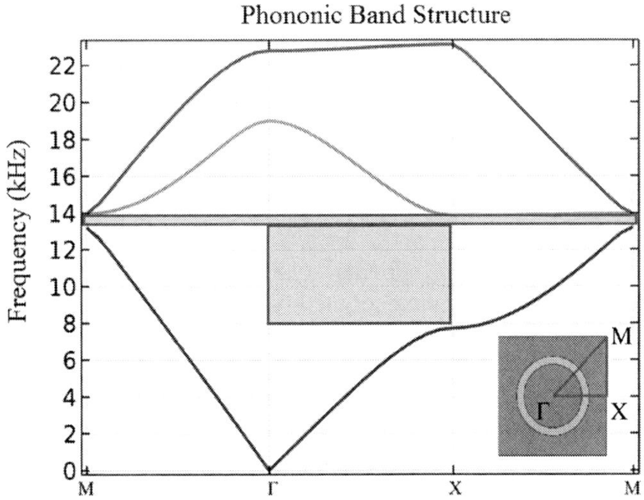

Figure 3. Acoustic band structure of PnC

In order to compare the band structure obtained by FEM and the sound attenuation obtained by PnC, transmission loss was obtained numerically with finite structure seen in Fig. 2 by FEM and experimental measurement. Fig.3 shows the numerical and experimental measurement results.

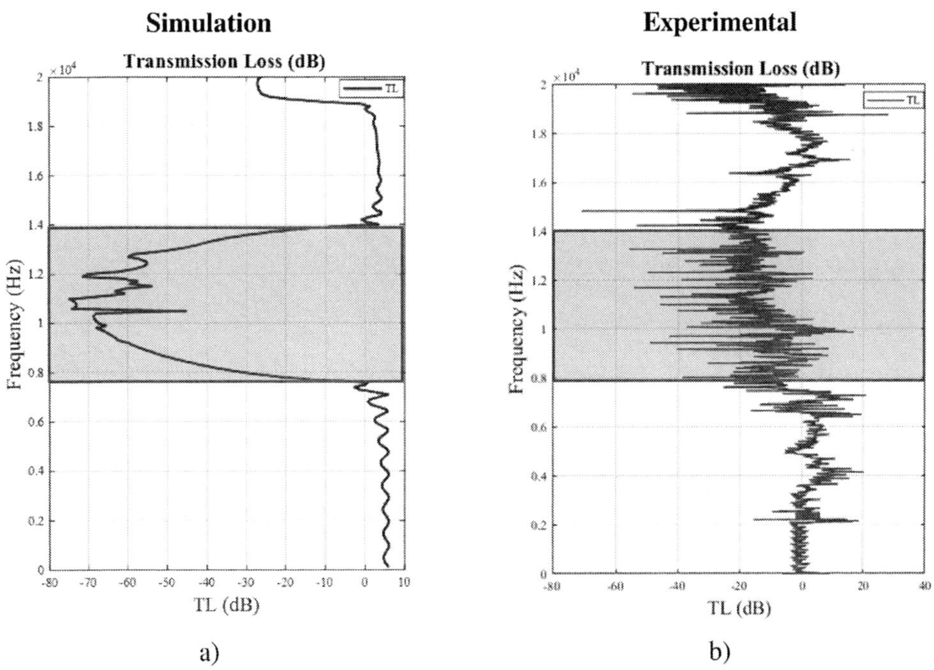

Figure 4. Transmission Loss of finite array a) FEM analysis b) measurement

According to the results of the transmission loss we obtained by FEM and experimental measurement, the sound level decreased between 8-14 kHz as in the band structure and the results are compatible.

We investigated the effect of wall thickness (r1-r2) on the band gap using FEM. as shown in table 1, the center frequency and bandwidth are reduced when the wall thickness is increased.

Table 1. Band relation with wall thickness

r_1-r_2[mm]	Central frequency [Hz]	Band Gap [Hz]
1	13543.5	661
1.5	13540.5	659
2	13538.5	657
2.5	13537.5	657
3	13537	656

4. Conclusions

In the present paper, a PnC with hollow cylinder columns was presented and investigated. The band structure was calculated using the finite element method. The band gap and frequency can be tuned by changing the inner radius of the hollow cylinders. It could be based on the design of the elastic acoustic wave guides that transmit at tunable frequencies using proposed structures. We have also shown the band structure of the 2D PnC that is compatible with the transmission losses obtained by numerical and experimental measurement.

References

[1] P Deymier 2013 Acoustic Metamaterials and Phononic Crystals *Springer-Verlag* Berlin Heidelberg

[2] M S Kushwaha, P Halevi 1996 *Applied Physics Letters* **69** (1) 31 https://aip.scitation.org/doi/10.1063/1.118108

[3] A Sukhovich, B Merheb, K Muralidharan, J O Vasseur, Y Pennec, P A Deymier, J H Page 2009 *Physical Review Letters* **102** (15) 154301 https://doi.org/10.1103/PhysRevLett.102.154301

[4] C He, H Zhao, R Wei; B Wu 2010 *Frontiers of Mechanical Engineering in China* **5** (4) 450 https://doi.org/10.1007/s11465-010-0105-y

IOP Conf. Series: Materials Science and Engineering **613** (2019) 012019 doi:10.1088/1757-899X/613/1/012019

Application of finite element method (FEM) to study stress-strain state and distribution of temperatures in cutting zone in turning of various structural materials by carbide tools with coatings of various composition and architecture

Alexey Vereschaka[1], Boris Mokritskii[2], Elena Mokritskaya[2], Oleg Sharipov[1] and Catherine Sotova[1]

[1] Moscow State Technological University STANKIN, Moscow (Russia)
[2] Komsomolsk-na-Amure State Technical University, Komsomolsk-na-Amure (Russia)

E-mail: ecotech@rambler.ru

Abstract. The paper considers simulation of cutting process conditions and state of a metal-cutting tool using the finite element method (FEM). The study considered the distribution of normal equivalent and tangential stresses in workpiece material during turning of aluminum-deformable alloy G-AlMg5 (DIN 1725) and copper C11000. Based on the calculated stress values, the values of plastic strains of the workpiece material were determined throughout the whole process of turning with the necessary time sampling of the process. The distribution of temperatures in cutting zone during turning of steel 5135 by carbide tools with coatings of different composition and thickness was studied. The obtained results were compared with the data obtained during the corresponding cutting tests. The paper considered the influence of such coating parameters as thickness, element composition and architecture on distribution of temperatures in a workpiece, cutting tool and chips during the cutting process.

1. Introduction

Simulation of the process of material-cutting using the finite element method (FEM) has a quite long history. Gunasekera et al [1] note that the finite element method (FEM) is actively used to simulate the metal-cutting processes. Simulation of the stress-strain state of a coated substrate, in general, for orthogonal cutting conditions [2-5, 11, 13]. Modeling of the friction and wear conditions and thermomechanical state for cutting tools with monolayer [6, 12, 14, 15] and multilayer [7-10, 13] coatings were conducted in a large number of studies.

Together with the development of new models for cutting with coated tools, modifying coatings of a new type multilayer, composite, gradient, and nano-structured are being created and studied. In particular, the study of the properties of multilayer wear-resistant coatings of various composition and architecture is considered in [16-22]. Meanwhile, real cutting tests are still an unchallenged and quite irreplaceable method to study the properties of a coated tool, and in some cases, it is reasonable to combine such studies with mathematical modeling of the process. Such a combination of methods makes it possible not only to obtain confirmation of the experimental data, but also to study much deeper the processes occurring in the cutting zone. The most important factors affecting tool life and reliability of a cutting tool are stress-strain state and temperature distribution in the cutting zone.

The temperature arising in the cutting system is generated as a result of the processes occurring in the zones of action of three main heat sources strain and two frictional ones from friction along rake and flank faces of a tool [23]. Given the weak possibilities of the influence of coatings on strain heat sources, it is of great interest to consider the possibilities of its influence on frictional heat sources along rake and flank faces of a tool. In the process of cutting, almost all mechanical energy is converted into thermal energy. Meanwhile, wear resistance of a tool is largely determined by temperature conditions of a cutting system. For a tool-workpiece system, regardless of the cutting modes, there is an optimum cutting temperature to minimize tool wear. That means that if during machining the cutting temperature is less than optimum, then it should be increased, and if the cutting temperature exceeds optimal, then it is necessary to create conditions for its reduction. Control over the cutting temperature and its maintenance at a pre-set level excludes the factor of negative effect of temperature on tool wear. Such control can be achieved due to the regulation of heat flows in a tool-workpiece-chips-environment system.

The conceptual role of a coating is in a dual function [16, 18], since it can not only sufficiently change the surface characteristics of tool material (frictional properties, thermal conductivity, propensity of tool material to physicochemical interaction with the material being machined, etc.), but it also can affect contact processes. Thus, the multifunctional performance of the intermediate technological medium like coating makes it possible to predict the possibility of directional control over cutting temperature [16-22].

In fact, on the one hand, a coating can significantly reduce the friction coefficient in the tool-workpiece system and reduce the power of frictional heat sources. On the other hand, the coating, characterized by screening function, can significantly reduce the intensity of heat flows into cutting wedge of a tool and thus increase the temperature threshold of the beginning of the adhesion interaction in the workpiece-tool system and reduce the intensity of interdiffusion between them [20-22].

2. Materials and experiments

Let us consider the challenge of kinematics for the process of turning of a workpiece [23].

The turning process was simulated based on the calculation of the stress-strain state of a workpiece, using a number of the following assumptions:

1. The coefficient of friction along rake face of a cutter is assumed as constant along its entire length and independent from tangential stresses.
2. The coefficient of friction along flank face is assumed as constant and equal to the coefficient of friction on rake face.

ANSYS Workbench software was selected as a simulation tool, with spatial mode as 3D and calculation type as dynamic. The calculation purpose is in the stress-strain analysis of a workpiece at the current moments of the cutting process. Finite element mesh is based on the SOLID186 element.

In a tool-workpiece interface, the heat from the cutting zone is distributed between a workpiece, tool, environment and chips, that is, some heat is transferred into the workpiece, some into the tool, and some into the chips [16, 23].

A directional change in the chemical heat, chemical and structural properties of the coating makes it possible to influence the distribution of heat flows in order to increase the volume of heat released into the chips (under the conditions of high-speed machining) or, on the contrary, its conservation (reclamation) in the cutting zone at low cutting speeds. By varying the thermophysical parameters and the tribological properties of the contact zone, it is possible to model the characteristics of wear-resistant coatings at the tool-workpiece interface. One of these parameters is the thermal conductivity of materials. In thermophysical analysis, the use of DEFORM and ANSYS software allows modeling and visualizing contact processes in the cutting zone at various thermal, physical, technological parameters. DEFORM 3D software was used to simulate the cutting conditions for a coated tool. Meanwhile, in DEFORM 3D, motion is applied to a rigid body modeling a cutting tool, and nothing but forces is applied to a workpiece. Therefore, the tool moves relative to the workpiece. The standard process of cutting was used, and the main cutting conditions were set up, including machining modes, configurations of a tool and a workpiece, tool coating parameters, etc. The following cutting modes were used in the calculations: n=1600 rpm, f=0.11 mm/rev, a_p=0.7 mm.

In computer simulation, it is possible to vary the thermophysical parameters and tribological properties of the contact zone by modeling the characteristics of wear-resistant coatings at the tool-workpiece interface. Such parameter is represented not only by thermal conductivity of the coating itself, but also by the initial conditions of the contact zone (for example, the coefficient of friction), which in turn influence the formation of heat sources and their intensity.

To assess the degree of influence of the above factors on the efficiency of the cutting process, the following experimental studies were performed:

• evaluation of tribological characteristics at different temperatures, carried out at adhesiometer using spherical indenters of carbide with TiAlN, Ti-TiN-TiAlCrN coatings; a friction pair was represented by special samples from aluminum deformable alloy G-AlMg5 (DIN 1725), copper C11000 and steel 5135 with hardness of 20 HRC;
• studies in longitudinal turning of steel 5135, alloy G-AlMg5 and copper C11000.

The coatings under study were obtained using the developed process of filtered cathodic vacuum arc deposition (FCVAD) at the VIT-2 station [16-22].

The studies of cutting properties of a tool made of carbide with developed coatings were conducted on a lathe CU 500 MRD in longitudinal turning.

3. Results and discussions

The temperature changes in the tool-workpiece system could be in situ measured by means of a high-speed infrared (IR) camera of type FLIR SC7600 (frame rate: 328 fps at 320 × 256 pixels, measuring range: -20 to +3000°C, accuracy:±1°C).

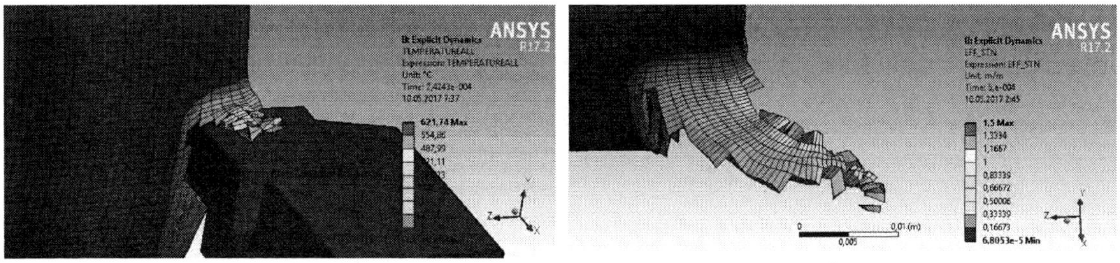

a b

Figure 1. Storyboard for tuning of AlMg6: chip formation moment (a), chip separation moment (b).

As a result of simulation of the turning process, it is possible to reveal the distribution of normal equivalent and tangential stresses in the workpiece material. From the known stress values, it is possible to determine the values of the plastic strains of workpiece material throughout the entire turning process. In these numerical simulations with a computer model of the turning process, the studies consider the features of the stress-strain state of workpiece material made of copper C11000. Based on the results of calculations, it is possible to view the formation of chips and their shapes at different moments of time (Fig. 2) and also determine the values of stresses and strains arising in the material against the entire process of machining of a workpiece.

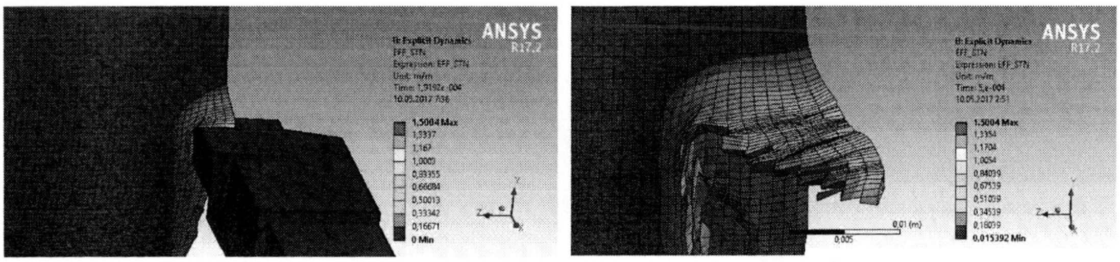

a b

Figure 2. Storyboard for turning of Cu material: start of chip separation (a), chip separation moment (b).

Fig. 3 shows the results of calculations in ANSYS at various coefficients of thermal conductivity of material of a workpiece, tool and coating. Meanwhile, coating material, which coefficient of thermal conductivity varies depending on the type of coating, is in direct contact with a workpiece. So, in Fig. 3a, coefficient of thermal conductivity of coating is $\lambda = 0.25$, and in Fig. 3b – $\lambda = 0.15$. It can be seen that the temperature in the chips increases as the coefficient of thermal conductivity of coating changes at the same contact temperature, that is, most of heat from the cutting zone is transferred to the chips.

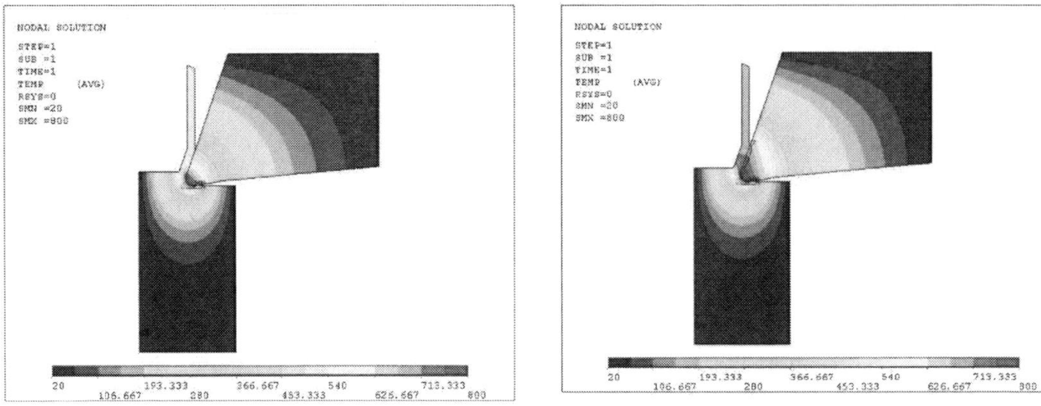

a b

Figure 3. Calculation of temperature fields in ANSYS.

With the use of this model, it is possible to trace the change in temperature fields in the cutting zone. In general, by changing the thermophysical properties of the friction surfaces, thickness and structural composition of the coatings for a cutting tool, it is possible to predict the thermal conditions and thereby provide the most favourable conditions to ensure higher wear resistance in blade cutting.

The most interesting results of information content were obtained during simulation in DEFORM environment (Fig. 4). In addition to thermal parameters, this software may additionally take into account cutting speed, presence of coating or CF, as well as wear of a cutting tool and their influence on the distribution of heat flows.

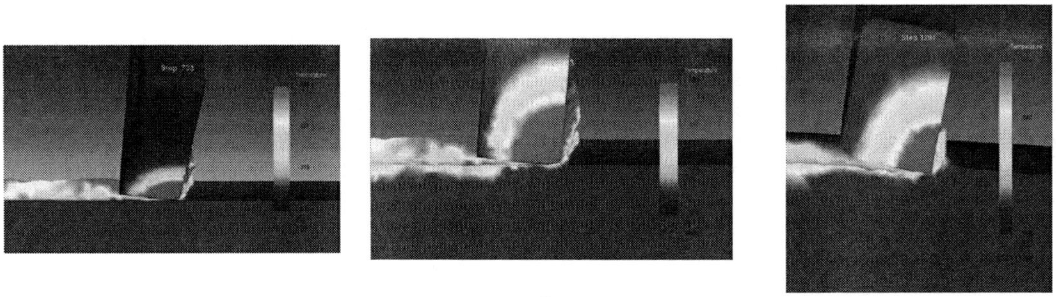

a b c

Figure 4. Distribution of heat flows in the contact zone of a cutting tool with Ti-TiN-TiAlCrN coating (a), uncoated tool(b) and TiAlN coating (c) at cutting speed V = 450 m/min (after 5 min and at VB = 0.10 mm).

4. Conclusions

The application of the finite element method (FEM) makes it possible to effectively simulate both the stressed-strained state of a cutting tool and the temperature distribution in the cutting zone. The application of the above method can significantly increase the efficiency of the process of designing a metal-cutting tool with coatings of new generation, while reducing its cost price.

The analysis of the obtained data makes it possible to note the following:

1. It has been found out that the highest temperature is generated during the machining of various materials with an uncoated tool.

2. The maximum decrease in the intensity of heat flows in the tool-workpiece system is provided by Ti-TiN-TiAlCrN coating with multilayer composite architecture.

3. The maximum improvement in the thermal state of a carbide tool cutting wedge is provided by three-layer Ti-TiN-TiAlCrN coating, and that is explained by favourable combination of layers in the multilayer coating architecture. In particular, for such a coating, upper layer of TiAlCrN coating provides maximum reduction of adhesion activity of carbide with regard to material being machined, intermediate layer TiN ensures strong adhesion between wear-resistant and adhesion layers, and lower

adhesion layer Ti efficiently shields heat flow action from frictional heat sources on rake and flank faces of a tool.

Acknowledgement
This research was financed by the Ministry of Education and Science of the Russian Federation (Leading researchers, project 16.9575.2017/6.7).

References

[1] J S Gunasekera, et al. 1982 *Computers in Mechanical Engineering* **1** 58
[2] B Haddag, M Nouari 2013 *Wear* **302** 1158 https://doi.org/10.1016/j.wear.2013.01.028
[3] K Komvopoulos. 1988 *Journal of Tribology* **110** 477 https://doi.org/10.1115/1.3261653
[4] H Djabella, R D Arnell 1993 *Thin Solid Films* **223** 98 https://doi.org/10.1016/0040-6090(93)90732-5
[5] B Zhang, A Bagchi 1994 *Journal of Engineering for Industry*. **116** 289
[6] F Du, M R Lovell, T W Wu, 2001 *Int J Solids Struct* **38** 4557 https://doi.org/10.1016/S0020-7683(00)00291-2
[7] T Mc Ginley, J Monaghan 2001 *J Mater Process Tech* **118** 293 https://doi.org/10.1016/S0924-0136(01)00969-4
[8] W Grzesik 2006 *J Mater Process Tech* **176** 102 https://doi.org/10.1016/j.jmatprotec.2006.02.025
[9] W Grzesik, P Nieslony 2003 *Int J Mach Tool Manu* **43** 1311 https://doi.org/10.1016/S0890-6955(03)00160-3
[10] W Grzesik, P Nieslony 2004 *Int J Mach Tool Manu* **44** 889 https://doi.org/10.1016/j.ijmachtools.2004.02.014
[11] W Grzesik, P Nieslony 2004 *Wear* **256** 108 https://doi.org/10.1016/S0043-1648(03)00390-9
[12] W Grzesik, M Bartoszuk, P Nieslony, 2004 *Int J Mach Tool Manu* **44** 1451 https://doi.org/10.1016/j.ijmachtools.2004.05.008
[13] S Zhang , Z Liu. 2008 *Int Commun Heat Mass* **35** 1311 https://doi.org/10.1016/j.icheatmasstransfer.2008.08.001
[14] Y-C Yen, A Jain, P Chigurupati, W-T Wu, T Altan 2003 Proceedings of the Sixth CIRP International Workshop on Modeling of Machining Operation 119
[15] B Haddag, H Makich, M Nouari 2014 *Tribol Int* **80** 58 https://doi.org/10.1016/j.triboint.2014.06.017
[16] Vereschaka AA, et al. 2018 *Wear* **416–417** 72 https://doi.org/10.1016/j.wear.2018.10.004
[17] AA Vereschaka, SN Grigoriev, NN Sitnikov, A Batako 2017 *Wear* **390–391** 209 https://doi.org/10.1016/j.wear.2017.07.021
[18] Vereschaka AA et al. 2019 *Surf. Coat. Technol.* **357** 218 https://doi.org/10.1016/j.surfcoat.2018.09.087
[19] Grigoriev SN, et al 2017 *Int. J. Adv. Manuf. Technol.* **90** 3421 https://doi.org/10.1007/s00170-016-9676-z
[20] Vereschaka AA, et al. 2017 *Int. J. Adv. Manuf. Technol.* **90** 27 https://doi.org/10.1007/s00170-016-9353-2
[21] Vereschaka AA, et al. 2018 *Int. J. Adv. Manuf. Technol.* **95** 2625 https://doi.org/10.1007/s00170-017-1325-7
[22] Vereschaka AA, et al. 2017 *Surf. Coat. Technol.* **332** 198 https://doi.org/10.1016/j.surfcoat.2017.10.027
[23] M F Poletika 2001 The theory of cutting. Part I. The mechanics of the cutting process: Textbook *Tomsk TPU*

Application of nanostructured Zr-ZrN-(Zr,Al)N and Zr-ZrN-(Zr,Cr,Al)N coatings for improvement of tool life and performance in end milling of carbides

Alexey Vereschaka[1], Maksim Oganyan[1], Yuri Bublikov[2], Nikolay Sitnikov[3], Konstantin Deev[4], Vladimir Pupchin[4] and Boris Mokritskii[5]

[1] Moscow State Technological University STANKIN, Moscow (Russia)
[2] IKTI RAN, Moscow (Russia)
[3] National Research Nuclear University MEPhI, Moscow (Russia)
[4] Central Aerohydrodynamic Institute named after N.E. Zhukovsky, Zhukovsky (Russia)
[5] Komsomolsk-na-Amure State Technical University, Komsomolsk-na-Amure (Russia)

E-mail: ecotech@rambler.ru

Abstract. The paper deals with the challenges of improving the tool life of end mills and performance in end milling of carbides through application of tools with the nanostructured Zr-ZrN-(Zr,Al)N and Zr-ZrN-(Zr,Cr,Al)N coatings, deposited using the filtered cathodic vacuum arc deposition (FCVAD) technology. The main mechanical characteristics of the above coatings (microhardness and critical fracture load) were determined. The cutting properties were investigated at mill rotation frequency of n = 1250, 1600, 2000 rpm. Samples were compared with uncoated tools and with tools with commercial coatings, including monolayer Ti-TiN and nanostructured Ti-TiN-(Ti,Al)N coatings. The studies have proved that the proposed coatings are not only able to improve the tool life, but they also allow milling at higher mill rotation frequency, and, consequently, with higher performance of milling. Meanwhile, tools with the Zr-ZrN-(Zr,Al)N and Zr-ZrN-(Zr,Cr,Al)N coatings are characterized by more balanced pattern of wear, without visible signs of brittle fracture.

1. Introduction

Due to the combination of a range of useful properties, titanium alloys are widespread in such areas of modern mechanical engineering as aerospace industry, engine manufacturing and a number of others. Meanwhile, titanium alloys belong to a group of hard-to-cut materials due to their high strength, toughness, oxidability and low thermal conductivity [1-3]. One of the obvious ways of improvement is a further increase in cutting speed, and that is associated with certain difficulties, since modern tool materials commonly used in the manufacture of end mills (mainly carbides) have reached their upper limit in terms of heat resistance [4].

There are a number of papers investigating the effect of different wear mechanisms on cutting tools and the ways for further improvement of the performance in end milling of carbides [5-12]. The use of wear-resistant coatings with multilayer architectures and multi-component compositions in milling titanium alloys can improve the tool life and cutting performance. In particular, the (Ti,Al)CN-VCN [13], (Ti,Al)N - TiN [14], TiN-TiC-TiCN [15], (Al,Ti)N-Si_3N_4 and (Al,Cr)N-Si_3N_4 [16], (Al,Ti)N and (Ti,Al,Cr)N [17], (Al,Ti)N-WN, (Al,Ti)N-MoN, (Al,Ti)N-CrN, (Al,Ti)N-VN and (Al,Ti)N-NbN [18], as well as (Al,Ti)N [19] coatings were considered. It was noted that coatings failed mainly because of the brittle fracture, plastic deformations and interlayer delaminations [15], while adhesive-fatigue, diffusion, and oxidation wear mechanisms were in general typical for coated tools [16, 18, 19].

There are two factors that influence decrease in the tool life:

- the presence of active processes of interdiffusion between the coating material and the material being machined, especially intensifying at high temperatures in the cutting zone;
- the formation of strong adhesive bonds between the material being machined (titanium alloy) and the titanium-containing coating.

Based on the above, it can be assumed that the use of wear-resistant coatings with high performance properties, containing no titanium in their compositions, will further increase the tool life in machining of titanium alloys. Multilayer nanostructured Zr-ZrN-(Zr,Al)N and Zr-ZrN-(Zr,Cr,Al)N coatings which showed their high performance in machining various materials [20-29] were used as coatings for cutting tools in end milling of titanium alloys. The tools with commercial Ti-TiN and (Ti,Al)N coatings and the uncoated tools were used as reference samples.

2. Materials and experiments

The coating was deposited at the VIT-2 (MSTU STANKIN-IKTI RAN, Moscow, Russia) unit using the technology of filtered cathodic vacuum arc deposition (FCVAD) [30-33]. A workpiece of VT20 titanium alloy (Al 6%, V-2%, Zr – 2%, Mo – 1.5%), widely used in aircraft engineering was machined. VT20 alloy is a typical alloy with a low content of isomorphic p-stabilizers within their solubility in the α-phase. The volume of the residual β-phase is insignificant. The cutting tests were carried out under the following cutting conditions: f_z = 0.11 mm/tooth, a_p = 1 mm, with the mill rotation frequency of n = 1250 rpm, n = 1600 rpm, and n = 2000 rpm. VB_{max} = 0.4 mm was assumed as a wear criterion. An end mill of R300-016B20L-08L with carbide inserts (R300-0828Z-PM 1130) (Sandvik Coromant, Sweden) was used.

3. Results and discussions

The monolayer TiN coating has a total thickness of about 4 μm. Nanostructured Ti-TiN-(Ti,Al)N, Zr-ZrN-(Zr,Al)N and Zr-ZrN-(Zr,Cr,Al)N coatings with three-layer architecture also have thicknesses of about 4 μm. The thicknesses of the nanolayers in the wear-resistant layer of the above coatings stay within the range of 60-100 μm (Fig. 1).

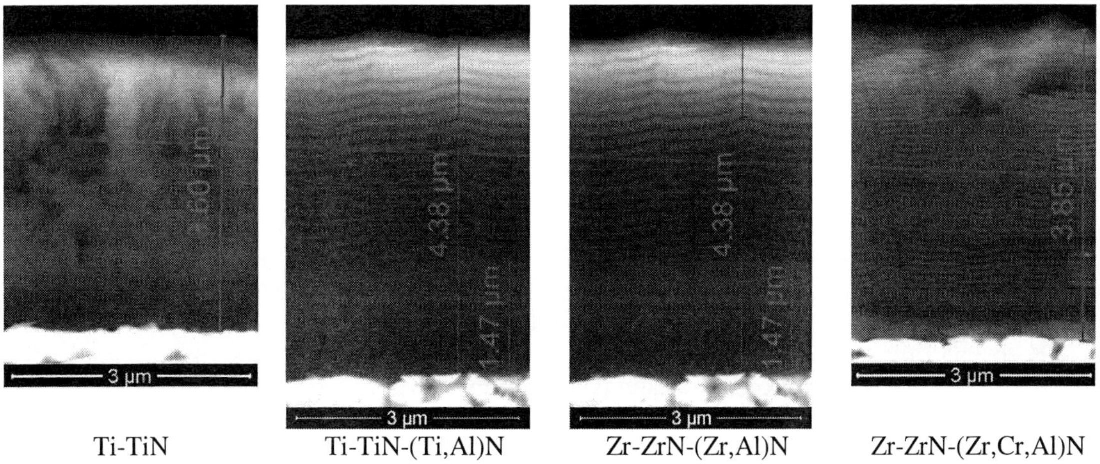

Figure 1. The microstructure of the coatings under study (SEM)

The results of the cutting tests at mill rotation frequency of n = 1250 rpm, n = 1600 rpm and n = 2000 rpm are presented in Fig. 2.

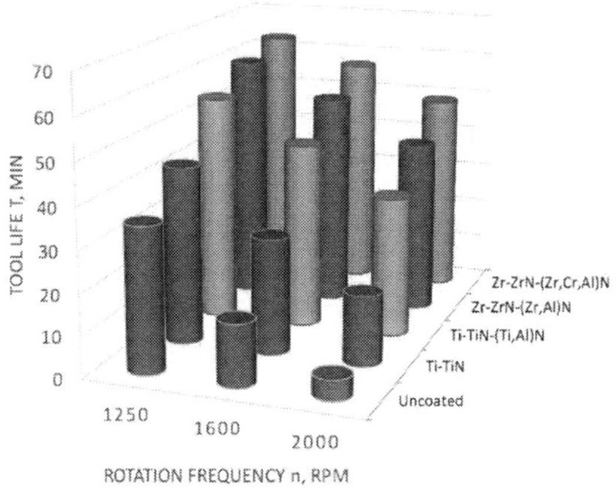

Figure 2. Relation between the tool life and the mill rotation frequency in end milling of VT20 carbide (f_z = 0.11 mm/tooth, a_p = 1 mm)

Following the analysis of the data obtained through the study of the cutting properties, it can be concluded that tools with the nanostructured coatings under study are not only characterized by a noticeably longer tool life compared to uncoated tools and tools with Ti-TiN coatings, but they also allow machining at a higher mill rotation frequency with higher performance. Meanwhile, the longest tool life at all mill rotation frequencies and the least intensive decrease in the tool life with an increase in mill rotation frequency was shown by the tool with the Zr-ZrN-(Zr,Cr,Al)N coating.

To understand the reasons for the high cutting properties of the tools with nanostructured coatings, let us consider the pattern of tool wear at mill rotation frequency of n = 1600 rpm. As known, failure of carbide tools occurs because of parallel mechanisms of abrasive, adhesive-fatigue, oxidation and diffusion wear. Let us consider the wear pattern on uncoated carbide tools and carbide tools with the proposed coatings after 12 min of cutting (Fig. 3).

Ti-TiN Ti-TiN-(Ti,Al)N Zr-ZrN-(Zr,Al)N Zr-ZrN-(Zr,Cr,Al)N

Figure 3. Wear patterns for contact pads of carbide inserts after 12 min of milling at n = 1600 rpm, f_z = 0.11 mm/tooth, and a_p = 1 mm

It can be seen that the development of wear centres on the tools with nanostructured coatings proceeds significantly slower compared to the uncoated tools and the tools with the Ti – TiN coating. Meanwhile, the uncoated tools and the tools with the Ti – TiN coatings showed signs of chipping of the carbide substrate, while no such signs were noticed on the tools with the proposed coatings.

Let us consider the nature of failure for the tools with the coatings under study in a plane passing perpendicular to the cutting edge, approximately in the centre of the cutting zone. The tool with the Ti – TiN coating (Fig. 4a) shows signs of brittle fracture with the formation of a series of longitudinal cracks. Such a pattern of cracking is usually typical for the adhesive-fatigue mechanism of wear [26]. The nature of failure for the Ti-TiN-(Ti,Al)N coating (Fig. 4b) with the nanolayer architecture is very close to the above considered nature of failure for the Ti – TiN coating. Here it is also possible to see an example of brittle fracture with extensive longitudinal cracks, in this case also combined with delaminations between the nanolayers. Such nature of failure also indicates the predominance of adhesive-fatigue processes. When considering the tool with the Zr-ZrN-(Zr,Al)N coating (Fig. 4c), it is possible to see noticeable differences in the nature of failure for this coating and the coatings considered above. This coating shows no pronounced brittle fracture. The interlayer delaminations formed in the coating structure indicate the presence of adhesive-fatigue wear processes, but these processes are less pronounced than those on the tools with the coatings considered earlier. Based on the pattern of wear, we can assume in this case the predominance of the mechanisms of abrasive, oxidation and diffusion wear. A similar pattern of failure is observed for a tool with the Zr-ZrN-(Zr,Cr,Al)N coating (Fig. 4d). The tool with this coating showed the longest tool life at all mill rotation frequencies. In this case, there are only minor delaminations, while no traces of chipping and brittle fracture of the tool are registered. The tool life is limited only by flank wear, which in this case is quite balanced. It can be concluded that the tool with the Zr-ZrN-(Zr,Cr,Al)N coating showed not only the longest tool life, but also the most favourable wear pattern in terms of predicting the cutting properties.

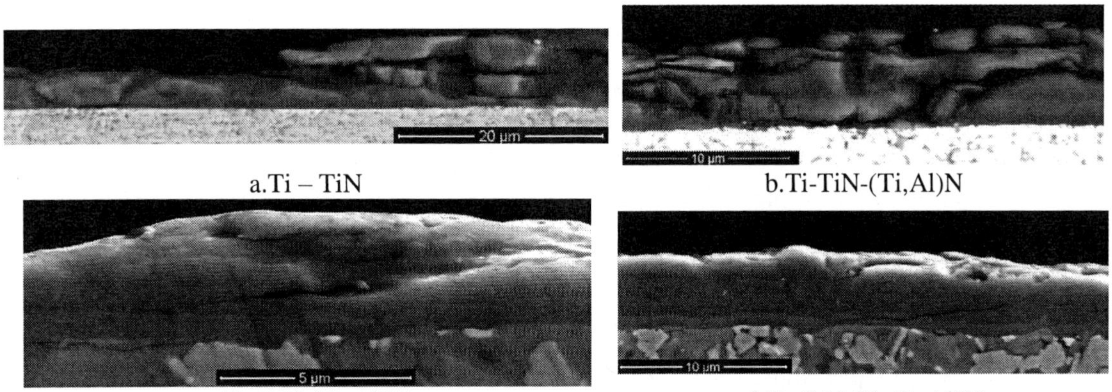

a.Ti – TiN b.Ti-TiN-(Ti,Al)N

c.Zr-ZrN-(Zr,Al)N d.Zr-ZrN-(Zr,Cr,Al)N

Figure 4. The pattern of wear on the tool with coatings

4. Conclusions

Thus, following the result of the study of the cutting properties and the patterns of wear for carbide end mills with the nanostructured Ti-TiN-(Ti,Al)N, Zr-ZrN-(Zr,Al)N and Zr-ZrN-(Zr,Cr,Al)N coatings, as well as the monolayer Ti-TiN coating, the following conclusions can be drawn:

- The use of wear-resistant coatings does not only contribute to significantly improvement of the tool life, but also enhances the cutting performance by increasing the mill rotation frequency.
- The tools with the nanostructured Zr-ZrN-(Zr,Al)N and Zr-ZrN-(Zr,Cr,Al)N coatings containing no titanium in their compositions showed the longest tool life at all mill rotation frequencies not only compared to the tool with the monolayer Ti-TiN coating, but also to the tool with the nanostructured Ti-TiN-(Ti,Al)N coating.
- The patterns of wear for tools with the nanostructured Ti-TiN-(Ti,Al)N coatings containing no titanium in their compositions are characterized by less pronounced adhesive-fatigue wear processes compared to those for the tools with the Ti-TiN and Ti-TiN-(Ti,Al)N coatings. This is manifested in the significant reduction in cracking and the formation of interlayer delaminations.

So, it can be argued that the use of tools with the nanostructured Zr-ZrN-(Zr,Al)N and Zr-ZrN-(Zr,Cr,Al)N coatings under study are not only able to improve the cutting performance of titanium alloys, but also provide a more balanced predictable pattern of wear. Thus, tools with the coatings under study can be successfully applied in computer-aided manufacturing.

Acknowledgement

This research was financed by the Ministry of Education and Science of the Russian Federation (Leading researchers, project 16.9575.2017/6.7).

References

[1] Rahman M, Wang Z G, Wong Y S 2006 *JSME Int J C* **49(1)** 11 https://doi.org/10.1299/jsmec.49.11
[2] Boyer RR 1995 *Adv Perform Mater* **2** 349 https://doi.org/10.1007/BF00705316
[3] Peters M, Kumpfert J, Ward CH, Leyens C 2003 Adv Eng Mater **5** 419 https://doi.org/10.1002/adem.200310095
[4] Ezugwu E O, Wang Z M. 1997 *J Mater Proc Tech* **68** 262 https://doi.org/10.1016/S0924-0136(96)00030-1
[5] Zhang S, Li J F, Sun J, Jiang F. 2010 *Int. J. Adv. Manuf. Technol.* **46** 69 https://doi.org/10.1007/s00170-009-2077-9

[6] Nouari M, Makich H. 2013 *Int J Refract Met H* **41** 259 https://doi.org/10.1016/j.ijrmhm.2013.04.011
[7] Sutter G, List G 2013 *Int J Mach Tool Manu* **66** 37 https://doi.org/10.1016/j.ijmachtools.2012.11.004
[8] Hou J, Zhou W, Duan H, Yang G, Xu H, Zhao N 2014 *Int. J. Adv. Manuf. Technol.* **70** 1835 https://doi.org/10.1007/s00170-013-5433-8
[9] Krishnaraja V, Samsudeensadhama S, Sindhumathia R, Kuppan P 2014 *Procedia Engineering* **97** 251 https://doi.org/10.1016/j.proeng.2014.12.248
[10] Cui D, Zhang D, Wu B, Luo M. 2017 *Int J Mech Sci.* **131–132** 613 https://doi.org/10.1016/j.ijmecsci.2017.07.027
[11] Hsu H-C, Wu S-C, Hsu S-K, Hsu C-C, Ho W-F 2016 *J. Mater. Eng. Perform.* **25** (5) 1986 https://doi.org/10.1007/s11665-016-2040-5
[12] Ramirez C, et al. 2017 *Wear* **390–391** 61 https://doi.org/10.1016/j.wear.2017.07.003
[13] Hovsepian P E, Ehiasarian A P, Petrov I 2010 *Surface Engineering*. **26** (8) 610 https://doi.org/10.1179/026708408X336337
[14] Safari H, Sharif S, Izman S, Jafari H 2015 *Int. J. Adv. Manuf. Technol.* **78** 651 https://doi.org/10.1007/s00170-014-6653-2
[15] Nouari M, Ginting A 2006 *Surf. Coat. Technol.* **200** 5663 https://doi.org/10.1016/j.surfcoat.2005.07.063
[16] Liu Z, An Q, Xu J, Chen M, Han S 2013 *Wear* **305** 249 https://doi.org/10.1016/j.wear.2013.02.001
[17] Fox-Rabinovich GS, et al. 2009 *Surf. Coat. Technol* **204** 489 https://doi.org/10.1016/j.surfcoat.2009.08.021
[18] Biksa A, et al. 2010 *Tribol Int* **43** 1491 https://doi.org/10.1016/j.triboint.2010.02.008
[19] Akmal M, et al. 2014 *Surf. Coat. Technol.* **260** 290 https://doi.org/10.1016/j.surfcoat.2014.06.066
[20] Jawaid A, Sharif S, Koksal S 2000 *J. Mater. Process. Technol.* **99** 266 https://doi.org/10.1016/S0924-0136(99)00438-0
[21] Ghafoor N, et al. 2013 *APL Mater.* **1** (2) 022105 https://doi.org/10.1063/1.4818170
[22] Lamni R, Sanjinés R, Parlinska-Wojtan M, Karimi A, Lévy F 2005 *J. Vac. Sci. Technol. A* **23** (4) 593. https://doi.org/10.1116/1.1924579
[23] Rogström L, Johansson MP, Ghafoor N, Hultman L, L Odén L 2012 *J. Vac. Sci. Technol. A* **30** (3) 031504 https://doi.org/10.1116/1.3698592
[24] Rogström L, et al. 2015 *J. Appl. Phys.* **118** (3) 035309 https://doi.org/10.1063/1.4927156
[25] Holec D, Rachbauer R, Chen L, Wang L, Luef D, Mayrhofer P H 2011 *Surf. Coat. Technol.* **206** 1698 https://doi.org/10.1016/j.surfcoat.2011.09.019
[26] Mayrhofer P H, Sonnleitner D, Bartosik M, Holec D 2014 *Surf. Coat. Technol.* **244** 52 https://doi.org/10.1016/j.surfcoat.2014.01.049
[27] Vereschaka A A, et al. 2018 *Wear.* **416–417** 72 https://doi.org/10.1016/j.wear.2018.10.004
[28] Vereschaka A A, Grigoriev S N, Sitnikov N N, Batako A 2017 *Wear* **390–391** 209 https://doi.org/10.1016/j.wear.2017.07.021
[29] Vereschaka A A, et al. 2017 *Surf. Coat. Technol* **332** 198 https://doi.org/10.1016/j.surfcoat.2017.10.027
[30] Kim Y J, et al. 2007 *Surf. Coat. Technol.* **201** 5547 https://doi.org/10.1016/j.surfcoat.2006.07.115
[31] Grigoriev S N, et al. 2017 *Int. J. Adv. Manuf. Technol.* **90** 3421 https://doi.org/10.1007/s00170-016-9676-z
[32] Vereschaka A A, et al. 2017 *Int. J. Adv. Manuf. Technol.* **90** 27 https://doi.org/10.1007/s00170-016-9353-2
[33] Vereschaka A A, et al. 2018 *Int. J. Adv. Manuf. Technol.* **95** 2625 https://doi.org/10.1007/s00170-017-1325-7
[34] Vereschaka A A et al. 2019 *Surf. Coat. Technol.* **357** 218 https://doi.org/10.1016/j.surfcoat.2018.09.087

IOP Conf. Series: Materials Science and Engineering **613** (2019) 012021 doi:10.1088/1757-899X/613/1/012021

Extra pure tellurium oxide for the growth of high quality paratellurite crystals

Mikhail Grishechkin, Anika Islam, Andrew Khomyakov, Marina Zykova, Elena Mozhevitina, Roman Avetisov, Ivan Ermochenkov and Igor Avetissov

D. Mendeleev University of Chemical Technology of Russia (Russia)

E-mail: aich@rctu.ru

Abstract. Tellurium oxide (IV) single crystals are in demand materials for manufacturing acousto-optical devices optical fibers for IR wavelength range. The initial material for production of active and passive optical materials must have purity better than 99.999 wt% and extremely low concentration of the residual water. The main method of synthesis of TeO_2 is oxidation in nitric acid followed by recrystallization from HCl is complicated, also amount of used high purity acids and waste are very high. Besides the residual water content is no less than 10 ppm. In the research we have developed the procedure for the synthesis of high-purity tellurium (IV) oxide without the usage of liquid oxidants. The process of oxidation of elementary Te was carried out in a flowing quartz glass reactor. Air oxygen was used as an oxidizing agent. The oxidation was conducted at 600-670°C during 24 hours. The produced material was a fine crystalline powder of acicular dendritic crystals of bright white color, identified as β-TeO_2 due to X-ray. It was as pure as 5N7 due to ICP-MS. The β-TeO_2 crystal grown form the synthesized water-free preparation has demonstrated better optical properties towards the crystals grown from commercial TeO_2, produced by the «wet» technology.

1. Introduction

Paratellurite crystal (β-TeO_2) is a demanded material in the technology of optoelectronics, acousto-optical and electro-optical converters [1,2]. TeO_2 also used as a component of promising glass-fibers transparent in IR wavelength region [3]. Both technologies need the use of high-purity TeO_2 (as pure as 99.999 wt% and better).

To the date the most widely used technology of high-purity TeO_2 based on an oxidation of elemental high-purity tellurium by nitric acid, and subsequent solution recrystallization from hydrochloric acid. This method is laborious and involves the consumption of a large number of high-purity mineral acids and the formation of equivalent amounts of liquid waste that require disposal. Besides, the produced crystalline product contains residual water or OH-groups which drastically influence on the optical properties of the paratellurite crystal and TeO_2 based glass fibers.

The goal of the research was the development of a methodology for the synthesis of high-purity TeO_2 without the use of liquid oxidants and the exclusive stage of recrystallization.

2. Experimental techniques

2.1. Method of TeO_2 synthesis

The concept of the proposed technique consists in oxidation of elementary tellurium by air oxygen under controlled conditions of heat and mass transfer. The oxidation process was carried out in a quartz reactor, sealed from one end, placed in a two-zone resistive furnace. The scheme of the setup is presented in Figure 1.

The preparation of the quartz parts of the setup was reduced to etching sequentially in nitric and hydrofluoric acids and subsequent repeated rinsing with bidistilled water. Liquid TeO_2 easily interacts with quartz glass forming the complex glass. But when solid TeO is inert to the quartz glass material which is known as a good and high-purity construction material for high temperature vacuum systems [4].

The original tellurium (99.9992 wt%, Figure 2), previously subjected to vacuum sublimation [5], was placed in a boat (5), which was then installed in a reactor (3). A quartz receiver tube – a condenser (6) was placed close to the boat to collect the synthesis product. A fungal seal (7) with a porous filter (4) was placed on the open end of the reactor. The filter (4) controlled the flow rate of oxygen entering the reaction zone and purified the entering gas from suspended solids.

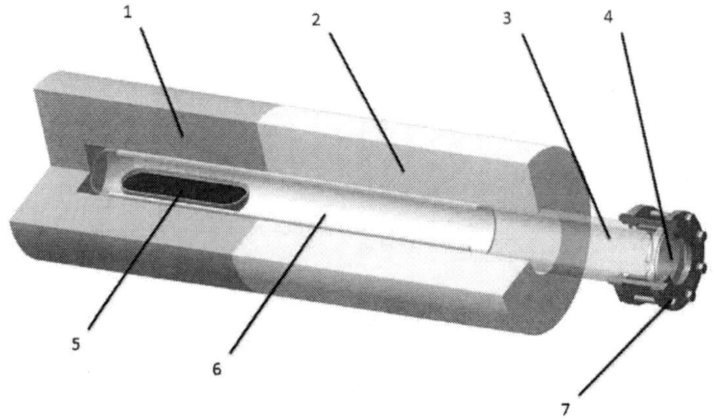

Figure 1. Scheme of the setup for direct synthesis of crystalline TeO_2:
1 – source heater, 2 – reaction zone heater, 3 – reactor, 4 – porous filter,
5 – boat with source tellurium, 6 – condenser, 7 – fungal seal.

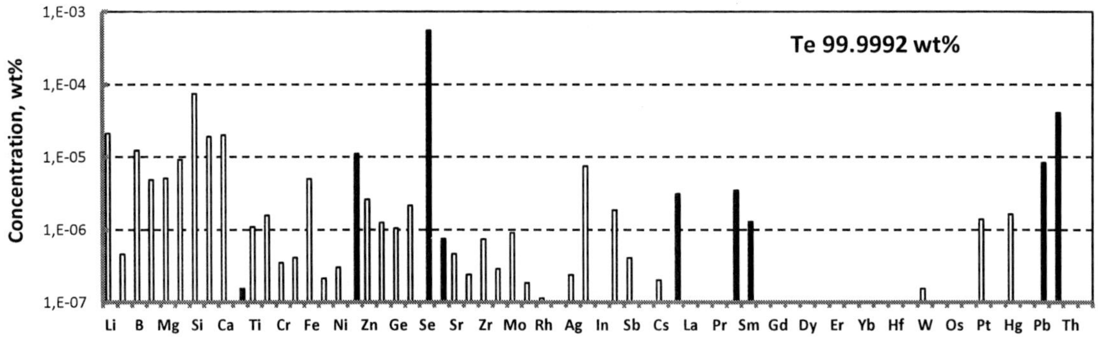

Figure 2. Impurity content in sublimated metal tellurium. The empty bars indicate the limit of detection (LD) of ICP-MS analysis. The concentrations of the non-presented elements were less 10^{-7} wt%

The reactor was heated and held for the time required to evaporate 90% of the loaded tellurium. The temperatures of the heaters (1) and (2) were selected in this way, on the one hand, to ensure the highest evaporation rate of the original tellurium, on the other hand, to prevent the interaction of TeO_2 with the material of the reactor walls and the internal quartz glass rigging.

After the end of the oxidation process, the reactor was cooled down to RT. The condenser (6) with the synthesized preparation was pulled out and the samples of the preparation were taken for the investigation of impurity and phase purity.

2.2. Impurity analysis technique

Samples of the tellurium (IV) oxide preparations were prepared for ICP-MS analysis in a laboratory room that met the ISO 6 cleanroom requirements.

Nitric (HNO_3) and hydrochloric (HCl) acids of 99.9999% purity (based on 68 elements) were produced by the surface distillation (BSB-929-IR, Berghof, Germany) of the respective pure-grade acids. The deionized water (H2O) was produced by an Aqua- MAX-Ultra 370 Series setup (Young Lin

Instruments, South Korea) and had 18.2 MΩ cm electrical resistance and 99.99999 wt% purity (based on 68 elements).

TeO_2 samples (250 mg) were dissolved in a mixture of nitric (HNO_3) and hydrochloric (HCl) acids at a HNO_3:HCl volume ratio of 1:3 in polytetrafluoroethylene (DAP-100, PTFE, Berghof, Germany) autoclaves using a SpeedWave Four microwave decomposition system (Berghof, Germany). The dissolution process was run for 40 min, and the temperature in the autoclave was maintained at 170°C. The dissolution product was transferred to a polypropylene (PP) test tube. A 1-mL aliquot taken from the resultant solution was transferred to a PP test tube, and then the solution was brought to 10 ml by water. The solution thus prepared was analyzed by inductively coupled plasma mass spectrometry (ICP-MS).

The measurements were performed using a NexION 300D ICP-MS system (PerkinElmer, USA) in collision mode (kinetic energy discrimination, KED) [7]. The measurement conditions are summarized in Table 1.

Table 1. ICP-MS analysis conditions and instrumental parameters

Nebulizer type	Concentric (Meinhard), PFA
Spray chamber	Scott double-pass chamber, PFA
Argon flow rate, L/min through the nebulizer plasma-forming auxiliary	 0.96 15 1.2
Generator power, W	1450
Collision gas (He) flow rate, L/min	14.6
Number of scan cycles	10

2.3. PXRD analysis

To confirm the phase of paratellurite crystal (β-TeO_2) powder X-ray diffraction analysis was carried out using an EQUINOX 2000 (INEL Corp., France). It is known that β-TeO_2 is tetragonal crystal system, space group P41212, cell parameters a = 4.812(5) Å, c = 7.615(9) Å. The X-ray diffraction analysis revealed no extraneous phases in the sample of the produced tellurium dioxide, except the β-TeO_2 phase (Figure 3), despite the fact that the zone of tellurium dioxide formation was close to the boat with the initial substance.

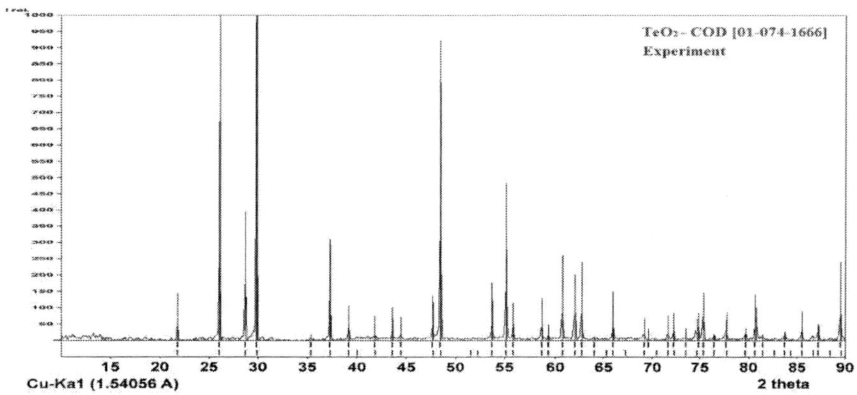

Figure 3. XRD pattern of polycrystalline β-TeO_2 produced by direct synthesis

2.4. IR transmission analysis

A comparative analysis of the IR absorption spectra of samples of a commercial preparation of tellurium oxide obtained by «wet» synthesis and a sample obtained by direct synthesis was carried out. To measure the absorption spectrum, the samples were crushed in an inert atmosphere and compressed into a tablet. Fourier transform infrared spectroscopy (FTIR) was performed on a Tensor 27 FT-IR spectrophotometer

(Bruker, Germany) in the wavelength range 6000-400 cm^{-} . The FT-IR absorption spectra of samples TeO_2 demonstrated two recent absorption bands around 775 and 656 cm^{-1}, which related to the vibration of the Te-O bond [6]. It was found that an absorption in the IR region of the (β-TeO_2 preparation which we made by the direct synthesis was two times less comparing to those for the commercial sample produced by the wet technique. In the case of the commercial sample we also observed vibrational peaks around 3500 cm^{-1} which are attributed to OH-group of the water (Figure 4).

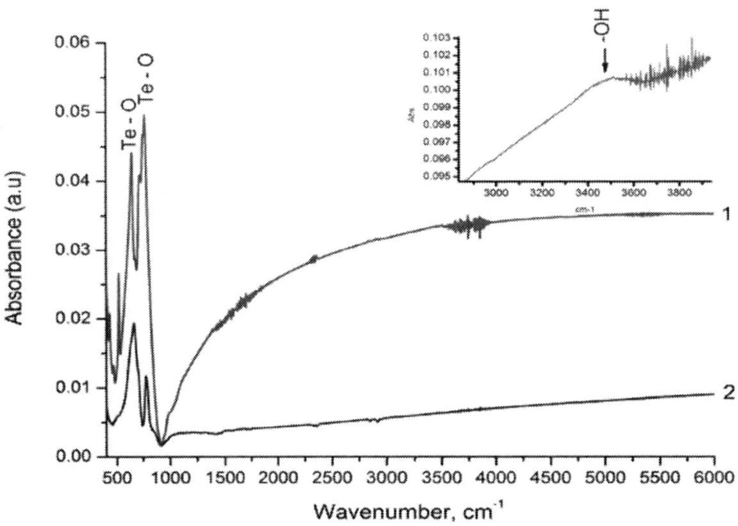

Figure 4. Normalize FT-IR absorption spectra of samples: 1 - commercial preparation («wet» synthesis); 2 – crystalline β-TeO_2 produced by direct synthesis

3. Results and discussions

3.1. TeO_2 purification

It has been experimentally established that when the synthesis is carried out at a temperature exceeding 690°C, the TeO_2 formed begins to interact with quartz glass, which leads to product contamination with silicon and rapid destruction of the reactor. Therefore, in the future, the synthesis process was carried out at a reaction zone temperature not exceeding 650°C. The resulting tellurium (IV) oxide was a mixture of fine powder and larger dendritic crystals (Figure 5).

According to X-ray phase analysis (Fig. 4), the content of the elemental tellurium phase in samples of tellurium (IV) oxide was below the detection limit (0.5 wt%). The cell parameters of the produced β-TeO_2 was slightly different from the reference value: a = 4.813(4) Å, c = 7.617(5) Å. We explained this difference by the deviation from stoichiometry in our preparation because during the synthesis we optimized the oxygen partial pressure to increase the yield of the final product which was 97%.

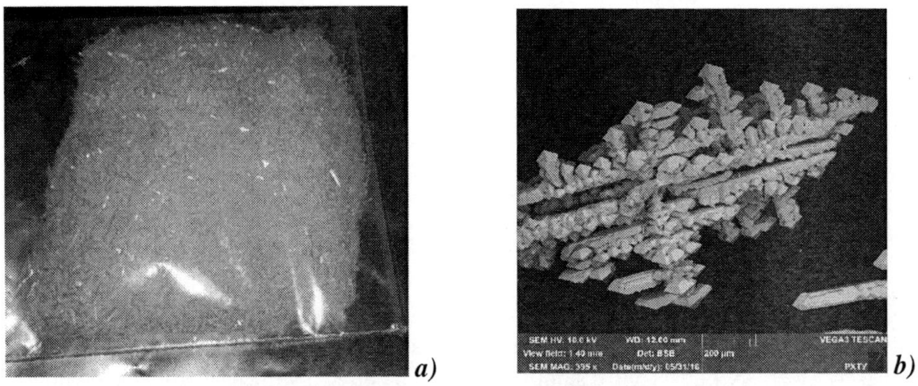

Figure 5. Habitus of the produced crystalline high-purity tellurium (IV) oxide: general image (a) and SEM image (b)

Analysis of the impurity content in the as-synthesized TeO_2 (Figure 6) showed that during the synthesis, the concentration of selenium impurity strongly decreases, while a slight contamination of the preparation with impurities of iron, cobalt, and nickel was observed.

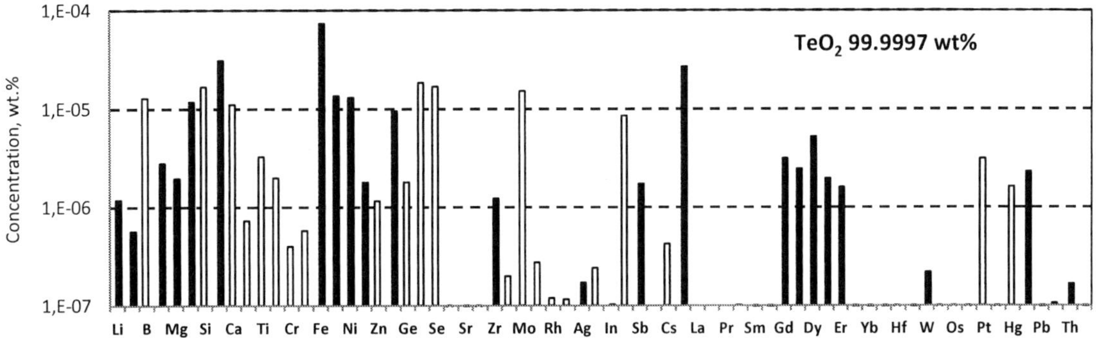

Figure 6. Impurity content in polycrystalline β-TeO_2 produced by direct synthesis. The empty bars indicate the limit of detection (LD) of ICP-MC analysis. The concentrations of the non-presented elements were less 10^{-7} wt%

The latter can be explained by secondary contamination during unloading of the preparation. The usage of Teflon devices for unloading improved the purity of the preparation. Finally, the total purity of the obtained polycrystalline β-TeO_2 preparation was 99.9997 wt%.

3.2. Paratellurite crystal growth

β-TeO_2 single crystal was grown in Czochralski configuration using a growth setup with controlled Ar-O atmosphere. P_{O2} pressure during the growth process was maintained at the level which had been found during the experiments for direct synthesis of β-TeO_2 from vapor. The pulling rate of the cylindrical part was 1 mm/h at 40 mm crystal diameter. The grown crystal demonstrated better optical properties towards the crystals grown from commercial TeO_2, produced by the «wet» technology (Table 2).

Figure 7. Impurity content in polycrystalline β-TeO_2 produced by direct synthesis. The empty bars indicate the limit of detection (LD) of ICP-MC analysis. The concentrations of the non-presented elements were less 10^{-7} wt%

Table 2. Values of specific optical rotation and Verdet constant of paratellurite crystals

λ, nm	Ref. [8]	Ref. [9]	Our crystal
ρ (633 nm), deg/mm	87	84.7±0.3	84.7±0.2
ρ (531 nm), deg/mm	143	146.0±0.3	147.2±0.2
ρ (488 nm), deg/mm	184	186.0±0.3	188.1±0.2
V (355 nm), red A^{-1}		1.7×10^{-4}	1.9×10^{-4}
Acoustic velocity [001] m/s (633 nm)	4260		4310
Figure of merit [010]>[100] 10^{-15} s3/kg	34.5		34.8

4. Conclusions

The developed technique makes it possible to produce high-purity (99.9997 wt%) polycrystalline β-TeO_2 with lowered residual water content. The preliminary crystal growth experiments showed that the usage of as-synthesized β-TeO_2 allowed decreasing the crystal adsorption in near IR region in about 30 rel.%. This is very promising result for both transmission IR devices (optical fibers) and high power acousto-optical and electro-optical converters. The β-TeO_2 crystal grown form the synthesized 5N7 water free preparation has demonstrated better optical properties towards the crystals grown from commercial TeO_2, produced by the «wet» technology.

Acknowledgement

The research was supported by the Russian Federation Ministry of Education and Science by the agreement with the unique identification code RFMEFI577716X0218.

References

[1] A E Kokh, V S Shevchenko, V A Vlezko, K A Kokh 2013 *J. Crys. Growth* **384** 1 https://doi.org/10.1016/j.jcrysgro.2013.08.027

[2] N F Declercqa, N V Polikarpova, V B Voloshinov, O Leroyd, J Degriecke 2006 *Ultrasonics* **44** e833 https://doi.org/10.1016/j.ultras.2006.05.113

[3] C Esin, I. Kabalci, M L Ovecoglu, G Ozen 2007 *J. Eur. Ceramic Soc.,* **27** (2,3) 1797 https://doi.org/10.1016/j.jeurceramsoc.2006.05.075

[4] J N Beaudry, S Grenier, S Amrate, M Mazzera, A Zappettini 2012 *Mater Chem. and Phys.* **133** (2,3) 804 https://doi.org/10.1016/j.matchemphys.2012.01.097

[5] M B Grishechkin, E N Mozhevitina, A V Khomyakov, M P Zykova, R I Avetisov, I Kh Avetissov **2017** *Rus. Microelectronics* **46** (8) 551 https://doi.org/10.1134/S1063739717080078

[6] F Arab, M Mousavi-Kamazani, M Salavati 2016 *RSC Advances* **6** (75) 71472 https://doi.org/10.1039/C6RA10770B

[7] S M Wilbut, Naoki Sugiyama and McCurdy **2010** *Spectroscopy suppl* 20 http://www.spectroscopyonline.com/optimizing-performance-collisionreaction-cell-icp-ms-system-operating-helium-collision-mode

[8] Handbook of Optical Materials, Eds. Marvin J Weber 2002 *CRC Press* NewYork ISBN 9780849335129 https://www.crcpress.com/Handbook-of-Optical-Materials/Weber/p/book/9780849335129

[9] A.I. Kolesnikov. A. Kaplunov, S.E. Ilyashenko, V.Ya. Molchanov, R.M. Grechishkin, M.A. Arkhipova, S.A. Tretyakov 2012 *Cystallography Reports* **57** (7) 909 https://doi.org/10.1134/S1063774512070115

The Diagnostics of Optical and Thermal Characteristics of Semitransparent Media at Modeling Phase Influence of Oscillating Convection and Thermoradiation

Fedor A. Zagumennov[1], Vladimir G. Merzlikin[2, 3] and Andrey V. Bystrov[2]

[1]Department of Instrument Production Techniques, Bauman Moscow State Technical Un, 2-ya Baumanskaya str., 5/1, Moscow, 105005, Russia
[2]Department of Industrial Economics, Plekhanov Russian Un of Economics, Stremyanny, 36, Moscow 117997, Russia
[3]Department of Technology and Equipment of Mechanical Engineering, Moscow Polytechnic Un, Bol'shaya Semenovskaya, 38, Moscow 107023, Russia

E-mail: sgw32@yandex.ru

Abstract. The paper suggests diagnostics method of non-stationary combined regime heating of semitransparent materials base on an evaluation of the interconnected complex of optical parameters (extinction – b_{rad}, scattering - σ and absorption κ indices) and thermophysical (thermal diffusivity a_T and conductivity coefs) ones using simulation the general temperature field $\theta(z,\omega t,\Delta\varphi_{conv}^{rad}) = \theta_{c+r}\left(z,\omega t,b_{conv},\Delta\varphi_{\theta_{c+r}}\right)+\theta_{rad}\left(z,\omega t,b_{rad},\Delta\varphi_{\theta_{rad}}\right)$ including the additive terms due to the convective and radiation fluxes with a phase shift $\Delta\varphi_{conv}^{rad}$ within the external oscillating (frequency ω) heat load. A one-dimensional problem of finding the temperature for a semitransparent material in the absence of aggregative transitions, neglecting long wavelength re-emission was considered in the analytic representation. The classical solution (of the homogeneous heat equation) and the particular one (of the inhomogeneous one with an internal thermal radiant source) defined summable terms with independent phases $\Delta\varphi_{\theta_{c+r}},\Delta\varphi_{\theta_{rad}}$ and different indices respectively for attenuation $b_{conv}=\sqrt{\omega/2a_T}$ (conductive heatwave) and for diffraction extinction $b_{rad}=\sqrt{\kappa^2+\kappa\sigma}$ (radiant heatwave). In-phase production of additive term $\theta_{rad}(z=0,t)$ simultaneously with external convection impact on the exposed surface allowed for the first time to establish interconnected complex $b_{rad}=\sqrt{\omega\cdot ctg(\Delta\varphi_{conv}^{rad})/a_T}$. The obtained formulas for specific temperature distributions allow you to use them to estimate values of optical and thermal parameters of a semitransparent material from experimentally recorded phase jumps of the temperature field during the action of alternate impact of initial phased convective and radiation components of the external heat flux. The results of the study can be used both for experimental diagnostics and to select the heat regimes for combined radial and convective processing of semitransparent ceramics, composites, and biomaterials.

1. Introduction

Traditionally in thermal physics it is important to analyze the heat regimes of materials under the influence of external oscillating convective and thermoradiation heat fluxes with different phase shift

[1]. Studies on the calculated and experimental estimates of the temperature fields for technical materials and natural environments are known [2-5].

But the interpretation of numerical and experimental estimates significantly complicated when predicting the impact of phased components of convective and radiant heating on formation of general temperature regime with predominance of conductive heating or alternative volumetric radiant overheating for semitransparent materials.

Therefore, obtaining an analytical solution is an urgent task allowing to evaluate an interconnected complex of optical and thermophysical characteristics of semitransparent material will depend on experimentally recorded temperature field characteristics, produced by the action of an external nonstationary convective-thermoradiant heat flux, changing according to a harmonic law with equal frequencies ω of its components and simulated phase shift [6-8].

2. Mathematical model of combined convective heating and volumetric radiant overheating

The ambient temperature $\theta_A(t)$ and probe radiation flux $q_{rad}(t)$ with phase shift $\Delta\varphi_{conv}^{rad}$ are set taking into account the time averages of the constants θ_{A0}, q_{rad0} and amplitudes θ_{A1}, q_{rad1} of harmonic oscillations:

$$\theta_A(t) = \theta_{A0} + \theta_{A1} \cdot \cos(\omega t), \quad (1a)$$

$$q_{rad}(t) = q_{rad0} + q_{rad1} \cdot cos(\omega t - \Delta\varphi_{conv}^{rad}). \quad (1b)$$

We are guided by the theory of radiative-conductive subsurface heating of semitransparent material with thickness $0 < z < H$ under the influence of radiant component overheating together with convective surface heating in one-dimensional approximation. Therefore the calculated temperature is determined not only by thermophysical parameters but also by optical properties.

The inhomogeneous heat equation should include the thermal source function $F(z, t, \omega)$ which is caused by penetrating radiation:

$$\frac{\partial\theta}{\partial t} = a_T \cdot \frac{\partial^2\theta}{\partial z^2} + \frac{F_{rad}}{c\rho}, \; F_{rad}(z,\omega t) = \frac{q_{rad}(t,\omega t,\Delta\varphi_{conv}^{rad}) \cdot (1-A) \cdot b_{rad}}{1 - A^2 \cdot \exp(-2b_{rad} \cdot H)} \left[e^{-b_{rad} \cdot z} - A \cdot e^{b_{rad} \cdot (z-2H)} \right], \quad (2)$$

where c – specific heat capacity, ρ – density, K_T – thermal conductivity, $a_T = K_T / c\rho$ - temperature diffusivity coefficient.

Boundary conditions on the exposed surface are determined by the influence of convection fluxes and long wavelength nonpenetrating radiation in the absence of aggregative transitions:

$$z = 0, \quad -K_T \frac{\partial\theta(0,t)}{\partial z} = \alpha_{turb} \cdot [\theta_{A1} \cdot cos(\omega t) - \theta(0,t)] + c_0 \cdot \varepsilon_{ef} \cdot \left\{ [\theta_{A1} \cdot cos(\omega t)]^4 - [\theta(0,t)]^4 \right\}, \quad (3)$$

where α_{turb} – coefficient of turbulent heat exchange, c_0 – Stephan-Boltzmann coefficient, ε_{ef} – effective emissivity coefficient.

Our examined samples of semitransparent material should be considered as scattering and weakly absorbing mediums and ensure the necessary reflection and transmission for fallen radiation. The methodology of calculative-theoretical estimations of the optical parameters is well known [4, 5], including version developed earlier by authors [6] which is used in this work.

The penetration of radiation is defined by volumetric coefficients of reflection, transmission and absorption, depending on scattering and absorption indexes of the given optically inhomogeneous environment.

Obtained numerical values of extinction b_{rad} and scattering σ indexes for example the ceramic layer are determined by experimental values of albedo A and absorption index κ. The given calculations were based on a functional relationship of these optical parameters [6]:

$$b_{rad} = \sqrt{\kappa^2 + p \cdot \kappa \cdot \sigma}, \quad A = \frac{b_{rad} - \kappa}{b_{rad} + \kappa}, \quad (4)$$

where b_{rad} - extinction index is considered as attenuation coefficient, A – albedo is reflection coefficient of semi-infinite layer with spherical scattering indicatrix ($p = 1$) in two-flux approximation of solution for radiative transfer equation.

For known samples of porous ceramics (based on silica, zirconium aluminum) $\kappa \sim 0.01\text{-}1$ m^{-1} and $\sigma \sim 1\text{-}3000$ m^{-1}, albedo changes in interval $A \sim 30\text{–}99\%$ with $b_{rad} < 10\text{-}30$ m^{-1}. It allows us to offer simple optical models for materials for solving engineering problems of complex heat transfer.

Thus it is possible to calculate the absorbed radiant energy function (2) within UV, visible and near IR spectral ranges for most artificial and natural radiant heat sources.

Considered physical and mathematical model of the radiant heat exchange is caused by technological conditions for convective (e.g. flame machining) or radiant (e.g. laser one) processing for materials. Intensive heat fluxes can exceed tens of MW/m^2. At temperatures of aggregative transitions up to ~2000-3000K for the refractory ceramics (type Al_2O_3, SiO_2, ZrO_2) the generated fluxes of self - IR radiation from expose surface may not exceed 1 MW/m^2.

Therefore, to obtain an analytical solution in the boundary conditions (3) authors will be neglected longwave re-emission on the external border and this thermal problem becomes linear:

$$z = 0, \quad -K_T \frac{\partial\theta(0,t)}{\partial z} = \alpha_{turb} \cdot [\theta_{A1} \cdot cos(\omega t) - \theta(0,t)]; \quad z \to \infty, \quad \theta(z,t) \to 0. \tag{5}$$

because of the accepted linear mathematical model for solving the problem total temperature field in semitransparent material can be considered as the result of independent additive contributions calculated temperature distributions $\theta_{c+r}(z, \omega t, \theta_A, q_{rad}, \Delta\varphi_{conv}^{rad}, b_{conv}, \Delta\varphi_{\theta_{c+r}})$ and $\theta_{rad}(z, \omega t, q_{rad}, \Delta\varphi_{conv}^{rad}, b_{rad}, \Delta\varphi_{\theta_{rad}})$, respectively (See Figure 1):

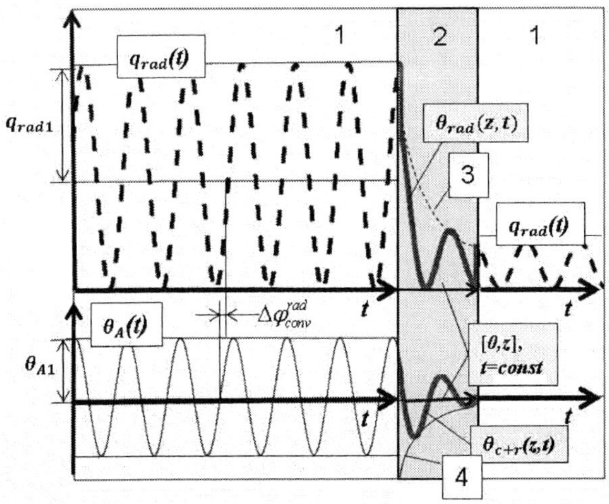

Figure 1. Block scheme of external (1) oscillating heat load on the semitransparent material (2) by components of convective surface heating $\theta_A(t)$ and subsurface radiant overheating $q_{rad}(t)$ (with a phase shift $\Delta\varphi_{conv}^{rad}$), which form and model the general temperature field $\theta(z,t)$ with additive terms $\theta_{c+r}(z,t)$ and $\theta_{rad}(z,t)$ with appropriate phases $\Delta\varphi_{\theta_{c+r}}(b_{conv}, b_{rad})$ and $\Delta\varphi_{\theta_{rad}}(b_{rad})$ with exponential conductive attenuation (Line *3*, with index b_{rad}) and radiative extinction (Line – 4, with index b_{conv}) for their heat waves

I - surface convective heating (1*a*) causing a heat wave with phase $\Delta\varphi_{\theta_{c+r}}$ and index attenuation $b_{\text{conv}} = \sqrt{\omega / 2a_T}$ at a conductive mechanism of heat propagation inside the material, but corrected by volumetric overheating, produced by penetrating radiation (1*b*);

II - volumetric overheating by penetrating thermal radiation (1*b*) with phase $\Delta\varphi_{\theta_{rad}}$ and extinction index b_{rad} (4) with conductive-radiate heat transfer.

Besides, for the constant components of the external temperature θ_{A0} (1*a*) and the incident radiation flux q_{rad0} (1*b*) the trivial exact solution is known [1], which will not be considered in our analysis of generated oscillating temperature fields.

Then for a total analytical solution for formed temperature fields we can limit ourselves to the influence only variable radiant $q_{rad1} \cdot cos(\omega t - \Delta\varphi_{conv}^{rad})$ and convective $\theta_{A1} \cdot \cos(\omega t)$ components of harmonic oscillations of an external heat load.

The formulated problem (1-5) admits an analytical solution.

Since the heat conductivity equation (2) is inhomogeneous with the desired temperature field $\theta(z,t)$, then it is possible to find separate solutions in the form of independent components:

1 - conductive component of homogeneous equation with oscillating convective heating at the boundary (5) and

2 - particular solution for the inhomogeneous equation (3) with thermoradiation subsurface overheating.

Then, taking into account the initial temperature, the general solution for (1-5) is [8]:

$$\theta(z,\omega t,\theta_A,q_{rad},\Delta\varphi_{conv}^{rad}) = \theta_{c+r}\left(z,\omega t,b_{conv},\Delta\varphi_{\theta_{c+r}}\right) + \theta_{rad}\left(z,\omega t,b_{rad},\Delta\varphi_{\theta_{rad}}\right) =$$

$$= \sqrt{c_1^2 + c_2^2} \cdot \cos\left(\omega t - \sqrt{\frac{\omega}{2a_T}} \cdot z - \Delta\varphi_{\theta_{c+r}}\right) \cdot e^{-\sqrt{\frac{\omega}{2a_T}} \cdot z} - \frac{q_{rad1} \cdot (1-A)}{c \cdot \rho} \cdot b_{rad} \cdot \frac{\cos\left(\omega t - \Delta\varphi_{\theta_{rad}}\right)}{\sqrt{\omega^2 + a_T^2 \cdot b_{rad}^4}} \cdot e^{-b_{rad} \cdot z},$$

$$\Delta\varphi_{\theta_{c+r}} = \arcsin\frac{c_2}{\sqrt{c_1^2 + c_2^2}}, \quad \Delta\varphi_{\theta_{rad}} = -\arcsin\frac{\omega}{\sqrt{\omega^2 + a_T^2 \cdot b_{rad}^4}},$$

$$c_1 = \frac{\left(K_T\sqrt{\frac{\omega}{2a_T}} + \alpha_{turb}\right) \cdot \Phi_1 + K_T\sqrt{\frac{\omega}{2a_T}} \cdot \Phi_2}{\Delta}, \quad c_2 = \frac{K_T\sqrt{\frac{\omega}{2a_T}} \cdot \Phi_1 - \left(K_T\sqrt{\frac{\omega}{2a_T}} + \alpha_{turb}\right) \cdot \Phi_2}{\Delta}, \tag{6}$$

$$\Phi_1 = \alpha_{turb} \cdot \theta_{A1} - \left(K_T b_{rad} + \alpha_{turb}\right) \cdot \frac{q_{rad1} \cdot b_{rad} \cdot (1-A)}{c \cdot \rho} \cdot d_1, \quad \Phi_2 = \frac{q_{r1} \cdot b_{rad} \cdot (1-A)}{c \cdot \rho} \cdot \left(K_T b_{rad} + \alpha_{turb}\right) \cdot d_2,$$

$$\Delta = \left(K_T\sqrt{\frac{\omega}{2a_T}} + \alpha_{turb}\right)^2 + K_T^2 \cdot \frac{\omega}{2a_T}.$$

On the surface of the medium, the temperature is defined by the formula:

$$\theta(z=0,t) = \sqrt{c_1^2 + c_2^2} \cdot cos\left(\omega t - \Delta\varphi_{\theta_{c+r}}\right) - \frac{q_{rad1} \cdot (1-A)}{c \cdot \rho} \cdot b_{rad} \cdot \frac{cos\left(\omega t - \Delta\varphi_{\theta_{rad}}\right)}{\sqrt{\omega^2 + a_T^2 \cdot b_{rad}^4}}. \tag{7}$$

Note also the case of maximum heating when the oscillations of the external temperature ($1a$) and the thermoradiation term in (7) can occur in the phase $\Delta\varphi_{\theta_{rad}} = 0$ (See formula 6).

This is done if the phase φ_{conv}^{rad} in the incident radiant flux ($1b$), the frequency ω and the optical-thermophysical parameters of the medium are related by interconnected specific complex (Figure 2):

$$b_{rad} = \sqrt{\kappa^2 + \kappa\sigma} = \sqrt{\omega \cdot ctg(\Delta\varphi_{conv}^{rad}) / a_T}. \tag{8}$$

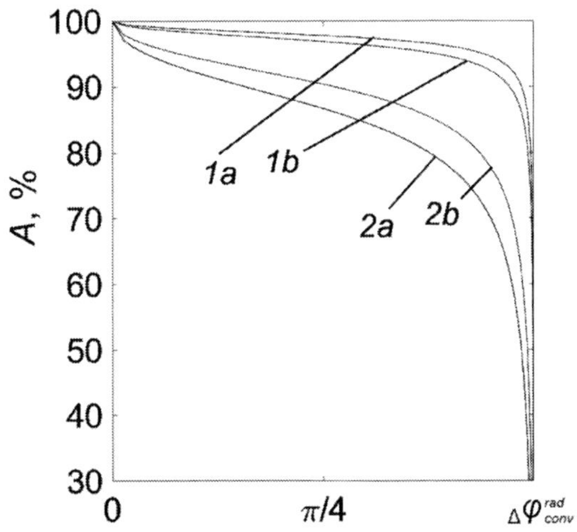

Figure 2. The dependence of semitransparent material albedo A vs the phase shift $\Delta\varphi^{rad}_{conv}$ (between convective and radiant components of the external oscillating heat flux with period T=3600 s) during maximum heating by surface convective impact (1a) with in-phase radiant overheating $\theta_{rad}\left(z,\omega t,b_{rad},\Delta\varphi_{\theta_{rad}}\right)$ (6) with $\Delta\varphi_{\theta_{rad}}=0$ for samples:

- porous ceramics (Lines #1a, #1b with a thermal conductivity K_{T1} =0.1 W/m·K and thermal diffusivity a_T = 2.2·10^{-7} m/s^2 coefs).
  Line # 1a for κ_{11} = 1 m^{-1}, σ_{11} = 450 m^{-1}. Line # 1b for κ_{12} = 2 m^{-1}, σ_{11} = 450 m^{-1}.
- monolithic ceramics (Lines #2a, #2b with a thermal conductivity K_{T2}=2 W/m·K and thermal diffusivity a_T = 4.4·10^{-6} m/s^2 coefs).
  Line # 2a for κ_{21} = 1 m^{-1}, σ_{21} = 45 m^{-1}. Line #2b for κ_{22} = 2 m^{-1}, σ_{21} = 45 m^{-1}.

Model parameters: θ_{A1}=±500 K, θ_{A0}=900 K; α_{turb}=100 W/m^2·K; q_{rad1}=±10 and q_{rad0}=10 kW/m^2

3. Conclusion

The obtained formulas for general solution $\theta(z,t)$ (6) with additive terms $\theta_{c+r}(z,t)$, $\theta_{rad}(z,t)$ allow you to use them to estimate values of optical and thermal parameters of a semitransparent material based on interconnected specific complex (8) from experimentally recorded phase jumps (7) of the temperature of the exposed surface at the alternating energy effect of initial phased convective and radiation components of the external heat fluxes.

The results of the study can be used both for experimental diagnostics and to select the heat regimes for combined radial and convective processing of semitransparent ceramics, composites, and biomaterials.

References

[1] H S Carslaw, J C Jaeger 1986 Conduction of Heat in Solids *Oxford Un Press* 255

[2] A K Pandey, M Kumar 2017 *Alexandria Eng. J.* **56** (1) 55 http://dx.doi.org/10.1016/j.aej.2016.08.035

[3] M A Azadi, M Baloo, G H Farrachi, S M Mirsalim 2013 *Int. J. of Autom. Eng* **3** (1) 305 http://www.iust.ac.ir/ijae/search.php?sid=1&slc_lang=en&auth=Farrahi

[4] L Wang, J I Eldridge, S M Guo 2013 *Acta Materialia* **64** 402 https://doi.org/10.1016/j.actamat.2013.10.053

[5] V A Petrov 2016 *High Temp* **54** (2) 186 https://doi.org/10.7868/S0040364416020150

[6] V Merzlikin 2007 *Proc. of 8th Int.* Conf *on Engines for Automobile - ICE2007* (*Capri, Napoly*) **24** 0066 https://www.sae.org/technical/papers/2007-24-0066

[7] V G Merzlikin, T V Zhubreva, A V Kostukov 2018 *Materials Engineering and Technologies for Production and Processing IV* **284** 1215 https://doi.org/10.4028/www.scientific.net/SSP.284.1215

[8] F Zagumennov et al 2018 Book of abstract (ic-cmtp5) *Igrex Ltd.* Miskolc (Hungary) 114 https://ic-cmtp4.eu/doc/book_of_abstract_iccmtp3.pdf

Structure and mechanical properties of ZrO_2-MgO composites with bimodal pore structure

A Buyakov[1, 2, 3], S Buyakova[1, 2, 3], S Kulkov[1, 2, 3]

[1]Tomsk State University, 36, Lenina ave., Tomsk, 634050, Russia
[2]Institute of Strength Physics and Materials Science of Siberian Branch Russian Academy of Sciences 2/4, Akademicheskii ave., Tomsk, 634021, Russia
[3]Tomsk Polytechnic University, 30, Lenina ave., Tomsk, 634050, Russia

E-mail: alesbuyakov@gmail.com

Abstract. Porous ceramic ZrO_2-MgO composite in a wide range of component content was studied. The effect of content and sintering parameters on pores size, its distribution, lattice structure, coherently diffracting domains and lattice microdistortion, tensile strength was investigated. It has been shown that the adding of pore-forming particles into the initial powder makes it possible to obtain a bimodal pore structure of composites. Sintering time and MgO content in ceramics are stipulated a changing of the average size of macro- and micropores and microstresses, which determine the macroscopic strength of the material.

1. Introduction

Ceramics, such as zirconia and alumina compounds, are widespread as dental and osteoprosthetics materials. These ceramics have identical type of chemical bonds with inorganic bone matrix, do not participate in electrochemical interaction with the body tissues and mechanical parameters do not degrade over time. However, their high strength, significantly exceeding similar characteristics of bone tissue, restricts using of ceramics as osteoreplacement material.

Significant prospects for the osteoreplacement materials creation may have the "Bio-inspired" approach, which consists in reproducing natural structures in artificially created materials. Bone tissue is a complex composite with heterogeneous structure and developed porosity. It is well known from studies that osteoreplacement materials with bimodal porosity have good conditions for bone tissue integration – macropores, with an average size of about 100 μm, play the role of niches for the osteoblast cells proliferation and micropores and micropore channels, with an average size of 30 μm, are necessary for the vascular system development and cell cultures nutrition [1]. Ensuring the structural and biomechanical identity with natural bone tissue will allow making a new kind of osteoreplacement materials.

Among ceramic materials, promising for osteoimplantation, materials based on ZrO_2 and MgO are of particular interest [2, 3]. Zirconia is included in the ISO registry (ISO 13356:2015) as a biocompatible material for prosthetics. Magnesium ions are participating in the DNA and RNA molecules synthesis and stabilization processes, creating favorable conditions for the bone tissue regeneration stimulating in the contact area [4]. However, currently knowledge of porous ceramics based on ZrO_2 and MgO system are not enough and require further study.

The purpose of this work is to study pore structure formation in the ZrO_2-MgO composite, depending on the composition and sintering parameters.

2. Materials and experiments

Porous ceramic materials based on partially magnesium-stabilized zirconium dioxide and magnesium oxide were studied. Samples were obtained by mechanical mixing of ZrO_2 and MgO powders in a wide range of components concentration followed by cold uniaxial pressing of ceramic mixtures and sintering in air. The pore structure was formed by adding 50 vol. % of ultrahigh molecular weight polyethylene particles (UHMWPE) with an average particle size of 80 μm. Samples sintering were carried out according to the following conditions: heating at a speed of 0.5 °C/min with holding at 300 °C for 3 hours to remove pore-forming particles and sintered at 1600 °C with holding time 10 minutes up to 10 hours.

The phase content, the average size of coherently diffracting domains and lattice microdistortion were estimated by the Williamson-Hall method [5] based on the X-ray diffraction patterns obtained under conditions of filtered Cu Kα radiation and an angular range of 20-80°.

Microstresses were evaluated by microdistortions and the Young's modulus values of the corresponding phase. The average value of microstresses was calculated by the mixture rule, according to the ceramic sample phase content.

The microstructure of ceramics was studied by using scanning electron microscopy Tescan. Porosity was measured by using the Archimedes method [6].

The tensile strength was estimated by the method of "Brazilian test" [7], with a constant traverse speed of 0.01 mm/min.

3. Results and discussions

The porous structure of ceramics is represented by two types of pores, fig. 1. Macropores inherited the structure of burnt UHMWPE particles and mostly have a regular spherical shape. The micropores did not have a clearly defined form and were formed due to the presence of interparticle voids formed during the powders compaction. Porosity of studied ceramics basically did not change and was 48 ± 2%.

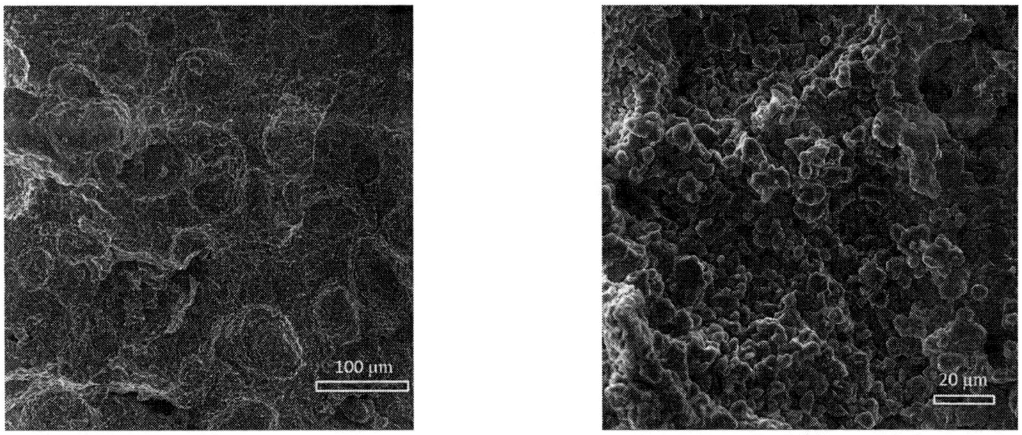

(a) (b)

Figure 1. The destruction surface of 50 wt. % ZrO_2 + 50 wt. % MgO ceramics after sintering at 1600 °C with a 1 hour holding time.

In ceramics sintered for 10 minutes, the average size of macropores decreases from 36 to 24 μm with increasing of MgO content. The average size of micropores decreases from 3.8 to 2.5 μm, Figure 2 (a). Increasing of sintering time up to 10 hours leads to decreasing of the macropores average size to 25 μm, due to the continuing volumetric shrinkage, however, the average size of micropores increases to 7 μm, which is a due to the pores coalescence mechanism [8], Figure 2 (b).

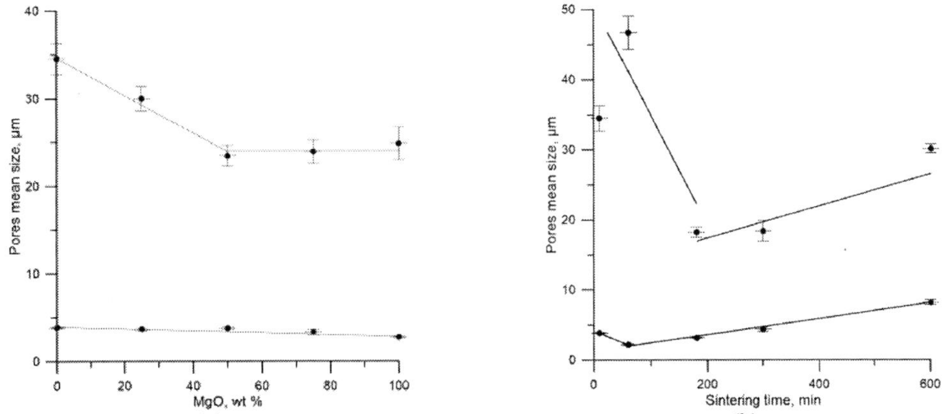

(a) (b)

Figure 2. The dependence of the average pore size on the MgO content after sintering for 10 minutes (a); dependence of the zirconia average pore size on the sintering time (b).

X-ray diffraction analysis showed that sintered zirconia is represented by the cubic phase with low-intensity peaks of monoclinic phase on the X-ray patterns of ZrO_2 samples sintered with holding more than 3 hours, which is corresponds with the ZrO_2 - MgO [9] state diagram. Increase of MgO content in the composite leads to a full stabilization of ZrO_2 in the cubic phase. Cubic zirconia lattice parameter was 5.090±0.017 Å, magnesia parameter was 4.213±0.009, which corresponds with JCPDS crystallographic base cards [Cubic ZrO_2 card # 27-997; Cubic MgO card # 4-829].

Figure 3 shows the changes of the coherently diffracting domains size and the lattice microdistortion of cubic ZrO_2 and MgO phases with varying sintering time. It can be seen that the cubic zirconia coherently diffracting domains size increases from 600 to 1600 Å, both with MgO content increasing and with time of sintering, figure 3 (a). The average size of MgO coherently diffracting domains varies from 900 to 1100 Å with increasing of MgO content in the composite and increases with the time of sintering, Figure 3 (b).

A cubic ZrO2 lattice microdistortion decreases from 14 to 4×10^{-4} with increasing of MgO content, Figure 1 (c). Increasing of sintering up to 10 hours leads to increasing of ZrO_2 lattice microdistortions, Figure 1 (d). MgO microdistortions decrease from 2 to $1 \cdot 10^{-4}$ with increasing of MgO content and time of sintering.

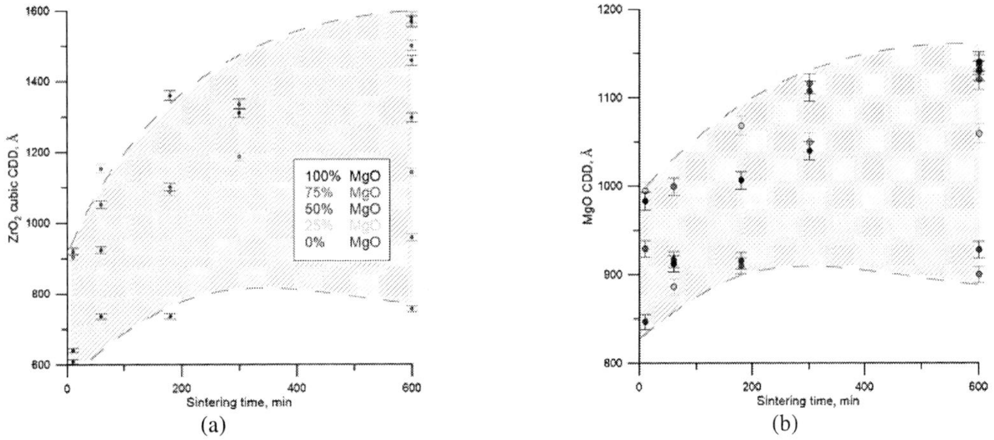

(a) (b)

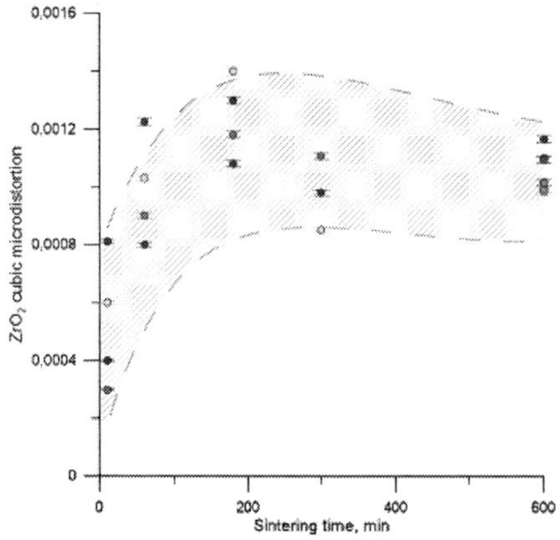

(c)

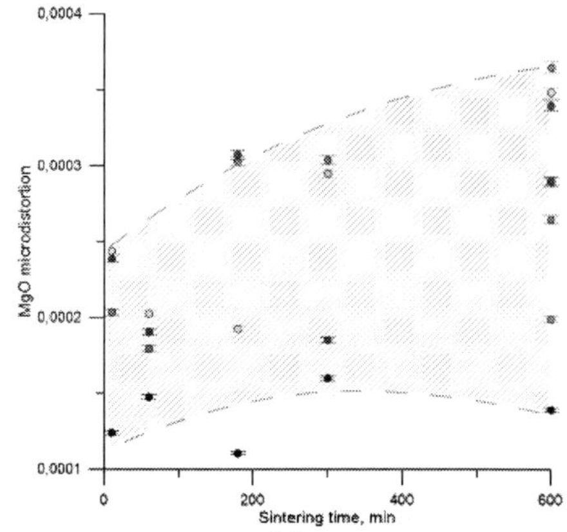

(d)

Figure 3. Cubic ZrO_2 CDD, the legend is valid for all the figures (a); MgO CDD(b); Cubic ZrO_2 microdistortion (c); MgO microdistortion (d).

Figure 4 shows the plotted data between the microstresses and macroscopic tensile strength. As one can seen there is a correlation between these values - microstresses, formed during sintering of the composite, determine the decreasing of macrostrength.

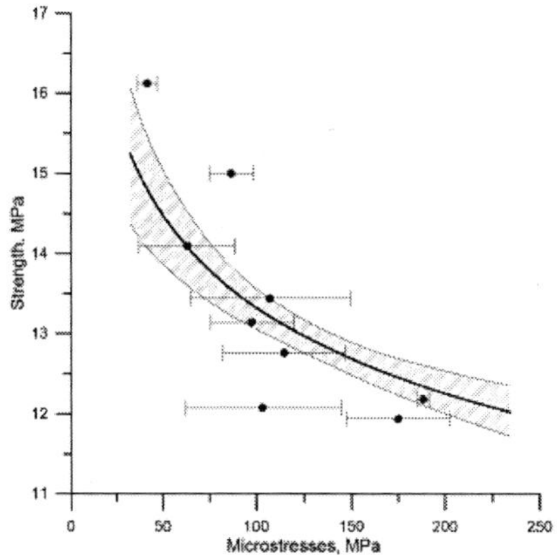

Figure 4. Dependence of average strength value on composite microstresses.

4. Conclusions

It has been shown that the adding of pore-forming particles into the initial powder makes it possible to obtain a bimodal pore structure of ceramics. Change of the sintering time and content of the ZrO_2-MgO leads to changing the average sizes pores, both macro- and micropores. Macro-strength of the material is determined by microstresses formed during sintering of the composite.

Acknowledgement

This work was carried out according to the Program III.23 of Fundamental Scientific Research of the State Russian Academy of Sciences for 2013-2020 and Tomsk State University competitiveness improvement program.

References

[1] Hadjicharalambous C, Prymak O, Loza K, Buyakov A, Kulkov S and Chatzinikolaidou M 2015 *Front. Bioeng. Biotechnol.* **3** 175 https://doi.org/10.3389/fbioe.2015.00175

[2] Kulkov S, Buyakova S, Chatzinikolaidou M and Kocserha I 2015 *Építőanyag-JSBCM* **67** (4) 155 http://dx.doi.org/10.14382/epitoanyag-jsbcm.2015.26

[3] Afanasy S Apkaryan A, Kulkov S and Gömze L 2014 *Építőanyag-JSBCM* **66** (2) 38 http://dx.doi.org/10.14382/epitoanyag-jsbcm.2014.8
[4] Anastassopoulou J and Theophanides T 2002 *Critical reviews in oncology/hematology* **42** 1 79 https://doi.org/10.1016/S1040-8428(02)00006-9
[5] Williamson G and Hall W 1953 *Acta Metall.* **1** 22 https://doi.org/10.1016/0001-6160(53)90006-6
[6] Dane J, Topp C, Flint A and Flint L 2002 *Methods of Soil Analysis: Part 4 Physical Methods* (Soil Science Society of America) 229
[7] Wang Q, Jia X, Kou S, Zhang Z and Lindqvist P 2004 *T Int. J. Rock Mech. Min. Sci.* **41** 245 https://doi.org/10.1016/S1365-1609(03)00093-5
[8] Xiong Y, Hu J and Shen Z 2013 *J. Eu. Cer. Soc.* **33** 11 https://doi.org/10.1016/j.jeurceramsoc.2013.03.015
[9] Grain C 1967 *J. Am. Ceram. Soc.* **50** 288 https://doi.org/10.1111/j.1151-2916.1967.tb15111.x

IOP Conf. Series: Materials Science and Engineering **613** (2019) 012024 doi:10.1088/1757-899X/613/1/012024

Studying effect of addition green inhibitor on compression strength of reinforced concrete

Shaymaa Abbas Abdulsada[1] and Tamás I. Török [1]

[1]Faculty of Materials Science and Engineering, University of Miskolc, Hungary.

E-mail: shaymaa.radhi@uokufa.edu.iq and fektt@uni-miskolc.hu

Abstract. In recent years the use of inhibitors in producing high-performance concrete has increased significantly to mitigate chloride and sulphate attacks. The present inhibitors in the market are toxic to the environment. Hence, the objective of the present investigation was to study a novel, eco-friendly and green waste plant (orange peel) extracts inhibitor and study its effects on compression strength of reinforced concrete. This inhibitor was added to the concrete mix with concentrations of 1% and 3% by weight of cement with two types of superplasticizers. Results showed that the tested penetrating type corrosion inhibitor as an admixture of the concrete samples did not impair on compressive strength, but with one of the superplasticizers showed higher (acceptable) values compared to the other type of superplasticizer due to the difference between their ability in reducing the ratio of water of the concrete samples.

1. Introduction

Corrosion of steel rebar in reinforced concrete structures is a big problem which might and cause significant human and economic losses. One of the most important reasons of corrosion in reinforced concrete is chloride attack. The main sources of chlorides are seawater and de-icing salts. Also, there are other sources of chlorides for example from gravels, sand, cement, water (being used for preparing the mixture of fresh concrete) and sometimes from (contaminated) steel rebar. In the subsistence of chloride, the protective passive stratum of steel is locally destroyed and the unprotected steel areas start dissolve. The formation of corroding products (rust) involves a substantial volume increase, i.e. the volume of corrosion products is greater than that of original steel bar. Therefore expansive stresses are induced around corroded steel bars causing possible cracking, spalling of concrete cover and loss of bond between steel/concrete (as shown in Fig. 1), and thus reducing the serviceability of concrete structures [1-8].

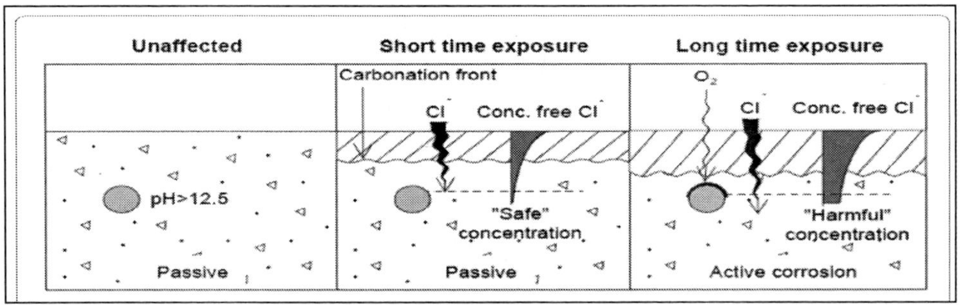

Figure 1. Corrosion by chloride ions and carbonation [8]

Among the many methods used to solve or reduce problems of corrosion in concrete, the use of corrosion inhibitors is gaining widespread acceptance due to its many advantages including ease of application, comparable lower costs and high effectiveness. Also, toxicity problems usually encountered from the use of inorganic chemical inhibitors are finding the solution via proposed alternatives of non-toxic organic chemicals or outright usage of green, environmentally-friendly extracts from natural plant materials, which are known to be rich in biocompatible organic chemicals [9]. The main purpose of this paper is to present the results of an experimental study on the compressive strengths to monitor and evaluate the effect of green inhibitor and superplasticizers of test samples prepared from concrete type XD3.

2. Materials and Methods

2.1 Materials and Samples Preparation

Portland slag cement CEM II/A-S 42.5R was used in this study conforming to the EN 197-1 [10] and it received from CRH Magyarország Kft. company in Miskolc, Hungary. Aggregates (fine and coarse) were used according standard EN 12620 [11] and it was also received from the CRH Magyarország Kft. company. Orange peels extract as a green inhibitor and two types of superplasticizers (Mapei Dynamon SR 31 and Oxydtron) were used. Tap water was used for both making and curing the specimens.

Concrete mixes were designed in accordance with the European mix design method (XD3 class) to have compressive strength C35/45 at age of 28 days. Two times three samples were prepared (with the two different plasticizers) and the two sets of three and three samples contained in 0%, 1%, 3% the green inhibitor as well as shown in Table 1. The concrete cubes so prepared for compressive strength testing had dimensions of 70x70mm. The molds were thoroughly cleaned and oiled before casting to avoid adhesion with the concrete surface. Mixing of materials was done manually after that water added to the mix with continued mixing, then the mix was put in the molds.

The specimens were taken out from the molds after 24 hours hardening, then were immersed in tap water for 28 day as in Fig.2.

Table 1. The concrete mixtures (specimens) prepared for this study

Type of Mixtures		
Symbol of Mix	Type of Admixture	Concentration of Green Inhibitor
A1 (Reference)	Mapei Dynamon SR 31	without
B1	Mapei Dynamon SR 31	1% by weight of cement
C1	Mapei Dynamon SR 31	3% by weight of cement
A2 (Reference)	Oxydtron	without
B2	Oxydtron	1% by weight of cement
C2	Oxydtron	3% by weight of cement

Figure 2. Curing of compressive strength specimens in tap water

2.2. *Compressive Strength Test*

The concrete compressive strength was measured by using a compressive strength testing machine (Kispesti Vas és Fém KTSZ) in the workshop of Institute of Ceramics and Polymers Engineering at Faculty of Materials Science and Engineering, University of Miskolc. The reported values are the average of three specimens for each age (nine for each mix).

3.Results and Disscusion

Fig.3 shows the results of compressive strength test at age 28 days for each mix.

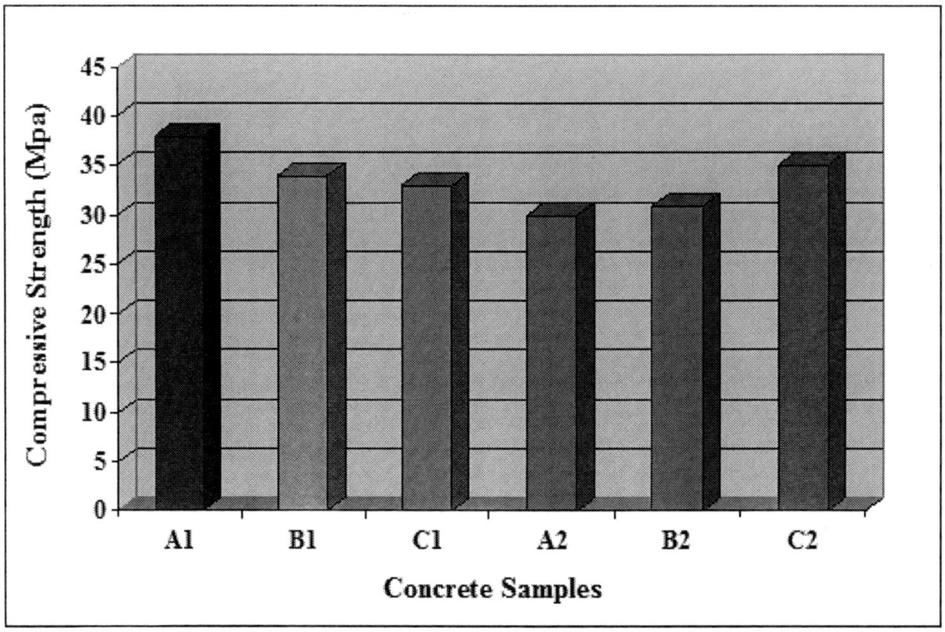

Figure 3. Compressive strength of concrete samples after immersion in tap water for 28 days

The results showed that: (1) For Dynamon SR 31 admixture in concrete there was not much reduction observed in the compressive strength. (Without this admixture the given type concrete should have compression strength of C35/45 MPa at age 28 days.)

By using the orange peels extract as corrosion inhibitor in conc. 1% and 3% by weight of cement, it reduced the compressive strength considerably by 10.5% for 1% addition and 13% for 3% addition, respectively. (2) For Oxydtron admixture with concrete causes reduction in compressive strength more than Dynamon SR 31 admixture in sample without green inhibitor. After using green inhibitor with 1% and 3% by weight of cement affect compressive strength with an increase 3.3% for 1% addition and 16.6% for 3% addition respectively. That's may be due to this inhibitor works as a retarder, it retards the action or effect of tricalcium silicate (C3S) or tricalcium aluminate (C3A) and these action modifies the compressive strength at the early ages of the concrete. The increase in compressive strength after addition green inhibitor because adding inhibitor mean decreasing the amount of cement (inhibitor adding to concrete by weight of cement%), so with increasing the concentration of inhibitor leads to decreasing the cement (decreasing the hydration components because it depends on water and cement) and already causing increasing the compressive strength.

Comparing the effect of superplastizers we found that Dynamon SR 31 caused reduction in water during the hardening process of concrete more than Oxydtron and these reduction in water caused increase in strength of concrete and also decreased the porosity as a consequence of a series of chemical reactions (called hydration) of cement with water to form the binding material. The hydration reaction of ordinary Portland cement (OPC) involves four major types of hydration components, i.e. tricalcium silicate (C3S), belite (C2S), tricalcium aluminate (C3A), and ferrite or brownmillerite (C4AF). More water in the concrete means more hydration components and decrease in the capillary porosity and finally decrease in strength of concrete because the strength depends also on the capillary porosity.

4.Conclusion

From compared effect of the two types of admixtures that used in this work we noted the compressive strength in general was a little bit higher with Dynamon SR 31 than with Oxydtron, that means the superplasticizer (Dynamon SR 31) caused greater reduction of water than Oxydtron and these reduction in water caused increase in strength of concrete and also must have decreased the porosity, but at some points it also might cause some risk because at these points it may introduce the risk of autogenous shrinkage and cracking.

Acknowledgement

We would like to thank everyone who helped us to complete this research especially the efforts of the Institute of Metallurgy and Institute of Ceramic and Polymer engineering, University of Miskolc in providing all the possibilities available to us in order to make the research successful.

Reference

[1] S A Abdulsada, T I Török and É Fazakas 2018 *Építőanyag-JSBCM* **70** (2) 48 https://doi.org/10.14382/epitoanyag-jsbcm.2018.10
[2] S A Abdulsada, É Fazakas and T I Török 2018 *International Journal of Corrosion and Scale Inhibition* **7** (1) 38 http://dx.doi.org/10.17675/2305-6894-2018-7-1-4
[3] S A Abdulsada, A I Al-Mosawi, A A Hadi, S Karrar and S Hassanein 2017 *Recent Trends in Civil Engineering & Technology* **7** (3) 53
[4] S A Abdulsada, A I Al-Mosawi and A A Hadi 2017 *Bioprocess Engineering* **1** (3) 81 https://doi.org/10.11648/j.be.20170103.14
[5] L Abosrra, A F Ashour and M Youseffi 2011 *Journal of Construction and Building Materials* **25** (10) 3915 https://doi.org/10.1016/j.conbuildmat.2011.04.023
[6] L Bertolini, B Elsener, P Pedeferri, E Redaelli and R B Polder 2014 Corrosion of Steel in Concrete: Prevention, Diagnosis, Repair *New York Wiley–Interscience* 2nd Edition
[7] J P Broomfield 2007 Corrosion of Steel in Concrete: Understanding, Investigation and Repair *London CRC Press Taylor & Francis* 2nd Edition
[8] M A Quraishi, DK Nayak, R Kumar and V Kumar 2017 *Journal of Steel Structures & Construction* **3** (1) 1 https://doi:10.4172/2472-0437.1000124
[9] O O Joshua, P I Abimbola and A L Cleophas 2017 *Energy Procedia* **119** 972 https://doi.org/10.1016/j.egypro.2017.07.130
[10] EN 197-1 2000 Cement-Part 1: Composition, specifications and conformity criteria for common cements *European Standard was Approved by CEN National Members*
[11] EN 12620 2002 Aggregates for Concrete *European Standard was Approved by CEN National Members*

Effect of composition and heat treatment on porosity and microstructures of technical ceramics made from kaolin and IG-017 additive

Emese Kurovics[1], László A. Gömze[1,2,3], Jamal Eldin F. M. Ibrahim[1] and Ludmila N. Gömze[2]

[1]Intsitute of Ceramics and Polymer Engineering, University of Miskolc, Hungary
[2] IGREX Engineering Service Ltd, Igrici, Hungary
[3]Tomsk State University, Russia

E-mail: femgomze@uni-miskolc.hu, mse205@gmail.com and igrex2009@yandex.ru

Abstract. Based on conventional kaolinite and IG-017 bio-original additives of IGREX Ltd. the authors have developed new ceramic composite materials for different industrial purposes. In this work, different powder mixtures of kaolinite and IG-017 bio-original additive were milled and uniaxially pressed at different compaction pressures into cylindrical discs, after compaction, the discs were sintered in electric kiln under oxidation atmosphere and in oxygen-free atmosphere. Using the oxygen-free sintering process, new high porosity ceramic materials were created. Through the examination of the microstructure of the produced specimens the authors have found that the ceramic structure is reinforced with carbon nanofibers.

1. Introduction

Recently, large amount of ceramic materials and ceramic matrices composites (CMC) are developed [1-11] and used in the industries because of their low density [12-15], hardness [17-20], wear resistance [21-29], toughness [30-34], mechanical strength [35-44], chemical [45-49], biological [50-52] and physical [53-60] properties. Many researches have been conducted to find low-cost solutions for the manufacture of porous ceramic composites, such as the combination of kaolin and alumina raw materials, to increase the toughness [61-63]. For these purposes the authors generally use highly expensive materials and costly technical equipment and processes [64-71]. Nowadays, it is necessary to produce high performance ceramic composites with reasonably priced complex materials for extreme consumptions and applications [72-85].

In this research work, our aims are to develop high porosity ceramic matrix materials on the basis of not expensive conventional kaolinite and **IG-017** additive powders and to find optimal material and morphological structures.

2. Materials and experiments

To develop new technical ceramic materials with high mechanical performance commercial kaolinite powders KKA-HB and IG-017 refined bio-original additives of IGREX Ltd were used. Figure 1 shows the grain size structures and distributions of used materials made by Malvern Mastersizer X Type Laser Granulometer.

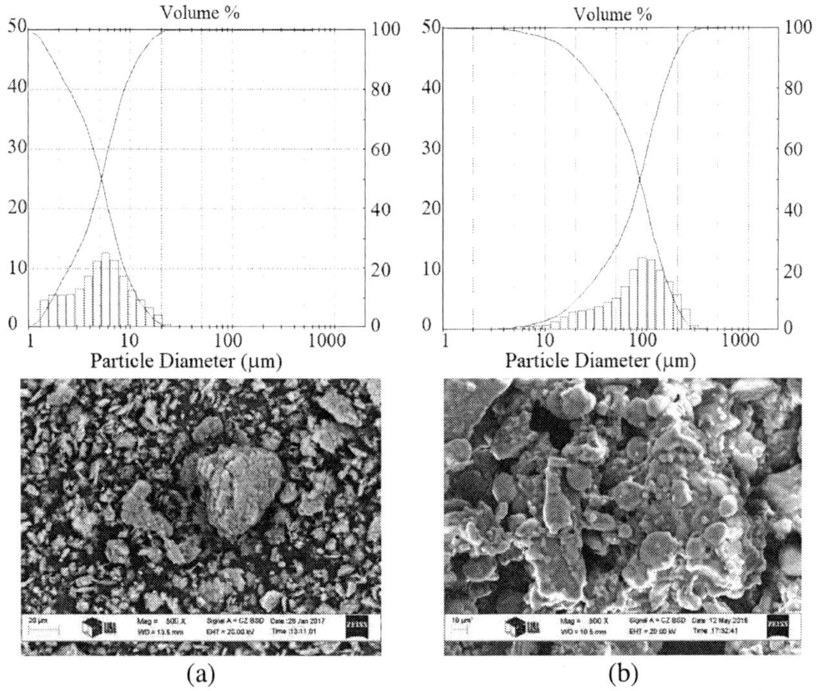

Figure 1. Morphology and grain size distribution of the used kaolinite (a) and IG-017 additive (b)

To optimize the heat treatment, sintering temperature curves and cycles, preliminary thermo-analytical tests were done for both components before their mixing and mechanical activating (Fig. 2). The thermo-analytical properties were measured by MOM Derivatograph-C. The achieved DTA curve of our high purity kaolinite powder was fully confirm the research works of Gabriel Varga and Igor Štubna [86, 87], where the dehydration process starts at about 452 °C and finish at 973 °C. In the present case the dehydration process is mostly intensive at 529.7 °C, meanwhile the IG-017 additive is highly loss the surface-absorbed moisture at 87.4 °C. At the same time the DTG curve shows that, the starting temperature of outgassing process of organic components is 242.8 °C, while the most intensive outgassing temperature is 281.6 °C. The organic component is mostly burned and generated organic gasses at 320.5 °C. Later, the temperature zones which achieved on the derivatograph were taken into the consideration during sintering of the uniaxially compacted cylindrical specimens.

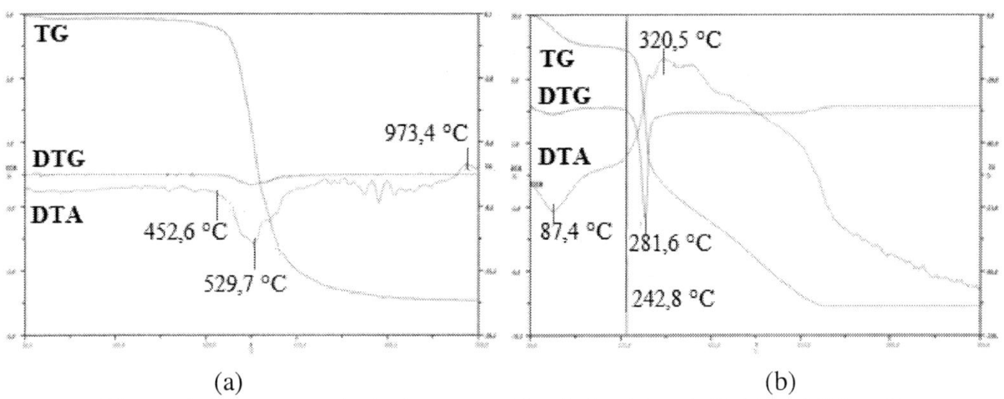

Figure 2. Thermo-analytical curves of kaolinite (a) and IG-017 additive (b)

7 different mixtures were prepared from the kaolin and IG-017 additives using the compositions in Table 1. The different prepared mixtures were mechanically activated in a planetary ball mill for 20 minutes at 200 rpm. After mechanical activation of the mixtures 24 cylindrical disc specimens with diameters of 20 mm for each mixture were compacted uniaxially using 100 kN mechanical pull-press.

Table 1. The mixtures compositions in m%

Number of mixtures	KKA-HB Kaolinite	IG 017 bio original material
K I	85	15
K II	75	25
K III	65	35
K IV	55	45
K V	45	55
K VI	35	65
K VII	25	75

Three different compaction forces (50 kN, 60 kN and 70 kN) were used to develop compression pressures of 156.21 MPa, 187.45 MPa and 218.69 MPa to produce cylindrical discs of 8 grams in each case. After compacting half of the specimens were fired (sintered) in normal (oxidation) atmosphere while the other half of the specimens were sintered in oxygen-free (quasi-reduction) atmosphere. In both cases the maximum sintering temperatures were 1250 °C on heating rate of 100 °C/hours. This sintering temperature was high enough for formation of mullite crystal phases from kaolinite, but not enough for creation of cristobalite crystals from the formed free SiO_2 component which is created from the degraded kaolinite at about 1080 °C [88]. In their previous work, the authors have already presented the color change, volume shrinkage, loss of mass and density of ceramic specimens depending on firing temperatures and atmospheres [89]. During the sintering processes the sintered specimens have changed the color (Figure 3), the specimens which sintered in normal (air) atmosphere give white color, while the specimens which sintered in oxygen-free (quasi-reduction) atmosphere give black color.

Sintered

IG-017 additive in m%

15 45 75

in normal (air) atmosphere

in oxygen-free (quasi-reduction) atmosphere

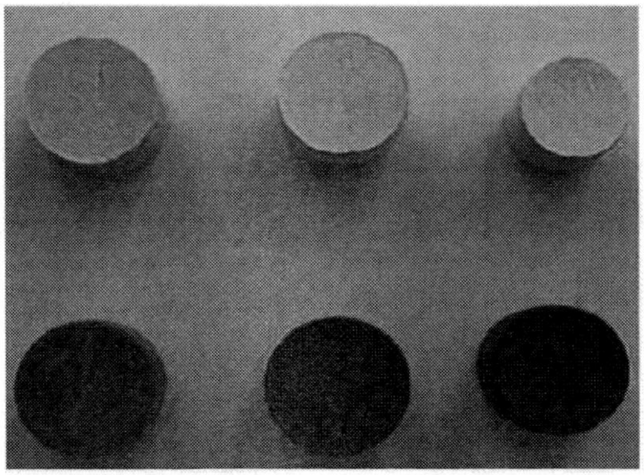

Figure 3. The specimens after the sintering

3. Results and discussions

The porosities of sintered specimens were determined based on water absorption capacity (Table 2) and the following formula was used for the calculations:

$$P = \frac{m_w - m_s}{V_s} \cdot 100\% \qquad (1)$$

Where P the porosity, m_s the mass of sintered specimens, m_w the mass of the tested specimens after boiling in distilled water for 2.5 hours and V_s the geometry volume of the sintered specimens.

Table 2. The porosities of the sintered specimens in %

Compacting pressures		**156.21MPa**		**187.45MPa**		**218.69MPa**	
Sintering in		**oxidation atm.**	**oxygen-free atm.**	**oxidation atm.**	**oxygen-free atm.**	**oxidation atm.**	**oxygen-free atm.**
Bio-original additive, m%	**15**	10.02	5.86	3.34	11.72	3.34	17.57
	25	6.45	16.64	6.47	13.87	9.71	16.64
	35	16.28	15.74	16.28	18.37	19.54	20.99
	45	26.75	20.42	23.41	20.42	30.10	22.97
	55	26.13	18.17	33.60	23.36	37.34	38.94
	65	31.83	21.99	40.93	27.48	45.48	41.23
	75	16.38	20.90	38.22	32.84	60.06	47.76

The porosities of sintered samples at the low mixing ratio of IG-017 did not depended greatly on the compacting pressures (Table 2 and Figure 4). In the case when the specimens have contained 55-75 m% bio materials, the degrees of porosity were significantly influenced by the applied compaction pressures.

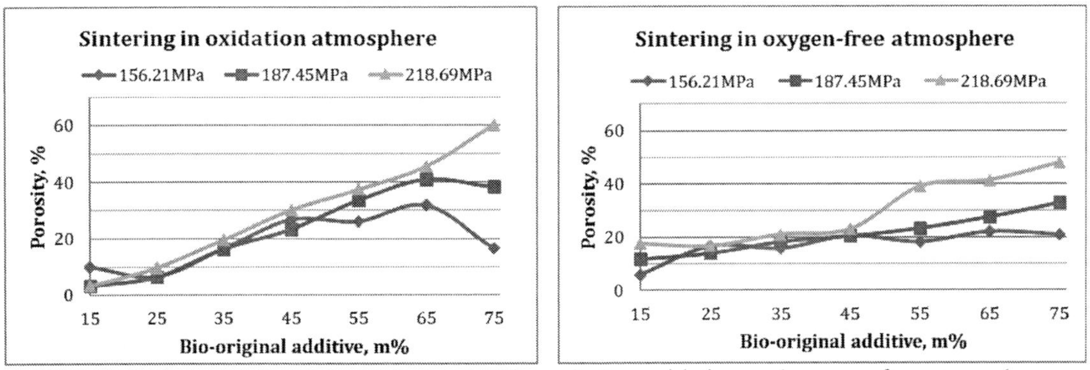

Figure 4. The porosity of the sintered test pieces in the oxidation and oxygen-free atmosphere

The surfaces of the sintered samples were examined by scanning electron microscopy to determine how the microstructures and the elemental compositions were changing depending on sintering conditions (atmosphere) and applied mixture compositions. In Figure 5 and in Table 3 big differences can be seen in the fracture surfaces due to the difference in the material compositions of specimens which were pre-sintered in normal atmosphere and in oxygen-free atmosphere. With increasing the volume of IG-017 additives in the mixtures, the porosity of the sintered samples has increased significantly.

The used amount of IG-017 additives have strongly influenced the carbon content of the sintered specimens in the oxygen-free atmosphere (23 % carbon in the case of samples with 75 % IG-017 content). Moreover, the different heat treatment conditions have a significant effect on the microstructure and elemental composition of the prepared and sintered ceramic specimens.

The chemical and the phase compositions of the sintered specimens were determined by XRD tests. In each cases the main phase was found to be mullite, together with a large amount of amorphous glass and small amount of quartz or cristobalite. During oxygen-free sintering, carbon fibers were formed (Table 4).

Sintering atmosphere | **normal (oxidation)** | **oxygen-free**

KI (15m% IG-017)

KIV (45m% IG-017)

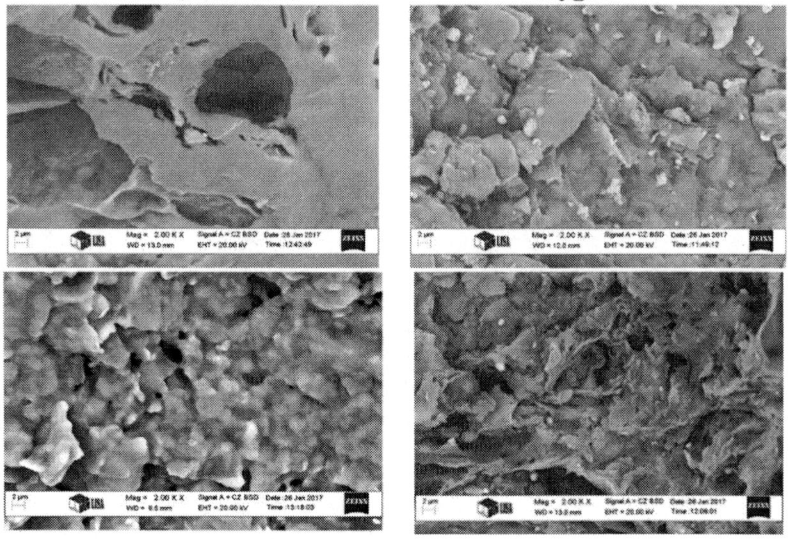

Figure 5. The microstructure of the sintered specimens

Table 3. The chemical compositions of the specimens from KI and KIV mixtures determined by EDAX

Mixture	Sintering atmosphere		C	O	Al	Si	K	Ti	Fe
KI (15m% IG-017)	normal (oxidation)	Wt %	4.62	34.96	24.22	32.19	1.89	0.91	1.21
		At %	8.17	46.47	19.09	24.37	1.03	0.40	0.46
	oxygen-free	Wt %	6.93	31.06	23.72	32.96	2.52	1.13	1.67
		At %	12.31	41.40	18.75	25.03	1.37	0.50	0.64
KIV (45m% IG-017)	normal (oxidation)	Wt %	1.51	34.62	25.67	33.79	2.13	1.18	1.11
		At %	2.77	47.63	20.94	26.48	1.20	0.54	0.44
	oxygen-free	Wt %	23.29	29.24	18.94	25.21	1.47	0.63	1.23
		At %	35.65	33.6	12.91	16.5	0.69	0.24	0.41

Table 4. The oxide- and phase composition of the sintered specimens from KIV mixture

Sintered atm.	Phase %	SUM	Mullite	Quartz	Cristobalite	Amorph glass	Carbon fiber
			$Al_6Si_2O_{13}$	SiO_2	SiO_2	SiO_2	C
Oxidation		**100**	**53**	**2**	**5**	**40**	**0**
	SiO_2	61.95	14.95	2	5	40	
	Al_2O_3	38.05	38.05				
Oxygen-free		**100**	**48**	**4**	**1**	**32**	**15**
	SiO_2	50.54	13.54	4	1	32	
	Al_2O_3	34.46	34.46				
	CO_2	54.96					54.96

4. Conclusions

In this research work, the authors have successfully developed and produced a high porosity ceramic matrix material using conventional KKA-HB kaolin and IG-017 bio-original additives. The volume of porosity is found to be very strongly depending not only on compacting pressures (156.21 MPa, 187.45 MPa and 218.69 MPa) and mixture ratio but also on sintering conditions as well. The samples which sintered in oxygen-free atmosphere using 156.21 MPa compression pressure give a porosity of 20V% while the same samples which compressed at 218.69 MPa give a porosity of 50 V% depending on quantity of the used additives. The reason of this larger porosity phenomena is that the generated gases from firing of IG-017 additive could not evaporate through the high dense surface of specimens during their sintering in oxygen-free atmosphere.

The carbon contents of specimens became insignificant when normal (oxidation) atmosphere was used for sintering, meanwhile using oxygen-free atmosphere and IG-017 additives lead to increase the quantity and volume of carbon contents of specimens up to 23.29 m%.

Acknowledgement

"The described study was carried out as part of the EFOP-3.6.1-16-2016-00011 "Younger and Renewing University – Innovative Knowledge City – institutional development of the University of Miskolc aiming at intelligent specialization" project implemented in the framework of the Szechenyi 2020 program. The realization of this project is supported by the European Union, co-financed by the European Social Fund."

References

[1] Dong Q, Yin S, Guo C, Wu X, Kimura Ti, Le T, Sakanakura T and Sato T 2013 *IOP Conf. Ser.: Mater. Sci. Eng.* **47** 012065 http://dx.doi.org/10.1088/1757-899X/47/1/012065
[2] Rocha-Rangel E, Hernández-Silva D, Terés-Rojas E, Martines-Franco E and Diaz-De La Torre S 2010 *Építőanyag-JSBCM* **62** (3) 75 http://dx.doi.org/10.14382/epitoanyag-jsbcm.2010.15
[3] Hakeem A S, Khan R M A, MAl-Malki M, Patel F, Bakare A I, Ali S, Hampshire S and Laoui T 2014 *Advancesin Science and Technology* **89** 63
[4] Bachar A, Mercier C, Tricoteaux A, Leriche A, Follet C and Hampshire S 2016 J. *Eur. Ceram. Soc.* **36** (12) 2869
[5] Kalatur E S, Kozlova A V, Buyakova S P and Kulkov S N 2013 *IOP Conf. Ser.: Mater. Sci. Eng.* **47** 012004 http://dx.doi.org/10.1088/1757-899X/47/1/012004
[6] Zharikov E V, Zaramenskikh K S, Popova N A, Faikov P P, Iskhakova L D, Gerke M N, Kutrovskaya S V and Nogtev D S 2011 *Glass and Ceramics* **68** 80
[7] O Kotova, A Shmakova and A Ponaryadov 2017 *IOP Conf. Ser.: Mater. Sci. Eng.* **175** 012002 https://doi.org/10.1088/1757-899X/175/1/012002
[8] A G Knyazeva, G A Pribytkov, Yu A Chumakov 2018 *Építőanyag-JSBCM* **70** (3) 74 http://dx.doi.org/10.14382/epitoanyag-jsbcm.2018.13
[9] M G Abd-Almutalib Al-Mosawy, E A Al-Mulla 2018 *Építőanyag-JSBCM* **70** (4) 116 http://dx.doi.org/10.14382/epitoanyag-jsbcm.2018.21
[10] Jr. D P Penaloza, T AP Seery 2018 *Építőanyag-JSBCM* **70** (5) 140 http://dx.doi.org/10.14382/epitoanyag-jsbcm.2018.26
[11] Jr. D P Penaloza 2019 *Építőanyag-JSBCM* **71** (1) 5 http://dx.doi.org/10.14382/epitoanyag-jsbcm.2019.2
[12] Katahira K, Watanabe Y, Ohmori H and Kato T 2002 *Int. Journal of Machine Tools and Manufacture* **42** 1307
[13] Carter C B and Norton M G 2007 *Ceramic Materials: Science and Engineering Springer* 1-716
[14] Gömze A L and Gömze N L 2009 *Építőanyag-JSBCM*, **61** (2) 38 http://dx.doi.org/10.14382/epitoanyag-jsbcm.2009.7

[15] Chuklina S G, Pylinina A I, Podzorova L I, Mikhailina N A, and Mikhalenko I I 2016 *Russian Journal of PhysicalChemistry* **90** (12) 2370
[16] Sadowski T and Golewski P 2012 *Comp. Mater. Sci.* **64** 285
[17] Skrovanek S D and Bradt R C 1979 *J. Am. Ceram. Coc.* **62** (3-4) 215
[18] Gubernat A, Stobierski L and Labaj P 2007 *J. Europ. Ceram. Soc.* **27** 781
[19] Z Koncsik, MB Maros, L Kuzsella 2010 *Materials Science Forum* **659** 313 https://doi.org/10.4028/www.scientific.net/MSF.659.313
[20] Liu J, Fu Z, Wang W, Zhang J, Wang H, Wang Y, Lee S and Niihara K 2014 J. *Eur. Ceram. Soc.* **34** 3095 http://dx.doi.org/10.1016/j.jeurceramsoc.2014.04.004
[21] Shinoda S, Aoba T, Suzuki T, Nakayama T, Suematsu H and Niihara K 2017 *Jpn. J. Appl. Phys.* **56** 019201 https://doi.org/10.7567/JJAP.56.019201
[22] Kulkov S and Savchenko N 2008 *Építőanyag-JSBCM* **60** (3) 62 http://dx.doi.org/10.14382/epitoanyag-jsbcm.2008.10
[23] Ershova N I and Kelina I Y 2009 *Építőanyag-JSBCM* **61**(2) 34 http://dx.doi.org/10.14382/epitoanyag-jsbcm.2009.6
[24] Urakov A, Urakova N, Reshetnikov A, Kasatkin A, Kopylov M, Baimurzin D 2016 *Építőanyag-JSBCM* **68** (3) 110 http://dx.doi.org/10.14382/epitoanyag-jsbcm.2016.19
[25] Vereschaka A, Kutin A, Sitnikov N, Oganyan G, Sharipov O 2016 *Építőanyag-JSBCM* **68** (3) 114 http://dx.doi.org/10.14382/epitoanyag-jsbcm.2016.20
[26] Isaev A, Grechishnikov V, Pivkin P, Ilyukhin Y, Kozochkin M, Peretyagin P 2016 *Építőanyag-JSBCM* **68** (2) 146 http://dx.doi.org/10.14382/epitoanyag-jsbcm.2016.8
[27] M Maros, AK Németh, Z Károly, E Bódis, Z Maros, O Tapasztó, K Balázsi 2016 *Tribology International* **93** (A) 269 https://doi.org/10.1016/j.triboint.2015.08.041
[28] Z Koncsik, MB Maros, L Kuzsella, A Kovács 2016 *IOP Conf. Ser.: Mater. Sci. Eng.* **123** 012033 https://doi.org/10.1088/1757-899X/123/1/012033
[29] MB Maros, AK Németh 2017 *Journal of the European Ceramic Society* **37** (14) 4357 https://doi.org/10.1016/j.jeurceramsoc.2017.05.005
[30] Izhevskyi V A, Genova L A, Bressiani J C and Bressiani A H 2000 *Cerâmica* **46** 297
[31] MB Maros, NK Helmeczi, BG Lenkey, P Arató 2005 *Key Engineering Materials* **290** 304 https://doi.org/10.4028/www.scientific.net/KEM.290.304
[32] Zerr A, Riedel R, Sekine T, Lowther J E, Ching W Y and Tanaka I 2006 *Adv. Mater.* 18 2933
[33] MB Maros, N Kaulics, P Arató 2007 *Ceramic Transactions* **199** 421 https://doi.org/10.1002/9781118144152.ch34
[34] Sadowski T and Golewski P 2015 *Archives of Metallurgy and Materials* **60** (4) 2737
[35] MB Maros, N Kaulics, J Dusza 2007 *Fractography of Glasses and Ceramics V: Ceramic Transactions* **199** 435 https://doi.org/10.1002/9781118144152.ch35
[36] MB Maros, NK Helmeczi, J Dusza 2008 *Materials Science Forum* **589** 73 https://doi.org/10.4028/www.scientific.net/MSF.589.73
[37] MB Maros, NK Helmeczi, P Arató, C Balázsi 2009 *Key Engineering Materials* **409** 338 https://doi.org/10.4028/www.scientific.net/KEM.409.338
[38] Gömze A L and Gömze N L 2011 *IOP Conf. Ser.: Mater. Sci. Eng.***18** 082001 http://dx.doi.org/10.1088/1757-899X/18/8/082001
[39] Yakushev V V, Zhukov A N, Utkin A V, Rogacheva A I and Kudakina V A 2015 *Combustion, Explosion and Shock Waves* **51** (5) 603
[40] Khasanov O, Reichel U, Dvilis E and Khasanov A 2011 *IOP Conf. Ser.: Mater. Sci. Eng.* **18** 082004 http://dx.doi.org/10.1088/1757-899X/18/8/082004
[41] Miranda-Hernández J G, Rocha-Rangel E and de la Torre S D 2010 *Építőanyag-JSBCM* **62** (1) 2 http://dx.doi.org/10.14382/epitoanyag-jsbcm.2010.1
[42] L A Gömze and L N Gömze 2017 *IOP Conf. Ser.: Mater. Sci. Eng.* **175** 012001 https://doi.org/10.1088/1757-899X/175/1/012001

[43] L A Gömze, S N Kulkov, E Kurovics, A S Buyakov, A Y Buzimov, M V Grigoriev, B I Kanev, T V Kolmakova, R V Levkov, S A Sitkevich 2018 *Építőanyag-JSBCM* **70** (1) 13 https://doi.org/10.14382/epitoanyag-jsbcm.2018.3
[44] M V Grigoriev, N L Savchenko, T Yu Sablina, E Kurovics, I N Sevostyanova, S P Buyakova, L A Gömze, S N Kulkov 2018 *Építőanyag-JSBCM* **70** (1) 18 https://doi.org/10.14382/epitoanyag-jsbcm.2018.4
[45] Tariq F, Rafique U, Yaqoob K 2017 *Építőanyag-JSBCM* **69** (3) 94 http://dx.doi.org/10.14382/epitoanyag-jsbcm.2017.16
[46] Ayub N, Rafique U 2017 *Építőanyag-JSBCM* **69** (3) 98 http://dx.doi.org/10.14382/epitoanyag-jsbcm.2017.17
[47] Tkachev G and Tkacheva O N 2009 *Steklo i keramika* **66** (2) 15
[48] A A Uspensky, S Yavshits, V Lipin, P Zhigalov and A Slobodov 2017 *IOP Conf. Ser.: Mater. Sci. Eng.* **175** 012018 https://doi.org/10.1088/1757-899X/175/1/012018
[49] R Gorshkova, D Khalikov, D Slobodova, A B Uspensky and A Slobodov 2018 *J. Phys.: Conf. Ser.* **1045** 012014 https://doi.org/10.1088/1742-6596/1045/1/012014
[50] Torres-Cadenas S, Bravo-Patiño A, Zarate-Medina J, Contreras-García M E 2017 *Építőanyag-JSBCM* **69** (1) 2 http://dx.doi.org/10.14382/epitoanyag-jsbcm.2017.1
[51] Torres-Romero A, Cajero-Juárez M, Contreras-García M E 2017 *Építőanyag-JSBCM* **69** (1) 8 http://dx.doi.org/10.14382/epitoanyag-jsbcm.2017.2
[52] E Kurovics, A Shmakova, B Kanev and L A Gömze 2017 *IOP Conf. Ser.: Mater. Sci. Eng.* **175** 012013 https://doi.org/10.1088/1757-899X/175/1/012013
[53] Zhukov I, Buyakova S P, Kulkov N S 2016 *Építőanyag-JSBCM* **68** (3) 74 http://dx.doi.org/10.14382/epitoanyag-jsbcm.2016.13
[54] Chigvinadze J G, Acrivos J V, Ashimov S M, Gulamova D, Machaidze T V and Uskenbaev D 2009 *Építőanyag-JSBCM* **61** (4) 104 http://dx.doi.org/10.14382/epitoanyag-jsbcm.2009.19
[55] Buyakova S P, Kalatur E S, Buyakov A S, Kulkov N S 2016 *Építőanyag-JSBCM* **68** (3) 70 http://dx.doi.org/10.14382/epitoanyag-jsbcm.2016.12
[56] Dedova E, Klevtsova E and Kulkov S 2013 *IOP Conf. Ser.: Mater. Sci. Eng.* **47** 012021 http://dx.doi.org/10.1088/1757-899X/47/1/012021
[57] A Shmakova, B Kanev, A L Gömze and O Kotova 2017 *IOP Conf. Ser.: Mater. Sci. Eng.* **175** 012015 https://doi.org/10.1088/1757-899X/175/1/012015
[58] E Kurovics, S N Kulkov, L A Gömze 2018 *Építőanyag-JSBCM* **70** (1) 3 https://doi.org/10.14382/epitoanyag-jsbcm.2018.1
[59] A Y Buzimov, W Eckl, L A Gömze, I Kocserha, E Kurovics, A S Kulkov, S N Kulkov 2018 *Építőanyag-JSBCM* **70** (1) 23 https://doi.org/10.14382/epitoanyag-jsbcm.2018.5
[60] A S Buyakov, R V Levkov, S P Buyakova, E Kurovics, L A Gömze, S N Kulkov 2018 *Építőanyag-JSBCM* **70** (1) 27https://doi.org/10.14382/epitoanyag-jsbcm.2018.6
[61] Li J, Yuan W J, Deng C J and Zhu H X 2013 *IOP Conf. Ser.: Mater. Sci. Eng.* **47** 012060 http://dx.doi.org/10.1088/1757-899X/47/1/012060
[62] Djangang C N, Kamseu E, Ndikontar M K, Nana G L L, Soro J, Melo U C, Elimbi A, Blanchart P és Njopwouo D 2011 *Materials Science and Engineering A* 528 (2011) 8311–8318 http://dx.doi.org/10.1016/j.msea.2011.07.006
[63] Ieva Zake-Tilhttp://dx.doi.org/an, Ruta Svinka, Visvaldis Svinka 2014 *Ceramics International* 40 3071–3077 http://dx.doi.org/10.1016/j.ceramint.2013.09.139
[64] MB Maros, JK Babcsan 2002 *Proceedings of the 11th Int. Conf. Ach. Mech. Mat. Eng. AMME* 303 ISBN:8391445879
[65] Tariq F, Rafique U, Yaqoob K 2017 *Építőanyag-JSBCM* **69** (3) 94 http://dx.doi.org/10.14382/epitoanyag-jsbcm.2017.16
[66] Bendoukha M, Mosbah M 2017 *Építőanyag-JSBCM* **69** (4) 122 http://dx.doi.org/10.14382/epitoanyag-jsbcm.2017.22

[67] Vunain E, Mishra S B, Mishra A K, Mamba B B 2017 Nanoceramic: Fundamentals and Advenced Perspectives, In Book: A K Mishra Sol-gel Nanoceramic Materials: Preparation, Properties and Applications, *Springer* pp.1-20 http://dx.doi.org/10.1007/978-3-319-49512-5_1
[68] Buyakov A S, Kulkov S N 2017 *IOP Conf. Ser.: Mater. Sci. Eng* **175** 012025 http://dx.doi.org/10.1088/1757-899X/175/1/012025
[69] Sadowski T, Golewski P 2017 *Arch. Metall. Mater.* **62** (3) 1577 http://dx.doi.org/10.1515/amm-2017-0241
[70] Torrisi A, Wachulak P W, Bartnik A, Fok T, Wegrzynski L, Fiedorowicz H, Mazzillo M, Sciuto A, Torrisi L 2017 *IEEE Transactions on Electron Devices* **64** (3) 1120 http://dx.doi.org/10.1109/TED.2017.2647780
[71] Eligiusz Postek, Tomasz Sadowski 2019 *Composite Structures* **213** 231 https://doi.org/10.1016/j.compstruct.2019.01.084
[72] Devaux X, Yu Tsareva S, Kovalenko A N, Zaramenskih K S, McRae E and Zharikov E V 2012 *Physica* E44 1028
[73] Zharikov E V, Duong T T T, Faikov P P, Popova N A and Sovyk D M 2016 *Inorganic Materials: Applied Research* **7** 163
[74] Gömze A L, Gömze N L 2013 *IOP Conf. Ser.: Mater. Sci. Eng.* **47** 012033 http://dx.doi.org/10.1088/1757-899X/47/1/012033
[75] Kulkov N S 2013 *IOP Conf. Ser.: Mater. Sci. Eng.* **47** 012042 http://dx.doi.org/10.1088/1757-899X/47/1/012042
[76] Hakuta Y, Nagai N, Suzuki Y-H, Kodaira T, Bando K K, Takashima H, Mizukami F 2013 *IOP Conf. Ser.: Mater. Sci. Eng.* **47** 012045 http://dx.doi.org/10.1088/1757-899X/47/1/012045
[77] Kovács E, Trencsényi R, Gulácsi Zs 2013 *IOP Conf. Ser.: Mater. Sci. Eng.* **47** 012048 http://dx.doi.org/10.1088/1757-899X/47/1/012048
[78] Yuan W, Yu C, Deng C, Zhu H 2013 *IOP Conf. Ser.: Mater. Sci. Eng.* **47** 012058 http://dx.doi.org/10.1088/1757-899X/47/1/012058
[79] Yu C, Deng C, Yuan W, Zhu H 2013 *IOP Conf. Ser.: Mater. Sci. Eng.* **47** 012059 http://dx.doi.org/10.1088/1757-899X/47/1/012059
[80] Teng X, Liu H, Huang C 2007 *Materials Science and Engineering A* 452–453 545–551 http://dx.doi.org/10.1016/j.msea.2006.10.073
[81] Radhi M M, Al-Dulimy W A G, Khalaf M S 2016 *Építőanyag-JSBCM* **68** (3) 90 http://dx.doi.org/10.14382/epitoanyag-jsbcm.2016.16
[82] Rodríguez Torres G M, Zarate Medina J, Contreras García M E 2016 *Építőanyag-JSBCM* **68** (4) 120 http://dx.doi.org/10.14382/epitoanyag-jsbcm.2016.21
[83] Smolin I Yi, Makorov P V, Eremin M O, Matyko K S 2016 *Procedia Structural Integrity* **2** 3353
[84] Gömze A L, Gömze N L 2010 *Építőanyag-JSBCM* **62** (4) 98 http://dx.doi.org/10.14382/epitoanyag-jsbcm.2010.18
[85] S Baba, T Goto, S Cho, T Sekino 2018 *Építőanyag-JSBCM* **70** (6) 195 http://dx.doi.org/10.14382/epitoanyag-jsbcm.2019.2
[86] Štubna I, Varga G, Trník A 2006 *Építőanyag-JSBCM* **58** (1) 6 http://dx.doi.org/10.14382/epitoanyag-jsbcm.2006.2
[87] Varga G 2007 *Építőanyag-JSBCM* **59** (1) 6 http://dx.doi.org/10.14382/epitoanyag-jsbcm.2007.2
[88] Kurovics E, Gömze A L 2017 *MultiScience - XXXI. microCAD International Multidisciplinary Scientific Conference* University of Miskolc https://www.researchgate.net/publication/316275265
[89] Kurovics E, Gömze A L 2017 *MultiScience - XXXI. microCAD International Multidisciplinary Scientific Conference* University of Miskolc https://www.researchgate.net/publication/318787305

IOP Conf. Series: Materials Science and Engineering **613** (2019) 012026 doi:10.1088/1757-899X/613/1/012026

Investigation of mechanical properties of porous alumina ceramics: experiment and simulation

V A Mikushina[1,2], I Yu Smolin[1,2] and A S Kulkov[1,2]

[1]National Research Tomsk State University, 634050 Tomsk, Russia
[2]Institute of Strength Physics and Materials Science SB RAS, 634055 Tomsk, Russia

E-mail: mikushina_93@mail.ru

Abstract. The paper presents the experimental data on elastic and strength characteristics of specimens made of porous Al_2O_3 ceramics in three-point bending. Ceramics specimens were prepared by the slurry casting of commercially pure Al_2O_3 powder and subsequent sintering at the temperatures 1400, 1500, and 1600 °C. The simulation results were reported for the mechanical behaviour of specimens of porous Al_2O_3-based ceramics at the mesolevel. We studied the mesovolumes with an explicit treatment of pores loaded in uniaxial tension. The pore structure of the mesovolumes was borrowed from experimental SEM images. The mesoscopic fracture was described using the fracture criterion which includes the Lode parameter to make allowance for the stress state type. Despite wide scatter, the experimental and calculated elastic and strength characteristics of the material agree rather well.

1. Introduction

Now porous ceramic materials are of interest for different technical applications, e.g., filters, membranes, insulators or biomedical implants. Porosity is one of the main microstructural parameters for ceramics. The effect of porosity on the variation of elastic and strength properties of porous materials have been studied by numerous researchers, using various mathematical models for describing inelastic deformation and fracture of these materials as well as different experimental techniques [1–5]. Wide use has been made of models of physical mesomechanics for the numerical study of the deformation and fracture features and to predict the mechanical behaviour of heterogeneous materials at different length scales [6–13].

The uniaxial tensile test for the brittle material is impossible because of their traits. Therefore the three-point bending flexural test is used to assess the mechanical characteristics of these materials intension. This method provides values for the modulus of elasticity in bending and flexural stress of the material.

In this study, the experimentally and numerically properties of the Al_2O_3-based porous ceramic specimens in three-point bending was investigated. The rectangular specimens with various values of porosity ranging from 17 to 33 % were chosen for the bending test.

2. The results of experimental investigation

Ceramic specimens were made by the slurry casting of commercially pure Al_2O_3 powder and subsequent sintering at the three temperatures 1400, 1500, and 1600 °C during an hour. The produced specimens had the form of rectangular parallelepipeds. The initial dimensions of the specimens were 65×7×5 mm. The specimens were tested in three-point bending to determine their mechanical properties. The porosity of specimens was 33, 26, and 17 % for the sintering temperatures 1400, 1500, and 1600 °C, respectively. The Image J program was used to determine the porosity of specimens. SEM images of the surface of Al_2O_3 specimens are shown in Figure 1.

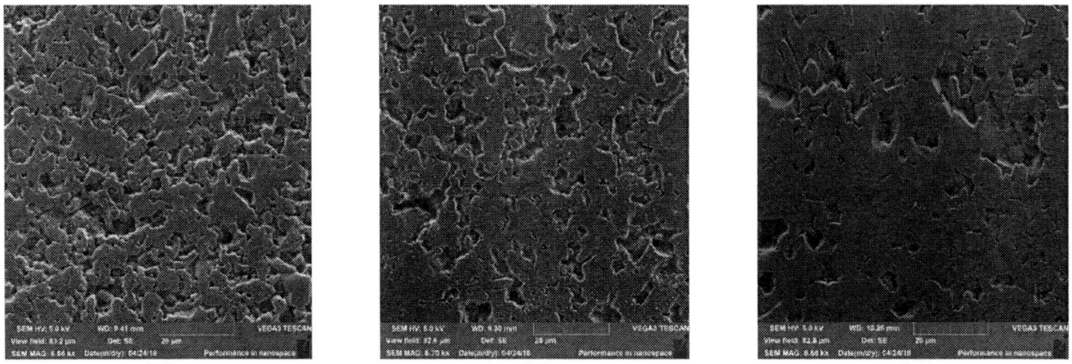

(a) (b) (c)

Figure 1. Scanning electron microscopy of the etched surface of Al_2O_3 specimens sintered at the temperatures 1400 (a), 1500 (b) and 1600 °C (c)

The ultimate strength and elastic modulus of ceramic specimens were determined in static three-point bending at the loading rate of 1 mm/min. Five specimens of the same batch were taken. The load-displacement curves were employed to evaluate the mechanical properties of the ceramic specimens. Using experimental data the values of the ultimate strength (σ_{exp}) and elastic modulus (E_{exp}) were calculated by the following equations:

$$\sigma_{\exp} = \frac{3Pl}{2bh^2},\ E_{\exp} = \frac{l^3 m}{4bh^3}$$

where P is the maximum load, (N); l is the support span, (mm); b is the width of the specimen, (mm); h is the thickness of the specimen, (mm); m is the gradient (i.e., slope) of the initial straight-line portion of the load-deflection curve, (N/mm). The calculated values of the ultimate strength and elastic modulus are presented in table 1.

Table 1. Mechanical and physical properties of Al_2O_3-based ceramics

Sintering temperature, °C	Porosity, %	σ_{exp}, MPa	E_{exp}, GPa	E_{calc}, GPa	σ_{calc}, MPa
1400	33±0.7	138±27	58±7	35±9	150±13
1500	26±0.6	243±14	80±13	85±23	236±7
1600	17±1	265±35	113±16	206±9	286±17

3. Simulation results and discussion

In three-point bending one of the parts of the specimen is in the tensile state, being most dangerous for porous ceramics. Consequently, uniaxial tension of mesovolumes with an explicit treatment of pores is modeled at the mesolevel in this work. Porous structures of the mesovolumes were borrowed from the electron microscopy data for the studied specimens (figure 2). Dimensions of mesovolumes in figure 2a and 2b are 100×100 μm. Three different digital mesovolume models of the same porosity were chosen for each sintering temperature.

The simulations were performed for problems in a two-dimensional formulation of plane-strain conditions using the finite difference method. We made use of constitutive equations that take into account the damage accumulation, which causes degradation of elastic properties [8]. The damage accumulation kinetic equation is built upon the effective stress by the Drucker-Prager material model and includes the Lode parameter to make allowance for the stress state type. To represent the mesoscopic fracture we took the critical damage criterion. Since the fracture criterion is fulfilled, the stresses are reduced to zero, and then the material stops to resist tension but not compression.

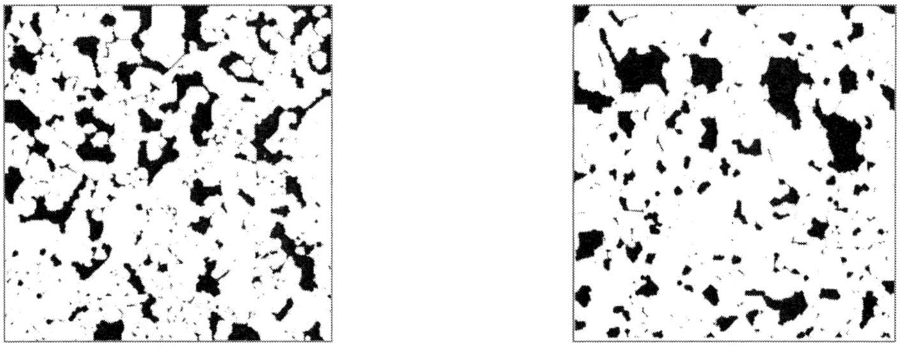

(a) (b)

Figure 2. Computer models of the ceramic structure with the porosity (a) 26 % and (b) 17 %

The results of the simulation at the mesolevel are shown in figure 3. The stress-strain curves averaged over the mesovolume are depicted in figure 3a. They appear as characteristic ones for brittle materials. We can notice that an increase in sintering temperature causes nonlinear growth of Young's modulus and strength of the porous material. Note, that the higher sintering temperature, the lower porosity. The calculated values of the elastic modulus (E_{calc}) and strength (σ_{calc}) are presented in table 1. It can be noted that the calculated strength values agree well the experimental ones. The calculated Young's modulus fits the experiment only for the sintering temperature 1500 °C, while it is lower than the corresponding experimental value at the sintering temperature 1400 °C and is higher than the experimental value at the 1600 °C. We mention that in general, the calculated elastic modulus and strength values roughly agree with the experimental data for the corresponding porosity [1].

Figures 3b and 3c display the fracture patterns in the two mesovolumes. One can see the cracks propagated from the potent stress concentrators specified by the pore shape and arrangement. The orientations of the cracks are mostly perpendicular to the tensile loading axis, which is typical for brittle materials fracturing.

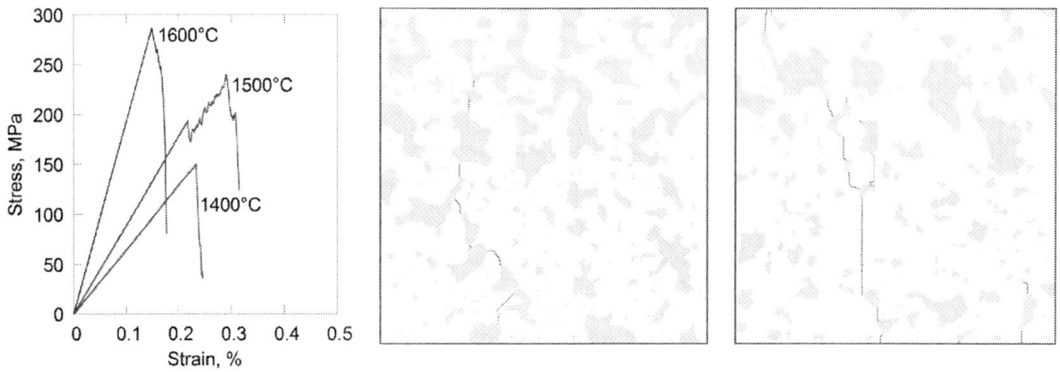

(a) (b) (c)

Figure 3. Mesoscale modeling results: (a) averaged stress-strain curves and fracturing patterns in mesovolumes of porous ceramics with the porosity (b) 25 % and (c) 17 %

4. Conclusion

Using the experimental results of the three-point bending test, we deduced the values of elastic modulus and tensile strength of the alumina ceramic specimens with the pore content varying from 17 to 33%. The experimental results evidence a decrease of the porosity with sintering temperature increasing, while elastic moduli and tensile strength increase nonlinearly. We mention in passing the wide scatter (in the range of 12–20 %) of the experimental data for both Young's modulus and strength, yet the dispersion of porosity does not exceed 2–6 % for each of the sintering temperatures. This is indicative of the profound effect of diverse parameters exemplifying the pore structure on the mechanical properties of porous materials (as also noted elsewhere [14]).

Here, we have analyzed how porous structure impacts on the features of mesoscopic fracturing and macroscopic stress-strain curves of the porous ceramics using our simulation results at the mesoscale. To perform modeling for the same total porosity, we used three different mesovolumes that vary in the pore shape and location specified by experimental photographs. The results obtained indicate that potent stress concentrators available in the mesovolumes determine the fracture nucleation site and influence the crack propagation. For all sintering temperatures, the calculated values of strength agree closely with experimental ones, while the calculated values of elastic modulus are distinct from the experimental ones at the sintering temperatures 1400 and 1600 °C, despite relatively large dispersion. The scatter of the calculated values is conditioned by the availability of heterogeneities of different shape and arrangement in the mesovolumes.

Acknowledgements
The work was carried out within the framework of the Fundamental Research Program of the State Academies of Sciences for 2013–2020, line of research III.23.

References

[1] Savchenko N L, Sevostyanova I N, Sablina T Yu, Gomze L and Kulkov S N 2014 *AIP Conf. Proc.* **1623** 547–550 https://doi.org/10.1063/1.4899003
[2] Grigoriev M V, Savchenko N L, Sablina T Y, Kurovics E, Sevostyanova I N, Buyakova S P, Gomze L A and Kulkov S N 2018 *J. Silic. Based Compos. Mater.* **70** 18–22 https://doi.org/10.14382/epitoanyag-jsbcm.2018.4
[3] Gömze L A and Gömze L N 2013 *Mater. Sci. Forum* **729** 43–48 https://doi.org/10.4028/www.scientific.net/MSF.729.43
[4] Orosz V and Gömze L A 2012 *Int. J. Microstruct. Mater. Prop.* **7** 205–219 https://doi.org/10.1504/IJMMP.2012.047500
[5] Corre Le, Brusselle-Dupend N and Moreaud M 2017 *Mech. Mater.* **114** 161–171 https://doi.org/10.1016/j.mechmat.2017.08.002
[6] Konovalenko Ig S, Smolin A Yu, Korostelev S Yu and Psakh'e S G 2009 *Tech. Phys.* **54** 758–761 https://doi.org/10.1134/S1063784209050272
[7] Mikushina V A, Smolin I Yu and Sidorenko Yu N 2017 *Key Eng. Mater.* **743** 172–175 https://doi.org/10.4028/www.scientific.net/KEM.743.172
[8] Smolin I Yu, Makarov P V, Eremin M O and Matyko K S 2016 *Proc. Struct. Integrity* **2** 3353–3360 http://dx.doi.org/10.1016/j.prostr.2016.06.418
[9] Romanova V A, Balokhonov R R and Emelyanova O S 2011 *Phys. Mesomech.* **14** 159–166. https://doi.org/10.1016/j.physme.2011.08.007
[10] Panin A V, Romanova V A, Balokhonov R R, Perevalova O B, Sinyakova E A, Emelyanova O S, Leontieva-Smirnova M V and Karpenko N I 2012 *Phys. Mesomech.* **15** 94–103 https://doi.org/10.1134/S1029959912010109
[11] Balokhonov R, Zinoviev A, Romanova V and Zinovieva O 2016 *Meccanica* **51** 415–428 https://doi.org/10.1007/s11012-015-0296-8
[12] Balokhonov R R, Makarov P V, Romanova V A and Smolin I Yu 1999 *Comput. Mater. Sci.* **16** 355–361 https://doi.org/10.1016/S0927-0256(99)00078-6
[13] Cherepanov O I, Smolin I Yu, Stefanov Yu P and.Makarov P V 1999 *Comput. Mater. Sci.* **16** 25–31 https://doi.org/10.1016/S0927-0256(99)00042-7
[14] Pabst W, Gregorova E and Ticha G 2006 *J. Eur. Ceram. Soc.* **26** 1085–1097 https://doi.org/10.1016/j.jeurceramsoc.2005.01.041

IOP Conf. Series: Materials Science and Engineering **613** (2019) 012027 doi:10.1088/1757-899X/613/1/012027

Computer Modeling of the Stress-Strain State of the Cervical Spine Segment

Yu A Rikun[1], T V Chaikovskaya (Kolmakova)[1, 2] and S N Kulkov[1, 2]

[1]Tomsk State University, Tomsk, Russia
[2]Institute of Strength Physics and Materials Science, Tomsk, Russia

E-mail: yulirik.93@mail.ru

Abstract. The article presents the results of the calculation of the stress-strain state of the C3-C4 segment of the cervical spine at flexion forward. The results of the calculations showed that there is the change in the deformation behavior of the segment and the change in the localization areas of the largest tensile and compression stresses at the increase in the elastic modulus of the intervertebral disk.

1. Introduction

Nowadays medicine is increasingly faced with the problem of optimal selection of an implant that can restore lost body functions. Adequate development and individual selection of implants play a decisive role in the treatment of a person. Improper selection of materials and design of implants can lead to deterioration of the bone tissue and the functioning of the spine as a whole. Despite the variety of modern materials, specialists whose professional activities is related to the development of methods for correction of human organs and structures and the creation of implants give their preference to ceramic materials. This is due to the fact that ceramics are identical to the inorganic bone matrix in the type of chemical bond [1-5].

The problems of mechanics arising in the creation and selection of implants of biological tissues are solved on the basis of studying the structure, mechanical behavior and properties of the biological tissues themselves and their interaction with implants. The use of computer modeling methods makes it possible to more deeply understand the patterns of functioning of the spine in norm, at pathologies, thus enhancing the development of recommendations on creation of implants replacing the whole organ or its part and effective methods for correction of the system in general.

The aim of this study is investigating of the stress-strain state of the cervical spine segment.

2. Model of the cervical spine segment

A vertebra geometrical model was built based on the literature data about experimentally obtained dimensions [6].

The simulation algorithm for the vertebra geometrical model was developed and implemented in the ANSYS system with the use of the APDL language. This algorithm allows an automatic rebuilding of the model when input parameters are changed.

Figure 1 shows the geometrical model of the cervical spine segment. The geometrical model includes the vertebrae C3 (1) and C4 (2), intervertebral disc (IVD) (3), facet joints (4), interspinous ligament (5), vertebral arches (6), spinous processes (7), transverse processes (8), and articular processes of vertebrae (9).

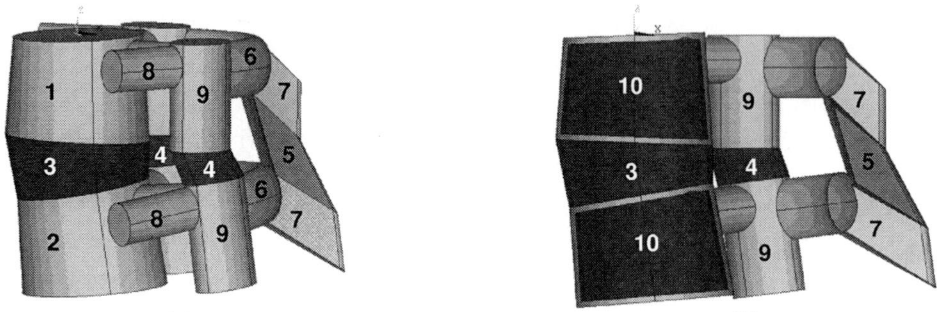

(a) (b)

Figure 1. Geometrical model of cervical spine segment: (a) lateral view, (b) sectional lateral view

The presence of the cortical and cancellous tissue in the vertebrae was taken into account. In Fig. 1b, the cancellous tissue is marked by 10, a thin layer of the cortical tissue [7] covers vertebral bodies. It is considered that vertebral arches and processes of the vertebrae fully consist of compact bone tissue. The Z axis of the coordinate system is located along the segment axis. The X axis is directed in the anteroposterior direction of the spinal segment.

Materials of the cortical and cancellous tissues of the vertebral bodies, materials of the intervertebral disc, facet joints, interspinous ligament, vertebral arches, and processes of vertebrae are considered as isotropic linear elastic materials. Degenerative changes of the IVD were simulated through a reduction of the disc height (h) from 6 to 4.5 mm and an increase of Young's modulus (E) from 2.5 to 98 MPa according to the previous research data [8,9], which in reality results from the water content reduction in the disc [10].

The task was resolved within linear theory of elasticity. The stress-strain state of cervical spine segment was calculated with the help of ANSYS software using the finite-element method. The lower surface of the vertebral body of C4 was rigidly fixed. The upper surface of the C3 vertebra was loaded by the pressure 1000 N. The bending moment 7.5 Nmm [11] was applied to the central point of the upper surface of the C3 vertebral body in the negative direction of the X axis in flexion of the spine segment. Specified load correspond to physiological flexion of the cervical spine segment.

3. Results of calculations

Figure 3 shows the stress fields σ_z in the segments of the spine with different heights and the Young's modulus of the intervertebral disc.

It can be seen from the figure that the compressive and tensile stresses are realized at segment flexion. At the value of the elastic modulus of the disk is 2.5 MPa the maximum absolute compressive stresses and maximum tensile stresses are localized at the sites of attachment of the legs of the C4 vertebra to the articular processes (Figure 2 (a, b)). With an increase in the elastic modulus of the disc, the maximum tensile stresses are localized in the upper part of the lateral surface of the vertebral body C3, and the maximum absolute compressive stresses in the central part of the upper surface of the vertebra C3 (Figure 2 (c, d)).

Such a change in stress localization is associated with a change in the character of the deformation of the cervical spine segment at increase the elastic modulus of the intervertebral disk (Figure 3).

Figure 3 shows the displacements fields in the Z axis direction together with the deformed shape of the segment.

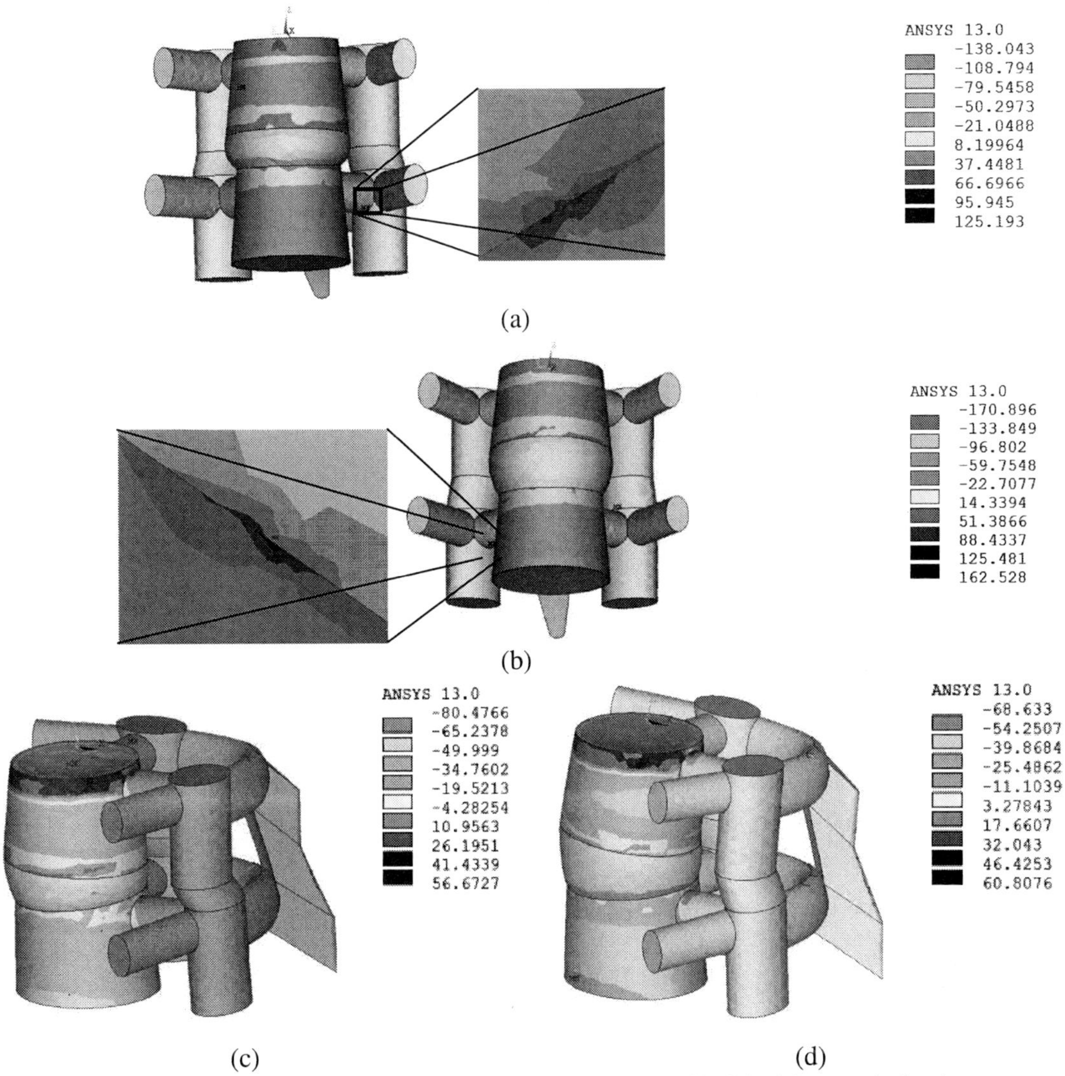

Figure 2. The stress σ_z distribution (MPa) in the segment C3-C4 of the cervical spine (a) h=4,3 mm, E=2,5 MPa; (b) h=6,0 mm, E=2,5 MPa; (c) h=4,3 mm, E=98 MPa; (d) h=6,0 mm, E=98 MPa

It can be seen from Figure 3 that under the Young's modulus of the disk 2.5 MPa, the greatest compression in the direction of the Z axis is observed in the anterior part of the vertebral body C3, affecting the upper part of the intervertebral disc, and greatest stretching is observed in the spinous process of the vertebra C3. As the Young's modulus of the IVD increases, the greatest compression region shifts toward the center of the upper surface of the vertebra C3. Decreasing the IVD height, as well as increasing its Young's modulus, leads to a decrease in the degree of compression and stretching of indicated regions in the direction of the Z axis.

4. Conclusion

Studies of the stress-strain state of the C3-C4 segment of the cervical spine at flexion forward showed that the increase in the Young's modulus of the intervertebral disc entails a change in the regions of localization of the greatest tensile stresses and the greatest absolute compressive stresses, which is associated with a change in the axial character of the spine segment deformation.

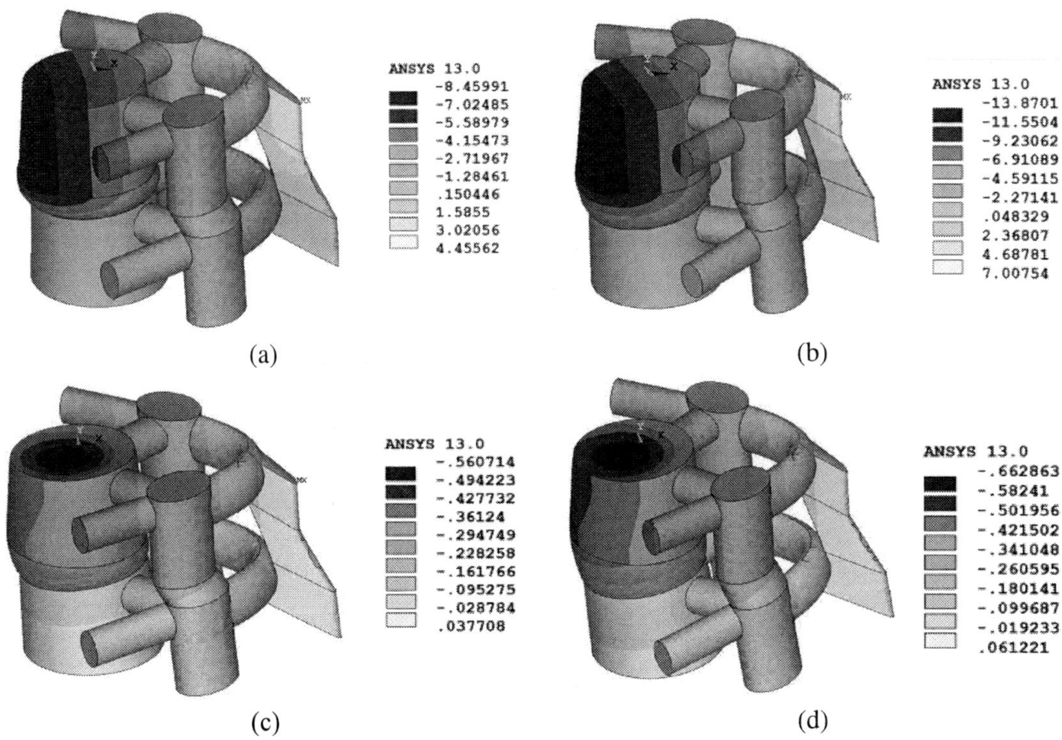

Figure 3. Displacements UZ (mm) and deformed shapes of the C3-C4 cervical spine segment (a) h=4,3 mm, E=2,5 MPa; (b) h=6,0 mm, E=2,5 MPa; (c) h=4,3 mm, E=98 MPa; (d) h=6,0 mm, E=98 MPa

Acknowledgements
The work was carried out within the framework of the Fundamental Research Program of the State Academies of Sciences for 2013–2020, line of research III.23.

References

[1] Gömze L A 2016 Applied materials science I. Compilation of Selected Scientific Papers *Published by IGREX Ltd* Hungary 189
[2] Gömze L A and Gömze L N 2017 *IOP Conf. Ser. Mater. Sci. Eng* **175** 012001 https://doi.org/10.1088/1757-899X/175/1/012001
[3] Kurovics E and Gömze L A 2017 *MultiScience - XXXI. microCAD Int. Multidiscip.Sci.Conf.* University of Miskolc, Hungary https://www.researchgate.net/publication/316275265
[4] Buyakov A S, Kulkov S N 2017 *IOP Conf. Ser.: Mater. Sci. Eng.* **175** 012025 https://doi.org/10.1088/1757-899X/175/1/012025
[5] Kulkov S N, Buyakova S and Gömze L A 2017 *J.Phis.: Conf. Ser.* **790** 012015 https://doi.org/10.1088/1742-6596/790/1/012015
[6] Tan S H, Teo E C and Chua H C 2004 *European Spine Journal* **13** 137
[7] Panzer M B 2006 Numerical modeling of the human cervical spine in frontal impact *Waterloo: University of Waterloo* 248
[8] Ng H W, Teo E C and Lee V S 2001 *Journal of Biomechanics* **23** 155
[9] Yang H, Jekir M G, Davis M W and Keaveny T M 2016 *Journal of Biomechanics* **49** 1134
[10] Wang Y, Chen H, Zhang L, Liu J and Wang Z 2012 *Chinese Journal of Traumatology* **15** 67
[11] Hueston S, Makola M, Mabe I and Goswami T 2012 Cervical spine anthropometric and finite element biomechanical analysis *Human Musculoskeletal Biomechanics* ed T Goswami (InTech, Rijeka, Croatia) 107

Unique microstructural and mechanical properties of Al-Zn alloys processed by high-pressure torsion

Nguyen Q. Chinh* and Zsolt Kovács

Department of Materials Physics, Eötvös Loránd University, H-1117 Budapest, Pázmány Péter sétány 1/A., Hungary.

*Corresponding author: Email: chinh@metal.elte.hu

Abstract: Al-Zn alloys with different Zn concentrations between 2 and 30 wt.% were processed by high-pressure torsion (HPT) to produce ultrafine-grained (UFG) materials. Microstructural and mechanical and properties of these UFG alloys were then investigated using depth-sensing indentations (DSI), focused ion beam (FIB), scanning electron microscopy (SEM) and differential scanning calorimetry (DSC). Emphasis was placed on the decomposition due to the HPT process, as well as on its effects on the mechanical properties of the UFG alloys. For low Zn contents, HPT gave strengthening due to grain refinement while for the highest Zn concentration the decomposition of the microstructure yielded an abnormal softening at room temperature. The microstructure decomposition led also to the formation of a Zn-rich phase, which wets the Al/Al grain boundaries and enhanced the role of grain boundary sliding with unusually high strain rate sensitivity. The occurrence of intensive sliding in these UFG alloys at room temperature is demonstrated by deforming micro-pillars, illustrating a potential for the effective application of these UFG materials in micro-devices
Keywords: *Al-Zn alloys, phase decomposition, grain boundary sliding, indentation, micro-pillars, strain rate sensitivity, ultrafine grains*

1. Introduction

It is well-known that ultrafine-grained (UFG) materials can be achieved by applying severe plastic deformation (SPD) techniques such as equal-channel angular pressing (ECAP) or high-pressure torsion (HPT) methods [1-3]. In general, the SPD-processed, UFG materials have reasonably saturated microstructures with steady-state dislocation density [4] so that the role of the grain boundaries is enhanced in the subsequent deformation processes. There are evidences suggesting that grain boundaries sliding (GBS) can occur easily even at room temperature in ultrafine-grained materials [5]. However, most of SPD-processed materials exhibit very limited ductility and low strain rate sensitivity (SRS) [6]. As the low ductility at ambient temperature restricts the use of these kind of materials (possibly having high strength) for structural applications, the enhancement of the ductility of UFG materials become as a topical subject of several investigations [7-10].

In this paper, some recent results are summarized describing the microstructural and the flow characteristics of the HPT-processed Al-Zn alloys, thereby giving insights into the unique microstructural and mechanical properties of these materials, especially into that of Al-30Zn alloy [11-15]. It should be emphasized that the Al-Zn alloys are probably the best known of all aluminum-based alloys. Some typical compositions, such as eutectic Al-95wt%Zn, eutectoid Al-70wt%Zn and solid solutions with Zn contents lower than 31.6%wt, have been extensively studied [11-23].

2. Microstructure and mechanical properties of Al-Zn alloys

a) *Unusual super-ductility microstructure-decomposition of HPT-processed Al-30Zn alloy.*
Supersaturated solid solution Al-30 wt.%Zn alloy, which is an important basic material in the aluminum industry, was processed by high-pressure torsion (HPT) [11] so that an ultrafine-grained microstructure developed which can be deformed by tensile for unusually high elongations up to 150% as shown in Fig. 1.

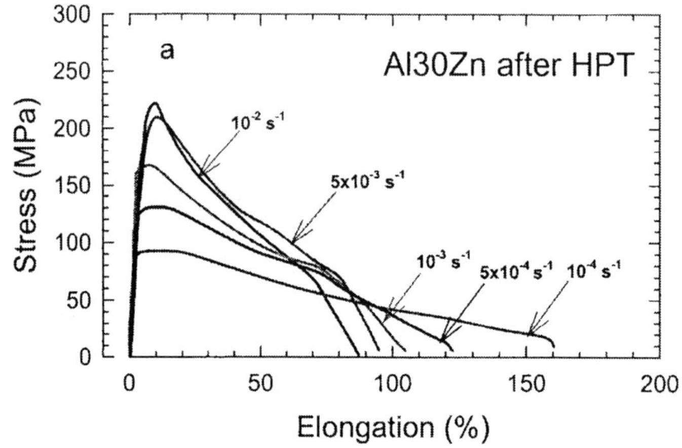

Figure 1 Stress-strain curves obtained at different strain rates on HPT-processed Al-30%Zn samples deformed by tensile test [11].

Together with the super-ductility of this HPT-processed Al30Zn alloy, microstructural investigations taken by using transmission electron microscopy (TEM) and energy dispersive spectroscopy (EDS) [14,20] have showed a decomposed structure which contained equiaxed ultrafine Al grains having a size of 300-400 nm and smaller Zn grains located at the triple junctions of the aluminium grains. Furthermore, more than 50% of the Al/Al grain boundaries is wetted by Zn-rich layers as shown in Fig. 2.

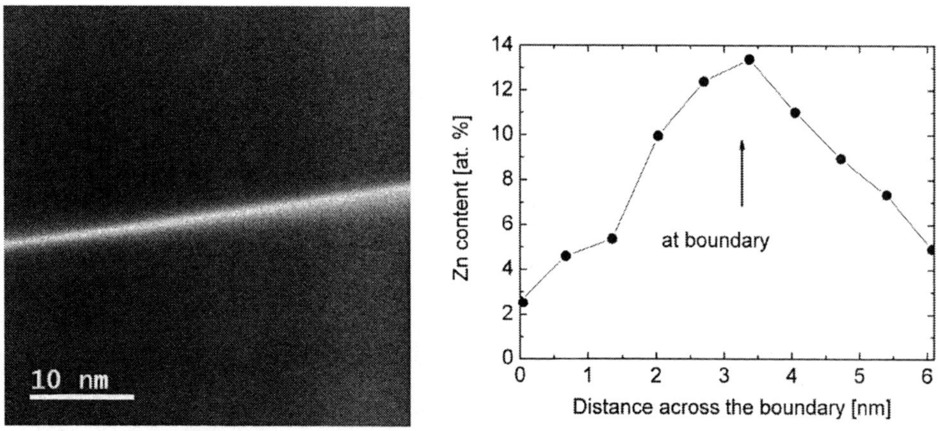

a) b)

Figure 2 Zn-rich layer at an Al/Al grain boundary in the HPT-processed Al-30Zn alloy, demonstrated by a) TEM and b) EDS results [14].

IOP Conf. Series: Materials Science and Engineering **613** (2019) 012028 doi:10.1088/1757-899X/613/1/012028

b) ***Effect of Zn content on the mechanical properties of Al-Zn alloys***

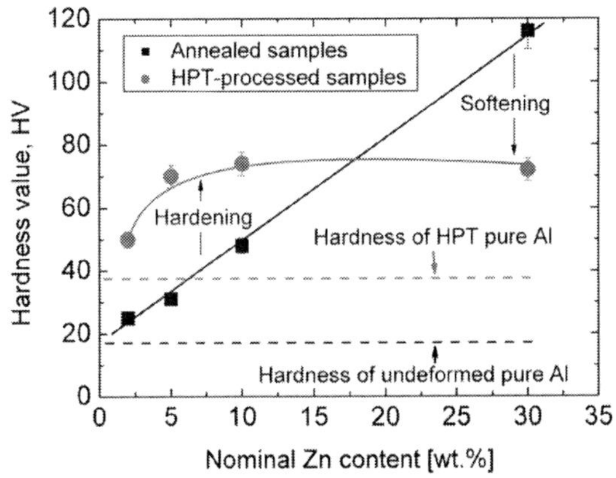

Figure 3 Effect of HPT processing on microhardness of different Al-Zn alloys [15]

Figure 3 shows the effect of Zn concentration on the hardness of the HPT-processed A-Zn alloys. In order to emphasize the significant changes, the hardness values of the annealed samples are also plotted in Fig.3. For Zn concentrations not larger than 10 wt%, the normal strengthening-effect of SPD can be observed as the hardness of the HPT-processed samples is higher than that of the annealed samples. In the case of the Al-30Zn alloy, however, there is a strong reduction in hardness, indicating an abnormal softening due to HPT. The abnormal softening of the high Zn-concentrated Al-30Zn alloy is a consequence of the mentioned strong decomposition of the annealed microstructure in this alloy [14,19].

c) ***The effect of the decomposed microstructures on the differential scanning calorimetry (DSC) thermograms***

Typical DSC thermograms taken on the HPT-processed Al-Zn samples having different Zn content can be seen in Fig. 4a. The thermogram obtained for the HPT-processed Al-30Zn sample is plotted together with that of the annealed state in Fig. 4b. It can be seen that the latter thermogram has well-defined endothermic indicating the dissolution of Guinier-Preston (GP) zones and the solid solution of Zn in Al [24]. However, for the HPT-processed Al-30Zn sample, the first DSC peak was not observed since SPD facilitated the development of Zn grains without the formation of metastable GP zones.

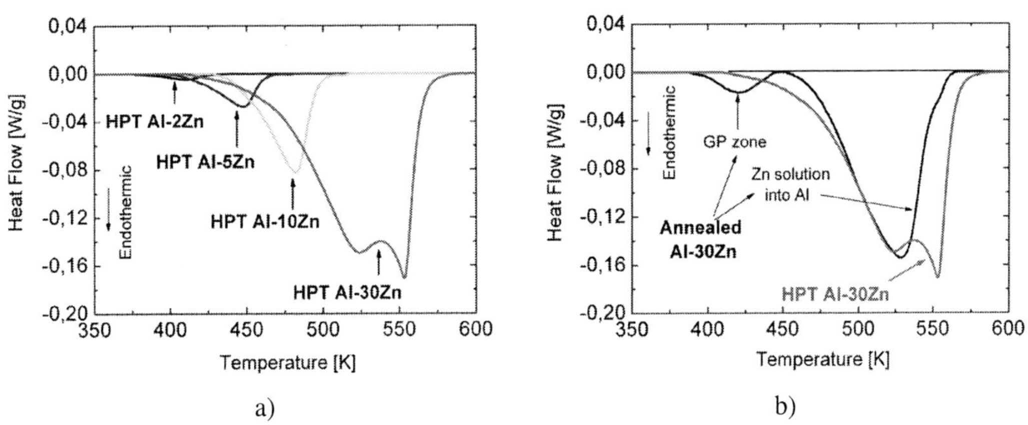

Figure 4 DSC thermograms taken at on a) HPT-processed Al–Zn alloys with different Zn contents and b) annealed and HPT-processed Al–30 wt.% Zn alloys [15].

Considering the HPT samples with different Zn contents (Fig. 4a), it can be seen that there is a significant difference between the Al-30Zn sample and the other three lower Zn-concentrated alloys, as a double-peak thermogram is observed only for Al-30Zn sample. The two peaks of this curve are located at T=525 K and 555 K. The peak at 525 K corresponds to the dissolution of Zn in Al, which is similar to the annealed state (see Fig. 4b). The other endothermic peak at 555 K most probably indicates the dissolution of the Zn-rich boundary layers. This second peak cannot be observed at lower Zn contents. Therefore, the present of this peak indicates the significant amount of the Zn-rich layers in the Al-30Zn alloy.

d) *The effect of the Zn-rich layers on the plastic behaviors of the HPT-processed Al-Zn alloys*

In order to study the plastic behavior of the HPT-processed samples, the investigation was focused on the determination of strain rate sensitivity (SRS) because it is well established that this parameter correlates to the ductility [25]. The SRS values were determined by using the indentation creep method developed in [26]. The obtained SRS values are listed in Table 1.

Table 1 The strain rate sensitivity of the HPT-processed Al-Zn alloys.

Sample	Al-2Zn	Al-5Zn	Al-10Zn	Al-30Zn
SRS value (± 0.02)	0.08	0.14	0.17	0.25

It can be seen that the SRS increases with Zn content and an unusually high value (0.25) was obtained for the Al-30Zn alloy. It is well known that UFG materials produced by SPD generally exhibit extremely low SRS of about 0.01-0.03 [6]. In the present case, the relatively high SRS is attributed to the Zn addition. Determining the SRS and activation energy characterizing the process of plastic deformation in HPT-processed Al-30Zn alloy [13] it was shown that the process of plastic deformation is controlled mainly by the Zn diffusion along Al/Al grain boundaries.

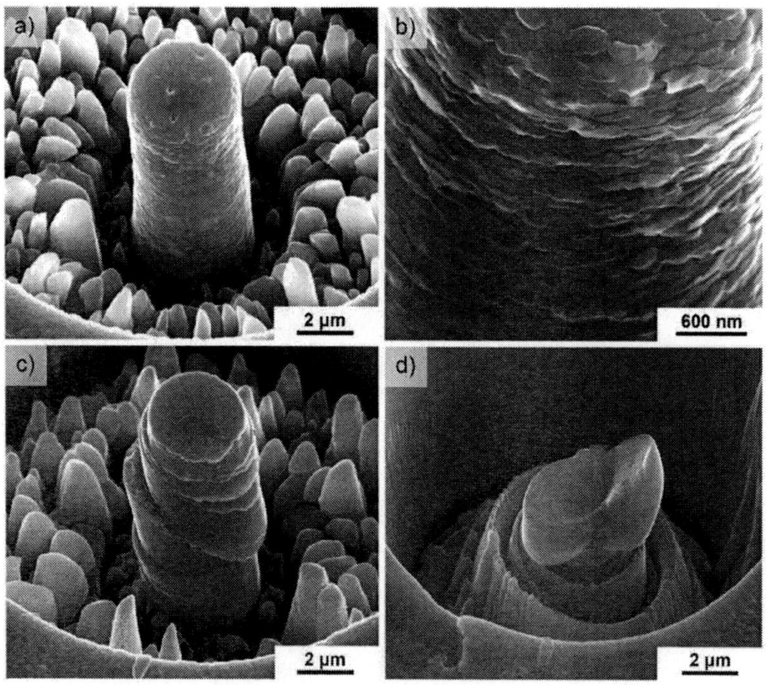

Figure 5 Surface morphologies of compressed micro-pillars for HPT-processed Al–30Zn in low and high magnifications shown in a) and b), respectively, c) HPT-processed Al-10Zn [15] and d) annealed, coarse-grained Al-30Zn [14] samples.

IOP Conf. Series: Materials Science and Engineering **613** (2019) 012028 doi:10.1088/1757-899X/613/1/012028

This mechanism is enhancing the role of the grain boundary sliding at room temperature. Thus, the unexpectedly high SRS obtained for the HPT-processed Al-30Zn sample is a consequence of the high fraction of Al/Al grain boundaries wetted by Zn-rich layers which leads to intensive grain boundary sliding and then to the mentioned super-ductility. In the case of lower Zn-content alloys, where the strain rate sensitivity is much smaller, the maximum elongation was lower the 50% [19].

In order to check the relationship between strain rate sensitivity and plasticity, as well as the effect grain boundary sliding in the investigated samples, micro-pillars were deformed by using nanocompression [14,15].

An examination of the surface morphologies of the compressed pillars by scanning electron microscopy (SEM) revealed significant differences between the basic mechanisms of plastic deformation in the different Al-Zn samples (see Fig. 5). In the case of the HPT-processed Al-30Zn sample (see Figs. 5a and 5b) the surface morphologies demonstrate clearly the occurrence of intensive grain boundary sliding, as individual ultrafine grains emerging from the pillar surface are visible [14]. The relatively high SRS of this sample is certainly a consequence of the high fraction of Al/Al grain boundaries wetted by Zn-rich layers, which lead to intensive grain boundary sliding. For the HPT-processed Al-10Zn sample having lower SRS, only strain localization and individual slip bands can be observed (Fig. 5c) [15]. In order to emphasize the super-ductility of the HPT-processed ultrafine-grained Al-30Zn sample, the surface morphology of the compressed pillar on the annealed Al-30Zn [14] can also be seen in Fig. 5d. In contrast with the UFG counterpart, the ductility of coarse-grained, annealed Al-30Zn sample having very low SRS of only 0.03 [13] seems to be very poor [14]. It can be seen that the deformation process of this sample consists of only few large strain fluctuations indicated by extreme slip bands in Fig. 5d. It has also been shown before that these large strain fluctuations lead to difficulties in plastically forming micrometer-scale single crystals because of the possibility of catastrophic failure. For this reason, therefore, micrometer-sized samples of coarse-grained metals are not suitable for use in the fabrication of micro-devices. On the other hand, the stable deformation by grain boundary sliding of the UFG micro-pillars, without the occurrence of any catastrophic avalanches, emphasizes the advantage of the sliding mechanism, suggesting an important potential for using these UFG materials in the fabrication of micro-devices.

3. Summary

Microstructures and mechanical properties of HPT-processed UFG Al-Zn alloys were studied by using transmission electron microscopy (TEM), tensile and microhardness tests, depth-sensing indentation (DSI), scanning electron microscope and focused ion beam (SEM/FIB), as well as by differential scanning calorimeter (DSC). The main results are the followings:

i) For low Zn contents normal strengthening was dominant whereas for the highest Zn concentration an abnormal softening was observed due to microstructural decomposition.

ii) The strong microstructure decomposition in high Zn-concentrated alloy leads also to the formation of Zn-rich grain boundary layers, which wet the Al/Al grain boundaries and enhance the role of grain boundary sliding in plasticity with an unusually high strain rate sensitivity.

iii) Occurrence of intensive grain boundary sliding at room temperature in the UFG Al-30wt%Zn alloy, leading to its super-ductility at room temperature.

iv) As a consequence of the role of the grain boundaries, the deformation process of UFG materials is relatively homogenous and this may have important practical implications for the electronics industry and especially for using these materials in micro-devices.

Acknowledgements

This work was presented in the 5th International Conference on Competitive Materials and Technology processes (ic-cmtp5, October 2018, Miskolc-Lilafüred, Hungary). The Hungarian-Russian bilateral Research program (TÉT) No. 2017-2.3.4-TÉT-RU-2017-00005 supported both the research and the participation in the Conference. This work was also financed by the Ministry of Human Capacities of Hungary within the ELTE University Excellence program (1783-3/2018/FEKUTSRAT).

References

[1] Valiev RZ, Islamgaliev RK, Alexandrov I V 2000 *Prog Mater Sci* **45** 103
[2] Valiev RZ, Langdon TG 2006 *Prog Mater Sci* **51** 881
[3] Zhilyaev AP, Langdon TG 2008 *Prog Mater Sci* **43** 893
[4] Chinh NQ, Csanádi T, Gubicza G, Langdon TG 2010 *Acta Mater* **58** 5015 http://dx.doi.org/10.1016/j.actamat.2010.05.036
[5] Chinh NQ, Szommer P, Horita Z, Langdon TG 2006 *Adv Mater* **18** 34
[6] Meyer MA, Mishra A, Benson DJ 2006 *JOM* **58** (4) 41
[7] Wang Y, Chen M, Zhou F, Ma E 2002 *Nature* **419** 912
[8] Zhao YH, Zhu YT, Liao XZ, Horita Z, Langdon TG 2006 *Appl Phys Lett* **89** 121906
[9] Zhao YH, Liao XZ, Cheng S, Ma E, Zhu Y T 2006 *Adv Mater* **18** 2280
[10] Horita Z, Ohashi K, Fujita T, Kaneko K, Langdon TG 2005 *Adv Mater* **17** 1599
[11] Valiev RZ, Murashkin MY, Kilmametov A, Straumal BB, Chinh NQ, Langdon TG 2010 *J Mater Sci* **45** 4718 http://dx.doi.org/10.1007/s10853-010-4588-z
[12] Chinh NQ, Győri T, Valiev RZ, Szommer P, Varga G, Havancsák K, Langdon TG 2012 *MRS Comm* **2** 75 http://dx.doi.org/10.1557/mrc.2012.11
[13] Chinh NQ, Csanádi T, Győri T, Valiev RZ, Straumal BB, Kawasaki M, Langdon TG 2012 *Mater Sci Eng A* **543** 117 http://dx.doi.org/10.1016/j.msea.2012.02.056
[14] Chinh NQ, Valiev RZ, Sauvage X, Varga G, Havancsák K, Kawasaki M, Straumal BB, Langdon TG 2014 *Adv Eng Mater* **16** 1000 http://dx.doi.org/10.1002/adem.201300450
[15] Chinh NQ, Jenei P, Gubicza J, Bobruk EV, Valiev RZ 2017 *Mater. Letters* **186** 334 http://dx.doi.org/10.1016/j.matlet.2016.09.114
[16] López GA, Mittemeijer EJ, Straumal BB 2004 *Acta Mater* **52** 4537
[17] Straumal BB, Baretzky B, Mazilkin AA, Phillipp F, Kogtenkova OA, Volkov MN, Valiev RZ 2004 *Acta Mater* **52** 4469
[18] Straumal BB, Valiev RZ, Kogtenkova OA, Zieba P, Czeppe T, Bielanska E, Faryna M 2008 *Acta Mater* **56** 6123
[19] Bobruk EV, Sauvage X, Enikeev NA, Straumal BB, Valiev RZ 2015 *Rev Adv Mater Sci* **43** 45.
[20] Sauvage X, Murashkin MYu, Straumal BB, Bobruk EV, Valiev RZ 2015 *Adv Eng Mater* **17** 1821 http://dx.doi.org/10.1002/adem.201500151
[21] Alhamidi A, Edalati K, Horita Z, Hirosawa S, Matsuda K, Terada D 2014 *Mater Sci Eng A* **610** 17 http://dx.doi.org/10.1016/j.msea.2014.05.026
[22] Mazilkin AA, Straumal BB, Borodachenkova MV, Valiev RZ, Kogtenkova OA, Baretzky B 2012 *Mater Letters* **84** 63 http://dx.doi.org/10.1016/j.matlet.2012.06.026
[23] Borodachenkova MV, Barlat F, Wen W, Bastos A, Gracio JJ 2015 *Inter J Plasticity* **68** 150 http://dx.doi.org/10.1016/j.ijplas.2014.01.009
[24] Mondolfo LF 1976 Aluminum Alloys: Structure and Properties *Butterworths* London
[25] Woodford DA 1969 *Trans ASM* **62** 291
[26] Chinh NQ, Szommer P 2014 *Mater Sci Eng A* **611** 333 http://dx.doi.org/10.1016/j.msea.2014.06.011

IOP Conf. Series: Materials Science and Engineering **613** (2019) 012029 doi:10.1088/1757-899X/613/1/012029

Effect of extrusion speed and ATO content on rheological properties of plasticised PVC

Ali I.Al-Mosawi [1] **and Kálmán Marossy** [1]

[1]Institute of Ceramic and Polymer Engineering, University of Miskolc, Hungary

E-mail: aliibrahim76@yahoo.com and polkal01@uni-miskolc.hu

Abstract. Rheological properties of antimony trioxide (ATO) modified plasticised PVC were measured at different extrusion speeds (20-100) rpm. The results showed a positive effect of the presence of ATO accompanied with increased extrusion speed, where the rheological properties and gelation grade of PVC have been improved.

1. Introduction

The extrusion speed one of the important processing parameters in addition to temperature, time, and pressure which significantly and directly affect the characteristics of the final product. The speed should not be low because the polymer will probably stick to the extruder barrel walls because the material will stay long in the extrusion cylinder in addition to the high temperature and the weakness of the flowability. Also, the speed should not increase significantly because this will generate cracks on the surface of the extruded material, and in both cases, the properties of the final product will be affected [1-4]. Due to polyvinyl chloride (PVC) considered one of the sensitive polymers; where its properties are a function of processing history, so, the extrusion speed should be controlled and balanced with high accuracy for achieving optimum properties accompany with a homogeneous structure which improves the quality of the product and reduces the deformations [5-7].

2. Methodology

The following materials have been used to produce samples: polyvinyl chloride (PVC) type Ongrovil® S-5070, dioctyl-phthalate (DOP or DEHP) as plasticizer, calcium-zinc type Newstab-50 as heat stabilizer, Wax-E type Licowax® E as lubricant, and antimony trioxide (ATO) with (0-5) wt.%. The extrusion machine type Göttfert was used to prepare the second batch, where the material extruded to the desired shape in one step by Ø 20 extruder rod diameter at uniform pressure, compression force, and temperature. The temperatures range of the machine heaters were 140, 150, and 160 °C at a uniform rate. PVC was extruded at different speeds (20-100) rpm with and without ATO additions. DSC samples were produced according to the ISO 11357-1 standard [8] with 1mm width and 10mm thickness, and DSC test was done by Mettler Toledo DSC823e instrument to finish this test for gelation grade (G.G) measurement. The temperature range was 30 °C-240 °C; and the heating rate was 20°C/min.

3.Results and Disscusion

The gelation grade (G.G) is affected by several factors, and the most important is speed and temperature of extrusion, and the change in the gelation grade (G.G) (increase or decrease) will also affect the rate of retraction, shear stress of the polymer, the presence of cracks, and the others properties of the PVC after extrusion, where they may be improved or reduced depending on the type of filler added to the polymer [1, 9]. From Figure.1 which represents the DSC traces of PVC, we see that the gelation grade increases as the speed increased, which considered as a function of extrusion speed. Figures 2, 3, and 4 which shows the behavior of the extruded PVC at different speeds after adding ATO in different percentages (1, 3, 5) %wt. respectively, where we noted that there are a significantly increments in gelation grade after adding this oxide, and the improvement rate depends on the percentage of ATO.

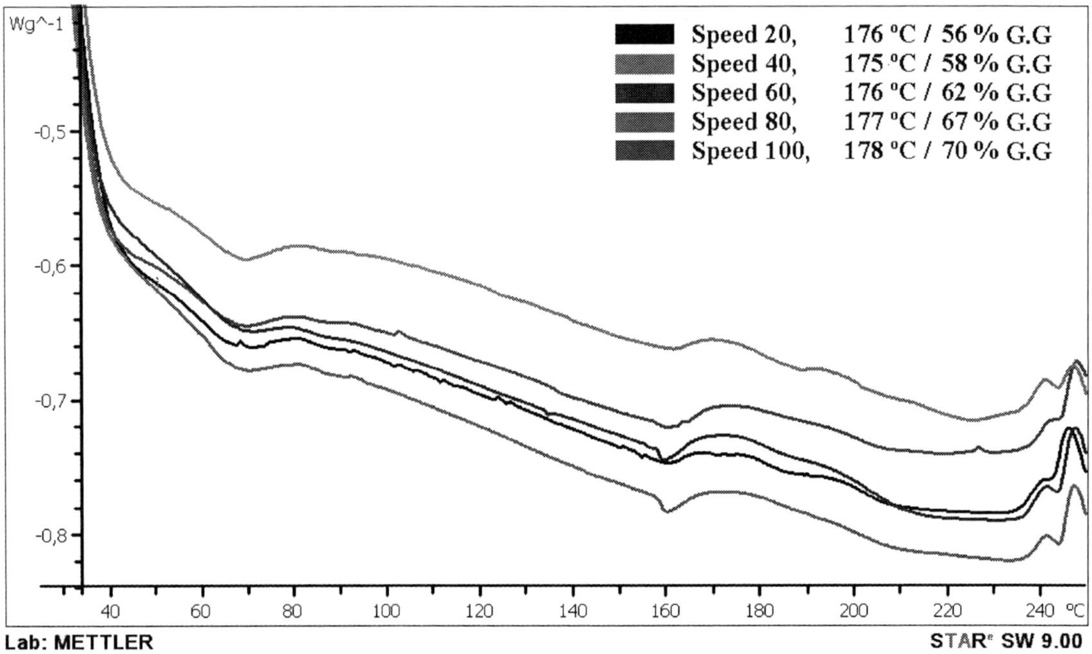

Figure 1. DSC traces of plasticised PVC at different extrusion speeds

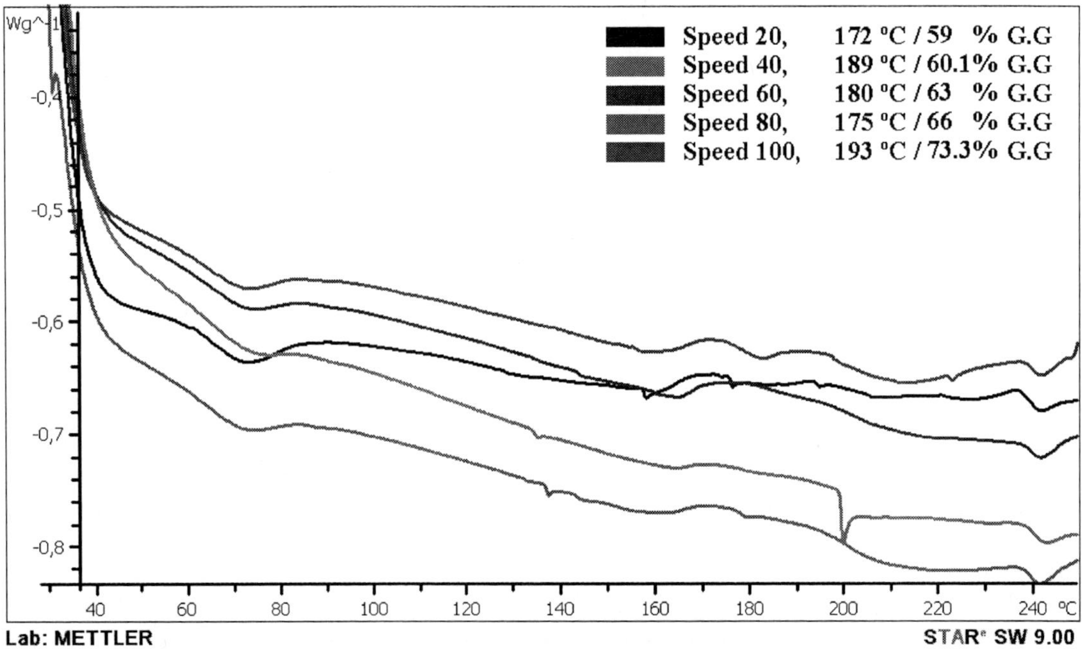

Figure 2. DSC traces of 1% ATO containing plasticised PVC at different extrusion speeds

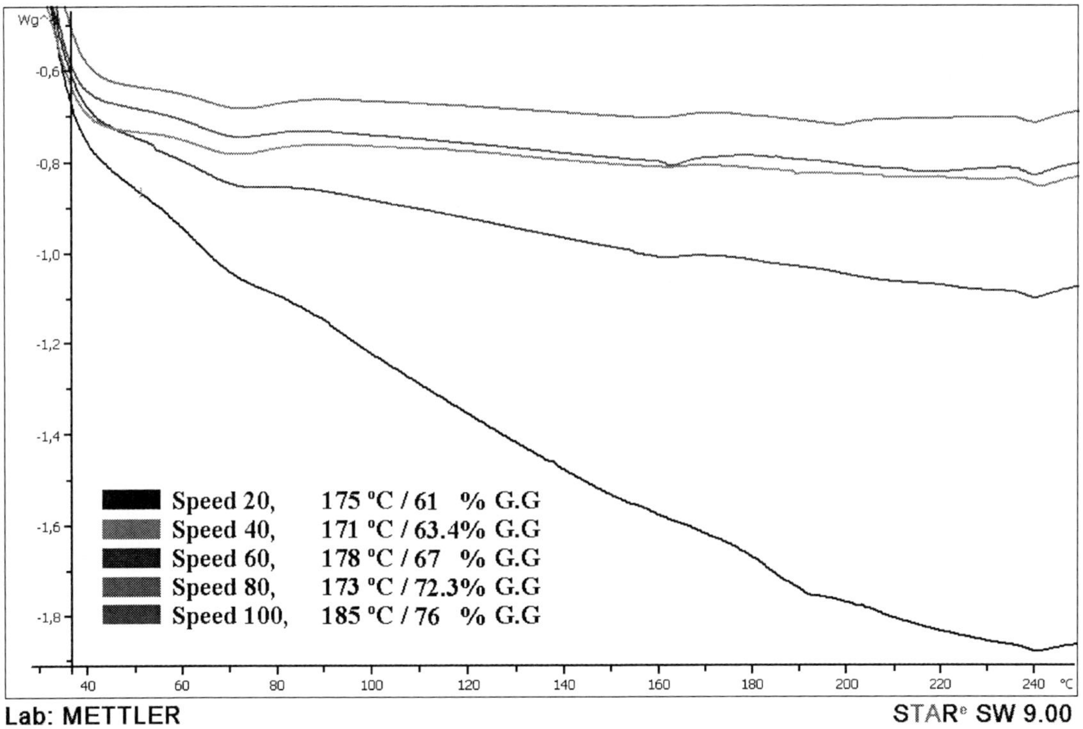

Figure 3. DSC traces of 3% ATO containing plasticised PVC at different extrusion speeds

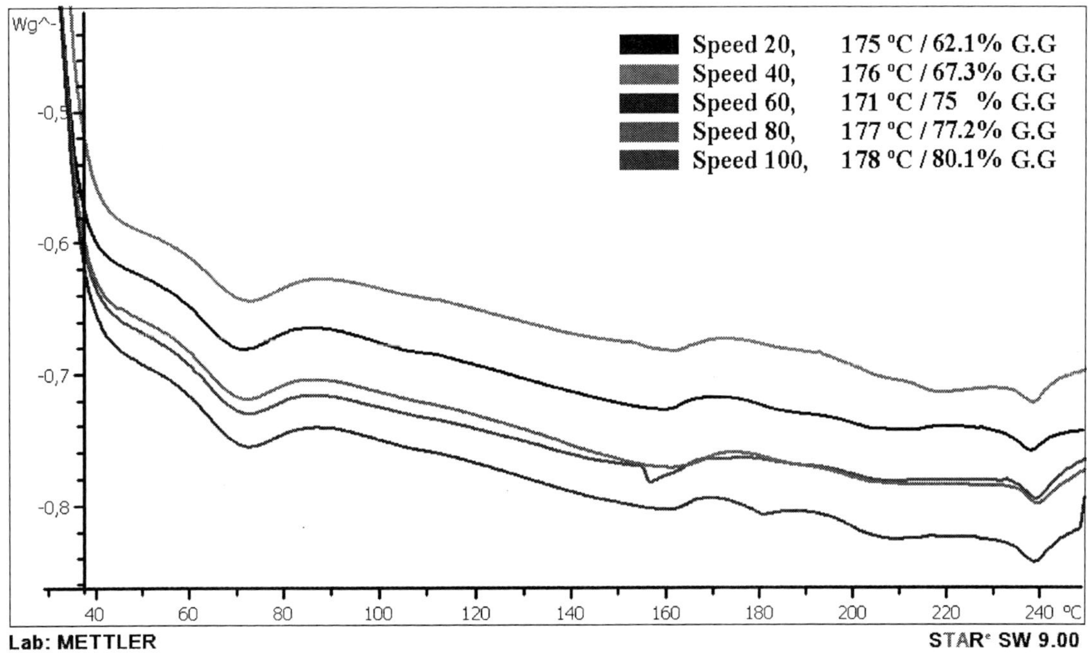

Figure 4. DSC traces of 5% ATO containing plasticised PVC at different extrusion speeds

4.Conclusion

1. The results prove that the DSC is a speed-dependent property (process dependent).
2. ATO additives improve DSC and the ratio of improvement depends on the percentage of ATO addition.
3. The best results obtained after adding ATO was at speed of 60 rpm, which are associated with maintaining the rest of the PVC properties.

Acknowledgments
We would like to thank Csaba Kónya, Gabriella Szemere, and Krisztina Tózsa at the BorsodChem Zrt., who helped us to complete this research.

References

[1] A I Al-Mosawi, K Marossy 2018 *MultiScience - XXXII. microCAD International Multidisciplinary Scientific Conference* 5-6 September University of Miskolc, Hungary http://doi.org/10.26649/musci.2018.017
[2] W V Titow 1990 PVC plastics properties, processing, and applications1st Edition *Springer* Netherlands http://doi.org/10.1007/978-94-011-3834-5
[3] D Ravoori, L Alba, H Prajapati, and A Jain 2018 *Additive Manufacturing* **23** 132 http://doi.org/10.1016/j.addma.2018.07.011
[4] Y Deng, P Yi, L Peng, X Lai and L Zhongqin 2015 *J. Micromech. Microeng.* **25** (6) 065004 https://doi.org/10.1088/0960-1317/25/6/065004
[5] K Marossy 2016 Formulation *International conference on the PVC industry* 5-7 April Cologne, Germany
[6] D M Burke 1994 *Journal of vinyl and additive technology* **16** (2) 102 https://doi.org/10.1002/vnl.730160208
[7] A I Al-Mosawi, K Marossy 2018 *Építőanyag-JSBCM* **70** (3) 86 http://doi.org/10.14382/epitoanyag-jsbcm.2018.16
[8] ISO 11357-1:2016 2016 Plastics differential scanning calorimetry (DSC) method *International Organization for Standardization*
[9] W Guoquan, C Yiaoting 1991 *Polymer Testing* **10** (4) 315 https://doi.org/10.1016/0142-9418(91)90025-S

Corrosion of glass used for radioactive waste disposal influenced by iron corrosion products

Ali AL DABBAS[1], **Katalin KOPECSKÓ**[2]

Faculty of Civil Engineering,
Budapest University of Technology and Economics,
Műegyetemrakpart 3, Budapest 1111, Hungary

E-mail: ali.al.dabbas@epito.bme.hu and kopecsko.katalin@epito.bme.hu

Abstract. Iron corrosion products of the steel canister in the nuclear waste container are highly silica sorptive materials and very much expected to delay formation of the protective layer (gel) on the nuclear glass interface. This study is focusing on the role of the iron carbonates (ankerite), which is probably one of iron corrosion products and already exists in the clay structure of the disposal depth. Simulating nuclear borosilicate glass samples (alternative to the R7T7-type) samples are planned to leach in different series of experimental models with dissimilar quantities of ankerite. Batch experiments all will be cultivated in synthetic groundwater; where different pH (4-9) values and temperatures (35, 50, 90 °C) are considered. In this paper the state of the art of the radioactive waste glass corrosion in the circumstances of deep geological disposal will focus on the role of ankerite as one of the iron corrosion products and clarify how it is expected to show delay in achieving the Si saturation status and accelerate glass alterations.

1. Introduction

Disposal of the High Level radioactive Waste (HLW) in deep geological formation in multi-layer containers is internationally highly recommended. After evaporation of the volatile radionuclides it is immobilized and vitrified by incorporating into borosilicate glass matrix (R7T7 type glass) [1, 2]. The glass matrix is formed to pellet shape and poured into stainless steel containers to facilitate its handling, storage and disposal operations. Finally it is embedded in a buffer i.e. concrete isolation in the clay-stones [3]. Experiments will be conducted with small polished simulating nuclear borosilicate glass chips, involving accelerated ageing of glass due to environment provided by the microbial activity to figure out the convenient and optimum conditions of the maximum Si uptake by ankerite. Ankerite is expected to show noticeable Si uptake, Iron-silicates precipitation is highly predicted to happen. They both are very much expected to sharing the glass corrosion role with the other important parameters mentioned in the literature. The experiments will show if the ankerite ratio will increase due to the microbial activity.

1.1. High level waste

Nuclear spent fuel mainly consists of (96-97%) uranium plutonium and (3-4%) fission products. Spent fuel is reprocessed (uranium and plutonium are separated) to generate new nuclear fuel used in economic reactors. The remaining fission products (^{129}I, ^{90}Sr, ^{99}Tc, ^{79}Se, ^{36}Cl, and ^{135}Cs) and other actinides are considered as HLW. In particular, some radioactive isotopes such as ^{99}Tc, ^{79}Se, and ^{129}I are mobile in water, requiring special storage medium that reduces their ability to move into the groundwater [4, 5].

1.2. Immobilization of the HLW
One of the most recommended immobilization methods of the HLW is the vitrification. It is calcination of the HLW and then incorporation it into borosilicate glass to be as glass matrix material then it is poured into stainless steel canisters. Finally, non-recoverable fission products and actinides became locked in the glass matrix insoluble in water and stable for hundreds of thousands of years [6].

1.3. Storage and disposal of HLW
Disposal of the non-recyclable HLW is a long-term storage for hundreds of thousands of years in a deep geological disposal located between 450 and 500 m depth [7, 8] in a multi-layer container, where tectonic stability is great and natural barrier are effective to contain that kind of waste.

2. Glass corrosion
Formation of alteration layer on glass in the geological formation has been studied as well as experimentally evaluated and determined by several studies. The long-term storage was experimentally estimated by evaluating the effects of the physico-chemical parameters on the short-term alteration of glass such as pH, temperature, glass composition and redox potential [7, 9].

2.1. Glass dissolution
Water saturated surrounding dissolution of glass is a result of various processes; First: glass hydration by water diffusion into the glass network followed by ion exchange between positive protons in the groundwater and alkali metals in the glass formation, which leads to glass hydration and formation of amorphous alteration layer (gel) on the glass surface. Second: hydrolysis of the ionic-covalent bonds of the most soluble elements by attacking the glass network (Si-O-X; X: Si, Al, Fe, Zr) bonds by groundwater. Third: condensation mechanism where the silica is reversely reacts and dissolves in the groundwater and condensate as a gel layer influenced by pH and temperature. In the experiment, the leaching solution should be saturated by Si as a condition necessary for the formation of the gel layer [7, 10, 11 and 12].

2.2. Function of the gel layer
The gel layer is a natural phenomenon important to reduce the direct contact between the groundwater and the glass to protect the glass from dissolving in the groundwater and so prevent potential release of the radioactive substances back to the biosphere [13, 14 and 15]. The Iron Corrosion Products (ICPs) in the solution will provide surfaces for Si sorption. Reducing Si content in the surrounding formation is expected to delay the creation of gel layer in the insufficient Si saturated solution thus motivate glass dissolution [16].

2.3. Iron corrosion products
Steel alteration in presence of bacterial activities is highly responsible for accelerating exponentially rising ICPs in the anaerobic conditions [8, 17]. Sulphate Reducing Bacteria (SRB) are considered as the main reason of that [17, 18]. They produce H_2 in the process formation of iron sulfide, FeS [19]. SRB exist widely in deep geological repository and they influence iron corrosion by releasing free Fe ions in the solution surroundings the glass, which are expected to create different kinds of iron carbonate minerals [20].

3. Silica sorption
Experiments have been performed with simulating glass samples to determine the chemical durability of R7T7 to better understand the glass corrosion and performance under potential disposal conditions. Silica saturation was achieved in 91days at 90 °C. Glass corrosion rate was measured 1000 times lower after Si saturation [21]. Nuclear waste is providing the higher temperature and microbial influenced corrosion mechanism is enhancing the lower pH by producing acidic by-products which are providing appropriate medium conditions for Si sorption mechanism on the ICPs, which were experimented and showed enhancing and accelerating effect on the corrosion process of the borosilicate glass [22, 23].

3.1. Power of ICPs to uptake silicon

Bacterial activities on the different carbon steel applications in both aerobic and anaerobic mediums have shown ICPs whether iron carbonates (i.e. siderite, $FeCO_3$) or iron oxides (i.e. magnetite, Fe_3O_4; goethite, α-FeOOH) in the bulk medium [9, 20 and 24]. Experiment was conducted on SON68 glass and ancient ICPs [25] has shown strong influenced alteration of the glass by the ICPs formed precipitation which lowered the Si concentration in the solution. Magnetite and goethite were added to the solution and showed deleterious effect on the glass chemical durability [11] justified by the power of Si sorption on the outer surface of the ICPs which showed up 5 times more of glass corrosion [21]. However, zero glass dissolution-rate was obtained due to clogging of the external gel porosity at the gel/Si saturated solution after about one month of leaching under certain conditions [26, 27].

3.2. Formation and role of ankerite

Ankerite is present in the clayey matrix and its dissolution showed goethite FeOOH formation [28]. In the carbonatites in the past siderite has approved to be ankerite, $Ca(Fe,Mg,Mn)(CO_3)_2$ [29]. Ankerite ($CaFe(CO_3)_2$) has the same crystal structure as dolomite ($CaMg(CO_3)_2$) [30] and considered as a member of the dolomite group of minerals and it shares siderite many of similarities [31]. Siderite, ankerite and dolomite have shown up significant identified precipitation in the sandstone, when Fe(II) source provider was exist there [30, 32]. MIC mechanisms on the steel canister release Fe(II), which most likely sufficient to precipitate ankerite. This study is predicting ankerite precipitation as one of the ICPs. However, Natural ankerite is already taking part in mineralogical composition of the French Callovo-Oxfordian host-rock [33, 34] and it is one of the main chemical compositions of the carbonates in the nuclear waste repository formations [35].

4. Experimental Parameters influencing the alteration of HLW glass

Experimental parameters will be special in composition and prepared in a standard methods to demonstrate the deep underground reactions.

4.1. Glass

The study is conducted over small polished simulating nuclear borosilicate glass chips/coupons, involving accelerated ageing of glass due to model environment. SON68 type glass and International Simple Glass ISG are inactive analogs of the French high-level waste glass (R7T7-type) and they are approved by many researches in the literature for studying the durability of the R7T7 glass [1, 2, 7, 12, 16, 25, 26, 27 and 36]. The composition of the SON68 glass is shown in Table 1.

Table 1. Chemical composition in weight of the inactive SON68 nuclear glass [7]

Oxide	Composition, wt.%	Oxide	Composition, wt.%	Oxide	Composition, wt.%
SiO_2	45.85	ZnO	2.53	Nd_2O_3	2.04
B_2O_3	14.14	P_2O_5	0.29	Pr_2O_3	0.46
Na_2O	10.22	SrO	0.35	Ag_2O	0.03
Al_2O_3	5.00	ZrO_2	2.75	CdO	0.03
CaO	4.07	MoO_3	1.78	SnO_2	0.02
Li_2O	1.99	Cs_2O	1.12	TeO_2	0.23
Fe_2O_3	3.03	BaO	0.62	Ce_2O_3	0.97
NiO	0.43	Y_2O_3	0.20	Others	0.39
Cr_2O_3	0.53	La_2O_3	0.93		

4.2. Synthetic Groundwater

To realize the groundwater in the nuclear waste repository formations; a synthetic groundwater is recommended. It is originally Ultra-Pure Water (UPW) under addition of some nutrients and organic carbon. Table 2 illustrates the components, which will be added experimentally to the UPW.

Table 2. Composition of the synthetic argillite water [8]

Compound	Concentration, mole/l	Compound	Concentration, mole/l
Fe	3.40E-05	Na	4.56E-02
Sr	2.00E-04	TIC	3.34E-03
K	1.03E-03	Si	1.80E-04
Mg	6.67E-03	Cl	4.10E-02
Ca	7.36E-03	NH_4	2.90E-04
Al	4.70E-09	Mn	4.70E-06
SO_4	1.56E-02	Lactate	8.00E-06

4.3. Clay
Scientific researchers are using drilled out clay from deep geological formation in selected countries where the HLW could be disposed. These include France (Callovo-Oxfordian Clayrock at Bure), Belgium (Boom Clay at Mol), Hungary (Boda Claystoneat Mecsek), Switzerland (Opalinus Clay at Mont Terri and Benken, and Palfris Formation at Wellenberg), United Kingdom (Oxford Clay) and Canada (Queenston Formation) [28], [33], [37].

4.4. Synthetic Ankerite
In our ongoing experiments synthetic ankerite will be added to the samples by different concentrations, however it is not planned to be added to the blank (reference) sample to figure out how does the ankerite role would affect the Si concentrations in the aqueous solution in the different samples.

5. Conclusions
High level radioactive wastes are immobilized and vitrified by incorporating into borosilicate glass matrix. This glass matrix is poured then into stainless steel containers. Finally it is embedded in a buffer i.e. concrete isolation in the clay-stones of deep geological disposal. In this paper the state of the art of the corrosion of radioactive waste glass in the circumstances of deep geological disposal were extensively studied. Microbial influenced corrosion is highly expected to be happening on the carbon steel canister of the nuclear waste in the anaerobic conditions deep in the geological repository formation. Steel alteration in presence of bacterial activities is proved as a basic factor responsible to accelerate the formation the iron corrosion products. Ankerite should be considered as one of the important iron carbonates, which is suggested to take part in the iron corrosion products in addition to its ubiquitous in soils. Therefore; it is highly expected to be additional factor influencing the glass alteration in the disposal repository formation in addition to the factors already studied by the literature. Ankerite could have multifunctional (direct and indirect) effects on the alteration of glass, which hypothesis should be experimentally proved.

Acknowledgement
Stipendium Hungaricum Scholarship Programme is highly acknowledged for supporting the PhD study and research work. Authors are acknowledged having the opportunity to present the paper at the 5th international conference on competitive materials and technology processes, Miskolc-Lillafüred, Hungary.

References
[1] Minet Y, Bonin B, Gin S, Frugier P 2010 *Journal of Nuclear Materials* **404** (3) 178 https://doi.org/10.1016/j.jnucmat.2010.07.015
Grambow B 2006 *Elements* **2** (6) 357 https://doi.org/10.2113/gselements.2.6.357
[2] Evaluation of the feasibility of a geological repository in an argillaceous formation, *Report Series*, Andra, France, 2005.
[3] Thompson L 2010 *Coursework for Physics* **240** Stanford University

[4] IAEA, Storage and disposal of spent fuel and high level radioactive waste. 2006 https://www.iaea.org/About/Policy/GC/GC50/.../English/gc50inf-3-att5_en.pdf (last visited 25/12/2017).
[5] Treatment and conditioning of nuclear waste *Word Nuclear Association* 2017
[6] Bouakkaz R, Abdelouas A, El Mendili Y, Grambow B, Gin S 2016 *RSC Adv* **6** 72616 https://doi.org/10.1039/C6RA12404F
[7] El Hajj H, Abdelouas A, Grambow B, Martin C, Dion M 2010 *Physics and Chemistry of the Earth Parts A/B/C* **35** (6-8) 248 https://doi.org/10.1016/j.pce.2010.04.007
[8] Damien Féron, Didier Crusset, Jean-Marie Gras 2008 *Journal of Nuclear Materials* **379** (1-3) 16 https://doi.org/10.1016/j.jnucmat.2008.06.023
[9] Gin S 2014 *Procedia Materials Science* **7** 163 https://doi.org/10.1016/j.mspro.2014.10.022
[10] Philippini V, Naveau A, Catalette H, Leclercq S 2006 *Journal of Nuclear Materials* **348** (1-2) 60 https://doi.org/10.1016/j.jnucmat.2005.09.002
[11] Burger E, Rebiscoul D, Bruguier F, Jublot M, Lartigue J E, Gin S 2013 *Applied Geochemistry* **31** 159 https://doi.org/10.1016/j.apgeochem.2012.12.016
[12] Rebiscoul D, Frugier P, Gin S, Ayral A 2005 *Journal of Nuclear Materials* **342** (1-3) 26 https://doi.org/10.1016/j.jnucmat.2005.03.018
[13] Vernaz E, Gin S, Jegou C, Ribet I 2001 *Journal Nuclear Materials* **298** (1-2) 27 https://doi.org/10.1016/S0022-3115(01)00643-2
[14] Gin S, Ribet I, Couillard M 2001 *Journal Nuclear Materials* **298** (1-2) 1 https://doi.org/10.1016/S0022-3115(01)00573-6
[15] Curti E 2003 *Technical Report* No. PSI-03-18, Paul Scherrer Institute, CH-5232 Villigen PSI (Switzerland) *INIS* **36** (3) 62.
[16] Kuang F, Wang J, Yan L, Zhang D 2007 *Electrochimica Acta* **52** (20) 6084 https://doi.org/10.1016/j.electacta.2007.03.041
[17] Sagues A A, Poor N D, Gaseres L, Akhoondam M 2009 *Final Report to Florida Department of Transportation* University of South Florida
[18] Beech I B 2003 *Microbiology Today* **30** (3) 115
[19] Little B J, Lee J S, Ray R I 2006 *Corrosion* **62** (11) 1006 https://doi.org/10.5006/1.3278228
Werme L, Bjorner I K, Bart G, Zwicky H U, Grambow B, Lutze W, Ewing R C, Magrabi C 1990 *J. Material Research* **5** (5) 1130 https://doi.org/10.1557/JMR.1990.1130
[20] Bart G, Zwicky H U, Aerne E T, Graber T H, Z'Berg D, Tokiwai M 1987*Symposium L - Scientific Basis for Nuclear Waste Management X* **18** (22) 459 https://doi.org/10.1557/PROC-84-459
[21] Mitsui S, Aoki R 2001 *Journal of Nuclear Materials* **298** (1-2) 184 https://doi.org/10.1016/S0022-3115(01)00595-5
[22] Jordan N, Marmier N, Lomenech C, Giffaut E, Ehrhardt J J 2007 *Journal of Colloid and Interface Science* **312** (2) 224 https://doi.org/10.1016/j.jcis.2007.03.053
[23] Dillmann P, Gin S, Neff D, Gentaz L, Rebiscoul D 2016 *Geochimica et Cosmochimica Acta* **172** 287 https://doi.org/10.1016/j.gca.2015.09.033
[24] Jollivet P, Angeli F, Cailleteau C, Devreux F, Frugier P, Gin S 2008 *Journal of Non-Crystalline Solids* **354** (45-46) 4952https://doi.org/10.1016/j.jnoncrysol.2008.07.023
[25] Jonathan P Icenhower, Carl I Steefel 2013 *Journal of Nuclear Materials* **439** (1-3) 137 https://doi.org/10.1016/j.jnucmat.2013.04.008
[26] Breitner D, Osán J, Fábián M, Zagyvai P, Szabó C, Dähn R, Fernandes M M, Sajó I, Máthé Z, Török S 2015 *Environmental Earth Sciences* **73** (1) 209 http://dx.doi.org/10.1007/s12665-014-3413-4
[27] Buckley H A, Woolley A R 1990 *Mineralogical Magazine* **54** 413 https://doi.org/10.1180/minmag.1990.054.376.06
[28] Chai L, Navrotsky A 1996 *American Mineralogist* **81** (9-10) 1141 https://doi.org/10.2138/am-1996-9-1012
[29] Amethyst Galleries' Mineral Gallery, Ankerite, http://www.galleries.com/Ankerite (last visited on 24/12/17).
[30] Xu T, Apps J A, Pruess K 2005 *Chemical Geology* **217** (3-4) 295

https://doi.org/10.1016/j.chemgeo.2004.12.015
[31] Gaucher E C, Tournassat C, Pearson F J, Blanc P, Crouzet C, Lerouge C, Altmann S 2009 *Geochimica et Cosmochimica Acta* **73** (21) 6470 https://doi.org/10.1016/j.gca.2009.07.021
[32] Lerouge C, Vinsot A, Grangeon S, Wille G, Flehoc C, Gailhanou H, Gaucher E C, Madé B, Altmann S, Tournassat C 2013 *Procedia Earth and Planetary Science* **7** 475 https://doi.org/10.1016/j.proeps.2013.03.133
[33] Tournassat C, Lerougea C, Blanc P, Brendlé J, Greneche J M, Touzelet S, Gaucher E C 2008 *Applied Geochemistry* **23** (4) 641 https://doi.org/10.1016/j.apgeochem.2007.11.002
[34] Collin M, Fournier M, Frugier P, Charpentier T, Moskura M, Deng L, Ren M, Du J, Gin S 2018 *Materials Degradation* **2** (4) https://doi.org/10.1038/s41529-017-0025-y
[35] Buocz I, Rozgonyi-Boissinot N, Török Á, Görög P 2014 *Pollack Periodica* **9** (3) 139 https://doi.org/10.1556/Pollack.9.2014.3.15

IOP Conf. Series: Materials Science and Engineering **613** (2019) 012031 doi:10.1088/1757-899X/613/1/012031

Time evolution of ZnO-CNT photoluminescence under variable ambient and temperature conditions

P Rauwel[1], A Galeckas[2] and E Rauwel[1]

[1]Institute of Technology, Estonian University of Life Sciences, Kreutzwaldi 56/2, Tartu (Estonia)
[2]Department of Physics, University of Oslo. P.O. Box 1048 Blindern, 0316 Oslo (Norway)

Email: protima.rauwel@emu.ee

Abstract. ZnO-CNT hybrid materials were prepared by non-aqueous sol-gel routes at 240°C. The morphology and defects have been studied by transmission electron microscopy and photoluminescence spectroscopy. The hybrid nanostructures manifest a broad luminescence emission covering the visible spectrum. Variations in photoluminescence with time are observed for the ZnO-CNT hybrid structures exposed to different ambient and temperatures. The studies show that photoluminescence from adsorbed oxygen dominates the emission emanating from other intrinsic defects and also affects the stability of the latter over time.

1. Introduction

ZnO and carbon-nanotube (CNT) hybrid materials have gained popularity over the past decades [1, 2] which is largely due to the influence that the carbon-based component exerts over the synthesis of ZnO nanoparticles, hence in determining the ultimate properties. In its general bulk form, ZnO is a semiconductor material with a wide band gap of 3.37eV, however, once the crystal dimensions are reduced to the nanometer scale, the presence of carbon-based materials can modify the crystallinity, defect states as well as the morphology of the ZnO nanoparticles (NPs). In turn, the photoluminescence (PL) of such hybrid structures is also modified as compared to the stand-alone ZnO NPs. Typical PL spectrum of ZnO exhibits two characteristic features - a relatively narrow peak in the UV region pertaining to the near-band-edge (NBE) emission and originating from exciton annihilation and recombination processes via shallow states [3, 4]. The other characteristic feature is a broad visible emission originating from intrinsic defects that introduce luminescent states within the band gap, hence commonly referred to as deep-level emission (DLE) band. The intensity ratio of NBE to DLE allows for qualitative assessment of the crystallinity of ZnO. The ZnO-CNT hybrid structures generally yield a high NBE to DLE ratio due to an enhanced surface plasmon mediated effect [5]. Also noteworthy is that the presence of carbon-based structures has a photocatalytic effect on ZnO, as it introduces defect states in the band gap of ZnO which then serve as efficient photoexcited charge separators [6].

Besides the ZnO NPs crystallinity, an equally important characteristic is the stability of the hybrid material properties, which can be assessed by monitoring the developments of luminescence properties over time under continuous UV exposure at different ambient and temperature conditions. In the case of ZnO-CNT hybrid structures, the time evolution of PL features to a large extent reflects the stability of the defects involved; these include point and surface defects of the ZnO as well as the interfacial defects between the ZnO and CNT. The charging effects of the surface states, their passivation and the energy transfer efficiency from ZnO to CNT along with the presence of impurities and defect sites, all these

IOP Conf. Series: Materials Science and Engineering **613** (2019) 012031 doi:10.1088/1757-899X/613/1/012031

factors can introduce significant variations in the charge carrier dynamics over time. In this regard, time-lapse PL measurements may provide an insight into the charge developments which is essential for determining the feasibility of ZnO–CNT hybrid structures in optoelectronic applications.

In the present work, we extend further our earlier studies of PL in air and in vacuum [7] and focus on the charge carrier behavior over time by monitoring time evolution of the PL features of ZnO-CNT hybrid structures prepared via a non-aqueous sol gel route at 240°C. Such an approach provides insights into the origin of luminescent states within the bandgap and their stability at different temperatures (10K and 300K) and ambient conditions (vacuum and air), which are among the key factors in determining the potential of these hybrid nanostructures in optoelectronic applications.

2. Experimental

2.1 Synthesis

The procedure for synthesizing ZnO NPs was carried out under air. In a typical synthesis, zinc acetate (3.41mmol) (99.99%, Aldrich) was added to 20 mL (183mmol) of benzylamine (≥99.0%, Aldrich). The reaction mixture was transferred into a stainless steel autoclave and carefully sealed. Thereafter, the autoclave was taken out of the glovebox and heated in a furnace at a temperature 240°C for 2 days. The resulting milky suspensions were centrifuged; the precipitates were thoroughly washed with ethanol and dichloromethane and subsequently dried in air at 60°C. In the case of the carbon nanohybrid synthesis, NANOCYL NC7000 MWCNT with an average diameter of 10nm and length of 1.5μm was used. For the synthesis of carbon-based ZnO nanocomposites, the MWCNTs were directly homogeneously dispersed into the solution of zinc acetate and benzylamine before transferring the solution into an autoclave for the reaction synthesis.

2.2 Characterization:

High-resolution transmission electron microscopy (HRTEM) was carried out on a probe-corrected Titan G2 80– 200kV operating at 200kV in TEM mode and providing a point to point resolution of 2.4Å. Photoluminescence (PL) measurements were carried out at 10K and 300K temperatures by employing the 325 nm wavelength of a cw He– Cd laser with an output power of 10mW as an excitation source. The emission was collected by a microscope and directed to a fiber optic spectrometer (Ocean Optics USB4000, spectral resolution 2 nm). Temperature dependent measurements were realized using closed-cycle He-refrigerator (Janis, Inc. CCS450). Room temperature PL measurements of the powder-like samples have been performed both in air and vacuum. The density of the compact powders is estimated to be approximately the same in all measurements.

3. Results and discussions

3.1 Morphological characterizations

Former x-ray diffraction studies demonstrated that ZnO nanoparticles are highly crystalline with a hexagonal wurtzite ($P6_3mc$) crystal structure (a = 3.25Å and c = 5.20Å) with an average diameter of 90nm [7]. The TEM image of figure 1 provides an overview of the nanoparticles synthesized at 240°C. ZnO particles of various shapes are illustrated. The nanotubes visible at the center of the image are linked to various ZnO particles, showing an intimate contact between them. The shape of the nanoparticle is an indication of the presence of polar surfaces in such materials. Hexagonal shaped nanoparticles contain 4 non-polar surfaces and 2 polar surfaces, where the former type is susceptible to oxygen vacancies and therefore also oxygen chemisorption [8]. The HRTEM study showed that the size of the ZnO nanoparticles is around 125nm. The arrows indicate strain in the ZnO structure implying stacking faults in the volume of the nanoparticle.

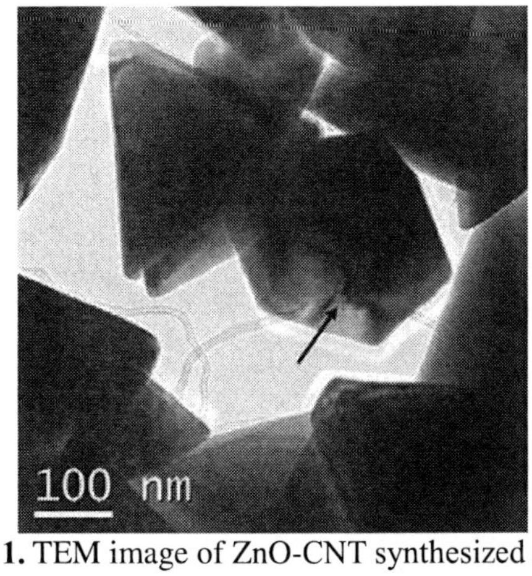

Figure 1. TEM image of ZnO-CNT synthesized at 240°C

3.2 Photoluminescence characterization

The time evolution of the ZnO-CNT photoluminescence was investigated as a function of different ambient (air versus vacuum) and temperature (room versus cryogenic) conditions. Figures 2, 3 and 4 summarize the time-lapse PL measurements with the key elements of emission represented by Gaussian deconvolution fits. One can observe a certain similarity of all PL plots featuring a repeatable set of the same key spectral components, as indicated by vertical markers positioned at around 3.1eV, 2.8eV, 2.5eV, 2.3eV, 2.1eV and 1.9eV. Another shared feature is a significant decrease of the PL intensity in the visible blue-green regions of spectra with the elapsed UV-exposure time. The dynamics of this process for each particular case of ambient and temperature is represented by a characteristic decay time constant (τ) estimated from the integrated PL intensity transients shown in the insets of figures 2, 3 and 4.

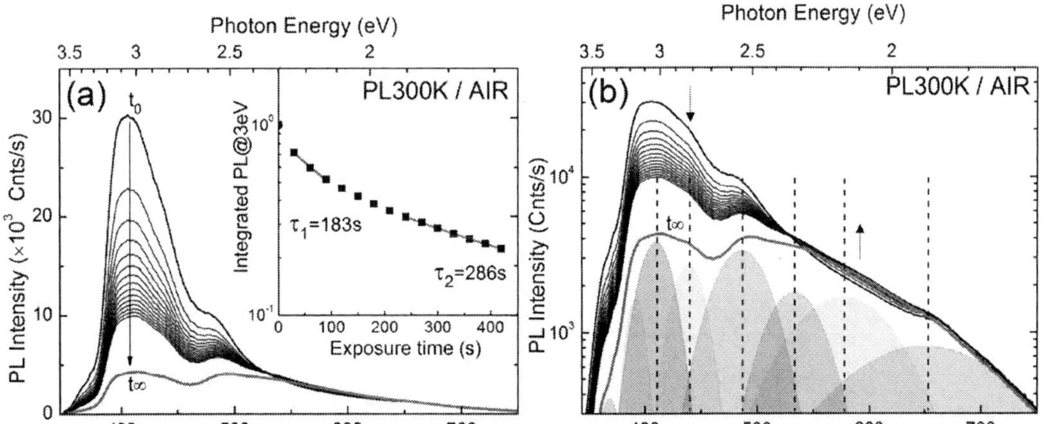

Figure 2. Time evolution of the PL measured at 300K in air ambient from the hybrid ZnO-CNT nanocomposites synthesized at 240°C: (a) time-lapse PL spectra as a function of UV-exposure time. Inset shows decay transient of the integrated PL intensity. (b) Semi-log plot of the PL developments with the key spectral components represented by Gaussian deconvolution fits (shaded areas); vertical dashed markers indicate the monitored spectral components; up/downward arrows point out PL intensity trends.

The dominant feature in all presented PL spectra is a broad blue band combining conventional NBE emission with two extra components at 3.1eV and 2.8eV, which are within spectral region known for optical manifestation of zinc vacancies (V_{Zn}) and interstitials (Zn_i) [9, 10]. The NBE emission adjoins and partially overlaps with the deep-defect related green luminescence comprised of two closely positioned bands centered at 2.3eV and 2.5eV. The latter, at 2.5eV, occurs when a singly ionized oxygen vacancy turns neutral (V_O^+/V_O) on capturing an electron from the conduction band and recombining with a valence band hole [11]. The other green emission component at 2.3eV is believed to originate from a singly ionized surface oxygen vacancy capturing a hole from the surface states and turning doubly ionized (V_O^+/V_O^{2+}), thus largely representing a surface phenomenon unlike the green emission at 2.5eV which is mostly volume-related. Finally, the red emission centered at 1.9eV is commonly attributed to accumulation of zinc vacancies, as separate V_{Zn} clusters or in association with extended defects, and apparently remains unchanged over time under any of varied ambient conditions (cf. figures 2, 3 and 4). The presence of the extended defects is directly evidenced by TEM image in figure 1, thus supporting the relation of the red emission with the accumulated V_{Zn} point defects.

Figure 2 shows PL of the hybrid ZnO-CNT measured in air ambient at 300K as a function of UV-exposure time. One can observe a substantial (~10 fold) and rapid PL intensity decrease (decay times 183s/286s) in the UV-visible spectral region. The underlying mechanism behind this dynamic phenomenon is attributed to surface band bending induced by adsorbents and their gradual desorption stimulated by UV-illumination. The actual band bending is a superposition of several opposed effects expected from trapping of H_2O and O_2 molecules, from hydroxyl (OH) termination or some residual surface contaminants. For ZnO-CNT hybrids prepared by non-aqueous route at 240°C, the most likely adsorbent in air ambient is molecular oxygen, which acts as negatively charged surface acceptor after electron capture from the conduction band and thus inducing upward band bending [12, 13]. Under UV-illumination, the photogenerated holes may react with surface acceptors and release the oxidizing agents, which in turn lead to increase of carrier concentration and downward band bending.

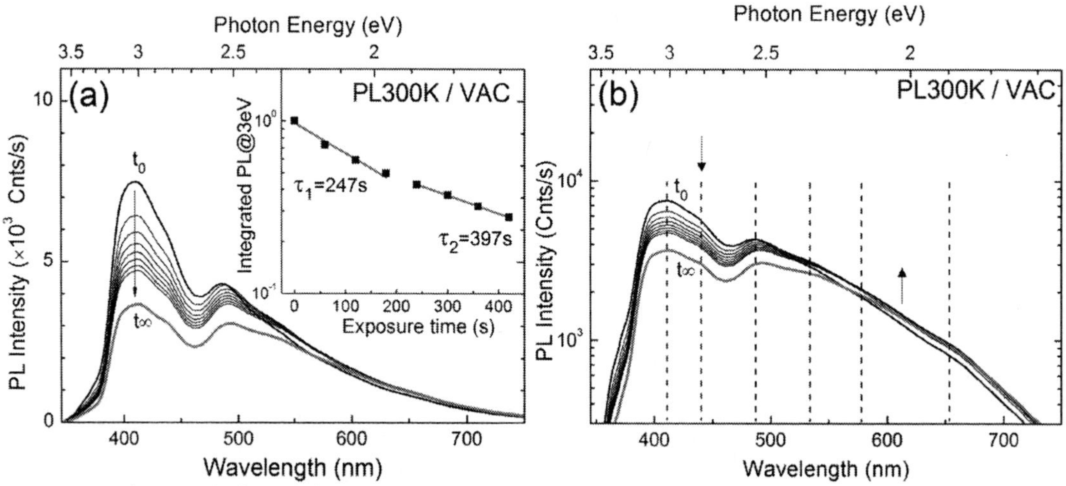

Figure 3. Time evolution of the PL measured from the hybrid ZnO-CNT nanocomposites at 300K in vacuum: (a) time-lapse PL spectra as a function of UV-exposure time. Inset shows decay transient of the integrated PL intensity. (b) Semi-log representation of the PL developments with the main spectral components indicated by vertical dashed markers.

In vacuum at 300K, the overall quantum efficiency is noticeably reduced (~4 fold) compared to that in air ambient (figure 3), as can be seen from the comparison of figures 3 and 2. Since there is no sustained oxygen supply in vacuum, the upward band bending is already reduced due to natural O_2 desorption (without UV-illumination) before the PL measurements even start. It is also noteworthy that the saturated PL spectra, i.e. those measured after prolonged UV-exposures, settle down at very similar levels in the case of air and vacuum, thus only the magnitude of PL changes appears as the decisive distinction. Figure

4 depicts qualitatively similar PL developments observed in vacuum at 10K, except for a considerably higher overall PL quantum efficiency, which is due to suppressed non-radiative recombination pathways at cryogenic temperatures.

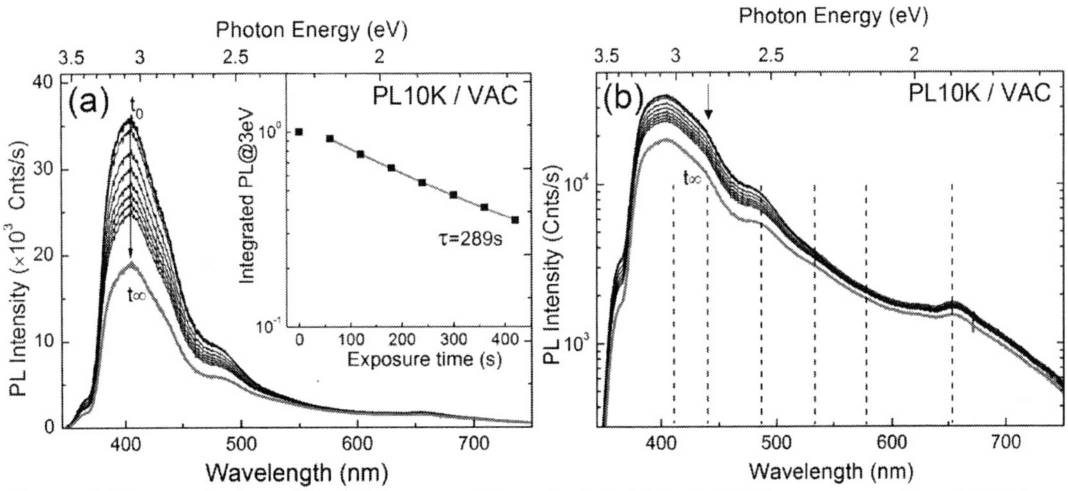

Figure 4. Time evolution of the PL measured from the hybrid ZnO-CNT nanocomposites at 10K in vacuum: (a) time-lapse PL spectra as a function of UV-exposure time. Inset shows decay transient of the integrated PL intensity. (b) Semi-log representation of the PL developments with the main spectral components indicated by vertical dashed markers.

The PL spectra indicate that ZnO nanoparticles are highly defective and are dominated by Zn vacancies. Extended defects formed by these Zn vacancies are rather robust and do not show much variation in the PL excitation-emission experiments performed by varying temperature and ambient. However, the Zn vacancies closer to the surface are greatly affected by ambient and temperature. It is to be noted that at 10K in vacuum the PL spectra lose some amount of intensity over time; this phenomenon is most likely due to charge transfer from ZnO to CNT or due to the band gap states becoming saturated.

4. Conclusion

We have synthesized ZnO-CNT hybrids at 240°C via non aqueous sol-gel routes. The PL spectra of the ZnO nanoparticles suggest a high level of Zn vacancies in the form of point and extended defects. The Zn vacancies close to the surface, like other surface defects, are very susceptible to the ambient conditions and tend to get dominated by adsorbed oxygen when measured over a period of time. Low temperature PL in vacuum helps in understanding the effect of suppressing the adsorbed oxygen in turn enhancing the PL emission from intrinsic defects which in this present case are mostly V_{Zn}. The observed optical sensitivity of hybrid ZnO-CNT nanocomposites to ambient may lead to new applications as gas sensors.

Acknowledgements

This research has been supported by the European Regional Development Fund project EQUiTANT (F180175TIBT). Research Council of Norway and the University of Oslo through the FUNDAMENT project (No. 251131) is gratefully acknowledged.

References

[1] D Eder 2010 *Chemical Reviews* **110** (3) 1348 https://pubs.acs.org/doi/10.1021/cr800433k
[2] P Rauwel, M Salumaa, A Aasna, A Galeckas, E Rauwel 2016 *Journal of Nanomaterials* 5320625 1 http://dx.doi.org/10.1155/2016/5320625

IOP Conf. Series: Materials Science and Engineering **613** (2019) 012031 doi:10.1088/1757-899X/613/1/012031

[3] B J Coppa, C C Fulton, S M Kiesel, R F Davis, C Pandarinath, J E Burnette, R J Nemanich, D J Smith 2005 *J. Appl. Phys.* **97** 103517 https://doi.org/10.1063/1.1898436
[4] E Rauwel, A Galeckas, P Rauwel, M Fleissner Sunding, H Fjellvåg 2011 *J. Phys. Chem. C* **115** (51) 25227 https://pubs.acs.org/doi/abs/10.1021/jp208487v
[5] X L Li, C Li, Y Zhang, D P Chu, W I Milne, H J Fan 2010 *Nanoscale Res. Lett.* **5** 1836 https://doi.org/10.1007/s11671-010-9721-z
[6] J Bae, J B Han, X-M Zhang, M Wei, X Duan, Y Zhang, Z L Wang 2009 *J. Phys. Chem. C* **113** 10379 https://pubs.acs.org/doi/abs/10.1021/jp901011u
[7] E Rauwel, A Galeckas, M Rosário Soares P Rauwel 2017 *J. Phys. Chem. C* **121** 14879 https://pubsdc3.acs.org/doi/abs/10.1021/acs.jpcc.7b03070?journalCode=jpccck
[8] H Zhang, J Sun, C. Liu, Y Wang 2015 *J. Catal.* **331** 57 https://doi.org/10.1016/j.jcat.2015.08.016
[9] A Janotti, C G Van de Walle 2007 *Physical Review B* **76** 165202 https://doi.org/10.1103/PhysRevB.76.165202
[10] H Zeng, Z Li, W Cai, P Liu 2007 *J. Applied Phys.* **102** 104307 https://doi.org/10.1063/1.2803712
[11] M Ghosh, A K Raychaudhuri 2008 *Nanotechnology* **19** 445704 https://doi.org/10.1088/0957-4484/19/44/445704
[12] J W P Hsu, D R Tallant, R L Simpson, N A Missert, R G Copeland 2006 *Applied Physics Letters* **88** 252103 https://doi.org/10.1063/1.2214137
[13] D Wang, N Reynolds 2012 *ISRN Condensed Matter Physics* 950354 1 http://dx.doi.org/10.5402/2012/950354

IOP Conf. Series: Materials Science and Engineering **613** (2019) 012032 doi:10.1088/1757-899X/613/1/012032

Degradation of methylene blue under UV and visible light illumination by yttria doped titania nanoparticles

J Grabis, Dz Jankoviča, I Sīpola

Riga Technical University, Institute of Inorganic Chemistry, Latvia

E-mail: jgrabis@rtu.lv

Abstract. Yttria doped TiO_2 nanoparticles were prepared by sol-gel method using the non-aqueous solutions in presence of urea and without it followed by calcination at 500–750 °C for 2h. Photocatalytic activity of the samples was determined by degradation methylene blue under Hg and solar light imitating illumination. Specific surface area, crystallite size, phase transition of TiO_2 and photocatalytic activity depend on content of Y_2O_3 and presence of urea during synthesis. The highest photocatalytic activity showed TiO_2 doped with 0.8–1.0 wt.% of yttria.

1. Introduction

Recently yttria doped titania was proven as promising active photocatalyst for degradation organic pollutants under ultraviolet and solar light illumination [1-3] and for the use as solar cell photoanode [4] due to lattice distortion, band gap narrowing and decrease of recombination rate of charge carriers.

Crystalline TiO_2 nanoparticles doped with yttria were prepared using combustion synthesis including drying of the precursors solution at 105 °C for 18 h followed by calcination at temperature of 300–800 °C for 20 h in air [1]. Samples with yttria content of 1 mol% showed the higher photocatalytic activity than commercial TiO_2 (P-25) but synthesis was very lasting. Active, partially X-ray amorphous TiO_2 films with Y_2O_3 content up to 12 at.% were prepared by reactive magnetron sputtering of titanium and yttrium target [2]. Highly active yttria doped titania nanosheets with specific surface area of 240–50 m^2/g were obtained by hydrothermal treatment of TiO_2 (P-25) in NaOH and $Y(NO_3)$ solution at 130 °C for 3 h [3]. The nanosheets with Y_2O_3 content of 0.75 wt.% exhibited the highest photocatalytic activity under solar light and UV-VIS light illumination. In similar way TiO_2 nanotubes doped with 0.5–2.0 at.% of yttria were prepared by microwave refluxing treatment of doped TiO_2 in presence of NaOH aqueous solution [5].

X-ray diffraction patterns analysis (XRD) of the prepared by known methods photocatalysts showed only characteristic TiO_2 peaks indicated to entering yttria into titania lattice. However, high-resolution electron microscopy studies showed that Y_2O_3 nanoparticles at least partially were deposited onto TiO_2 nanosheets [3]. The yttria doped TiO_2 samples show enhanced photocatalytic activity but optimal content of yttria differs and, obviously, it depends on preparation method.

The aim of the work was preparation of crystalline yttria doped TiO_2 nanoparticles by sol-gel method using a non-aqueous solutions and determination of their parameters and photocatalytic activity in dependence on used precursors, calcination temperature and content of yttria.

2. Experimental

TiO_2 and TiO_2/Y_2O_3 nanoparticles were prepared by sol-gel method using titanium isopropoxide (TTIP, 97%, Merck), acetyl acetone (AcAc, 99%, Merck), yttrium nitrate (Merck), nitric acid and ethanol as precursors. TTIP was dissolved in ethanol and AcAc at 40 °C. Definite amount of yttrium nitrate solution in ethanol and HNO_3 was added drop-wise with constant stirring for 30 min (route 1). In separate experiments urea solution was added (route 2). The temperature was increased up to 80 °C. The formed gel was calcinated at temperature range of 500–750 °C for 2 h. The samples prepared by both methods (with urea and without it) contained Y_2O_3 in the range of 0. 25–3.0 wt.%.

Phase and chemical composition of the prepared samples were determined by X-ray diffraction analysis (Advance D8, Bruker AXS), and X-ray flourescence spectroscopy (PioneerS4, Bruker AXS). The crystalline size was calculated from broadening of diffraction maxima using software Topas3.

The photocatalytic activity of the prepared samples was investigated by degradation of methylene blue (MB) solution with concentration of 7.2 mg/L. The photocatalyst (0.1 g/L) was dispersed into MB solution in quartz cylinder equiped with magnetic stirrer. Before illumination by Hg lamp (120 W) or sunlight imitating Osram Vitalux lamp (300 W) the solution was stirred for 30 min in dark to ensure adsorption-desorption equilibrium. The degradation of MB was determined by measuring of the light absorption by solution after removing solid particles at wave length of 662 nm using Jenway-6300 spectrometer. The MB degradation rate constant was calculated by using a pseudo-first kinetic model.

3. Results and discussion

XRD patterns of the pure or doped TiO_2 samples calcinated at 500 °C showed only intensive broad TiO_2 anatase peaks independently on the content of Y_2O_3 in the range of 0.25–3.0 wt.% and preparation route. Increase of the calcination temperature in the range of 550–750 °C led to narrowing of the diffraction maxima due to crystallization and to change phase composition of TiO_2 in dependence on the preparation route and Y_2O_3 additives

Anatase phase transition to rutile of bare TiO_2 samples started at temperature above 500 °C and process intensity depended on the use of urea during synthesis, especially at calcination temperatures in the range of 500–570 °C. The TiO_2 samples prepared with or without urea after calcination at 570 °C contained 57% and 30% anatase respectively (Fig. 1).

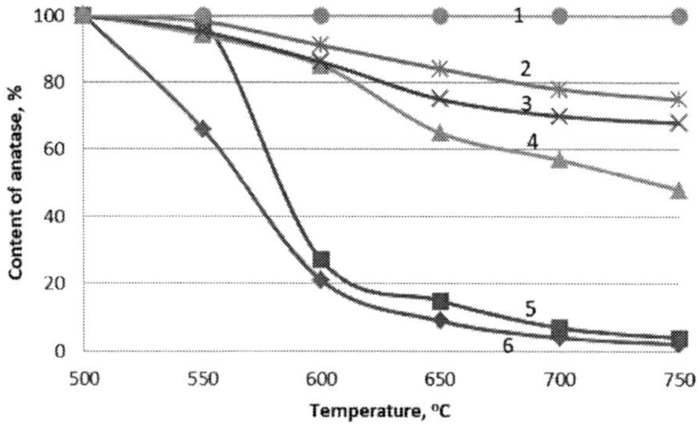

Fig. 1. Content of anatase in samples prepared by route 1 (3, 4, 6), by route 2 (1, 2, 5);1, 3 – content of yttria 2.0 wt.%; 2, 4 – content of yttria 0.8 wt.%; 5, 6 – bare titania.

The phase transition temperature of the samples was increased by TiO_2 doping with Y_2O_3. Among yttria doped TiO_2 samples powders prepared by route 2 showed the highest resistance against rutile formation. The TiO_2 particles containing 2.0 wt.% of Y_2O_3 and prepared by route

1 or route 2 contained 100% and 68% anatase respectively after calcination at 750 °C. The influence of Y_2O_3 and urea on the TiO_2 phase composition was close to literature data [6]. Accordingly, to [1] the phase transformation of the anatase into rutile phase in yttria doped TiO_2 occurs above 700 °C. It is well-known that the use of urea in sol-gel synthesis of TiO_2 leads to incorporating nitrogen (0.3 wt.%) into titania and urea prevents phase transformation from anatase to rutile phase at least at temperature below 600 °C [6]. In our case combined influence of yttria and urea on the TiO_2 phase composition increased phase transformation temperature of TiO_2/Y_2O_3 sample prepared in presence of urea up to 750 °C.

Some additional effect can be connected with different crystallite size of 8–10 nm and 15–20 nm and specific surface area of the samples (Fig. 2) prepared by route 2 and by route 1 respectively.

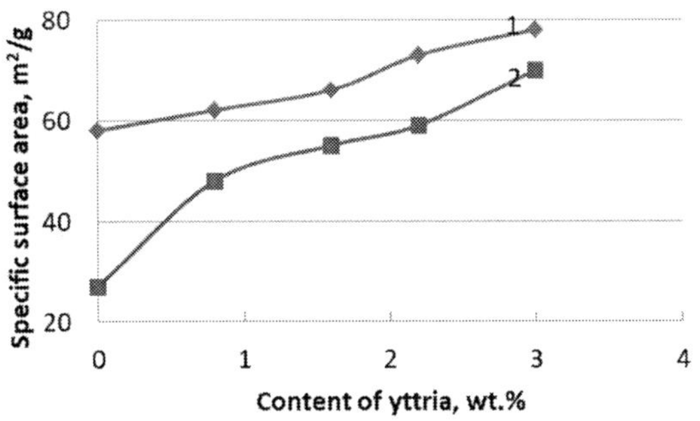

Fig. 2. Specific surface area of doped TiO_2 samples calcinated at 550 °C versus content of Y_2O_3: 1 – prepared by route 2; 2 –by route 1.

The specific surface area of the bare TiO_2 samples prepared by the route 2 approximately twice exceeded that of the samples prepared by route 1. The specific surface area of doped samples prepared by both routes gradually increased with content of Y_2O_3. The specific surface area of all samples decreased with calcination temperature but some influence of preparation route on surface area retained.

Degradation of MB solution by Y_2O_3(0.8 wt.%) doped TiO_2 and bare TiO_2 depended on preparation route and calcination temperature (Fig. 3).

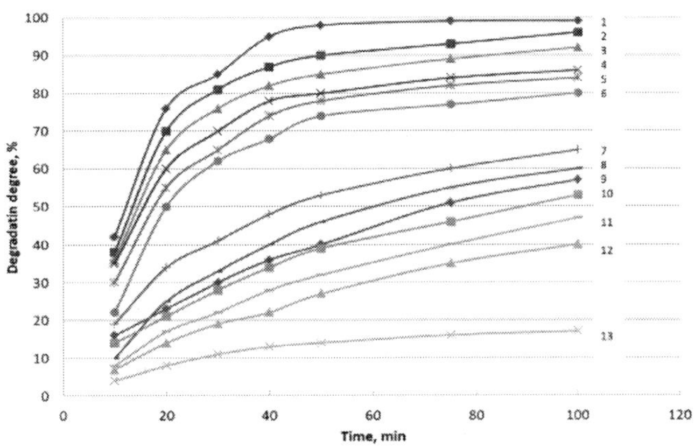

Fig. 3. Degradation degree of MB solution by Y_2O_3(0.8 wt.%) doped TiO_2 (1-6) and by pure TiO_2 (7-12) particles calcinated at 550 °C (2, 6, 8, 10), 600 °C (1, 4, 7, 9), and 750 °C (3, 5, 11, 12), and prepared by route 2 (1, 2, 3, 9, 10, 11) and by route 1 (4, 5, 6, 7, 8, 12); blank solution (13).

The photocatalytic activity of the samples increased with calcination temperature up to 600 °C. It should be explained by increase of crystallinity and by the presence of rutile phase (10–15%). On the other hand calcination at 750 °C led to increase content of rutile phase at least up to 52% and reduction of specific surface area, and therefore to decrease of the sample photocatalytic activity. From data followed that activity of doped TiO_2 remarkably exceeded that of bare TiO_2. The highest activity showed Y_2O_3(0.8 wt.%) doped titania and bare TiO_2 prepared using route 1 and route 2 respectively. Similar MB degradation curves under solar and Hg illumination were obtained for TiO_2 samples with Y_2O_3 content in the range of 0.25–3.0 wt.% and degradation rate constant were calculated (Fig. 4).

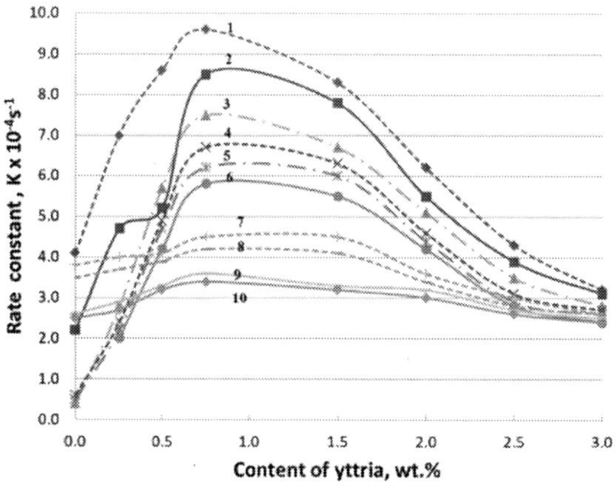

Fig. 4. MB solution degradation rate constants. TiO_2/Y_2O_3 nanoparticles prepared by route 2 (4, 5, 6, 8, 10) and by route 1 (1, 2, 3, 7, 9), calcinated at 550 °C (2, 6, 9, 10), 600 °C (1, 4, 7, 8) and 750 °C (3, 5) under solar light (1-6) and Hg lamp illumination (7-10).

Photocatalytic degradation rate constant reached maximal value at Y_2O_3 content about 0.8–1.0 wt.% and it was higher under solar light imitating illumination due to higher power of the lamp with respect to Hg lamp.

4. Conclusions

Sol-gel synthesis of yttria doped titania nanoparticles in presence of urea ensured increase specific surface area of the powders and eliminated rutile formation during calcination up to 750 °C but led to decrease of their photocatalytic activity under solar light imitating illumination with respect to the particles prepared without using urea.

The photocatalytic activity of the yttria doped titania depended on the content of the dopant, calcination temperature and phase composition.

The highest photocatalytic activity in degradation of MB solution under solar light simulated illumination showed TiO_2 doped with 0.8–1.0 wt.% of Y_2O_3 calcinated at 600 °C and containing 85% anatase and 15% rutile.

Acknowledgements. The research was supported by ERDF project No. 1.1.1.1/16/A/079 «Development of active under sunlight immobilized TiO_2-ZnO based photocatalysts"

5. References

[1] Young J W, P S, Dae L G, Sig L M, Gang-Woo L, and Seong-Soo H 2010 *Res J Chem Environ* **14** 17
[2] Zhang W, Wang K, Zhu S, Li Y, Wang F, He H 2009 *Chem Eng J* **155** 83 http://dx.doi.org/10.1016/j.cej.2009.06.039

IOP Conf. Series: Materials Science and Engineering **613** (2019) 012032 doi:10.1088/1757-899X/613/1/012032

[3] Zhao X, Wu P, Liu M, Lu D, Ming J, Li C, Ding J, Yan Q, Fang P 2017 *Appl Surf Sci* **410** 134-144 http//:dx.doi.org/10.1016/j.apsusc.2017.03.073.
[4] Wang W, Liu Y, Sun J, Gao L 2016 *J Alloys Compd* **659** 15 http://dx.doi.org/10.1016/j.jallcom.2015.10.254.
[5] Rao J, Xue H, Zhang W, Li X, You X, Xing Z 2016 *Int. J Electrochem Sc* **11** 2408
[6] Cheng P Changsheng D, Gub M, Daia X 2008 Mat Chem Phys **107** 77 http://dx.doi.org/10.1016/j.matchemphys.2007.06.051

Influence of nanodiamond loading on properties of poly(vinyl alcohol) nanocomposite membranes.

Tomáš Remiš, Jaroslav Kadlec, Tomáš Kovářík

New Technologies - Research Center, University of West Bohemia, Plzeň 306 14, Czech Republic

tremis@ntc.zcu.cz

Abstract. Carbon nanomaterials are well-known for their unique physical properties. They have attracted interest as reinforcing fillers because of their superb mechanical properties. We report the excellent reinforcement properties of polymer nanocomposites by the incorporation of nanodiamond (ND). ND has been expected to offer polymer nanocomposites optimal properties because of its smooth surface and excellent optical, mechanical, and thermal properties, which can approach the values of single diamond crystal. We prepared advanced nanocomposites membranes based on poly(vinyl alcohol) (PVA) and ND in a single step using the solution casting method from aqueous medium and achieved the high dispersibility of ND in the PVA matrices. The thermal, mechanical and rheological properties of the membranes were investigated by means of thermogravimetry (TGA), dynamic mechanical analysis (DMA) and Ares G2 rheometer. The resulting nanocomposites had excellent properties derived both from ND and PVA. Thermal and mechanical properties increased dramatically with increasing ND content, indicating a strong interaction between ND and PVA. We anticipate that ND will be able to compete as a nano filler against conventional carbon-based nano fillers for polymer composites, and it is possible their reinforcement properties will be extended in the future.

1. Introduction

Polymer nanocomposites have attracted much attention not only in scientific fields but also in industrial applications due to the excellent abilities of nano-sized fillers. Nanocomposites have been reported to provide remarkable improvements in materials properties, including mechanical, thermal, and barrier properties, compared with those of polymer itself and micro-sized fillers composites.[1] Carbon-based nanomaterials are well-known for their incredibly high mechanical properties, and they have attracted interest as reinforcing fillers for polymer nanocomposites.[2] In recent years, NDs has been produced by several procedures, such as detonation shock-wave synthesis (from various starting materials), and it has been exploited in vast fields of nanotechnology.[3] In this work, we selected PVA, a watersoluble and biodegradable polymer, as a polymer matrix. PVA membranes are applied in many fields, especially in organic/inorganic separations, pervaporation, biomedical applications, catalysis, controlled drug release.[4-6] We prepared PVA+ND nanocomposites by single-step solution casting method from an aqueous suspension, and we investigated the structure and thermo-mechanical properties of the nanocomposites.

2. Experimental

2.1. Material
PVA (hydrolyzed 99+%, M_w = 86000–98000) was obtained from Sigma-Aldrich (USA). Powder of grade G01 detonation nanodimanods (NDs) was purchased from PlasmaChem GmbH. All the chemicals were of analytical grade and used as received. Deionized water was utilized for all the membrane preparation experiments.

2.2. Membrane preparation
Nanocomposite PVA membranes were prepared using a solution casting method. Surface of NDs have been modified via thermal treatment in air atmosphere to provide them negatively charge nanodiamonds (NDs^-), which were obtained by annealing at 450 °C for 5 h. Then, the annealed NDs^- powder was dispersed in deionized water (DI) at different concentration from 0.05 to 1.9 %, expressed as weight percent in mentioned solvent (Figure 1). As a polar protic solvent was used DI with resistivity 14.3 MΩ·cm at 25°C. Bandelin Sonopuls - Ultrasonic homogenizer HD 3200 was applied to obtain required dispersion of NDs^-. Sonication for 3 h with continuous stirring were applied to acquire grey-colored nanodispersion. Afterwards, the solutions were centrifuged three times at 4000 rpm (2325g) for 1 hour. Next a 5 wt% solution of PVA in NDs solutions was prepared by continuous stirring at 90 °C until the complete dissolution was achieved. The prepared solution was poured into the automatic film applicator (TQC, Germany) equipped with a glass plate and was fabricated into membranes with uniform thickness and dried at 40 °C.

Figure 1. The visual appearance of NDs dispersions in different weight concentrations 0.05, 0.25, 1 and 1.9 wt% for DI water.

2.3. Characterization
The rheological studies of the prepared PVA and PVA+ND solutions were measured on a rheometer ARES G2 from TA Instruments. The polymer solutions were inserted between two paralel plate geometries (d = 25 mm). For each measurement, the shear rate from 0.1 to 100 s^{-1}, for a duration 30 s was carried out at room temperature.

TGA was used to evaluate the relative thermal stability of modified PVA membranes. The TGA curves were obtained with a TGA Q500 thermogravimetric analyser (TA Instruments, USA). Samples of about 5 mg were heated at a rate of 10 °C/min in N_2 atmosphere. The weight of the samples was measured as a function of temperature.

Tensile properties were determined by DMA using a DMA Q800 instrument (TA Instruments, USA). The initial length of the specimen was 20 mm, and the extension rate was 1 mm/min. For measurement was used room temperature.

3. Results and discussion

3.1. Rheology

The viscosity of the PVA solution changed after the incorporation of NDs⁻ (Figure 2). Viscosity increase with NDs⁻ content. At the shear rate 10 s-1 the viscosity increase from ~0.5 to ~10 Pa·s for pure PVA and PVA+1.9ND, respectively. Highest viscosity for PVA+ND solutions is caused by the presence of NDs⁻ which operate as crosslinking agent (Scheme 1).[7]

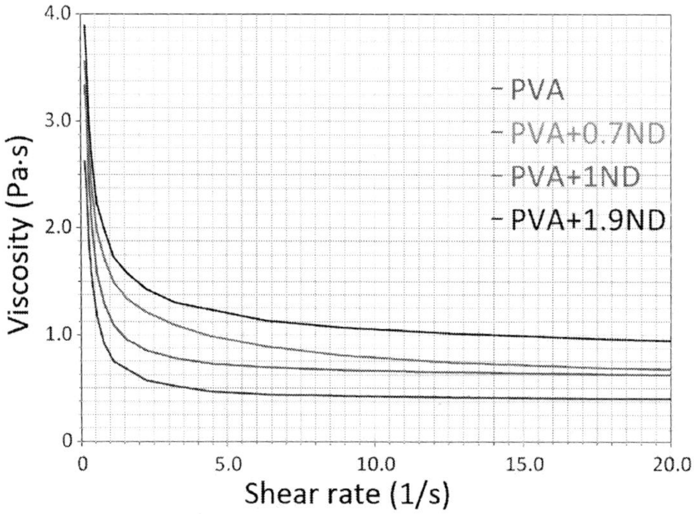

Figure 2. Rheological spectra of pure PVA solution and different NDs loadings of PVA+ NDs⁻ solutions.

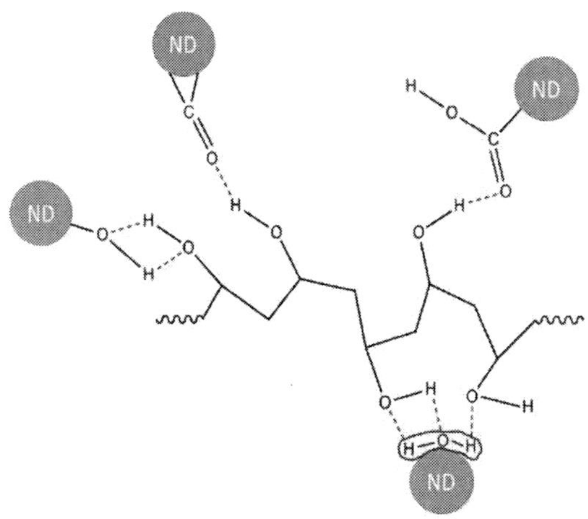

Scheme 1. Schematic illustration of the interaction between PVA and NDs⁻ particles.

3.2. TGA

The TGA traces of the PVA membrane and the PVA+ND nanocomposites is shown in Figure 3. Thermal stability of nanocomposites increase with NDs⁻ content. This indicates that the nanodispersion of the NDs⁻ particles can act as a barrier to hinder the volatile decomposition products throughout the composites.[7]

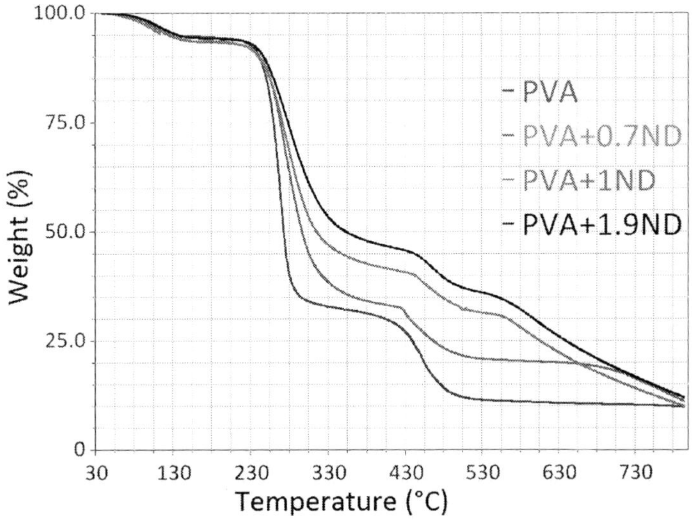

Figure 3. TGA thermograms of pure PVA and nanocomposite PVA+ND membranes.

3.3. DMA

Figure 4 shows the stress (σ) - strain (ε) curves of pure PVA membrane and the PVA+ND nanocomposites with different NDs⁻ contents. The PVA showed the typical σ-ε curve for a conventional polymer film exposing elastic limit and maximum tensile strength (σ_{max}). σ_{max} increased drastically with the NDs⁻ content, from 25,7 to 63,3 MPa for pure PVA and PVA+1.9ND, respectively. Thus, we can conclude that the NDs⁻ loading clearly improved the mechanical strength of the membranes.

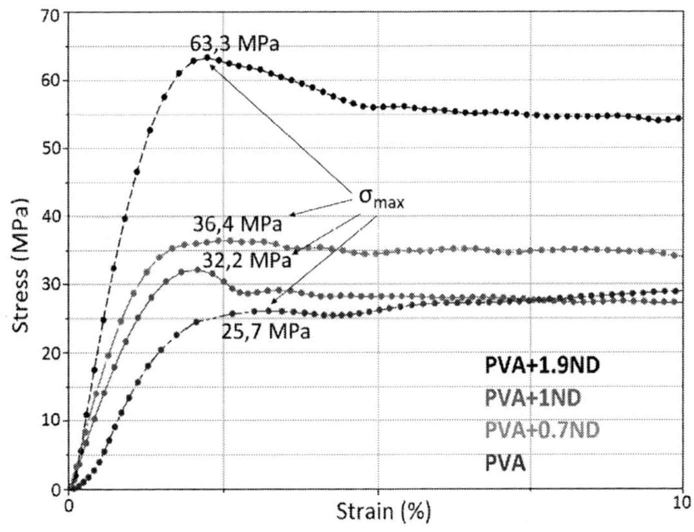

Figure 4. Stress (σ) - strain (ε) curves of pure PVA and the PVA+ND nanocomposites.

4. Conclusion

Nanodispersion of the NDs⁻ particles in the PVA matrix was achieved by using a solution casting method from aqueous medium. The mechanical properties of the PVA+ND nanocomposites increased remarkably by the incorporation of NDs⁻ particles. The σ_{max} values of the nanocomposites increase

IOP Conf. Series: Materials Science and Engineering **613** (2019) 012033 doi:10.1088/1757-899X/613/1/012033

with NDs content. Therefore, the ideal reinforcement effect of NDs was revealed. Thermal properties of the PVA+ND nanocomposite increased with increasing content of NDs particles. Thus, we created advanced nanocomposites by using NDs$^-$, which will be able to compete as a nanofiller against conventional nanofillers for polymer composites.

Acknowledgements
The result was developed within the CENTEM project, reg. no. CZ.1.05/2.1.00/03.0088, cofunded by the ERDF as part of the Ministry of Education, Youth and Sports OP RDI programme and, in the follow-up sustainability stage, supported through CENTEM PLUS (LO1402) by financial means from the Ministry of Education, Youth and Sports under the "National Sustainability Programme I.

References

[1] Koo, J. H. 2006 *Polymer nanocomposites* McGraw-Hill Professional Pub
[2] Wang, Z., et al 2013 *Pol. Compos* **34** 1735 https://doi.org/10.1002/pc.22577
[3] Mochalin, V. N., et al 2015 *Diamond and R. Mat.* **58** 161 https://doi.org/10.1016/j.diamond.2015.07.003
[4] Ceia, T. F., et al 2014 *Catal. T.* **236** 98
[5] Yang, D., et al 2007 *Carbohyd. Polym.* **69** 538
[6] Yadav, R., et al 2013 *Mat. Lett.* **110** 130 https://doi.org/10.1016/j.matlet.2013.07.109
[7] Morimune, Seira, et al 2011 *Macromolecules* **44** 4415 https://doi.org/10.1021/ma200176r

IOP Conf. Series: Materials Science and Engineering **613** (2019) 012034 doi:10.1088/1757-899X/613/1/012034

Formation of iron sulfide photocatalytic colloidal dispersion via pulsed laser ablation in liquids

Lukáš Vála[1], Veronika Vavruňková[1], Věra Jandová[2], Michal Pola[1], Tomáš Křenek[1]*

[1] University of West Bohemia, Research Centre New Technologies 30614 Pilsen, Czech Republic

[2] Institute of Chemical Process Fundamentals, Czech Academy of Sciences, Rozvojova 135, 160 00 Prague 6, Czech Republic

E-mail: tkrenek@ntc.zcu.cz

Abstract. Pulsed laser irradiation of iron sulfide in water and ethanol allows laser ablation and generation of FeS nano/micro particles. Measurement of the size distribution reveals 100, 1000 and 5000 nm sized particles in water and 180 nm sized particles in ethanol. The values of zeta potential: 13.2 mV for colloid in water confirm incipient stability and tendency for coagulation whereas the value: -40.6 mV shows good stability of ablatively achieved nanoparticles in ethanol. SEM analyses of particles obtained by evaporation of solvents on Ta substrate revealed shapeless, roundshape and sheet-like morphology of agglomerates whose size span from units up to tens of μm. Also spherical particles sized around tens of nm were detected. EDX shows Fe/S ratio ~3.25 and ~ 1.2 for particles ablated in water and ethanol respectively. Raman spectroscopy indicates the formation of mackinawite ($Fe_{1+x}S$) and pyrrhotite ($Fe_{1-x}S$) phase. The photocatalytic effect of prepared water colloid was tested in methylene blue (MB) degradation under the daylight.

1. Introduction

Ferrous sulfide represents an attractive material for solar cell application [1] which exhibits also interesting semiconducting [2], magnetic [3] and biocatalytic [4] and photocatalytic [5] properties. Also, hydrogen evolution reaction (HER) electrocatalytic activity of ferrous sulfide has been confirmed [6]. Moreover, some iron sulfides such as mackinawite (FeS) and pyrite (FeS_2) have been considered as excellent materials for mercury adsorption due to their high affinity towards Hg^{2+} [eg.7] and for ability to capture also other toxic elements, including Cu, Cr, Cd, Pb, and As [8-11]. FeS offers abundance, non-toxicity and low price. The iron sulfide (FeS) belongs between considerably polymorphic compounds with complex phase diagram including seven phases: pyrite (cubic- FeS_2), marcasite (calcium chloride structure-FeS_2), pyrrhotite-IT ($Fe_{1-x}S$), pyrrhotite-4M (Fe_7S_8), Fe_9S_{10}, greigite (cubic spinel- Fe_3S_4), troilite-2H (FeS) and mackinawite ($Fe_{1-x}S$) [12-15]. Except stable pyrite (FeS_{2p}) and pyrrhotite phases are metastable or unstable. Pyrite phase of iron sulfide is suitable as an absorbing material for thin film solar cells because of its band gap (Eg = 0.95 eV) and high absorption coefficient (~105 cm^{-1}) [16].

FeS stoichiometric iron sulfide has troilite structure which possesses antiferromagnetic properties at room temperature. Above 120°C, troilite transforms to the NiAs-type structure composing of ($Fe_{1-x}S$) and Fe_7S_8 pyrrhotites [17]. Vacancies of Fe cause that many pyrrhotites ($Fe_{1-x}S$) create compositions with interesting magnetic and electrical properties [18, 19]. The non-stoichiometric $Fe_{1-x}S$ shows different morphologies, including nanorods [20], whiskers [21], and U-shaped microslots [22]. Nanoparticles of iron sulfide have been obtained via high energy mechanical milling [23] and by hydrothermal methods [eg. 24]. Pulsed laser deposition of nanostructured FeS in the vacuum has been described in our recent studies [25, 26]. To best of our knowledge there is not published any information about LAL of iron sulfide. Laser ablation in liquid (LAL) has been developed into an important method to prepare metal, semiconductor, and even polymer colloidal dispersion of nanoparticles. As a technique, LAL is somewhat different from the other laser ablation approaches operate in vacuum or gaseous environments because the liquid medium not only provides some effective controlling parameters for fabrication, but also greatly affects the morphology and microstructure of the products [27].

Here we report the first study on LAL of iron sulfide. LAL of FeS results in generation of less stable colloid in water and good stable colloid in ethanol. The morphology of ablatively achieved particles is roundshape and sheet-like and Raman spectroscopy suggests formation of mackawite and pyrrhotite phase. Photocatalytic efficiency of the aqueous colloid for methylene blue (MB) degradation has been shown.

2. Experimental

A 3rd harmocnic of pulsed Nd:YAG laser with base wavelength 1064 nm (model Q SMART 850, wavelength: 355 nm , energy: 180 ± 5 mJ per pulse, pulse duration: 10 ns, repetition rate: 10 Hz, pulse full width at half maximum: 23 ns) was focused by lens (f = 15 cm) on the spot area of 0.02 cm^2 on FeS target. Simple tubular Pyrex reactor (70 mL in volume) reactor was furnished with a borosilicate glass windows and filled with DEI water or ethanol. The duration of irradiation was 2 hours.

The irradiated target of FeS pellet with diameter 8 mm and height 5 mm was positioned vertically in the center of the reactor. After the irradiation the prepared colloid was characterized by MALVERN Zetasizer for the measurement of the size of particles and Zeta potential of the colloid and part of colloid has been evaporated on Ta substrate in order to measure SEM-EDX and Raman spectroscopy of nanoparticles/agglomerates.

Size distribution of the particles in prepared colloids and its zeta potential were measured by Dynamic Light Scattering – Zetasizer Nano (Malvern).

A SEM (Scanning Electron Microscope, Tescan Indusem) with mounted EDS was used for the composition evaluation of the particles obtained by evaporation of the solvents of the colloids. Particles were catching on Ta substrate and analyzed by SEM and EDS at acceleration voltage 15 kV.

Raman spectra were obtained using a DXR Raman microscope with Diode-pumped solid state laser emitting at 532 nm using high resolution gratings working in the range of 50 – 1800 cm^{-1} and spectral resolution 2 cm^{-1} FWHM.

The FeS pellet was made at 100 atm. on a hydraulic press from a commercially available iron sulfide powder (FeS, 99% Fe, Aldrich).

The photocatalylic oxidation of MB by FeS nanoparticles dispersed in water was performed as follows: 1ml of 0.05mM solution of methylene blue and 2.5 ml of freshly prepared aqueous colloid. This mixture was exposed to daylight (without external light source) for 180 min. Intensity of the light was $\Phi \sim 800$ lm. The depletion of MB was measured each 30 minutes by UV spectrometer (RED TIDE USB650 UV).

3. Results and discussion

Highly focused pulsed irradiation of FeS target in water/ethanol results in generation of colloids based on FeS NPs/agglomerates. The color of solvents (DEI water, ethanol) changed into slightly brownish. The distribution of NPs and the progress of agglomeration were measured by Malvern Zetasizer analyzer. Fig. 1 A depicts size distribution of the particles obtained by LAL of FeS in DEI water were several fractions have been revealed. The size distribution around 100 nm, 1000 nm and 5000 nm has been detected. The measuring of size distribution depending on time shows slowly continuing agglomeration of the particles. The intensity of the peak which belongs to 100 nm sized particles decreases and the peaks attributed to 1000 and 5000 nm sized agglomerates become more intensive.

Zeta potential of the colloid measured immediately after irradiation represents value 13.2 mV which suggests incipient instability of the colloid. It corresponds with agglomeration process. The FeS particles generated via LAL in ethanol exhibit higher stability compared to water colloid. Fig. 1 B gives information about one size fraction varies around 180 nm where the values are not changing in dependence on time. Also, the value: -40.6 mV of zeta potential confirms good stability of the prepared colloid.

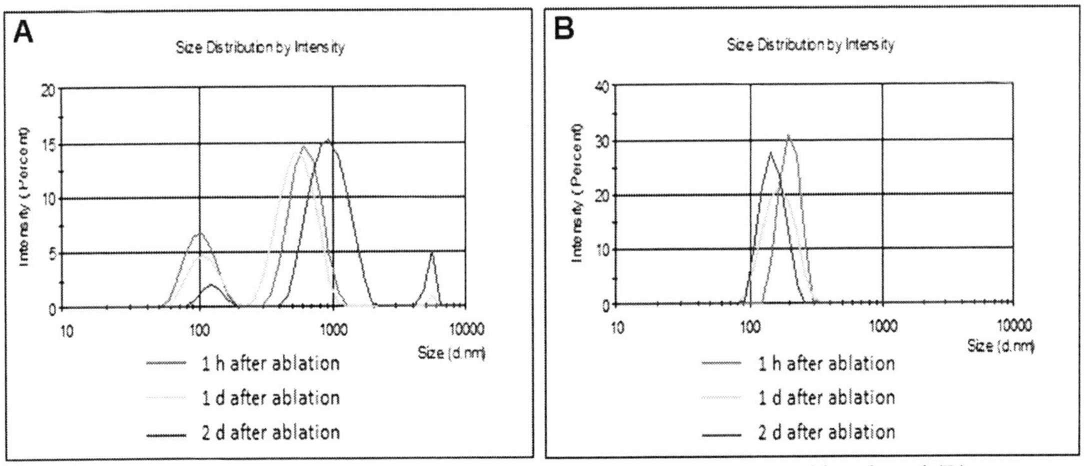

Fig. 1 Size distribution of particles generated in water (A) and in ethanol (B)

In order to search phase and chemical composition of prepared FeS particles the colloids were evaporated in small cuvette and achieved particles were deposited on Ta substrate. The SEM (Fig. 2) revealed shapeless, roundshape and needle-like/sheet-like morphology in case of colloid prepared by laser ablation in the water (Fig. 2 A). More detailed view shows also smaller spherical particles whose size varies around tens of nm. The particles obtained by evaporation of the ethanol exhibits shapeless agglomerates with sheet-like units with random orientation. Similar flower-like morphology was observed eg. in case of FeS nanosheets synthesized by a poly(vinyl alcohol)-assisted precipitation method [28] or after thermal decomposition of the organometallic precursor in oleylamine [29]. Also in case of ethanol colloid, spherical particles sized in order of tens of nm are present.

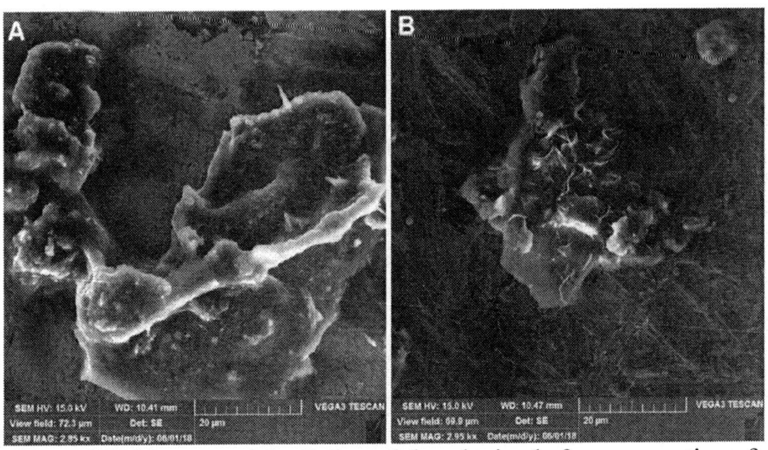

Fig. 2 SEM images of the FeS particles obtained after evaporation of water (A) and ethanol (B)

EDX analyses allow assessment of the chemical composition of the particles. The agglomerates achieved using LAL in water show Fe/S ratio of at.% around 3.25 and in ethanol Fe/S ratio ~ 1.2. These values of S-deficient objects indicate some decomposition of FeS during LAL. The decomposition is more significant in water environment where iron sulfide can undergo some hydrothermal processes [30].

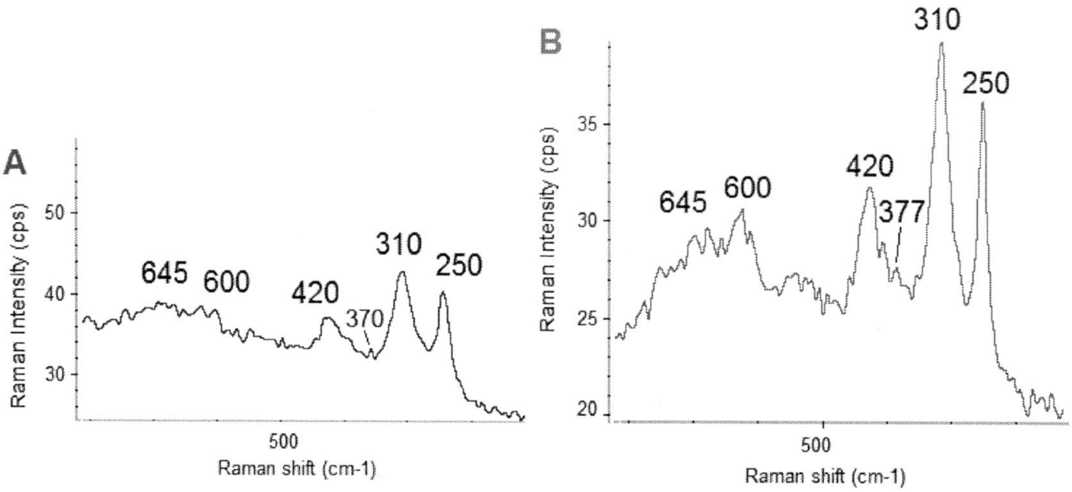

Fig. 3 Raman spectroscopy of FeS particles obtained by LAL of FeS target in water (A) and in ethanol (B)

Raman spectroscopy of FeS particles obtained via evaporation of water and ethanol reveals similar features (Fig. 3). For both spectra sharp peaks positioned at 250 and 310 cm^{-1} followed by less intensive broader peak at 420 cm^{-1} and broad peaks at 600 and 645 cm^{-1} have been detected. The peaks at 310 and 250 cm^{-1} are assignable to mackinawite ($Fe_{1+x}S$) phase [eg. 31, http://rruff.info reference R070302]. For mackinawite phase is typical also peak situated around 370 – 380 cm^{-1} [31] which could be demonstrated by less intensive peak at 370 (Fig. 3 A) and 377 cm^{-1} (Fig. 3 B). The broad peaks positioned at 600 and 645 cm^{-1} corresponds with the peaks of pyrhotite ($Fe_{1-x}S$) [31] and in accordance with the reference R060388 in the RRUFF database these peaks coincident also with mackinawite phase. In agreement with the reference R060440 spectrum (RRUFF database) peak at 420 cm^{-1} can be attributed to pyrhotite phase. The absence of bands at 206 and 280 cm^{-1}, which are characteristic for amorphous or weakly crystalline FeS phase [31], shows highly crystalline structure. In summary, Raman spectroscopy suggests formation of two-phase structure composed of mackinawite and pyrhotite phases. The phase transition from mackinawite into pyrrhotite is line with experimental data in aqueous solution [eg.32].

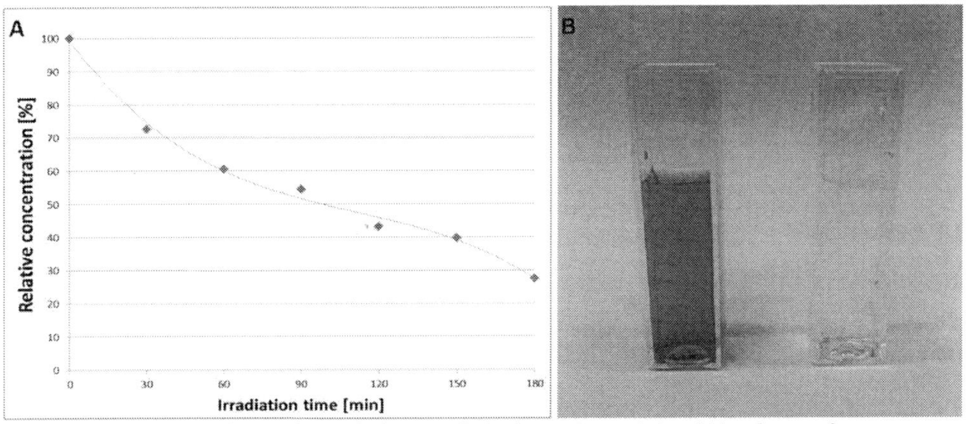

Fig. 4 Progress of MB degradation on light by FeS particles (A), photo of aqueous colloid before and after irradiation (180 min) at Φ ~ 800 lm of luminous flux.

The photocatalytic effect of aqueous FeS colloid was tested in terms of MB degradation. The mixture of 1ml of 0.05mM solution of MB and 2.5 ml of FeS aqueous colloid was exposed on daylight and the depletion of MB was measured by UV spectroscopy. Fig. 4 A shows significant MB degradation proceeding under daylight irradiation (without any external light source). After 180 minutes of irradiation ~27 % of original MB concentration remains. Fig. 4 B illustrates the image of the cuvettes with the mixture before (blue one) and after 180 minutes of daylight exposure (transparent one).

4. Conclusion

Laser ablation of FeS traget in the water and ethanol results in laser ablation and formation of nanoparticles and their clusters. Different colloid stability has been revealed for FeS colloid in water and ethanol. Size distribution of 100, 1000 and 5000 nm of particles in water together with the value of zeta potential 13.2 mV suggest tendency for agglomeration and incipient stability. In ethanol, one size of particles around 180 nm and the value of zeta -40.6 mV have been detected. It confirms good stability of this colloid. The morphology of obtained FeS particles was measured after evaporation of the liquids on Ta substrate. Shapeless, roundshape and sheet-like agglomerates of microparticles and spherical nanoparticles were shown by SEM. EDX revealed Fe/S ratio ~3.25 for particles obtained in water and ~ 1.2 for particles ablated in ethanol. Two-phase structure consisting of mackinawite ($Fe_{1+x}S$) and pyrrhotite ($Fe_{1-x}S$) was assigned by Raman spectroscopy. The photocatalytic activity of aqueous FeS colloid was tested by MB degradation under the daylight exposure.

Acknowledgements

The result was developed within the CENTEM project, reg. no. CZ.1.05/2.1.00/03.0088, cofunded by the ERDF as part of the Ministry of Education, Youth and SportsOPRDI programme and, in the follow up sustainability stage, supported through CENTEM PLUS (LO1402) by financial means from the Ministry of Education, Youth and Sports under the National Sustainability Programme I. Computational resources were provided by MetaCentrum (LM2010005) and CERIT-SC (CZ.1.05/3.2.00/08.0144) infrastructures.

References

[1] Hofmann W K, Birkholz M 1990 *Sol. Energy Mater* **20** 149
[2] Lehner S W, Savage K S, Ayers J S 2006 *J. Cryst. Growth* **286** 306
[3] Takahashi T 1973 *Solid State Commun.* **13** (9) 1335
[4] Wang H, Bigham J M, Tuovinen O H 2007 *Hydrometallurgy* **88** 127
[5] Bhar S K, Sumanta Jana S, Anup Mondal A, Nillohit Mukherjee N 2013 *Journal of Colloid and Interface Science* **393** 286 https://doi.org/10.1016/j.jcis.2012.10.049
[6] Villalba M, Peron J, Giraud M, Tard C 2018 *Electrochemistry communications* **91** 10 https://doi.org/10.1016/j.elecom.2018.04.019

[7] Sun Y, Lv D, Zhou J, Zhou X, Lou Z, Baig S A, Xu X 2017 *Chemosphere* **185** 452 https://doi.org/10.1016/j.chemosphere.2017.07.047
[8] Borah D, Senapati K 1999 *Nurs. Case Manag. Manag. Process Patient Care* **4** 85
[9] Bostick B C, Fendorf S 2003 *Geochim. Cosmochim.* **67** 909
[10] Bostick B C, Fendorf S, Helz G R 2003 *Environ. Sci. Technol.* **37** 285
[11] Doyle C S, Kendelewicz T, Bostick B C, Gordon Jr E B 2004 *Geochim. Cosmochim.* **68** 4287
[12] Soon J M, Goh L Y, Loh K P 2007 *Appl. Phys. Lett.* **91** 084105
[13] Vaughan D J, Lennie A R 1991 The iron sulphide minerals: their chemistry and role in nature, *Sci. Prog.* **75** 371 (Edinburgh)
[14] Rao C N R, Pisharody K P R 1975 *Prog. Solid State Chem.* **10** 207
[15] Goodenough J B 1978 *Mater. Res. Bull.* **13** 1305
[16] Bessergenev V 2004 *J. Phys. Condens. Matter* **16** S531
[17] Gosselin J R, Townsend M G,Tremblay R J 1976 *Solid Stat. Commun.* **19** 799
[18] Nakazawa H, Morimoto N 1971 *Mater. Res. Bull.* **6** 345
[19] Horwood J L, Townsend M G, Webster A H 1976 *J. Solid State Chem.* **17** 35
[20] Barnad A S, Russo S P 2007 *J. Phys. Chem.* **111** (31) 11742
[21] Almond M J, Redman H, Rice D A 2002 *J. Mater. Chem.* **10** 2842
[22] Ma X, Xu F, Wang X, Du Y, Chen L, Zhang Z 2005 *J. Cryst. Growth* **277** 314
[23] Chin P, Ding J, Yi J B, Liu B H 2005 *J. Alloys Compd.* **390** 255
[24] Morales-Gallardo M V, Ayala A M, Pal M, Cortes Jacome M A, Toledo Antonio J A, Mathews N R 2016 *Chemical Physics Letters* **660** 93 https://doi.org/10.1016/j.cplett.2016.07.046
[25] Pola J, Urbanová M, Pokorná D, Bezdička P, Kupčík J, Křenek T 2014 *RSC Adv.* **4** 11543 https://doi.org/10.1039/C3RA46580B
[26] Křenek T, Medlín R, Karatodorov S, Mihailov V, Pola M, Reshak A H 2017 *Journal of Alloys and Compounds* **723** 689 https://doi.org/10.1016/j.jallcom.2017.06.229
[27] Zeng H, Du X-W, Singh S C, Kulinich S A, Yang S, He J, Cai W 2012 *Adv. Funct. Mater.* **22** 1333 https://doi.org/10.1002/adfm.201102295
[28] Zhang Q, Yao X, Mwizerwa J P, Huang N, Wan H, Huang Z, Xu X 2018 *Solid State Ionics* **318** 60
[29] Wang X, Zhou W-H, Zhou Z-J, An Y-Q, Wu S-X 2013 *Materials Science in Semiconductor Processing* **16** 530
[30] Richard D, Luther G W 2007 *Chem. Rev.* **107** 514
[31] Mendili Y E, Abdelouas A, Hajj H E, Bardeau J-F 2013 *RSC Adv.* **3** 26343
[32] Rickard D T 1969 *Stockholm Contrib. Geol.* **26** 67

IOP Conf. Series: Materials Science and Engineering **613** (2019) 012035 doi:10.1088/1757-899X/613/1/012035

The Effect of Heat Input Parameters on Residual Stress Distribution by Numerical Simulation

Mahmood Hasan Alhafadhi[1], Gyorgy Krallics[1]

[1]Intsitute of physical Metallurgy, Metal Forming, and Nanotechnology, University of Miskolc (Hungary)

E-mail: mahmoodhs199@gmail.com

Abstract. This paper deals with Numerical Simulation to analyze the behavior of residual stresses in the plate. The residual stresses at the surface of some weld specimen by using the finite element technique. The results of the numerical simulation analysis were compared with three cases data to evaluate the accuracy of Goldak model (double ellipsoidal heat distribution model). Based on this study, a modeling procedure with reasonable accuracy was investigated. The developed finite element modeling was used to investigate the effects of welding heat input on magnitude and distribution of welding residual stresses in a welded plate made of steel. The different directions of residual stresses in the surface of the plate of 25 mm thick for V-groove shape were studied. It is shown that the welding heat input parameters have a significant effect on the magnitude and distribution of residual stresses.

1. Introduction

Welding is reliable and efficient metal joining processes used in almost all industries. Arc welding is extensively used for welding palates. The finite element method applied to simulate the welding processes has focused mainly because of the increase the capacity of computers and availability of commercial numerical simulation code. The numerical simulation has been the most common for analysis and simulates the wedding problems which include thermal and mechanical with each other. When two-part of plates are welded together residual stress is created in the vicinity of weld due to the arc welding of the plates. The elastic-plastic response of creating the residual stresses by thermal cycle can be responsible for many problems in weld such as cracking or failures in weld structures. [1-5].

Generally, tensile residual stresses are increasing weld fatigue damage, stress cracking and fracture. It has been observed in near the weld fusion on surfaces of plates for example. The distribution of residual stresses changes in different positions on welding, so residual stresses plays a very important role to measure weld quality. The first studies used numerical simulation of welding process were employed by Friedman (1975) who studied the thermal and mechanical FEM model to determine the temperature and stresses distribution. Finite Element simulation has become a common tool for the prediction of welding residual stresses. A substantial amount of simulation and experimental work and studies focusing on welding with emphasis on plates welding is available in the literature [6-13]. Morrow and Sinclair [14] studied the prediction of residual stress relaxation based on mean residual stress relaxation observed in the axial fatigue test. Dong (2003) [15] described thermal-mechanical analysis provides accurate results and simplifies the numerical solution. El-Ahmar (2007) [16] explained that the effect of finite elements with simulations is not significant in the welding process application. Ziolkowski and Brauer (2009) [17] used the combination of Gaussian distribution on the surface and along the thickness with 3D distribution, by applying the heat source model. Wu et al. (2009) [18] studied the Goldak model in double ellipsoid with spherical and cylindrical volumes through the thickness. The study aim to investigate the mechanism of residual stress with different parameters Goldak model and analytical model for under various cases in 2D FE simulation of a single pass weld joint geometry is performed using Marc software.

2. Simulation model

Two plate halves 25 x 500 x 1000 mm are butt-welded together. A double V-groove is made in the plates. The plates are welded from the top using an arc welding process. No weld is made in the lower groove as shown in fig.1 (a) is considered for the analysis. Due to symmetry conditions, only one of the plates is analyzed using the same conditions. The meshed model of the plate and the welding local coordinate system is also shown in fig. 1 (b). The material used is steel and it's mechanical and thermal properties with varying temperature are shown in fig.2.

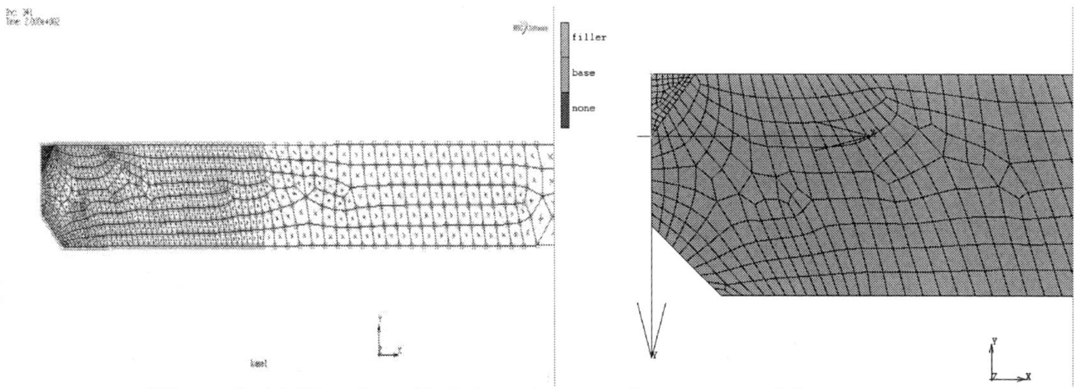

Figure 1. (a) Plate is welded from the top using an arc welding process
(b) FE Mesh and Weld Coordinate System used for Butt-Welding Process

Both the weld filler and base metal are assumed to be made of the same steel material. The initial temperature for the base metal is taken at 30°C.

Table 1. Welding parameters (Mechanical and Thermal properties)

Mechanical Properties	
Young's Modulus	$2.0x10^5$ MPa
Poisson's Ratio	0.35
Mass Density	7850 kg/m^3
Yield Stress	$3.0x10^2$ MPa
The coefficient of thermal expansion	$1.0x10^{-5}$/°C.
Thermal conductivity	40 W/m/ °C
Specific heat	500 J/kg/ °C

The welding parameters mechanical and thermal properties shown in **Table 1.** Solid-Solid phase transformations in the steel during heating and cooling are not considered in the current study. The solid-liquid transition is accounted for by providing a latent heat of fusion of 250 kJ/kg with a solidus temperature of 1100°C and a liquidus temperature of 1200°C. Symmetry boundary conditions are applied at one end of the structure and applied at the other end. A volumetric weld flux is applied to all elements of the base metal. Based on the dimensions given for the welding flux, Marc automatically determines which base elements actually receive the flux. A convective film boundary condition is applied to all the exposed edges of the base metal. The heat transfer coefficient is taken as 12 W/m^2/°C and the ambient temperature is taken as 30°C.

Simulation process

An analysis is done with the plate is symmetrical, half of the plate is only modeled. By using mesh optimization technique fine mesh generated around and in weld zone and comparatively coarse mesh far from the weld line. The governing equation for transient heat transfer during welding is given by

$$pc\frac{\partial T}{\partial t}(x,y,z,t) = -\nabla.q(x,y,z,t) + Q(x,y,z,t) \quad (1)$$

Where p is the density of the materials, c is the specific heat capacity, q is the heat flux vector, T is the temperature, Q is the internal heat generation rate, x, y and z are the coordinates in the reference system, t is time and ∇ is the spatial gradient operator. To perform the analysis, the heat distribution model is modified in order to accommodate cylindrical coordinates. The plate geometry is generated, and welding torch is applied using different the welding parameters given in Table 2., Fig. 2 shows the various weld parameters, in a double ellipsoidal distribution proposed by Goldak et al. [19].

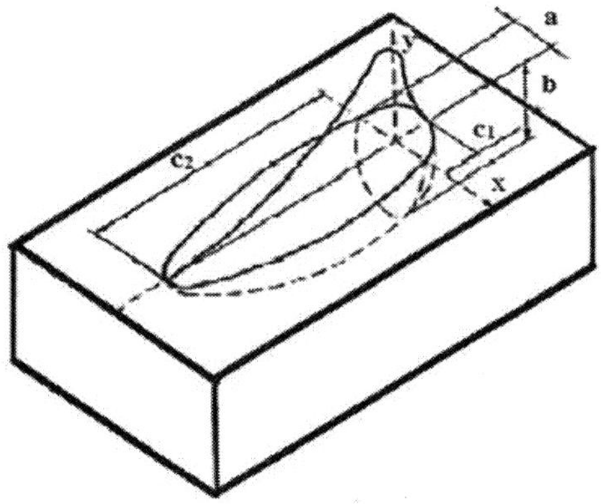

Figure 2. Weld parameters [20]

Table 2. FE model with different parameters

Specimen	Forward length (mm)	Rear length (mm)	Width (mm)	Depth (mm)
	5	5	2.5	10
Case 1	5.5	5.5	2.75	11
Case 2	6	6	3	12
Case 3	7	7	3.5	14

The parameters a, b, c_1 (forward length) and c_2 (Rear length) are related to the characteristics of the welding heat source. To compare and analyze the result the different parameter (Goldak model) as shown in Table 2. The heat from the moving source is applied as a double ellipsoidal distribution (Goldak model) is explained by the following equations:

$$Q(x, y, z.t) = \frac{6\sqrt{3}f_i\eta VI}{abc\pi\sqrt{\pi}} e^{\left[-3\left(\frac{x-vt}{c}\right)^2 -3\left(\frac{y}{b}\right)^2 -3\left(\frac{z}{a}\right)^2\right]} \quad (2)$$

Where x, y, and z are the local coordinates of the double ellipsoid model. fi is the fraction of heat deposited in the weld region. V and I are the applied voltage and current respectively. η is the arc efficiency for the TIG welding process, v is the speed of torch travel in mm/s and t is the time in seconds. The parameters a, b and c are related to the characteristics of the welding heat source. The parameters of the heat source are chosen according to the welding conditions. This study is concerned with simulations of temperatures, displacements, stresses and strains in the elded structure without solid-state phase transformations using Marc software. In mechanical properties is same as used in the hermal analysis.

3. Results and discussions

In first calculation were studied at the same model to get an accurate result of temperature and residual stress in the welded plate. The main difficulty of the thermal field simulation in a welding process is Goldak model. Several cases of different parameters of heat distribution, heat input, and plate thickness have had their weld pool geometries analyzed and compared with each other as shown in figure 3. It is obvious some different from case to another. A 2D model is formulated in present work. The temperature dependent on thermos-physical and mechanical properties. The finite element is developed to validate the result. The predicted results are shown some differences in various cases. In

IOP Conf. Series: Materials Science and Engineering **613** (2019) 012035 doi:10.1088/1757-899X/613/1/012035

the numerical simulations, Figure 4-8 show the residual stress distribution of the welded plate in x, y and z direction with different cases.

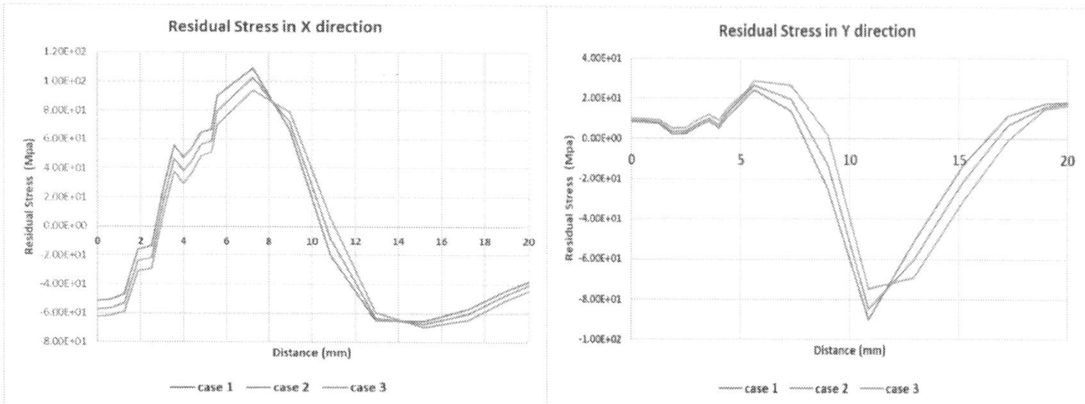

Figure 3. Residual stresses 2D plate in X, Y direction

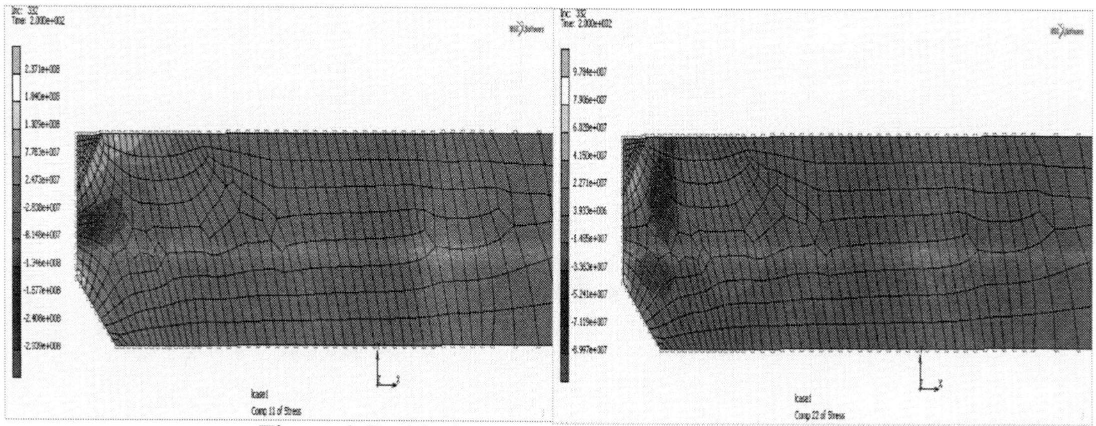

Figure 4. Residual stresses 2D plate in X, Y (Case 1)

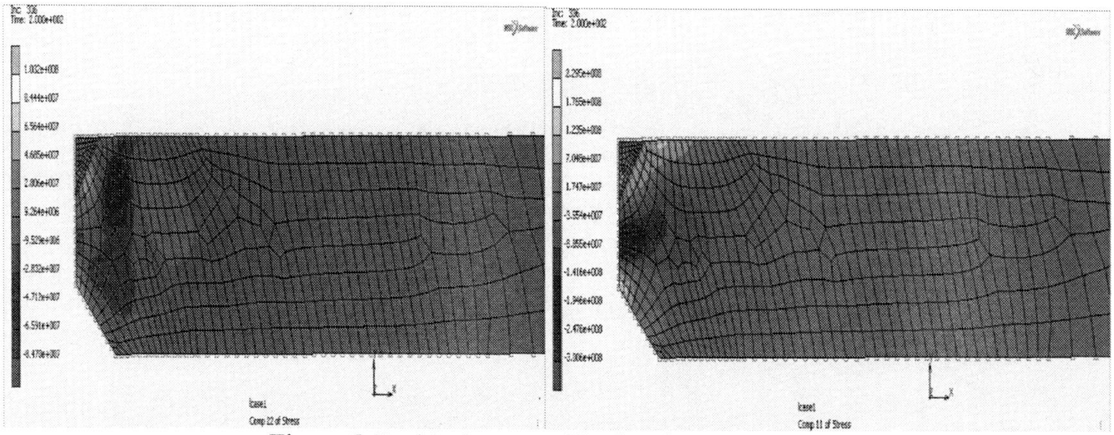

Figure 5. Residual stresses 2D plate in X, Y (Case 2)

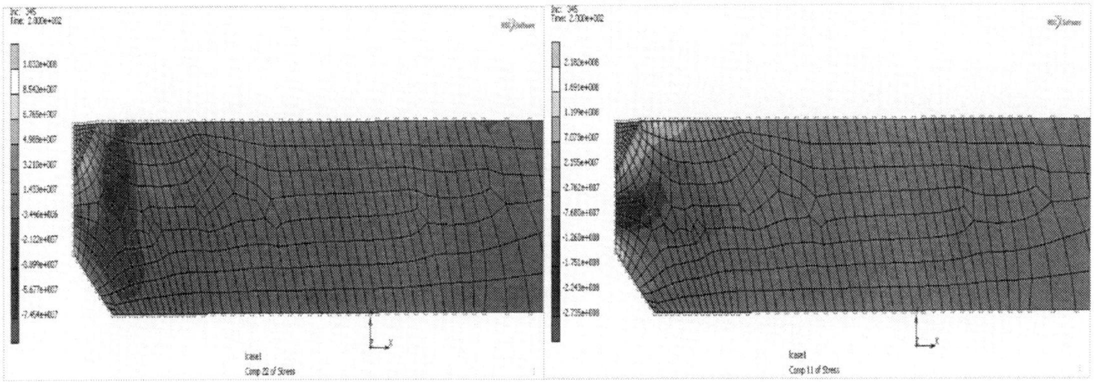

Figure 6. Residual stresses 2D plate in X, Y (Case 3)

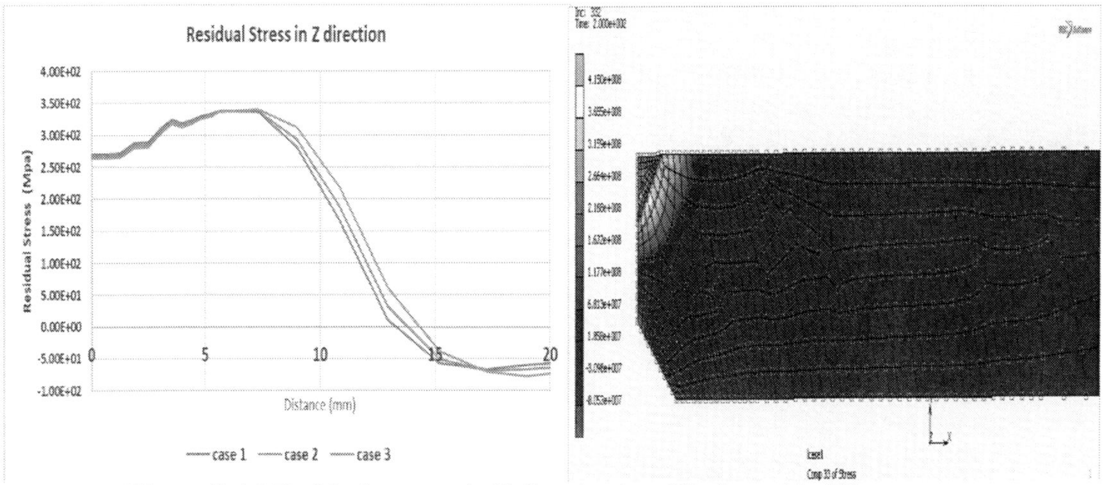

Figure 7. (**a**) Residual stresses in Z direction in a 2D plate with different cases
(**b**) Residual stresses in the Z2 D plate direction (Case 1)

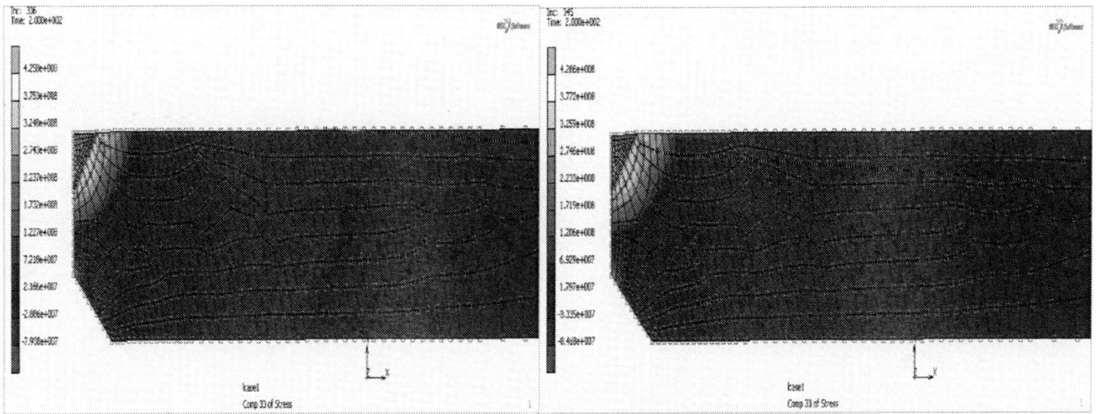

Figure 8. (**a**) Residual stresses in Z direction (Case 2)
(**b**) Residual stresses in Z direction (Case 3)

4. Conclusions

In this study, the accuracy of the Goldak model parameters was investigated. Generally, the numerical simulation of the welding process is the prediction of residual stress. These cases showed that the methodology based on the finite element technique and Goldak model is a good proposal for welding cases when welding processes are employed. The parameters of the Goldak model have a significant influence on residual stress.

IOP Conf. Series: Materials Science and Engineering **613** (2019) 012035 doi:10.1088/1757-899X/613/1/012035

References

[1] R I Karlsson, B L Josefson 1990 *Journal of pressure vessel technology* **112** (1) 76 https://doi.org/10.1115/1.2928591
[2] S Fricke, E Keim, J Schmidt 2001 *Nucl Eng Des* **206** (2) 139 https://doi.org/10.1016/S0029-5493(00)00414-3
[3] M Siddique, M Abid, H F Junejo, R A Mufti 2005 *Mater Sci Forum* **13** 491 https://iopscience.iop.org/article/10.1088/0965-0393/13/6/010/meta
[4] M Jonsson, L Karlsson, L E Lindgren 1995 *AWS Weld J, Weld Res Suppl* **64** (10) 301.
[5] P Dong 2003 *Welding in the world* 60 (2) https://link.springer.com/article/10.1007/s40194-015-0286-4
[6] Ruben L, Marina C, María Á and Pedro M 2017 *Metals* **7** 136 https://doi.org/10.3390/met7040136
[7] Wahab, M A Painter, M J 1998 *Journal of Materials Processing Technology* **77** (1) 233 https://doi.org/10.1016/S0924-0136(97)00422-6
[8] B Brickstad, B L 1998 *Int. J Pres Vessels Piping*, **75** (1) 11 https://doi.org/10.1016/S0308-0161(97)00117-8
[9] E M Qureshi, A M Malik, N U Dar 2015 *Advance in Mechanical Engineering* **1** 22 https://doi.org/10.1155%2F2009%2F351369
[10] E F Rybicki, P A McGuire, E Merrick, J Wert *ASME J Press Vessel Technol* https://doi.org/10.1115/1.3264205 **104** (3) 204
[11] E F Rybicki, R B Stonesifer 1979 *Trans ASME J Press Vessel Technol* https://doi.org/10.1115/1.3454614 **101** (2) 149
[12] L Karlsson, M Jonsson, L E Lindgren, L M Nasstrom 1989 *American Society of Mechanical Engineers* http://urn.kb.se/resolve?urn=urn%3Anbn%3Ase%3Altu%3Adiva-38656
[13] I Sattari Far, and Y Javadi 2008 *International Journal of Pressure Vessels and Piping* **85** (4) 265 https://doi.org/10.1016/j.ijpvp.2007.07.003
[14] Z Wyman, Gary Zhuanga, R Halford 2001 *International Journal of Fatigue* **23** (1) 31 https://doi.org/10.1016/S0142-1123(01)00132-3
[15] Y Dong 1997 *Welding Journal* **76** (10) https://www.osti.gov/biblio/562021
[16] W El-Ahmar 2007 https://www.theses.fr/2007ISAL0023
[17] M Ziolkowski, H Brauer 2009 *The International Journal For Computation And Mathematics In Electric And Electronic Engineering* **28** (1) 140 https://doi.org/10.1108/03321640910918940
[18] C S Wu, Q X Hu, J Q Gao 2009 *Computational Material Science* **46** (1) 167 https://doi.org/10.1016/j.commatsci.2009.02.018
[19] J Goldack, A Chakravarti and M Bibby. **15** (2) 1984 299 https://link.springer.com/article/10.1007/BF02667333
[20] D Klobcar, J Tusek, B Taljat 2004 *J. Comp. Mater. Sci.* **31** (3) 368 https://doi.org/10.1016/j.commatsci.2004.03.022

IOP Conf. Series: Materials Science and Engineering **613** (2019) 012036 doi:10.1088/1757-899X/613/1/012036

Local mechanical properties of advanced skutterudites processed by various routes

Jiří Buršík[1], Vilma Buršíková[2], Gerda Rogl[3,4] and Peter Rogl[3,4]

[1] Institute of Physics of Materials, Czech Academy of Sciences, Žižkova 22, CZ-61662 Brno, Czech Republic
[2] Institute of Physical Electronics, Faculty of Science, Masaryk University, Kotlářská 2, CZ-61137 Brno, Czech Republic
[3] Institute of Materials Chemistry, University of Vienna, Währingerstrasse 42, A-1090 Wien, Austria
[4] Christian Doppler Laboratory for Thermoelectricity, Wien, Austria

E-mail: bursik@ipm.cz, vilmab@physics.muni.cz, gerda.rogl@univie.ac.at and peter.franz.rogl@univie.ac.at

Abstract. Skutterudites are an important class of thermoelectric p- and n-type materials and they have already achieved fair efficiencies for the conversion of heat to electricity. Nevertheless researchers try to further enhance the figure of merit, ZT, by various ways. In this work we compare the microstructure and mechanical properties of an industrial p-type $DD_yFe_3CoSb_{12}$ skutterudite powder processed by several routes containing various combinations of ball milling, hot pressing, cold pressing and high pressure torsion. Details of the resulting microstructure are studied by means of analytical electron microscopy as well as by means of local mechanical testing.

1. Introduction

Skutterudites are known as excellent thermoelectric (TE) materials and have already achieved good efficiencies for the conversion of heat to electricity. Researchers try to further enhance the figure of merit, ZT, by various ways [1-4]. Among several groups of state-of-the-art TE materials, filled skutterudites continue to draw much scientific interest and have also been identified as one of the top candidates for real-life thermoelectric applications, particularly in the 300–800 K range. Didymium (DD) represents a natural double-filler composed of about 5 wt. % Pr + 95 wt. % Nd, being cheaper than rare earth elements. The TE performance may be enhanced by nanostructuring procedures resulting in submicron crystallites.

In this work we compare the microstructure and mechanical properties of an industrial p-type $DD_yFe_3CoSb_{12}$ skutterudite powder processed by several routes of thermomechanical processing leading to compacting and nanostructuring of the material. Overall characterization of these materials including measurements of temperature dependent electrical resistivity, Seebeck coefficient, power factor, and figure of merit as well as measurements of hardness and elastic moduli was completed and is summarized in [5].

2. Materials and experiments

The material studied was a commercially produced p-type $DD_yFe_3CoSb_{12}$ skutterudite powder (Treibacher Industries AG, TIAG, Austria). It was further either *(i)* hot pressed (Ar atmosphere, 700 °C, 56 MPa, 30 min, HP), or *(ii)* cold pressed (resulting in pellets 1 cm in diameter and about 1 mm in height, CP) followed by high pressure torsion (Ar atmosphere, 375 °C, applying 4 GPa and 1 revolution, HPT), or *(iii)* ball milled (BM) followed by HP and HPT; for details see [5].

IOP Conf. Series: Materials Science and Engineering **613** (2019) 012036 doi:10.1088/1757-899X/613/1/012036

Thin cross sectional lamellae were prepared by focussed ion beam (FIB) in a Tescan LYRA 3XMU SEM×FIB scanning electron microscope (SEM). A Philips CM12 STEM transmission electron microscope (TEM) and a FEI Titan Themis 60-300 cubed high resolution TEM with high sensitivity EDX system (0.7 srad solid angle) were then used to study details of the microstructure.

The mechanical properties were studied using nanoindentation techniques by means of a Hysitron TI950 (Bruker) triboindenter. The nanoindentation tests were carried out using a Berkovich diamond tip with a diameter less than 50 nm. A nanoscale measuring head with resolution of 1 nN and load noise floor less than 30 nN was used for this study. Two testing modes were employed in the range of indentation loads from 0.1 to 10 mN, namely quasistatic nanoindentation and quasistatic tests with several unloading segments. The quasistatic indentation tests were carried out in a load controlled regime at a constant loading rate of 0.2 mN/s. Both the partial unload regime and the nanodynamical analysis were carried out in a constant strain rate regime. The standard procedure proposed by Oliver and Pharr [6] was used for the evaluation of the hardness and elastic modulus. The hardness and effective elastic modulus maps were obtained on areas of 5×5 μm^2. The effective elastic modulus can be expressed as

$$E_{eff} = \frac{E_s}{1-\nu_s^2} = \frac{E_{ind}E_r}{E_{ind} - E_r\left(1-\nu_{ind}^2\right)}, \tag{1}$$

where E_s, ν_s, E_{ind} = 1141 GPa and ν_{ind} = 0.07 are the Young's moduli and Poisson ratio values of samples and the diamond indenter, respectively. E_r is the reduced modulus value.

Moreover, nanodynamic mechanical analysis (nanoDMA) in the range from 0.1 to 300 Hz served to study the local mechanical properties of the samples prepared. The modulus mapping capability was applied on areas 5×5 μm^2 and 1×1 μm^2 to obtain quantitative maps of the storage and loss stiffness and the storage and loss modulus. A sharp CubeCorner diamond tip with diameter less than 40 nm was used for the modulus mapping studies.

3. Results and discussion

According to the SEM observations the HP, BM-HP-HPT, CP-HPT and the annealed CP-HPT samples showed substantial differences in their grain size. The HP sample consisted of grains in the size of 10μm. The BM-HP-HPT sample exhibited a bimodal structure (mixture of coarse grains with diameter ≤10 μm and fine grains) similarly to the CP-HPT sample which also showed grains in the size of almost 10 μm mixed with fine and nano-sized grains. The annealed CP-HPT sample exhibited a relatively homogeneous grain structure.

Mechanical properties of $DD_yFe_3CoSb_{12}$, prepared by different technological procedures, were studied using both quasistatic nanoindentation with a maximum load of 5mN and quasistatic nanoindentation with 20 unloading segments and a maximum load of 10mN. The results obtained with these techniques were in good accordance.

Table 1. Summary of mechanical properties of $DD_yFe_3CoSb_{12}$ samples. HV is the Vickers hardness, E^{MI} is the Young's modulus received from microindentation tests (maximum load of 1 N, loading rate of 0.1 N/s and a loading time of 10 s) and ν is Poisson's ratio as reported in ref. [5]. H_{IT} is the indentation hardness and E_r is the reduced elastic modulus, E_{eff} is the effective elastic modulus and E^{NI} is the Young's modulus determined on the basis of the nanoindentation tests.

$DD_yFe_3CoSb_{12}$ processing	HV	E^{MI} [GPa]	ν	H_{IT} [GPa]	E_r [GPa]	E_{eff} [GPa]	E^{NI} [GPa]
CP-HPT	525 [5]	140 [5]	-	7±2	134±9	140±10	-
CP-HPT-ann.	506 [5]	136 [5]	-	8±3	140±9	160±10	-
BM-HP-HPT	522 [5]	-	0.23 [5]	7±3	130±8	147±9	139±8
HP	510 [5]	139 [5]	0.225 [5]	7.4±0.3	142±4	162±5	154±4

It was found that the bimodal grain size distribution of the samples also manifested itself in the local mechanical properties. In each sample it was possible to find areas with relatively high hardness (around 9 GPa) and regions with a lower hardness (around 6 GPa). The CP-HPT sample exhibited a low amount of soft grains with hardness (H_{IT}) of 5.5 GPa and effective elastic modulus (E_{eff}) of 120 GPa. The remaining part of the CP-HPT sample was characterized by H_{IT} of up to 8.5 GPa and E_{eff} of 155 GPa. The annealed CP-HPT sample exhibited grains with H_{IT} of up to 10 GPa and E_{eff} up to 170 GPa together with softer grains (H_{IT} ~ 6.5 GPa and E_{eff} ~ 150 GPa). In case of the BM-HP-HPT sample, H_{IT} values were around 8 and 6.5 GPa and the E_{eff} values were around 160 and 135 GPa in harder and softer regions, respectively. The HP sample exhibited relatively homogeneous mechanical properties (H_{IT} ~ 7.4 GPa and E_{eff} ~ 160 GPa). Moreover, soft particles (~ 1.5 GPa and E_{eff} ~ 40 GPa) with a size up to 100 nm were found in the samples studied.

Mechanical properties of $DD_yFe_3CoSb_{12}$ prepared by different technological procedures using both micro- and nanoindentation methods are compared in table 1. The microindentation results were obtained at a maximum load of 1N [5]. The nanoindentation results were obtained by means of quasistatic nanoindentation tests using a sharp Berkovich indenter (tip radius R<50 nm). The indentation hardness H_{IT} and the reduced elastic modulus E_r were calculated according to the Oliver&Pharr method [6]. Results obtained from 25 indentation tests from an area of 50×50 μm^2 were averaged. The effective elastic modulus was calculated according to formula (1). E^{NI} is the Young's modulus, estimated from nanoindentation tests as $E^{NI}=E_r(1-\nu^2)$, where ν was used from ref. [5]. The microindentation and nanoindentation data show some differences because nanoindentation gave localized data from different grains whereas microindentation provided the average hardness over a larger area containing several grains.

In the further study thin lamellae were prepared by FIB for TEM observations of fine details of the microstructure. Long term high magnification EDX maps were collected in TEM to see at least qualitatively the elements' distribution. As it can be seen on EDX maps in figure 1, precipitates of (Nd+Pr)-rich phases were often found both inside the skutterudite grains and at the grain boundaries. Possibly a small amount of the fillers was squeezed out of the icosahedral Sb-cage during HPT processing under a pressure of 4 GPa and formed rare earth oxides and antimonides.

The modulus mapping capability was applied to obtain quantitative maps of the storage and loss modulus. In figure 2 examples of storage modulus maps obtained at an oscillation frequency of 300 Hz and oscillation amplitude of 2 μN are shown. The loss modulus was negligible for both samples studied. In modulus maps the (Nd+Pr)-rich phases were observed as nanometre sized areas with a low elastic modulus around 50-80 GPa.

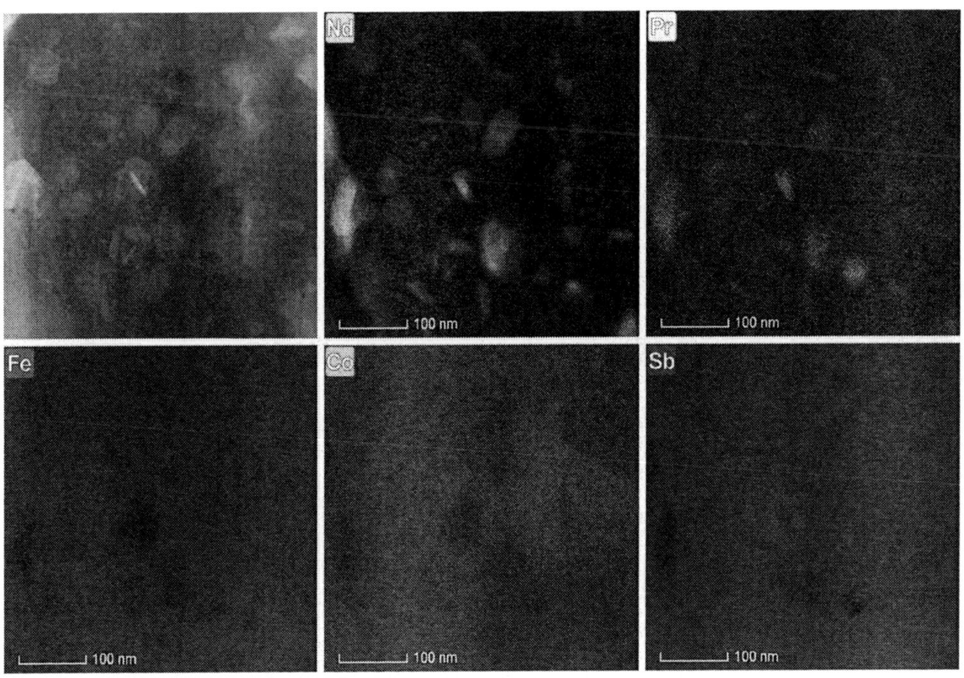

Figure 1. EDX maps showing (Nd+Pr)-rich precipitates in the grain interior.

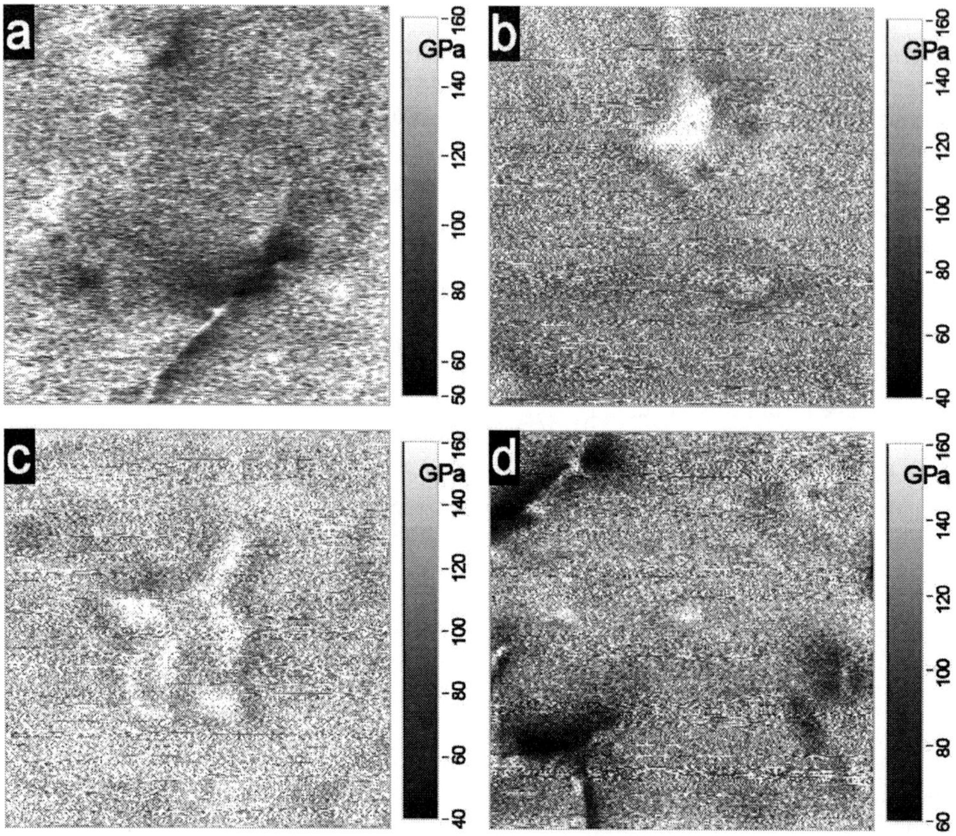

Figure 2. Storage modulus mapping results obtained on an area of 1×1 μm^2 of $DD_yFe_3CoSb_{12}$ samples prepared using CP-HPT (a), annealed CP-HPT (b), BM-HP-HPT (c) and HP (d).

4. Conclusions

The processing route influenced both the microstructure and the local mechanical properties of $DD_yFe_3CoSb_{12}$ samples. Samples which underwent high pressure torsion showed a bimodal grain structure and mechanical properties. According to nanoindentation and EDX results, the samples contained soft (Nd+Pr)-rich precipitates inside the grains and at the grain boundaries.

Acknowledgement

The work was supported by the Czech Science Foundation as Project 17-12844S. CEITEC Nano research infrastructure (Brno, Czech Republic) is acknowledged for granting access to experimental equipment.

References

[1] Rogl G and Rogl P 2017 *Current Opinion in Green and Sustainable Chemistry* **4** 50 http://dx.doi.org/10.1016/j.cogsc.2017.02.006

[2] Rogl G, Grytsiv A, Failamani F, Hochenhofer M, Bauer E and Rogl P 2017 *J. Alloys Compd.* **695** 682 http://dx.doi.org/10.1016/j.jallcom.2016.10.114

[3] Rogl G, Grytsiv A, Bursik J, Horky J, Anbalagan R, Bauer E, Mallik R C, Rogl P and Zehetbauer M 2015 *Phys. Chem. Chem. Phys.* **17** 3715 http://dx.doi.org/10.1039/c4cp05230g

[4] Rogl G, Bursik J, Grytsiv A, Puchegger S, Sopronyuk V, Schranz W, Yan X, Bauer E and Rogl P 2018 *Acta Materialia* **145** 359 https://doi.org/10.1016/j.actamat.2017.12.001

[5] Rogl G, Grytsiv A, Anbalagan R, Buršík J, Kerber M, Schafler E, Zehetbauer M, Bauer E and Rogl P 2018 *Acta Materalia* **159** 352 https://doi.org/10.1016/j.actamat.2018.08.020

[6] Oliver W C and Pharr G M 2004 *J. Mater. Res.* **19** 3 https://doi.org/10.1557/jmr.2004.19.1.3

IOP Conf. Series: Materials Science and Engineering **613** (2019) 012037 doi:10.1088/1757-899X/613/1/012037

The effect of the synthetic fibre reinforcement on the fracture energy of the concrete

K P Juhász

Assistant lecturer, Department of Mechanics, Materials & Structures, Budapest University of Technology and Economics, Budapest, Hungary

E-mail: juhasz@szt.bme.hu

Abstract. Fibre reinforced concrete belongs to the group of composite materials, where matrix means the concrete, and fibres mean the various shaped (straight, hooked, wavy, etc.) and material (steel, polypropylene, glass, etc.) fibres, which are short compared to the size of concrete elements. Fibres are mixed in fresh concrete which supposedly provides a perfect and uniform mix. Although material of the fibres can be different (steel and synthetic polypropylene are the most widespread), material models in literature are primarily for steel fibres. Elastic modulus of steel and synthetic fibres differs from each other in terms of scale (steel 210 GPa, synthetic 8-12 GPa), despite this, however, similar results are received frequently at bending beam tests. The different result lies in the energy absorption of the fibres pulling out from the concrete matrix. In the present paper, this different behaviour is examined and a suggested material model for fibre reinforced concrete is presented. This material model is based on the modification of the fracture energy.

1. Introduction

Synthetic macro fibre reinforced concrete is becoming more and more widespread in the industry and more and more accepted among engineers as well [1]. Their applicability was questioned primarily because of their low elastic modulus (steel: 210 GPa, synthetic: 8-12 GPa) [2]. There are numerous other arguments against their applicability (their low melting point, the creeping of the synthetic polypropylene material), but they have one significant advantage over steel fibres, which is their corrosion resistance. When a structure is exposed to a corrosive environment, it is worth considering the use of synthetic macro fibres.

Material model of fibre reinforced concrete has plenty of recommendations in the literature, however in most cases these only apply to steel fibres [3,4,5], or do not distinguish between steel and synthetic materials [6]. The difference is apparent at long-term loading due to the creeping of synthetic material [7], or at temperature loading due to the low melting point. These qualities of the material are not manifested at normal temperature short-term static loading. Examining the energy absorption of steel and synthetic fibres, a new and simplified material model is presented in the current paper, which models fibre reinforced concrete material by modifying fracture energy of the concrete. As fibres have a low impact on the flexural tensile strength of plain concrete, material model of plain concrete has to be modified only after cracking. Fracture energy of concrete is estimated based on the formulas in the literature, which then is modified by added fracture energy depending on fibre type and fibre dosage. With this approach, a simple yet accurate material model can be defined.

IOP Conf. Series: Materials Science and Engineering **613** (2019) 012037 doi:10.1088/1757-899X/613/1/012037

2. Fracture energy of the concrete

Concrete is a quasi-brittle material, which means that after the cracking of the material it is capable of taking decreasing stress depending on the opening of the crack. Due to the tensile stresses in the concrete a thin band with micro-cracks will appear: this is called the fracture process zone [8]. Increasing the stress the concrete reaches its tensile strength, and leads to merging of the micro-cracks, and finally a macro crack is created and is visible. After this point the tensile capacity of the concrete will decrease, the cracks will bypass or cross the aggregates and then slowly the entire section will be crossed by the crack, see figure 1.

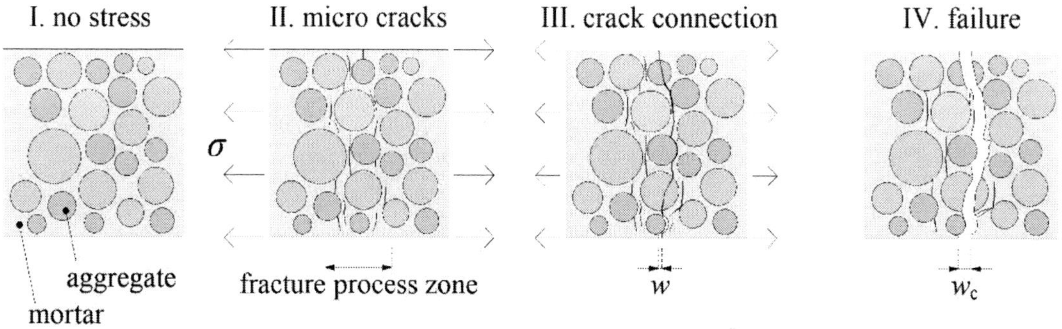

Figure 1. Crack propagation in the concrete

Measuring the residual tension stress and crack width a stress-crack width (σ-w) diagram is obtained, area below this curve is the fracture energy of concrete: G_F (Nmm), see figure 2.

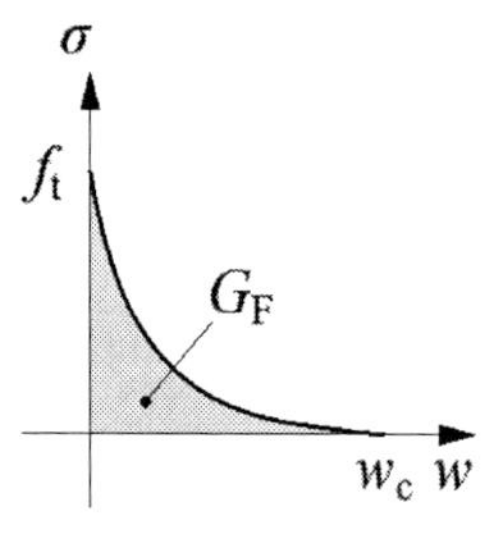

Figure 2. Fracture energy of the plain concrete

The fracture energy of the concrete is influenced by a number of factors which are clearly not related to the concrete strength class. Most of the existing design methods neglect the fracture energy of the concrete and do not pay much attention to the tensile strength. There are numerous references in the literature about how to specify the fracture energy of concrete, most frequently in function of strength class and maximum aggregate size [9,10,11]. According to CEB-FIP Model Code the fracture energy can be estimated as following:

$$G_F = G_{F0}\left(f_{cm}/f_{cm0}\right)^{0.7} \tag{1}$$

where:
f_{cm0} = 10 MPa

G_{F0} is base value of fracture energy, depends on the maximum aggregate size, according to table 1.

Table 1. Base values of fracture energy G_{F0}

d_{max} (maximum aggregate size) [mm]	G_{F0} [N/mm]
8	0.025
16	0.030
32	0.058

Some recommendations of the stress-crack width diagram can be seen in figure 3, and some formulas in table 2.

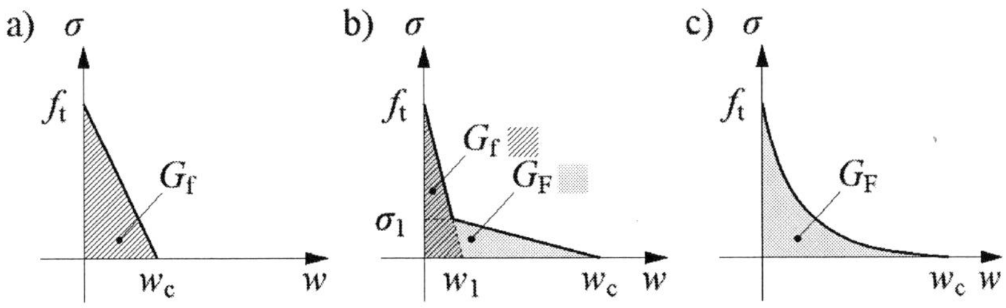

Figure 3. Recommendations of the stress-crack width diagrams
a) Linear [8] *b)* Bilinear [21] *c)* Power curve [22]

Table 2. Recommendations of σ-w diagram

Name	Expression	
Linear curve [8]	$\sigma = f_t\left(1-\frac{w}{w_c}\right)$ $w_c = \frac{2G_f}{f_t}$	(2)
Bilinear curve [21,23,24]	$\sigma = \begin{cases} f_t-(f_t-\sigma_1)\frac{w}{w_1} & \text{if} \quad w \le w_1 \\ \sigma_1-\sigma_1\frac{w-w_1}{w_c-w_1} & \text{if} \quad w_1 > w \end{cases}$ $w_1 = \frac{2G_f}{f_t}$ $w_f = \frac{2}{\psi f_t}\left(G_F-(1-\psi)G_f\right)$	(3)
Power curve [22]	$\sigma = f_t\left(\left(1+\left(c_1\frac{w}{w_c}\right)^3\right)\exp\left(-c_2\frac{w}{w_c}\right)-\frac{w}{w_c}\left(1+c_1^3\right)\exp\left(-c_2\right)\right)$ $c_1 = 3$ $c_2 = 6.93$ $w_c = 5.14\frac{G_F}{f_t}$	(4)

3. Energy-absorbing fibre-matrix mechanisms of steel and synthetic fibres

The fibres start to work effectively after the crack of the matrix and increase the residual strength and the ductility of the material, thus increasing fracture energy of the concrete. According to Zollo [12] the energy-absorbing function of the fibres could be 1) fibre bridging, 2) fibre pull-out and 3) fibre failure (figure 4). The bridging and pull-out of fibres could give the highest added ductility for the concrete, whilst fibre failure will decrease it. For steel fibres, the fibre pull-out will have a major role even from small crack opening, while synthetic fibres function more through a bridging mechanism.

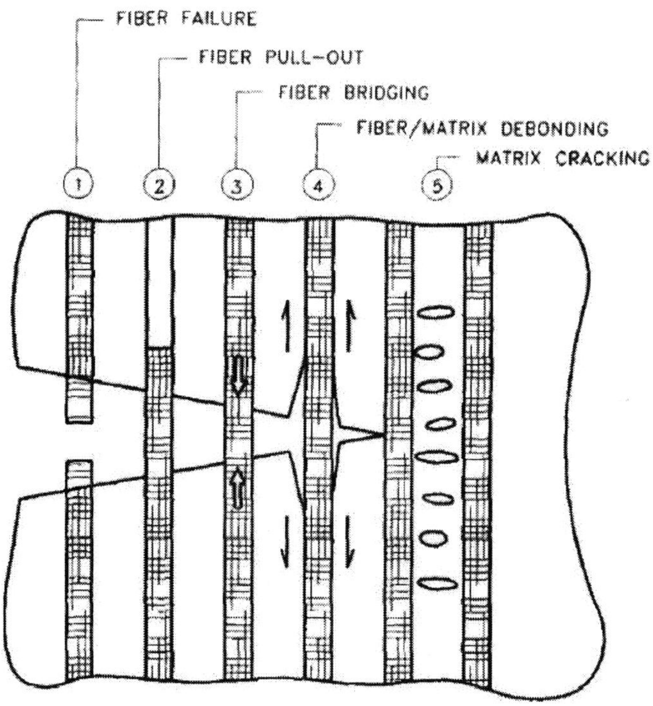

Figure 4. Energy-absorbing fibre/matrix mechanism [12]

The tensile strength of the fibre reinforced concrete will be the same as plain concrete, the effect of the fibres in the linear phase is limited. The fibres start to work when the micro cracks start appearing, first the fibres detaching from the concrete matrix. This is an important phase of the mechanism, because if this cannot happen the fibres will rupture. This may justify a reduction in ductility of steel fibre reinforced concrete in time, called as embrittlement according to Bernard [13]. After the detachment there is the pull-out phase (figure 5).

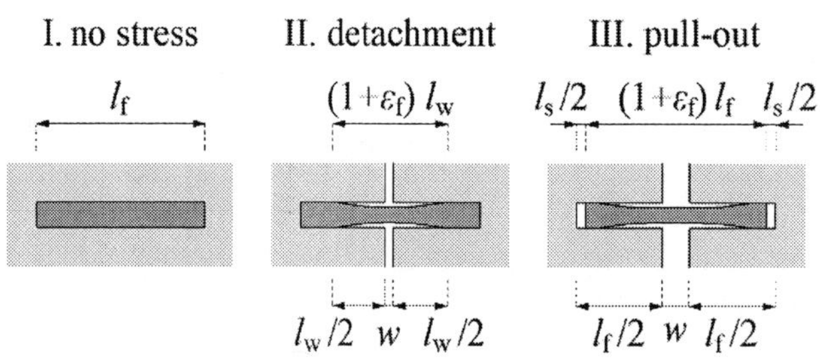

Figure 5. Mechanics of the fibre

Thus the following formula for crack opening can be written as:

$$w = l_w \varepsilon_f + l_s \tag{5}$$

where:

l_f	length of the fibre
l_w	working length of the fibre ($l_w \leq l_f$) as given in equation (6)
l_s	slip distance as given in equation (7)

$$l_w = \begin{cases} \dfrac{w}{\varepsilon_f} = \dfrac{wE_f}{\sigma_f} & \text{if} \quad \dfrac{wE_f}{\sigma_f} \leq l_f \\ l_f & \text{if} \quad \dfrac{wE_f}{\sigma_f} > l_f \end{cases} \tag{6}$$

which means that it is assumed that the fibres will pull-out if the working length (l_w) reaches the fibre length (l_f).

$$l_s = w - l_w \varepsilon_f \tag{7}$$

which will give zero if $l_w < l_f$, i.e. only some parts of the fibre are working.

In this model the influence of the unsymmetrical anchoring and different orientation of the fibre were neglected. This model is made only for comparing the steel- and synthetic fibre mechanics during the crack opening. The fibres give added fracture energy via detachment, bridging and pull-out. The detachment depends on the connection of the mortar and the surface of the fibre, the bridging process on the detachment, while the pull-out phase depends on the surface of the fibre and the end-hook configuration. Fibre reinforced concrete beams were tested according to RILEM recommendations [14] by three point notched beam test and the load and the CMOD were measured. The beams were reinforced with 20 kg/m^3 steel and 5 kg/m^3 synthetic macro fibres. From these series steel and synthetic fibre reinforced concrete beam test results were selected where the load-CMOD results are close to each other (figure 6).

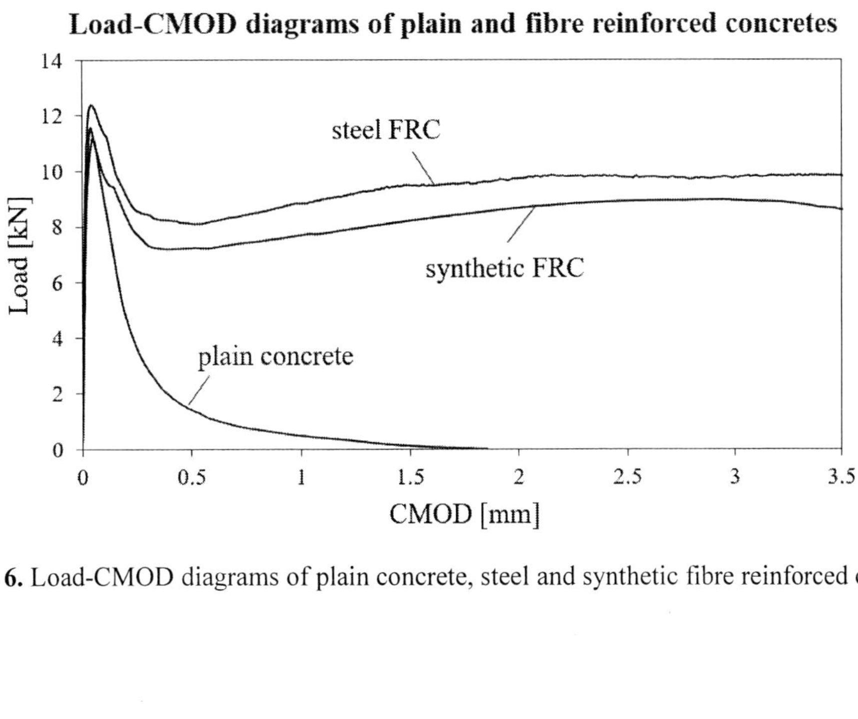

Figure 6. Load-CMOD diagrams of plain concrete, steel and synthetic fibre reinforced concrete

The stress-strain and stress-crack width material law was defined according to the RILEM recommendation [4]. Due to this it was possible to find out the stress distribution on the section. If location of the fibres is known on the section then the stresses in the fibres (σ_f), the working length of the fibre (l_w) and the slip distance (l_s) could be calculated. As the beams were reinforced with different fibres, steel and synthetic, but the load-CMOD diagrams were almost the same, it can be estimated that the stress-crack width material law is also affined for both materials. Having the same stress-crack opening law for the same load-CMOD curve can be estimated (figure 7). The method of calculation is the following: the value of the CMOD and the stress distribution on the section is known. From this the crack opening (with linear estimation) and the stress in the individual fibres (according to the distribution of the fibres) can be calculated. The working length and the slip can be calculated according to equation (5) and (6).

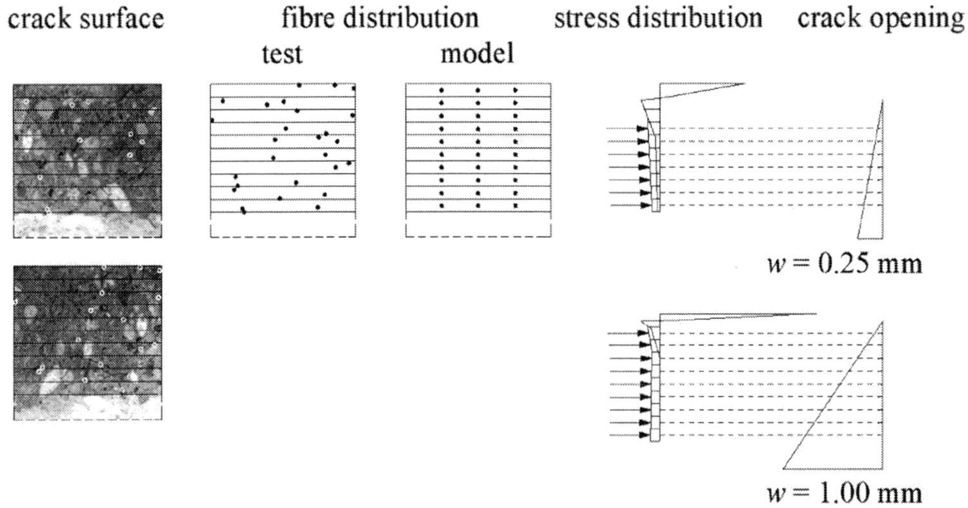

Figure 7. Modelling the stress, strain and slip in the individual fibres

With these approximations the strain and slip of the fibres were calculated at CMOD = 0.25 mm and 1 mm in the case of steel- or synthetic fibres. The results are shown in figure 8.

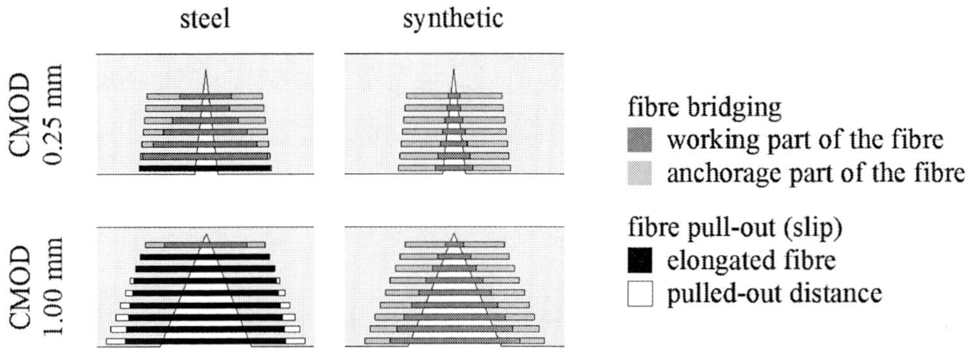

Figure 8. Fibre bridging and fibre pull-out of steel and synthetic fibres

As shown in figure 8, because of their small strain capacity, the steel fibres must start slipping at small crack opening, while the synthetic fibres are still in bridging phase. If the material is still in its elastic range after unloading it will want to close the crack. At bridging, fibres are equal to the whole crack opening, i.e. they want to close the whole crack, while at slipping, fibres could close only with their strain energy and the slipping cannot be reversed.

4. Added fracture energy of the fibre reinforced concrete

Fracture energy of concrete can be estimated by many recommendations. Although the added fibre increases fracture energy, only a few instances occur when fibre reinforced concrete is modelled by considering modification of fracture energy [15] however it would be plausible. Added fibres increase this residual strength and the ductility of the material, thus increasing fracture energy of concrete. The various types of fibres exert their primary effect at different crack width [7]. Steel fibres with higher elastic modulus are able to take high stress even at low crack width, however by the increase of crack width fibres tear or pull out from the matrix, thus resulting in the decrease of residual stress. However, synthetic fibres of low elastic modulus can work even at higher strain than steel fibres, meaning that synthetic fibres behave efficiently in the case of higher crack width [13].

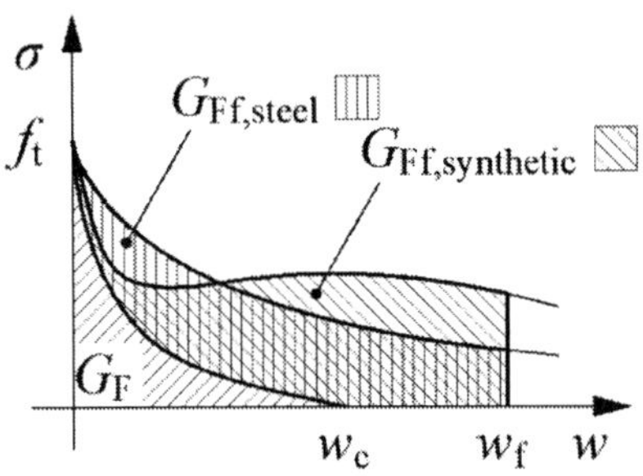

Figure 9. σ-w diagram of steel- and synthetic fibre reinforced concrete and interpretation of fracture energy

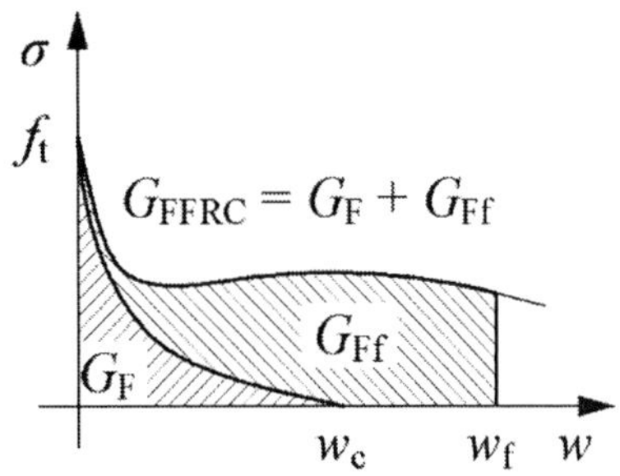

Figure 10. Fracture energy of concrete and fibre reinforced concrete, and added fracture energy

Crack width at which stress terminates between the crack surfaces of the concrete is called final crack opening width marked by w_c [15]. In the case of fibre reinforced concrete this value is significantly higher, so it can be taken into account during calculation with a given value, according to guidelines. Thus, considered maximum crack opening width of fibre reinforced concrete is marked by w_f. Value of w_f depends on the type and length of the fibre, its recommended baseline value is w_f =3 mm. Fracture energy of the concrete is the area below the stress-crack width diagram of the concrete, marked by G_F, while for fibre reinforced concrete it is similarly the area below its curve, marked by G_{FFRC}. Fracture energy added by fibres is the difference of the two, thus: $G_{Ff} = G_{FFRC} - G_F$ (figure 10).

Based on both my own experimental results and the publications found in the literature [16,17,18,19] it can be stated that post-crack residual strength of synthetic fibres takes an approximately constant value. A simplified constitutive law is usually provided in the guidelines, where post-crack tension strength is modelled by a constant value [20,5]. Based on the fib model code 2010 value of the post-crack tension strength in the case of rigid-plastic model is the following:

$$f_{Ftu} = \frac{f_{R3}}{3} \tag{8}$$

where:

f_{Ftu} post-crack residual tension strength
f_{R3} residual flexural tensile strength corresponding to CMOD = 3 mm, according to equation 9.

$$f_{R3} = \frac{3F_3 l}{2bh_{sp}^2} \tag{9}$$

where:

F_3	is the load corresponding to CMOD = 3 mm
l	span length, 500 mm
b	specimen width, 150 mm
h_{sp}	distance between the notch tip and the top of the specimen, 125 mm

From the above the following constitutive law can be defined according to figure 11. Value of f_f can be defined based on the guidelines, thus $f_f = f_{Ftu}$ based on the fib model code demonstrated above. The overlapping part of the G_F and G_{Ff} at the numerical diagram is negligible.

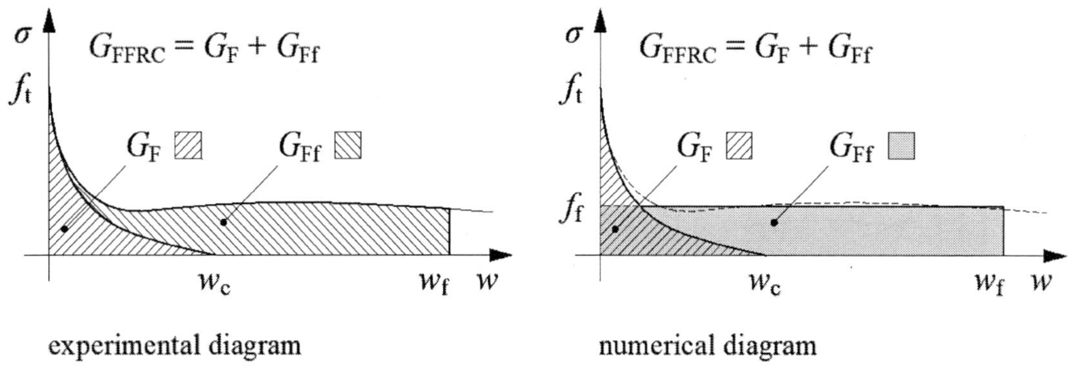

Figure 11. Experimental and numerical diagram of fibre reinforced concrete

Based on the above model added fracture energy (G_{Ff}) can be defined as follows:

$$G_{Ff} = f_f \, w_f \tag{10}$$

5. Summary

Elastic modulus of the synthetic fibres (8-12 GPa) is much smaller than the elastic modulus of the steel fibres (201 GPa), despite this fact with the proper dosages similar load-CMOD diagram is received at beam tests. The reason for this is the different energy-absorption mechanism of the fibres: while steel fibres pull-out from the matrix, synthetic fibres bridge the crack. These two kinds of behaviours were analysed with beam tests.

Effect of fibres can be defined by analysing fracture energy of fibre reinforced concrete, which can be divided into two parts: fracture energy of the concrete and added fracture energy of fibres. There are numerous recommendations in literature defining the value of fracture energy of plain concrete, of which the most relevant models were introduced. By knowing the fracture energy of the concrete and the added fracture energy of the fibres the material model of the fibre reinforced concrete can be easily defined.

References

[1] Bernard E S 2015 *Tunn. and Undergr. Space Techn.* **49** 241
[2] BEKAERT. Steel Fiber Reinforced Concrete FAQ [Internet] [cited 2 June 2018] Accessed from: https://www.bekaert.com/en/product-catalog/content/building-products-pages/steel-fiber-reinforced-concrete-faq#20
[3] ACI Committee 544. Standard 544.2R-89 - Measurement of Properties of Fiber Reinforced Concrete [Internet]. 1998 [revised 4 February 2017, cited 4 February 2017]. Available from:

IOP Conf. Series: Materials Science and Engineering **613** (2019) 012037 doi:10.1088/1757-899X/613/1/012037

https://www.concrete.org/store/productdetail.aspx?ItemID=544289&Format=DOWNLOAD&Language=English

[4] Vandewalle et al., 2003 RILEM TC 162-TDF: Test and design methods for steel fibre reinforced concrete – σ-ε design method *Mat. and Struct.* **36** 560

[5] ÖVBB, 2008 *Richtlinie Faserbeton* (Vienna)

[6] CNR-DT. Standard 204/2006 - *Guide for the Design and Construction of Fiber-Reinforced Concrete Structures* [Internet]. 2006 [revised 4 February 2017, cited 4 February 2017]. Available from: https://www.cnr.it/en/node/2640

[7] Juhász, K P 2014 *Építés-Építészet.* **42** 57

[8] Hillerborg A, Modéer M and Petersson P E 1976 *Cem. and Conc. Res.* **6** 773

[9] CEB 1991 *CEB-FIP Model Code 1990* (London: Thomas Telford Services Ltd.) pp 36-37.

[10] Hilsdorf, H K and Brameshuber W 1991 *Int. J. of Fract.* **51** 61

[11] Papanikolaou V K and Kappos A J 2007 *Int. J. of S. and Str.* **44** 7021

[12] Zollo R F 1997 *Cem. Concr. Comp.* **19** 107

[13] Bernard E S 2008 *Shotcrete* **10** 16

[14] Vandewalle L. et al. 2000 *Mat. and Struct.* **33** 3

[15] Park K, Paulino G H and Roesler J 2010 *Cem. and Concr. Res.* **40** 956

[16] Li V C, Chan Y W and Wu H C 1994 *Proc. of Int. Symp. on Br. Mat. Comp. 4 (Warsaw)* (Warsaw: Ike and Woodhead Publishing) 7

[17] Bernard E S 2009 *Shotcr. for Underg. Supp. XI, ECI Sym. Ser.* P11 (Accessed 15 June, 2013) Retrieved from: http://dc.engconfintl.org/shotcrete/14

[18] Carnovale D. Behaviour and analysis of steel and macro-synthetic fibre reinforced concrete subjected to reversed cyclic loading: a pilot investigation. [M.A. Sc. Thesis on the Internet]. [Toronto]: University of Toronto. [cited 5 May 2018]. Available from: https://tspace.library.utoronto.ca/bitstream/1807/42702/1/Carnovale_David_J_201311_MASc_Thesis.pdf

[19] Di Maida P, Radi E, Sciancalepore C and Bondioli F 2015 *Const. and B. Mat.* **82** 39

[20] fib 2012 *Model Code 2010 (Final Draft vol 1)* (fib bulletin 65) 234

[21] Roelfstra R E and Wittmann F H 1986 Fracture Toughness and Fracture Energy in Concrete ed F H Wittmann *Amsterdam Elsevier Science* 163

[22] Hordijk D A. Local approach to fatigue of concrete. [doctoral thesis on the Internet]. [Delft]: Delft University of Technology. [cited 5 May 2018]. Available from: https://repository.tudelft.nl/islandora/object/uuid:fa87147b-8201-47ed-83d7-b812b09c5fbb/

[23] Bazant Z P 2002 *Eng. Fr. Mech.* **69** 165

[24] Roesler J, Paulino G H, Park K and Gaedicke C 2006 *Cem. & Co. Com.* **29** 300

Experimental Study of the Technology Parameters Affect in the Laser Welded Joints

Tünde Anna Kovács[1], **Péter Pinke**[2], **Enikő Bitay**[3]

[1,2]Óbuda University, Bánki Donát Faculty of Mechanical and Safety Engineering, Department of Materials Technologies, 1081 Hungary, Budapest, Népszínház u., 8
[3]Sapientia Hungarian University of Transylvania, Faculty of Technical and Human Sciences, 540485 Romania, Târgu-Mureş, Op. 9., Cp. 4

E-mail: kovacs.tunde@bgk.uni-obuda.hu

Abstract. The laser welding technology is improving immediately rapidly nowadays. It has many advantages again the traditional welding technologies. The heat affected zone in case of the laser welding is very thin and the penetration of the joint is depth. This high energy technology work by a focused high energy laser light accompanied by protective gases. For welding process well known the different laser devices (gas laser, solid-state laser, disc laser). The effect of the gases and the technological parameters are not well known yet. In this work, we wanted to find some relationship between the welding parameters and the joint mechanical properties. To find this relationship it was welded the joint by different technologies parameters and the samples experimented the joint mechanical properties with different parameters and controlled by the experimental way.

1.Introduction

The welding technology is a rapidly developed joining technology. Laser welding is a fusion welding technology, what use heat. The melted materials solidification prepares the joint and occur a cohesion join between parts during the process [1-3]. Laser technology is a special technology because the used heat occurred from light. The concentrated light energy means the laser is very useful in case of many different technologies like thermal cutting, welding, engraving.

The laser welding technology process effects aren't well understood yet. The laser machining setup is the base of this technology. Between the welding fusion welding technologies, this process causes a small heat affected zone (HAZ). The quality of the welded joint depends on the welding technology parameters and the affected heat and the metal heat transfer. The key to this technology process the knowledge of the effect of the used parameters. The precedent way the practice based results to acquire this knowledge.

The laser technology important parameters are like the welding gas (Ar, CO_2, He, etc. and their mixes, and percentages), the welding speed, the focus point and the used power are also important. The used alloys (chemical composition), the sheet thickness and the joint type also influence the welded joint quality [2,4]. It can specify the optimal welding process parameters based on the practice and quality control results (destructive and nondestructive inspections).

2. Welding processes and Materials

In the material the laser waves effect a melted metal surrounded depth plasma canal (keyhole welding). The essential requirement is to produce and sustains the plasma canal during the keyhole welding. In case of welding processes very useful the laser welding (gas laser, solid-state laser or disc laser). The concentrated heat during the welding process melt the filler metal and the base metal and cause a very thin heat affected zone.

2.1. CO_2 laser welding

In the practice, the most applied gas laser type is the CO_2 laser technology. In case of this gas laser technology, the used laser material is a mix of the Carbon-dioxide, Nitrogen and Helium, and the active material is the CO_2 gas. In the gas volume, it works in continuous laser emission but it can work in impulse mode too. A CO_2 laser wavelength is λ=10,6 μm, that means the infrared area. In this way, the power can be more than 100 kW and the average emission part is 20% [5].

2.2. YAG laser welding

An optically pumped solid-state laser which is constructed of components held in association by a support structure which is configured to receive the components and Yttrium-Aluminium-Garnet (YAG). ND: YAG lasers are widely used in a variety of applications [5-8].

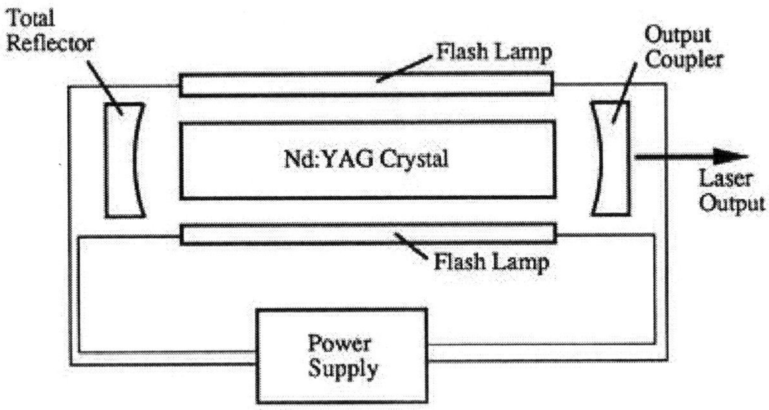

Figure 1. Nd: YAG laser setup [6].

2.3. Used materials

It was used two different steel in this study. It was joined an austenite stainless steel and low carbon steel. The chemical composition of the used steels showed in Tab.1-2.

Table 1. The chemical composition of the stainless steel beside irons in weight %

steel sign: X2CrNiMoN17-11-2								
C %	Si %	Mn %	Ni %	P %	S %	Cr %	Mo %	N %
0,03	1,0	2,0	10	0,035	0,015	16,5	2	0,12

In case of low carbon steel (S355JR) welding process, the preheating is not a requirement because of the low carbon equivalent (the carbon equivalent can calculate from the chemical composition) and the small samples sizes. In the case of the austenitic stainless steel, the preheating is usually forbidden and about the samples size isn't required.
The prepared joint was between two rings. The welded sample is showed in Fig. 2.

Table 2. The chemical composition of the low carbon steel beside irons in weight %

steel sign: S355JR						
C %	Si %	Mn %	P %	S %	N %	Cu %
0,21	0,6	1,4	0,045	0,045	0,014	0,6

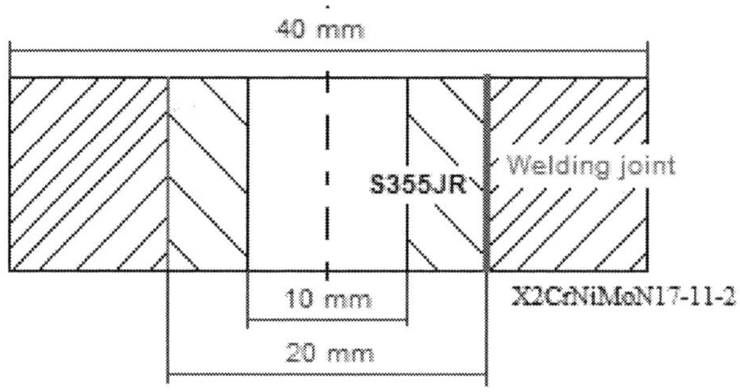

Figure 2. The welded sample

3. Experiments

In this work, it was made the welding by two different laser welding process. In case of the first group, it was weld 3 samples by CO_2 laser and Nd: YAG laser welding with the same parameters.

3.1. CO_2 welding tests

These welding process parameters were: welding rate 2000 mm/min, laser power 3250 W. The used gas was the M20 signed working gas, the volume of the gas 7 l/minute, pressure 7 bar. The welded joint depth is 4 mm (shown in Fig. 3.). The joint not symmetric because it was welded two different steel with different heat transfer properties. The joint chemical composition is a mix of these different steel.

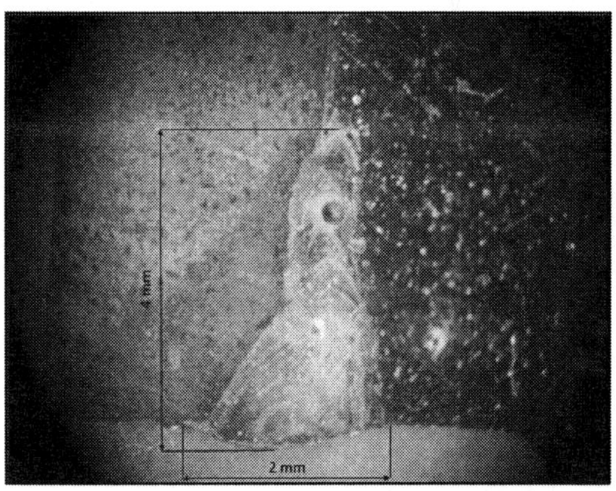

Figure 3. CO_2 laser welded joint

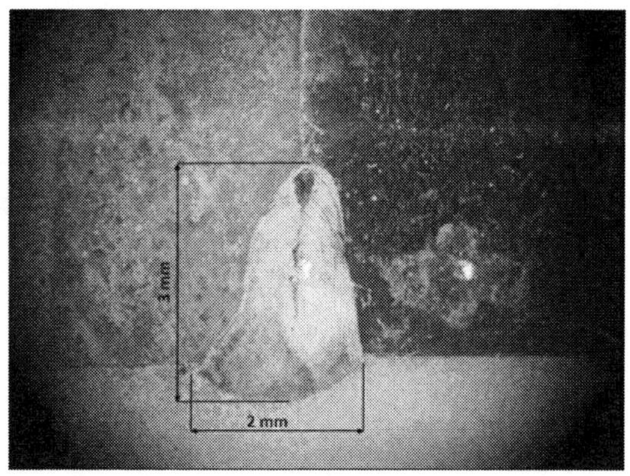

Figure 4. Nd:YAG laser welded joint

3.2. Nd:YAG laser welding tests

The Nd: YAG solid-state welding parameters were: welding rate 390 mm/min, laser power 2700 W. The used shielding gas was CO_2, volume 20 l/minute, pressure 5 bar. The joint depth was 3 mm (shows in Fig.4.). In the middle of the joint find in every tested joint a longitudinal crack.

4. Results and Discussion

Mechanical properties of the experimented samples

Table 3. The hardness in the tested joint and the HAZ (HV_{10})

	Hardness test location		
State	**X2CrNiMoN17-11-2**	**Joint**	**S355JR**
Original	223	-	239
CO2 welded	222	354	240
Nd:YAG welded	230	382	281

The welded joint and the heat affected zone (HAZ) was tested by microscopy and hardness test (HV_{10}). To determine the hardness in the welded joint and in the heat affected zone it was used a traditional Vickers hardness tester the applied load is 10 kg. The tests results are shown in the Tab. 3. and on the Fig. 5-6. The hardness of the steels before and after the welding process shows in the Tab.3.

In case of the welded joint and the HAZ the required hardness need to be under 350 HV_{10}.

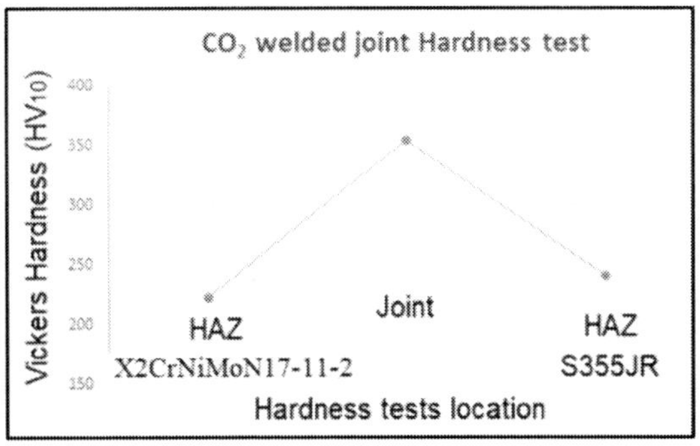

Figure 5. CO_2 laser welded joint hardness

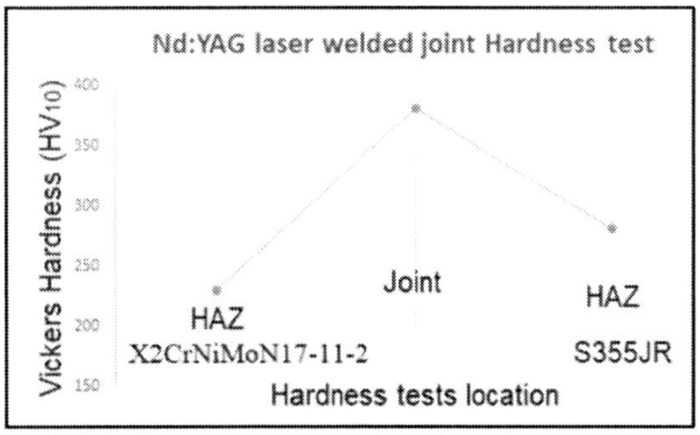

Figure 6. Nd:YAG laser welded joint hardness

5. Conclusion

It can conclude that the CO_2 laser welded joint the measured hardness showed the required results, under 350 HV10.

I. In the CO_2 laser welded joint it wasn't found crack propagation, but in the Nd: YAG laser welded joint it was detected this failure in case of every tested sample.
II. The hardness test results showed that in the joint was the hardness the highest. These results cause the welded joint chemical composition because of the low carbon steel and the austenitic stainless steel mixed during the fusion welding process.
III. The joint hardness was lower in case of the CO_2 welded joint.
IV. It can suggest that in the case of this samples (with the used chemical composition and geometrical properties) the followed process is the CO_2 laser welding besides the used technological parameters.

Acknowledgement
We acknowledge the financial support of this work by the Hungarian State and the European Union under the EFOP-3.6.1-16-2016-00010 project.

References

[1] Bitay Enikő, Dobránszky János 2014 XIX. Fiatal Műszakiak Tudományos Ülésszaka – FMTÜ, *EME*, ISSN 2067 – 6 808 Kolozsvár 77 http://hdl.handle.net/10598/28203
[2] Buza Gábor 2012 Lézersugaras technológiák I. *EDUTUS* http://www.tankonyvtar.hu/en/tartalom/tamop412A/2010-0017_50_lezersugaras_technologiak_I/index.html
[3] T. Kovács 2018 *Procedia Manufacturing* **22** 147 https://doi.org/10.1016/j.promfg.2018.03.023
[4] E Bitay, T Kulin, Zs Puskás, Gy György, J Dobránszky 2014 *Biomaterials cutting technology* 405 In: Gáti József (ed.) *27. Welding Conference* 2014. Óbuda University, Budapest, Hungary
[5] John H Clark, Dennis L Werth 1988 *Solid state laser,* Patent number: US 4731795 A Amoco Corporation
[6] http://www.aml.engineering.columbia.edu/ntm/level1/ch05/html/l1c05s07.html (10.01.2018.)
[7] Gy Bagyinszki, E Bitay 2013 *Danger of the laser industrial engineering works* XVIII. FMTÜ, EME, Kolozsvár 51
[8] www.lasermarking.trumpf.com (15.12.2017.)

IOP Conf. Series: Materials Science and Engineering **613** (2019) 012039 doi:10.1088/1757-899X/613/1/012039

Surface Modification Innovation for Wear Resistance Increasing

Tünde Anna Kovács[1], Umesh Mhatre[2], Zoltán Nyikes[3], Bitay Enikő[4]

[1]Óbuda University, Bánki Donát Faculty of Mechanical and Safety Engineering, Department of Material Science, 1081 Hungary, Budapest, Népszínház u., 8
[2]Surface Modification Technologies Pvt. Ltd. Research and Development Lab, India, Sativali Road Vasai Road – (East), Dist. Palghar, Maharashtra - 401208.
[3]Óbuda University, Doctoral School on Safety and Security Sciences, 1081 Hungary, Budapest, Népszínház u., 8
[4]Sapientia Hungarian University of Transylvania, Faculty of Technical and Human Sciences, 540485 Romania, Târgu-Mureş, Op. 9., Cp. 4

E-mail: kovacs.tunde@bgk.uni-obuda.hu

Abstract. The surface modification is a suitable technology to increase the surface wear resistance of steels. In this way, the used substrate metal can be a low alloyed, cost-effective chemical composition steel. After the substrate metal heat treating, hardening and annealing, it needs surface modification.

It well knew some surface heat treating technologies, like carburizing, nitridation, nitrocarburizing, and surface quenching. The surface coating is also very useful technologies when by the chemical or physical process it built a new layer on the surface.

In this work, we compare different modification processes results made by the way of the traditional and new technologies advantages. The chemical composition and the structure determine the surface mechanically and wear resistance properties. The available chemical composition determines the surface modification technology [1-3].

It was prepared with different surface layers on the same substrate steel by a different process and inspected by different testing methods. On the base of the tests results needs to specification a new surface modification process for the high-level surface resistance layer.

1. Introduction

The steels are widely developed by the steel industry in the last decade. For the manufacturing of the high strength steels and the new high strength metal alloys is necessary the tool steels innovation. The challenge is the increasing of the wear resistance and rises up the lifetime of the tools. The new materials need new hardening technologies and surface treatings. The innovation is realizable by the modernization of traditional technologies, the invention of new technologies and combinate the traditional and modern heat treating technologies [4].

The traditional hardening technology for steels is well known, but in the case of the new tool steels considering the chemical composition (special and high amount alloy content) and the primary manufacturing technology (f.ex. powder metallurgy) caused microstructure the heat treating technology can be different.

The required mechanical properties of the tool steels are the high wear resistance, the high strength and the long lifetime. The tool steels surface requirements are the high hardness and wear resistance [4]. The increasing of the tool steels productivity insurable by surface treatments. The traditional surface treating technologies are well known.

The modern surface coatings technologies are the Physical Vapor Deposition (PVD) and the Chemical Vapor Deposition (CVD) technologies these are suitable to establish a new, thin, high hardness and high strength layer or multilayer on the substrate surface. By the traditional surface treatment, the surface chemical composition can modify (nitriding, boriding, etc.) or can modify the surface microstructure with martensite formation (induction surface hardening, laser surface hardening, etc.) or can combinate the two process (nitrocarburizing, etc.) [2].

The lifetime and wear resistance prediction in case of the treated tool steels is not simple. The industries use the hardness test because the hardness and the wear resistance show a strong correlation. Also usable the comparative experiments by model test (pin on disc test, ball cratering, etc.). In our innovation work, we used the comparative experiments method when we used the same test parameters and after the heat treating, we prepared the tested samples in the same way.

2. Used material and heat treating processes

The used tool steel is H13 (ISO 683-13, W.n. 1.2344, X40CrMoV5-1) heat treatable steel, chemical composition shown in Tab.1.

Table 1. Chemical composition (in wt. %) and the remain is the iron (Fe)

C	Mn	P	S	Si	V	Cr
0.32-0.45%	0.2-0.6%	max. 0.03%	max.0.03%	0.8-1.25%	0.8-1.2%	4.75-5.5%

Heat treatment was the hardening, austenitization and quenching by rapid cooling, the heat treatment parameters shown in Tab.2. The heat treating and the surface coating made in India supported by Surface Modification Technologies Pvt. Ltd. Research and Development Laboratory.

Table 2. Heat treating of the samples

Treatments	**Parameters**
Austenization temperature	1080 °C
Cooling media	Oil cooled
Plasma Nitriding temperature	480 °C
Nitriding time for 60 μm depth	5 h
Nitriding time for 120 μm depth	12 h

3. Wear coefficient determination theory

The wear coefficient is calculated by the following modified Archard equation (1) from the constant parameters (N_c, f, t) and the crater depth (h) [5]:

$$K = \frac{h^2}{2 \cdot N_c \cdot f \cdot t} \tag{1}$$

IOP Conf. Series: Materials Science and Engineering **613** (2019) 012039 doi:10.1088/1757-899X/613/1/012039

Where:
R the wear ball radius [10 mm]
h the depth of the wear crater (calculated) [mm]
t=5 min time of the wear process
N_C=0,86 [N] normal load
f=570 [1/min] revolution per min
K wear coefficient [mm^2/N]
e-diameter of the crater [mm]
S length of the wear [mm]

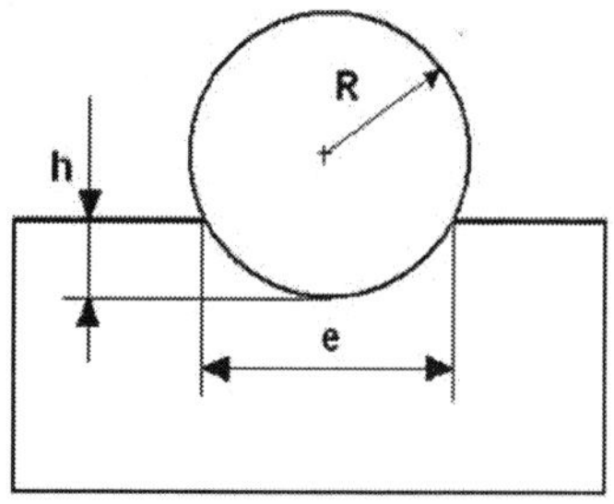

Figure 1. The wear crater parameters

4. The experiments and results

For the wear resistance determination well known several experimental setup, the ball cratering tests are suitable to investigate comparative wear tests [6-7]. The ball cratering wear tests were investigated in the laboratory of the Óbuda University by a modified ball cratering experimental setup. Disc test samples size were Ø15x3 mm. It was tested all samples 5 times under constant parameters (load, time and instruments) without any lubricant. The wear crater diameter was measured by microscopy (Neophot 21, supported by computer visual analysis software) one crater in case of every treated surface is shown in Fig. 2-5.

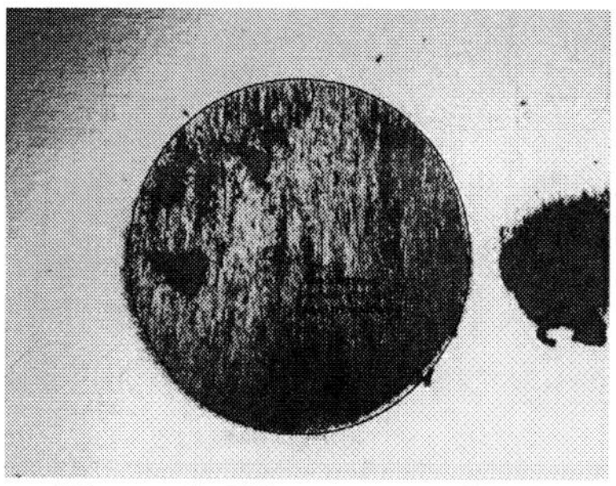

Figure 2. Wear crater of N.1. sample

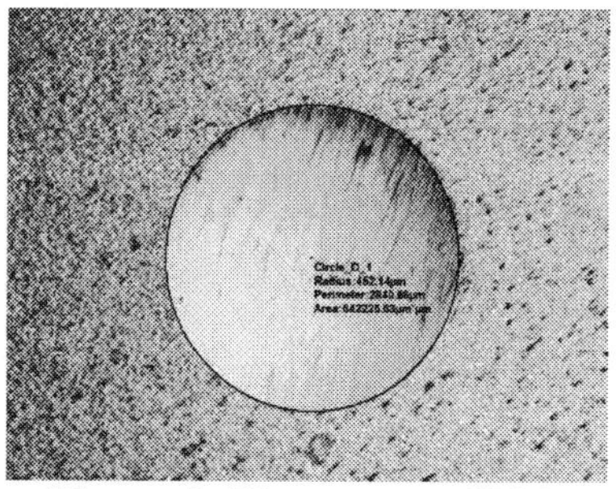

Figure 3. Wear crater of N.1. sample

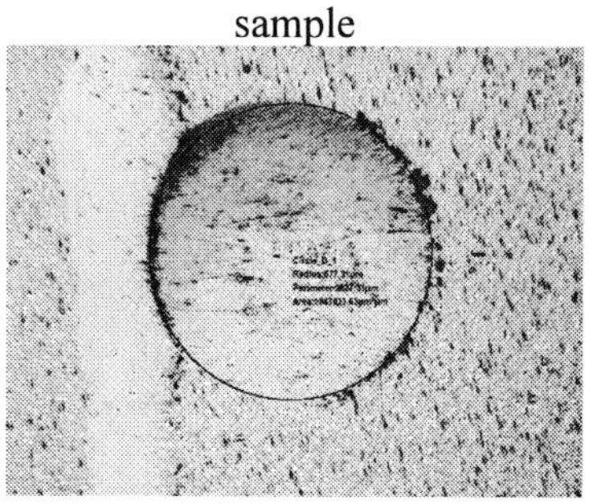

Figure 4. Wear crater of N.1. sample

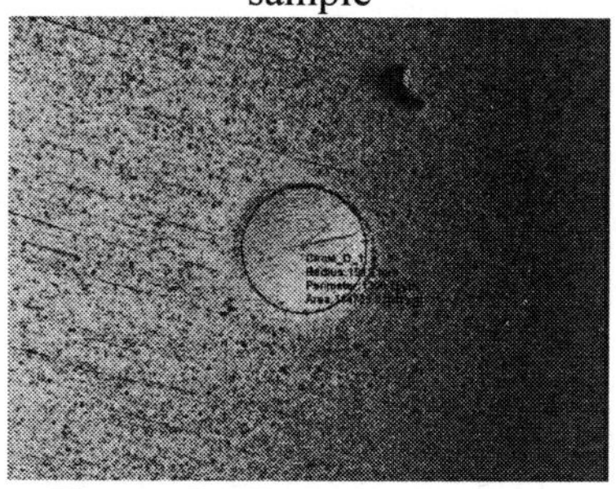

Figure 5. Wear crater of N.1. sample

The Tab.3. shown the average value of the hardness (HV) and the wear coefficient of the measured test samples. The hardness in case of the plasma nitrided surface samples (N.2. and N.3.) it can not find the relevant difference even that the layer thickness is different. The hardened and PVD coated surface hardness is the highest, more than then 5 times harder than the used tool steel hardened value. The special metal (chromium) and ceramics layers (titan nitride, zirconium tungstate, etc.) hardness can be incredible high because of the special crystal structure [8].

The wear coefficient was calculated by the (1) equation from the measured wear crater radius. The Fig. 2-5 shown some typical crater and their radius.

Table 3. The test samples properties

N.	Heat and surface treatment	Treated layer thickness (μm)	Hardness HV	Wear coefficient K (mm^2/N)
1	Hardening	0	549	$6{,}32 \cdot 10^{-9}$
2	Hardening + Plasma nitriding	50 μm	1130	$1{,}37 \cdot 10^{-9}$
3	Hardening + Plasma nitriding	200 μm	1140	$1{,}95 \cdot 10^{-9}$
4	Hardening + PVD coating Cr/CrN	2,67 μm	2775,4	$4{,}23 \cdot 10^{-11}$

The wear coefficient K which characterizes the wear performance of treated surfaces were calculated, the Fig.6 shows the wear coefficient in case of the different treated surfaces as a function of the hardness.

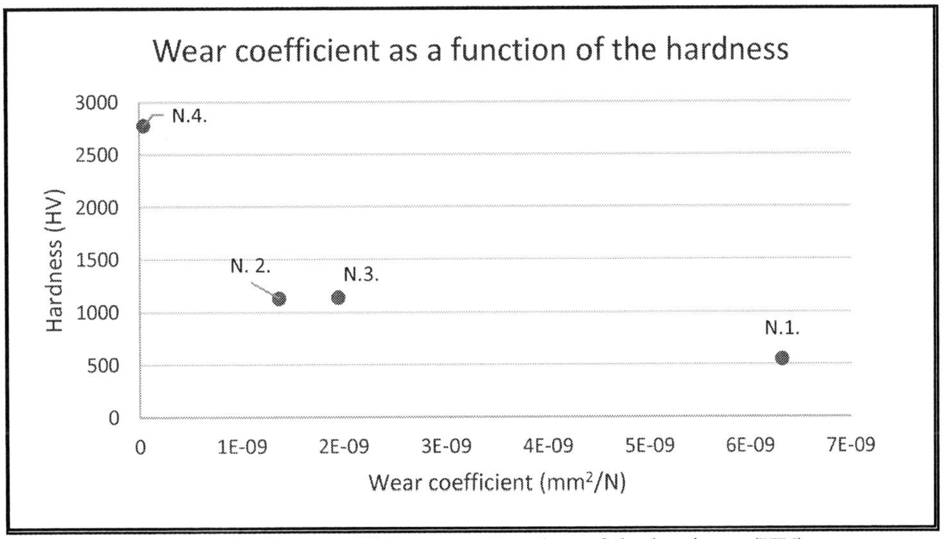

Figure 6. Wear coefficient as a function of the hardness (HV)

5. Conclusion

The correlation between the hardness and the wear resistance is well known we waited for the comparative results also on the base of this knowledge. We concluded after the experimental results evaluation :

I. The PVD Cr/CN coated surface showed the highest wear resistance under the used experimental methods and parameters.

II. The plasma nitride surfaces even that the thickness of the treated layer was significant (4 times) showed same hardness, even that the wear resistance in case of the thinner layer given a higher value. To justify this result we need to know the real surface roughness, because this parameter was not tested only we supposed the surfaces roughness equality.

III. We concluded, that by the traditional hardening heat treating process the used tool steel hardness and wear resistance much lower than the surface teated values.

IV. The used tool steel hardening and wear resistance can increase by surface treating technologies.

Acknowledgement
We acknowledge the financial support of this work by the Hungarian State and the European Union under the EFOP-3.6.1-16-2016-00010 project.

References

[1] George E Totten et all 2004 Surface Modification *Marcel Dekker* Inc. U.S.A.
[2] Karthikeyan Rajan, Vinod Joshi, Arindam Ghosh 2013 *Journal of Surface Engineered Materials and Advanced Technology* **3** (3) http://dx.doi.org/10.4236/jsemat.2013.33023
[3] L Tóth 2018 *European Journal of Materials Science and Engineering* **3** (3) 165
[4] L Tóth, T Réti 1994 *Bulletin of The Polytechnic Institute of Jassy XL* **1-2** 377
[5] T Kovács-C and P Pinke 2011 *Periodica Polytechnica Mechanical Engineering* **55** (1) 39 https://doi.org/10.3311/pp.me.2011-1.05
[6] Gee M G, Gant A, Hutchings I, Bethke R, Schiffman K 2003 *Wear* **255** 1 https://doi.org/10.1016/S0043-1648(03)00091-7
[7] Stachowiak G W, Batchelor A W, Stachowiak G B 2004 *Tribology series* **44** 25 https://doi.org/10.1016/S0167-8922(04)80019-1
[8] E S Dedova, E S Shutilova, R Géber, L A Gömze, S N Kulkov 2016 *IOP Conference Series: Materials Science And Engineering* **140** 012007 https://doi.org/10.1088/1757-899X/140/1/012007

IOP Conf. Series: Materials Science and Engineering **613** (2019) 012040 doi:10.1088/1757-899X/613/1/012040

Investigation of structures and properties of oxide-hydroxide systems after mechanical activation

R V Levkov[1,3], **S N Kulkov**[1,2,3]

[1]Institute of Strength Physics and Materials Science SB RAS, Akademicheskii ave. 2/4, 634055, Tomsk, Russia
[2]Tomsk State University, Lenin ave. 36, Tomsk, 634050, Russia.
[3]Tomsk Polytechnic University, Lenin ave. 30, Tomsk, 634034, Russia

E-mail: kulkov@ms.tsc.ru

Abstract. It has been studied Al_2O_3 and ZrO_2 with $Al(OH)_3$mixtures after mechanical activation in a planetary ball mill.Measurement of the specific surface by the BET method has shown that an increase of mechanical treatment time leads to an increase in the specific surface in all studied powders. In the initial states a large agglomerate of the order of 50-150 μm and isolated particles of 0.05-0.1 μm were observed. Mechanical treatment leads to fracture of agglomerates into separate particles. It has been revealed that the amount of agglomerates decreased with the increasing of the mechanical activation time, and the number of isolated particles increased. The average size of the agglomerates of Al-based powders varies in the range of 50-15 μm. In the case of powders based on ZrO_2, it varies in the range of 10-14 μm.

1. Introduction

Porous ceramic materials based on Al_2O_3 and ZrO_2 are widely used in various fields of technology, due to the unique combination of high strength, wear resistance, thermal conductivity and heat resistance [1].A disadvantage of the technology for producing composite ceramics based on Al_2O_3 and ZrO_2 is the presence of large agglomerates in the powder, which leads not only to heterogeneity in the material structure but also to different compaction kinetics in microvolumes [2], which in turn leads to the formation of structural defects. In the literature [3,4] were shown that the mechanical processing of the powder makes it possible to achieve uniform distribution of particles in the powder, which in turn inevitably leads to improved compaction and uniformity in the volume of the ceramic material. One of the methods of mechanical processing of powders is the treatment of the powder in a planetary mill.

The aim of this work is to study the effect of mechanical treatment in a planetary mill on the structure and morphology of Al_2O_3, $Al(OH)_3$, ZrO_2 (MgO), ZrO_2 (Y_2O_3) powder mixtures.

2. Experiment

Al_2O_3, $Al(OH)_3$, ZrO_2(MgO), ZrO_2(Y_2O_3) powders were investigated. The powders were subjected to mechanical activation for 0-600 s in a planetary ball mill.

The average particle size of the $Al(OH)_3$ powder was 70μm, Al_2O_3 - 60μm, ZrO_2(MgO) and ZrO_2(Y_2O_3) - 5μm. Mechanical activation of the polymer binder powder was performed in a planetary ball mill AGO-2. Determination of the crystal structures of Al_2O_3, $Al(OH)_3$, ZrO_2(MgO), ZrO_2(Y_2O_3)

powders was carried out on a X-ray diffractometer with CuKα radiation. The determination of the CDD size was performed using the Hall-Williamson method. Morphological studies of the powders were performed using Tescan Vega-3scanning electron microscope.

3. Results and Discussion

On figures 1 are shows the dependence of the average particle size after mechanical activation with different time. As one can see in all cases a decreasing of particle sizes are observed and with milling time increase leads to appearance of dependences saturation up to 2-4 μm of particle sizes.

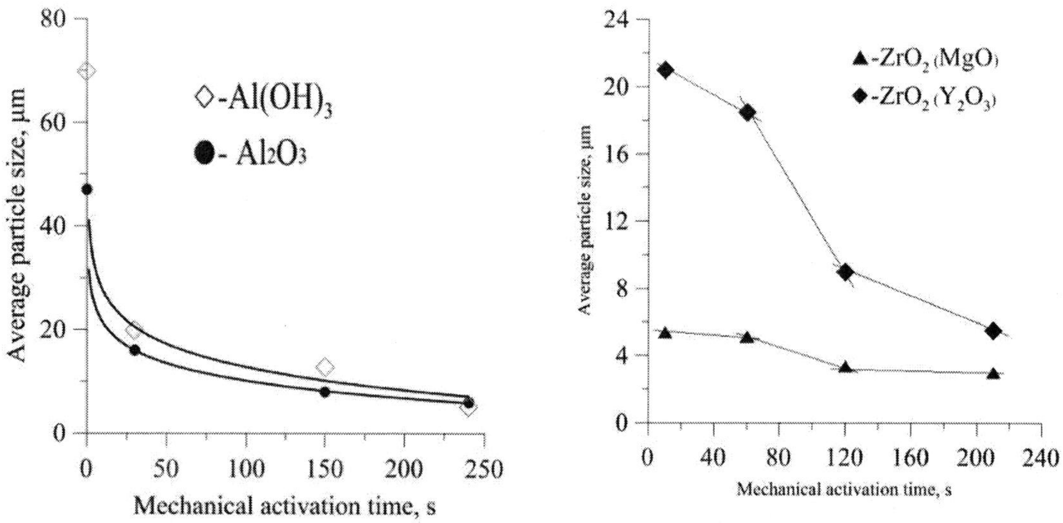

Figure 1. An average particle sizes with increasing of mechanical activation time

On figure 2 are shown specific surface of powders with increasing of milling time. The dependences for ZrO_2 powders and aluminum oxide and hydroxide are differing each others. Specific surface area of $ZrO_2(MgO)$ and $ZrO_2(Y_2O_3)$ reaches maximum value at 100-180 s of mechanical activation while subsequent increase of activation time leads to a slow decrease of the specific surface area.

These behavior looks like very similar to theoretical one [5] and means that on first step agglomerates are destroyed and on the next step there are formed another aggregates.

Specific surface area of Al_2O_3 and $Al(OH)_3$ powder only increase with milling time activation, this is probably that we didn't reach a maximum and necessary more time for destroy agglomerates in these powders.

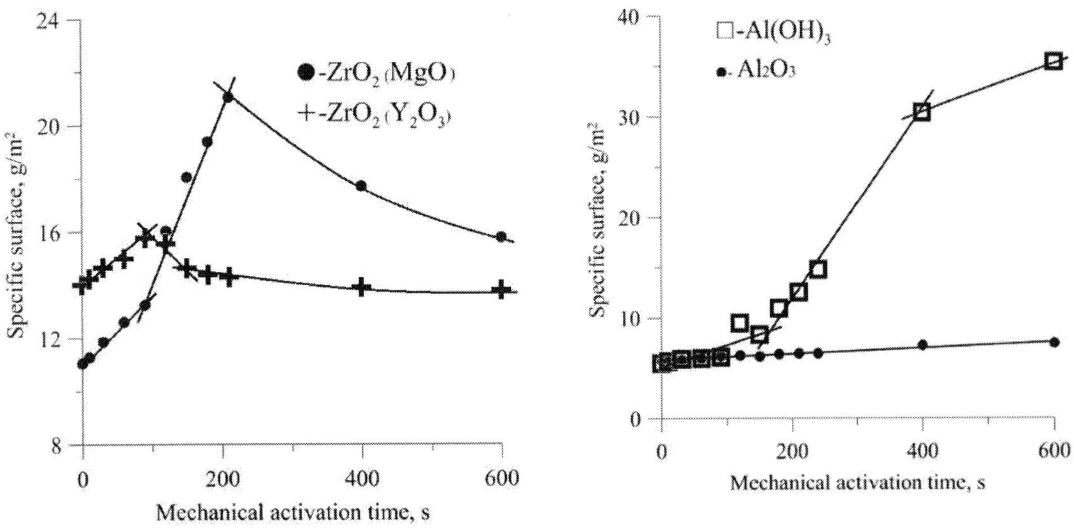

Figure 2 - Specific surface changes with mechanical activation time for powders

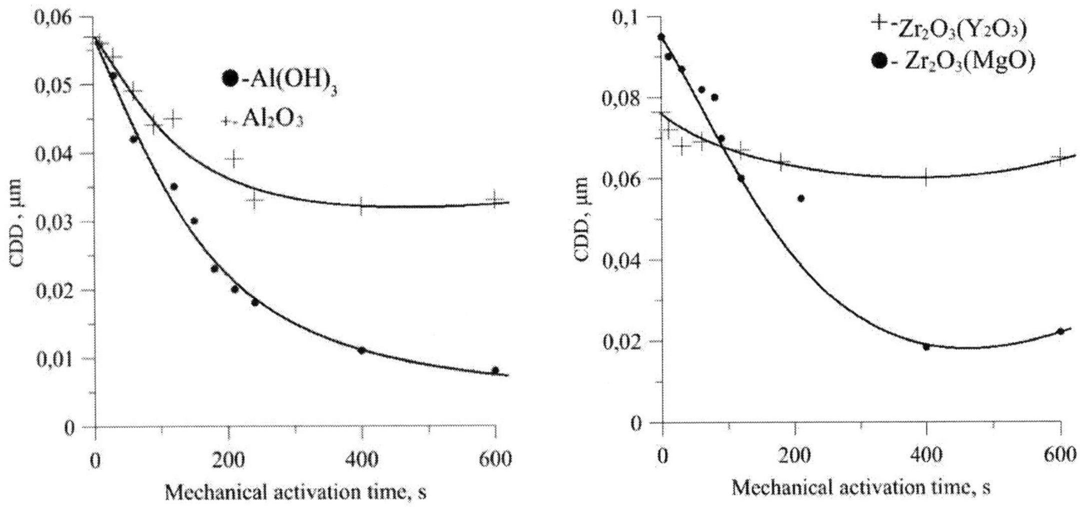

Figure 3 - Graph of CDD versus mechanical activation time.

A calculations of CDD from X-ray data, fig. 3, are shown that of Al_2O_3, $Al(OH)_3$, $ZrO_2(MgO)$ and $ZrO_2(Y_2O_3)$ powders showed that these values are decreased with increasing mechanical activation time and reached sizes about 10-20 nm which coincide with [6,7].

After comparison of the data for the average particle size, specific surface area and CDD size obtained from XRD analysis one can conclude that all studied powders has a polycrystalline structure which, in turn, became smaller during mechanical treatment.

4. Conclusions

After mechanical activation of powders particle size distribution is uni-modal. With mechanical activation time increase the average size of agglomerates decreases and size of particles in these agglomerates decreases as well.

The average size of the agglomerates of alumina-based powders varies in the range of 15-50 μm. In the case of powders based on zirconia, it varies in the range of 10-15 μm.

IOP Conf. Series: Materials Science and Engineering **613** (2019) 012040 doi:10.1088/1757-899X/613/1/012040

It has been shown that the average particle size determined from the measurement of the specific surface showed that the average particle size decreases with increasing time of mechanical activation.

XRD analysis showed that the CDD size of aluminum hydroxide powder during mechanical activation decreases with increasing of mechanical activation time and all studied powders has a polycrystalline structure.

Acknowledgement

This work was carried out with financial support of project 14.584.21.0026 (RFMEFI58417X0026 / 000000007518PRY0002)

References

[1] A Nylund, I Olefjord 1994 *Surface and Interface Analysis* **21** (5) 283
[2] A Egesz, L A Gomze 2015 IOP Journal of Physics: Conference Series **602** 012023 https://doi.org/10.1088/1742-6596/602/1/012023
[3] Z Y Deng, T Fukasawa, M Ando, G J Zhang, T Ohji 2001 *Journal of the American Ceramic Society* **84** (3) 485
[4] W H Tuan, R Z Chen, T C Wang, C H Cheng, and P S Kuo 2002 *Journal of the European Ceramic Society* **22** (16) 2827
[5] M V Korobenkov, R V Levkov 2018 *AIP Conference Proceedings* **2051** (1) 020135 https://doi.org/10.1063/1.5083378
[6] I Seyssiec, S Veesler, R Boistelle and J M Lamerant 1998 *Chemical engineering science* **53** (12) 2177
[7] P J Eng, T P Trainor, G E Brown Jr, G A Waychunas, M Newville, S R Sutton, and M L Rivers 2000 *Science* **288** (5468) 1029

IOP Conf. Series: Materials Science and Engineering **613** (2019) 012041 doi:10.1088/1757-899X/613/1/012041

The termal deformation reducing in sheet metal at manufacturing parts by CNC cutting machines

Aleksandr A. Petunin[1,2], Efim G. Polyshuk[1], Pavel A. Chentsov[1,2], Stanislav S. Ukolov[1] and Vladimir I. Krotov[3]

[1] Ural Federal University, Yekaterinburg (Russia)
[2] Institute of mathematics and Mechanics, UBr RAS, Yekaterinburg, Russia
[3] LLS „Technocomproject", Yekaterinburg (Russia)

E-mail: a.a.petunin@urfu.ru

Abstract. In various industries for the manufacturing of parts (workpieces) from sheet metal the CNC machines for thermal cutting (laser/plasma/gas cutting) are used. During process of thermal cutting, various deformations of the metal can occur, that cause the distortion of the geometric forms and sizes of the cut parts. These distortions are caused both by the uneven distribution of temperature in the sheet, and by certain geometric characteristics of the cutting process. Geometric characteristics are determined by the set sequence (order) of cutting of figured parts, as well as by selecting the piercing points of sheet material, i.e. the places of inserting the tool into the material. Previously, the authors formulated heuristic geometric rules, that allow reducing the value of geometric deformations of parts produced by the CNC cutting machines. To comply with these rules, users of Computer-Aided Design (CAM) systems that are used to generate NC programs, have to use interactive design methods. One of these rules (so-called "Part Hardness Rule") refers to the selecting the location of points for inserting (piercing) a CNC machine tool into the material. The paper describes a model based on the use of the heat conduction equation, which makes it possible to calculate the temperature of the sheet material at each moment of cutting process. The computational experiments showed that the points, which according to the "Part Hardness Rule" can't be used for piercing, have a higher material temperature than the points that satisfy this rule. In the paper the results of the computational experiments are given. Another rule (socalled "Sheet Hardness Rule"), which also reduces the deformation of the material during thermal cutting, refers to the order of parts cutting from the sheet. A heuristic algorithm that automatically generates the order of cutting parts that satisfy the rules of "Part Hardness" and of "Sheet Hardness" is developed. Developed software allows us to abandon the interactive methods of the NC programs generation for the thermal cutting of figured parts on CNC machines.

1. Introduction

Development of NC programs for the CNC sheet cutting machine includes two main functions (procedure):

1) Nesting: the placement of figured parts onto plate;
2) The tool path routing and choosing the tool path components.

Both procedures have to execute according to some technological constraints and optimization criteria (Nesting problem [1-2], Tool path problem [3]). In this work we are discussing the second problem. The basic components of tool path are showed on Figure 1.

Usually in non-series production for the manufacturing of parts (workpieces) from sheet metal the CNC machines for thermal cutting (laser/plasma/gas cutting) are used. During process of thermal cutting, various deformations of the metal can occur, that cause the distortion of the geometric forms and sizes of the cut parts. Figure 2 shows an example of distortion of the geometric forms and sizes of the long rectangulare plate.

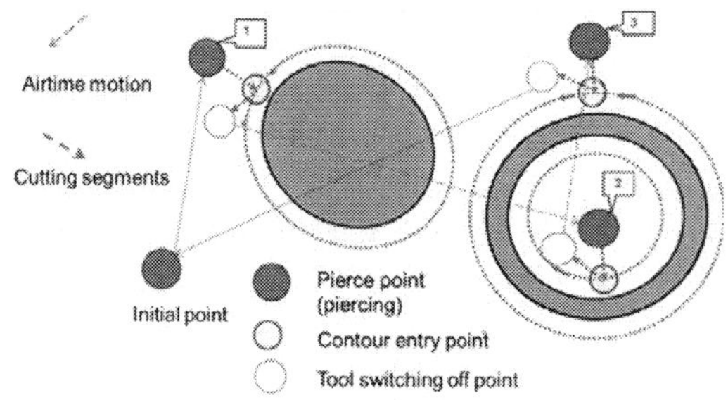

Figure 1. Basic components of tool path

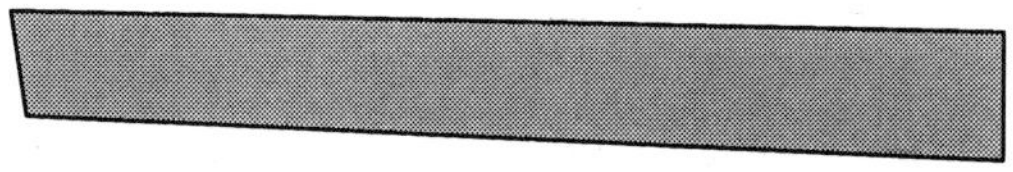

Figure 2. An example of the geometric forms and sizes distortion the long rectangular plate produced by CNC flame cutting machine

The distortions are caused both by the uneven distribution of temperature in the sheet, and by certain geometric characteristics of the cutting process. Geometric characteristics are determined by the set sequence (order) of cutting of figured parts, as well as by selecting the pierce points of sheet material, i.e. the places of inserting the tool into the material. Previously, the authors formulated heuristic geometric rules, that allow reducing the value of geometric deformations of parts produced by the CNC cutting machines, are described in [4-5]. To comply with these rules, users of Computer-Aided Design (CAM) systems that are used to generate NC programs, usually use interactive design methods. One of these rules (so-called "Part Hardness Rule") has been formalize [6-7] (see Figure 3) that allowed us to develop algorithms and software for optimizing of the tool path, taking into account this rule.

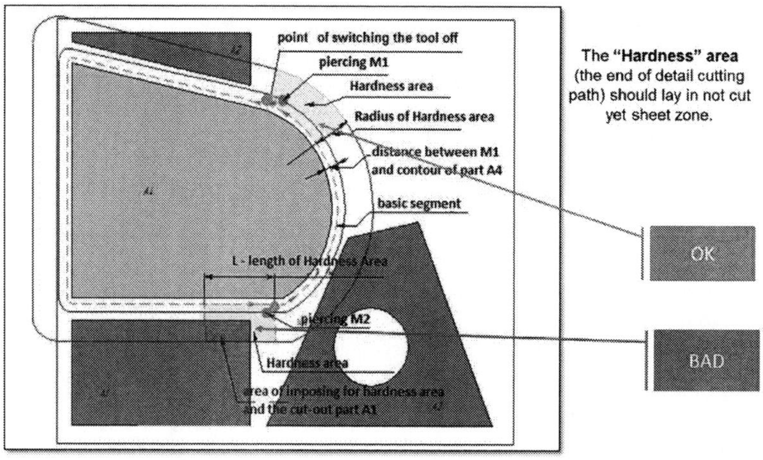

Figure 3. Illustration of „The Part Hardness Rule"

In [8] next Hypothesis is formulated: If the piercing and the zone of the completion cutting (a Hardness area) are satisfies the Part Hardness Rule then the temperature of the material in the Hardness area is lower than at using of the bad piercing. To test the hypothesis, the special software has been developed that makes it possible to calculate the temperature of the sheet material at each moment of cutting process.

2. Modeling of the thermal fields calculating and computing experiments

For each contour of parts in succession we consider the problem of finding the temperature $\theta(t,x)$ (t - time moment, x - point in the region), satisfying the following conditions:

- the heat equation $c\rho\frac{\partial\theta}{\partial t} = k\Delta\theta + N, \quad x \in \Omega$

- the initial condition $\theta(t_0, x) = \theta_0(x)$, $x \in \Omega$
- the boundary condition $-k\dfrac{\partial\theta}{\partial n} = M(\theta - \theta_*)$, $x \in \partial\Omega$

t lies in the interval $[t_0, t_1]$, x is a point in the region $\Omega \in R^3$.

Here, t_0 and t_1 are, respectively, the starting and ending time of contour cutting, Ω is the part of the plate left after the removal of all the regions bounded by the previous contours, $\partial\Omega$ is the boundary of region .

Here, c is the specific heat per unit mass, ρ is the density, k is the thermal conductivity, $N(t,x)$ is the density of heat source, M is the heat transfer coefficient, $\theta_0(x)$ is the current temperature field prior to cutting a given contour, θ_* is the air temperature. The density of heat sources N is as follows. Let p = w/($\pi r^2 h$) be the power density of the heat "ray" and *m(t)* is the position of the heat "ray" axis at time *t*. Then $N(t,x)$ = p at the points lying from the line *m(t)* at a distance smaller than r and 0 at the remaining points.

Approximation to the problem:
The process of recalculation of the temperature field $\theta(t,x)$ during contour cutting is split into small time intervals $[t_{r-1}, t_r]$ of length Δt and for the calculation of $\theta(x) = \theta(t_r, x)$ one considers the problem

$$c\rho\frac{\theta(x) - \theta_0(x)}{\Delta t} = k\Delta\theta(x) + N(x),$$

$$-k\frac{\partial\theta}{\partial n} = M(\theta - \theta_*),$$

where $\theta_0 = \theta(t_{r-1}, x)$ $\quad N(x) = N(t_{r-1}, x)$.

The region Ω is split into tetrahedra, the functions $\theta(x), \theta_0(x), N(x)$ are piecewise linear, they are determined by their values at the knots (vertices of tetrahedra).
The solution of this problem delivers the minimum to the following functional

$$\mathrm{I}(\theta) = \frac{1}{2\Delta t}\int_\Omega c\rho(\theta - \theta_0)^2 dx + \frac{1}{2}\int_\Omega k\,|\,grad\theta\,|^2\, dx - \int_\Omega N\theta dx + \frac{1}{2}\int_{\partial\Omega} M(\theta - \theta_*)^2 dS$$

The minimum of this functional is searched by the relaxation method [9]. The choice of this method is based on the following considerations.

The radius *r* of the "heat ray" is small, and hence any tetrahedron of the partition must have a small size (in calculations the side length was 2 mm) and hence their number is very large. The temperature still has the original value at the knots located far from the points already visited by the "heat ray" and where the heat change could not reach by the given time instant. The relaxation method allows one to skip such knots, which reduces the running time.

Remark. To decrease the number of tetrahedra (knots) in designing the cuts of the successive contour K_i one considers not the entire plate, but rather a piece $\Omega_i \supset \Omega_{i-1}$.

A piece is supposed to be sufficiently large in order that the temperature outside Ω_i would not change by the time of completion of cutting of contour K_i. Once the cutting of the contour is complete, the tetrahedra inside the contour are thrown away, their total number becomes smaller. Next, mesh is enlarged until we reach the next piece Ω_{i+1}.

Specialized FEM-module was developed based on the above mathematical model. Using it, a number of numerical experiments were conducted with the following parameters: sheet size 1000 x 1000 mm, beam diameter 2 mm, beam power 1500 W, cutting speed 10 m/min. Figure 4 shows the difference in average temperature in the hardness area while cutting the last of eight parts in case of selecting pierce point in compliance with part hardness rule (b) and (a) in disagreement with that rule. Temperature difference is about 120 degrees.

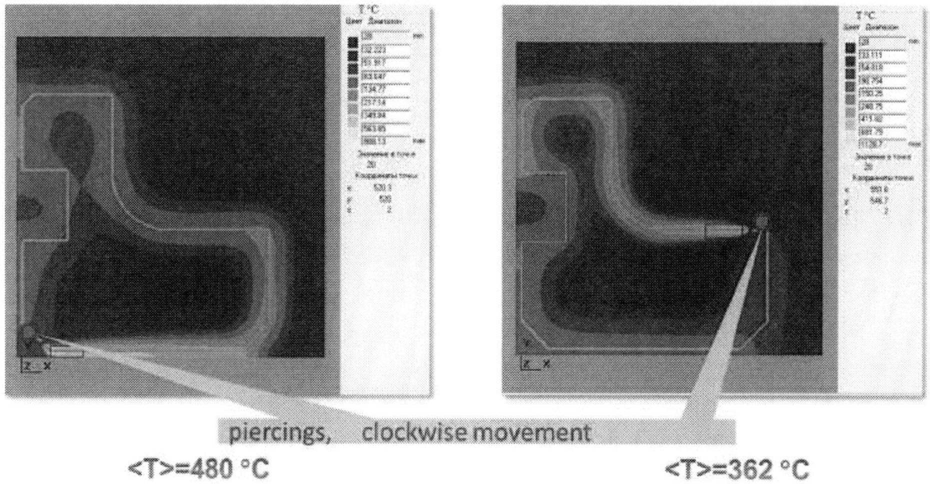

Figure 4. Thermal distribution evaluation according to the part hardness rule

3. Results and discussions

18 numerical experiments were conducted with different initial nestings and cutting orders. Obeying the part hardness rule always leads to decrease in average temperature in the part hardness area. Average temperature decrease was about 63%. Next two examples illustrate results.

An example 1

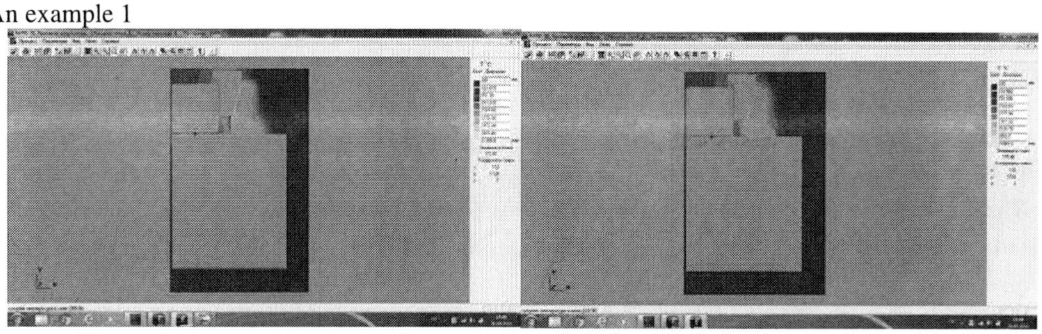

Figure 5. Average temperatue decreases from 309 degress (left; pierce point position not complying to the part hardness rule) to 120 degrees (right)

An example 2

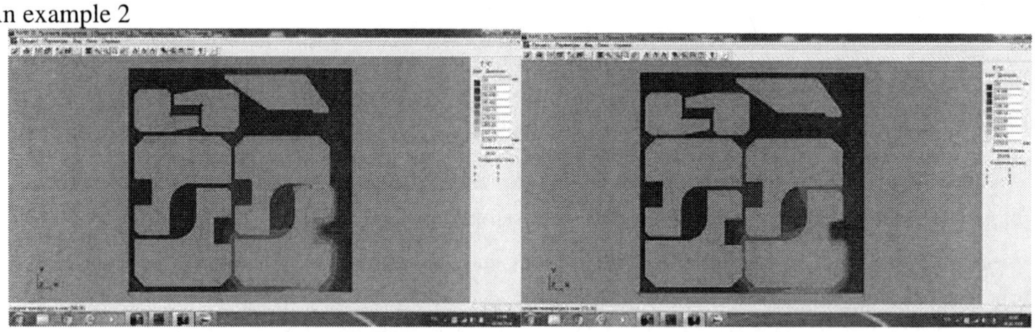

Figure 6. Temperature decrease from 550 degrees (left) to 130 degrees (right)

4. Conclusions

It has been shown that analysis of temperature fields during thermal sheet cutting on the CNC cutting machine can be used as additional consideration for tool path optimization in order to decrease thermal deformations. The results obtained are in good compliance with empirical "part hardness rule".

Acknowledgement

The work was supported by the Russian Foundation for Basic Research (grant 19-01-00573)

IOP Conf. Series: Materials Science and Engineering **613** (2019) 012041 doi:10.1088/1757-899X/613/1/012041

References

[1] G Wäscher, H Haußner and H Schumann 2007 *European Journal of Operational Research* **183** (3) 1109 https://doi.org/10.1016/j.ejor.2005.12.047
[2] M A Carravilla, C Ribeiro, J F Oliveira 2003 *International Transactions in Operational Research* **10** 651 https://doi.org/10.1111/1475-3995.00434
[3] Hoeft J, Palekar U S 1997 *IIE Transactions* **29** (9) 719 https://doi.org/10.1023/A:1018582320737
[4] Petunin A A 2009 *Ser. Upravlenie, vychislitelnaya tekhnika i informatika* **13** 280 (in Russian) http://journal.ugatu.ac.ru/index.php/Vestnik/article/view/1144
[5] Petunin A, Krotov V 2015 *Mashinostroenie, Energetika* 466 (in Russian)
[6] Chentsov P A, Petunin A A 2016 *Ifac papersonline* **49** (12) 645 http://dx.doi.org/10.1016/j.ifacol.2016.07.762
[7] Petunin A A, Stylios C 2016 *IFAC-PapersOnLine* **49** (12) 23 http://dx.doi.org/10.1016/j.ifacol.2016.07.544
[8] Petunin A A, Polishuk E G, Chentsov A G, Chentsov P A, Ukolov S S 2016 *AIP Conference Proceedings* **1789** 060002 http://dx.doi.org/10.1063/1.4968494
[9] Glowinski R, Lions J L, Trémolières R 1967 *Analyse numérique des inéquations variationnelles.* **1-2** Dunod Paris

IOP Conf. Series: Materials Science and Engineering **613** (2019) 012042 doi:10.1088/1757-899X/613/1/012042

Biogeosorbents for solving ecological problems

T Shchemelinina[1], O Kotova[2], M Harja[3] and E Anchugova[1]

[1] Institute of Biology, Komi Science Center, Ural Branch of RAS, Syktyvkar, Russia
[2] Institute of Geology, Komi Science Center, Ural Branch of RAS, Syktyvkar, Russia
[3] «Gheorghe Asachi» Technical University of Iasi, Iasi, Romania

E-mail: tatyanakomi@mail.ru

Abstract. Among the variety of the ecotoxicants the petroleum products are the most environmentally dangerous, since the intensive migration of pollutants from them can occur because of the solubility of ionic components. Zeolites have high adsorption and cation exchange capacity and can be used as adsorbents for hydrocarbons polluting water bodies and soil. The disadvantage of their application is that after the adsorption of pollutants the secondary waste requiring costly disposal methods would be formed. Developing biogeosorbents via immobilization of petroleum oxidizing microorganisms on glauconites can provide the solution to the problem of immobilization of hydrocarbons polluting water bodies and soil. The obtaining zeolite-based biogeosorbents through immobilization, rather than freeze drying, solve the problem of sorbing material utilization, moreover, of increasing the efficiency cell retention, cell viability and extracellular enzyme activity. In extremely poor environmental conditions zeolite carrier provides transport of bacteria, allowing to exist in state suspended animation for up to 10 years.

1. Introduction

Petroleum products are one of the most dangerous ecotoxicants for the environment, because migration of pollutants from them can be the most intensive, because compounds in their composition are in molecular and ion-soluble state [1]. Zeolites have high adsorption and cation exchange properties and can be used as adsorbents of petroleum products polluting water bodies and soil. The problem of their use is that adsorption of petroleum products results in the secondary pollution, which requires application of expensive utilization methods. The development of biogeosorbents by immobilizing oil-oxidizing microorganisms on glauconites can be the response to this problem. Glauconite has such unique properties as a high dispersity and physical-chemical activity. Its cohesive-adhesive properties and ability to regenerate are very important [7]. The adsorption activity of glauconite is primarily related to ion-exchange properties. The ion exchange takes place both at a level of interlayer cation and hydroxyl exchange. The oil adsorbed by glauconite will be further degraded by cells of microorganisms.

In addition, biogeosorbents from zeolites solve problems of not only utilization of sorbing materials, but also of replacement of freeze drying for immobilization of microorganisms, which increases preservation of their vitality and activity of extracellular enzymes. The freeze drying process of microorganisms has a negative impact on the structure of their surface proteins, activity of adhesion, and also leads to destruction of valuable metabolites [12]. In environmentally unfavorable conditions, zeolite sorbents serve as a base to transport bacteria, which allows them to remain suspended for up to 10 years [2].

The aim of this work was to study the change in the sorption, destructive and biochemical properties of GEOLEX biogeosorbent, depending on storage conditions and age since manufacture.

2. Materials and methods

In the experiments we used:

1. Glauconite of Bondarskoe deposit from Tambov region (Ionsorb ™) – mineral ionite obtained from a complex of aluminosilicates according to the author's technology [11].
2. Strains of hydrocarbon-oxidizing microorganisms (HOM):
   - bacteria Pseudomonas yamanorum VKM B-3033D [9];
   - yeast Rhodotorula glutinis, VKM Y-2998D [4];
   - microalgae Chlorella vulgaris Beijer. f. globosa V. Andr. A1123 [6].

   BIOTRIN biopreparation comprises all these strains [3].
3. GEOLEX biogeosorbent, consisting of glauconite with immobilized strains of microorganisms of BIOTRIN biopreparation [5].
4. Oil contaminated water for experiment 1 as a Control: 0.5 g of oil added to 100 cm^3 of sugarless Czapek medium was aerated in a shaker at 180 rpm during 4 days under normal conditions. Then water was filtered off.

Experiments were conducted as follows:

Experiment 1. In sterile tap water 50 cm^3 we added oil (0.25 g), biogeosorbent or glauconite 0.5 g. The control was a sample of oil contaminated water. The experiment was performed in triplicate within 3 days in the shaker (180 rpm) under normal conditions. The intensity of oil destruction was assessed according to the residual total petroleum hydrocarbons (TPH) content.

Experiment 2. The contaminated water prepared according to the above was bottled in 250 ml flasks and 1 g of glauconite and GEOLEX biogeosorbent were added. The experiment was carried out during 4 days in the shaker (180 rpm) under normal conditions. The residual TPH content in water and sorbents were separately determined.

Experiment 3. 35 g of glauconite or GEOLEX biogeosorbent suspended in 1 dm3 of tap water were applied per 1 m2 contaminated soil of the railroad track. Tilling to a depth of 15 and sampling after 30 days for residual TPH content were conducted. Untreated soil samples before glauconite and biogesorbent application were analyzed as a control.

Experiment 4. The oil polluted sand-gravel mixture (SGM) that coated the concrete railway platform was treated with GEOLEX biogeosorbent until complete coating, then mixed and sampled for residual TPH content after 30 minutes and after 6 days. The SGM sampled before the biogeosorbent application was investigated as the control.

In the Experiment 5 biogeosorbent samples of different age from manufacture and the temperature conditions of storage were used. Biogeosorbents applied were: freshly prepared GEOLEX (G-fp), GEOLEX stored at room temperature for 1 year (G-rts), GEOLEX stored in frozen form (G-ffs). 55 mg of oil were suspended in 50 ml of sterile tap water, and then biogeosorbents (G-fp, G-rts, G-ffs) in the amount of 0.5 g were added. A sample of oil contaminated water was used as a control. The experiment was carried out under carefully controlled conditions such as room temperature, plant grow light illumination, aeration in the shaker for 3 days (180 rpm). The experiment was performed in triplicate. At the end of the experiment, the residual TPH content was determined.

We determined the dehydrogenase activity of the samples according to the procedure [10].

The TPH content in the samples was analyzed fluorometrically [8].

To assess the effectiveness of adsorptive immobilization, the samples were examined microscopically with a scanning electron microscope VEGA3 TESCAN, × 5000 (Fig. 1).

IOP Conf. Series: Materials Science and Engineering **613** (2019) 012042 doi:10.1088/1757-899X/613/1/012042

3. Results and discussions

The study's first phase identified the possibility of fixing microbial cells on a carrier, the sorption and destructive properties of raw glauconite and glauconite based biogeosorbent (GEOLEX).

The composition and structure of Ionsorb™ allow using it as an effective carrier for immobilization of specialized oil oxidizing microorganisms. The BIOTRIN biopreparation strains cultivated were immobilized on Ionsorb™ sorbent. Accumulation of cells in the body of the carrier was shown under the microscope (Fig. 1). In this case the carrier was not only a sorbent of hydrocarbons, but also a source of biogenic elements.

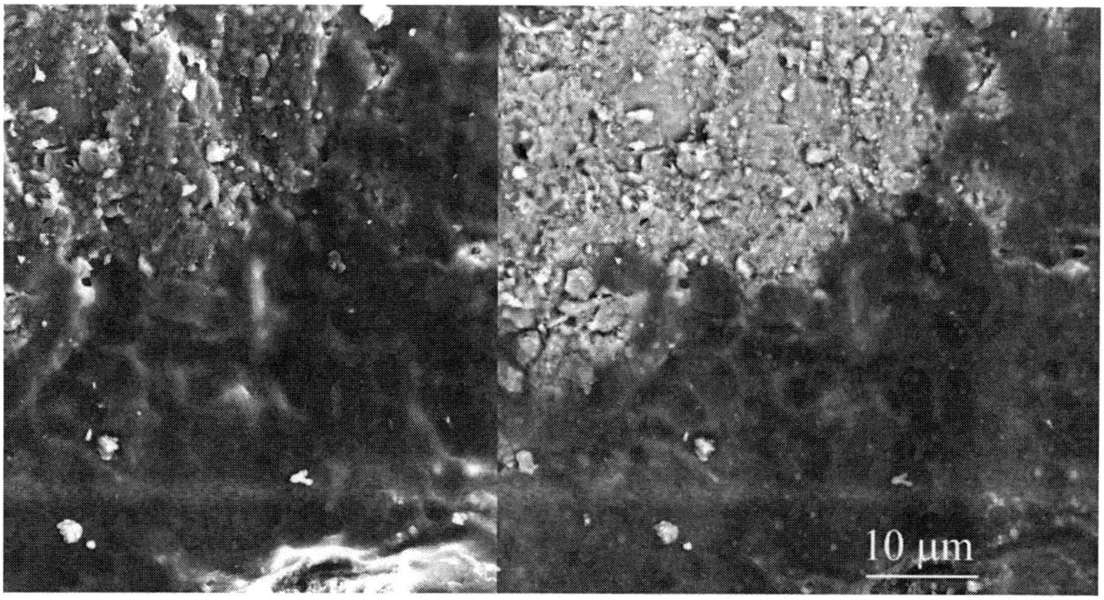

Figure 1. Ionsorb™ with immobilized cells of BIOTRIN biopreparation.

In the experimental water, when glauconite and GEOLEX biogeosorbent were applied, adhesion of petroleum hydrocarbons was observed. It should be noted that sorption properties of glauconite declined 1.2-2.0 times after immobilization of strains (Fig. 2, Table 1). This was the result of a partial decrease in sorption capacity because of decline of porous permeability and sorbent specific surface area by microorganisms adhered.

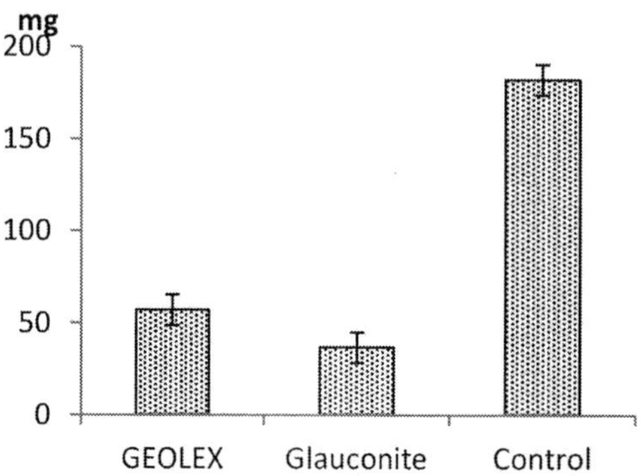

Figure 2. TPH content in the experimental water after application of biogeosorbent and glauconite.

When studying the destructive properties of the biogeosorbent in the experimental water, it was determined that 62 % of hydrocarbons sorbed had been oxidized by immobilized microorganisms (Table 1).

Table 1. Residual TPH content in the Experiment 2.

GEOLEX	Glauconite	Control
$\frac{0.045 \pm 0.016}{34 \pm 14}$	$\frac{0.035 \pm 0.012}{90 \pm 40}$	0.12±0.041

*Note: the numerator – TPH content in the experimental water, mg/dm^3, the denominator – TPH content in the sorbent and biogeosorbent after the experiment, mg/g.

In the bulk of the railroad ballast, the reduction of TPH content due to glauconite sorption was 8.3 % after 30 days (Fig. 3). GEOLEX biogeosorbent application reduced the pollutant content by twice in 30 days. The efficiency of soil clean up was 60 %.

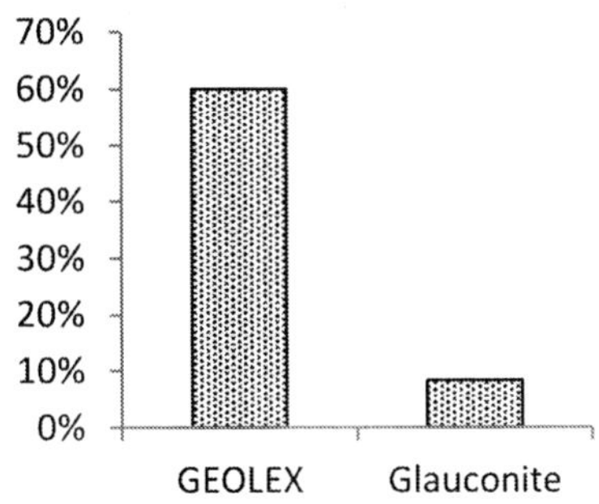

Figuer 3. Effectiveness of TPH reduction in the contaminated soil.

Evaluations of cleaning of SGM on a concrete site by GEOLEX biogeosorbent have found that sorption efficiency of hydrocarbons in 30 minutes was 14 % from the reference value. HOM immobilized on the geosorbent, activate metabolic processes and after 6 days the efficiency of purification at the concrete site was 45 % (Fig. 4).

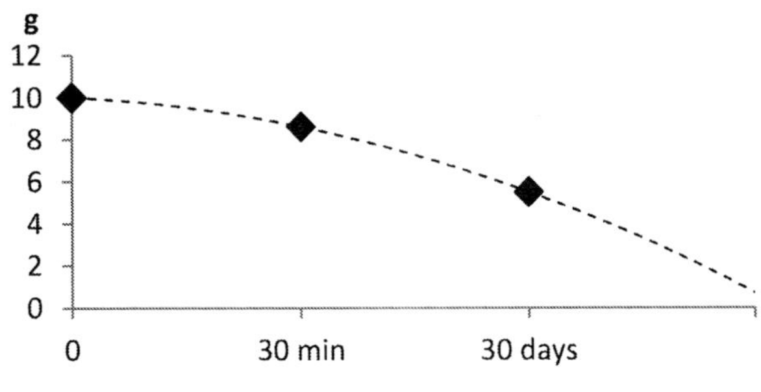

Figure 4. TPH Content in SGM.

The study's second phase revealed sorption, destructive properties of GEOLEX biogeosorbent depending on the conditions of its storage and age since manufacture (Table 2).

Table 2. Storage conditions of GEOLEX biogeosorbent of various age since manufacture

Sample	Storage temperature range, °C	Age since manufacture
G-ffs	+25…–35	12 months
G-rts	+23…+25	12 months
G-fp	+23…+25	7 days

There is evidence in the scientific literature [13] that biochemical methods, such as viability, metabolic and synthetic activity are used for the analysis of various functional parameters of immobilized cells. We chose a method of dehydrogenase activity determination, because the HOM cultural fluids have a high DA level. As a result of the immobilization on glauconite, the cells of microorganisms, as well as their various enzymes, stabilize and remain active for a long time. So DA of the biogeosorbents confirms the presence of viable cells in them.

The DA of the biogeosorbent G-fp is 1.5 to 4 times higher than that of the biosorbents G-rts and G-ffs, i.e. after 12 months of storage GEOLEX biogeosorbent reduces the activity by 2.5 times in average. Dehydrogenation processes do not depend on storage conditions. The microorganisms, immobilized on the biogeosorbent G-ffs in 60 °C temperature swing do not decrease their activity; moreover, DA was higher than in biogeosorbent stored at room temperature (Fig.5). Thus, immobilization on glauconite increases cell thermoresistance.

Despite decreasing dehydrogenase activity in the biogeosorbents stored for 12 months (G-ffs, G-rts) in comparison with freshly prepared one (G-fp), the microorganisms, immobilized on glauconite, remain viable and start multiplying in favorable conditions. Thus, during the test of GEOELEX biogeosorbent, the decrease in TPH content in water varied within the error margin, which proved that both the storage temperature factor and the age since manufacture of biogeosorbent practically do not influence the oxidizing properties of the microorganisms immobilized (Table 7). The effectiveness of hydrocarbon destruction using GEOLEX biosorbent amounted to: G-fp - 27.7 %, G-rts - 25.6 %, G-ffs - 17.7 %. Based on that glauconite is a prolonged action carrier increasing the thermal stability of the microorganisms.

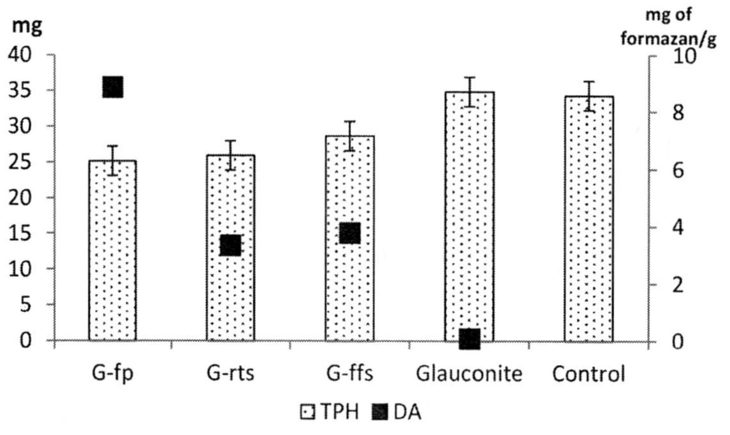

Figure 5. The effect of different age since manufacture and temperature conditions of storage on residual TPH content an DA of biogeosorbents.

4. Conclusion

Our studies revealed that the sorption properties of glauconite after application of microorganism cells were reduced by 1.2-2 times.

We determined that strains of microorganisms immobilized oxidize oil hydrocarbons by 62 % in water within 4 days and in the bulk of a railroad ballast to 50 % within 30 days, in SGM of concrete site by 45 % within 6 days.

The age since manufacture of the biogeosorbent affects its dehydrogenase activity. We observed decreasing dehydrogenation processes by 2.5 times in average.

The effectiveness of the TPH reduction by GEOLEX does not depend on the term and storage conditions investigated.

Acknowledgment

The research was carried out with the financial support of the State task "Development of biocatalytic systems based on enzymes, microorganisms and plant cells, their immobilized forms and associations for the processing of plant raw, production of biologically active substances, biofuels, remediation of contaminated soils and wastewater treatment" No. AAAA-A17-117121270025-1. The work was carried out with a partial financial support of UB RAS Programs (project 18-5-544).

References

[1]. C Courbe, B Velde and A Meunier 1981 *Clay Miner* **16** (3) 231 https://doi.org/10.1180/claymin.1981.016.3.02
[2]. T N Shchemelinina, O B Kotova, M Harja, E M Anchugova, Y Pelovski, I Cretescu 2017 *Vestnik IG Komi SC UB RAS* **36** 40 https://doi.org/10.19110/2221-1381-2017-6-40-42
[3]. Conclusion on the toxicological and hygienic assessment of "BIOTRIN" consortium of strains of oil-oxidizing microorganisms (in Russian).
[4]. Application for a patent of the Russian Federation No. 2016126521
[5]. Application for a patent of the Russian Federation No. 2018120922
[6]. Application for a patent of the Russian Federation No. 2018120704
[7]. M L Levchenko, A M Gubaidullina 2009 *Drilling and oil* **4** 56 (in Russian)
[8]. Federal environmental regulations PND F 16.1.21-98 Quantitative chemical analysis of soils. Methods of measurement of mass concentration of petroleum products in the soil samples using Fluorat-02 fluid analyzer 1998 15 (in Russian)
[9]. S M Meshkelo, T N Shchemelinina, E M Anchugova, M Yu Markarova, S V Zheludkova 2017 Strain of bacteria *Pseudomonas yamanorum* VKM B-3033D for activation of biodegradation of oil and oil products in water, as well as in oil soils on sections of the railway RU Patent 2 615 458 (in Russian)
[10].SNiP 2.04.03-85 Water-supply and sewerage outdoor networks and structures. The USSR State Committee for Construction USSR 1986. (in Russian)
[11].Prize of the Russian Geological Society (ROSGEO) and the Federal Agency for mineral resources for merits in the field of science and innovative technologies in the geological study of the mineral resources of Russia 2010 (in Russian)
[12].I V Solovyeva, A G Tochilina, I V Belova, E I Efimov, N A Novikova, T P Ivanova 2012 *Bulletin of the Nizhny Novgorod University named after N.I. Lobachevsky* **2-3** 85 (in Russian)
[13]. N Mallick 2006 Immobilization of microalgae Immobilization of enzymes and cells Ed J M Guisan (Totowa, New Jersey: *Humana Press Inc.*) 373

Hypersound excitation of three-layer magnetic film by microwave field

L N Kotov[1*], M Yu Dianov[1], V S Vlasov[1], V A Ustyugov[1], S V Shilov[1]

[1]Department of Radio Physics and Electronics, Pitirim Sorokin Syktyvkar State University (Russia)

E-mail: kotovln@mail.ru

Abstract.
The work deals with the definition of elastic and magnetic oscillations exited by microwave magnetic field in the three-layer structure. The frequency dependences for the amplitudes of the magnetic and elastic vibrations of the film layers with different values of the material parameters and amplitudes of the external fields are calculated.

1. Introduction

In the last decade, the control of magnetization dynamics under the influence of elastic waves at frequencies up to 500 GHz is a rapidly developing area of magnetoacoustics [1, 2]. This paper continue above mentioned works and is devoted to solving the problem of magnetic and elastic dynamics in a three-layer film in the most general nonlinear case. Based on the results of studies of the dynamics of a three-layer magnetic structure, a generator of hypersonic vibrations can be constructed. This generator can be used in various electronic devices, for example, in medicine for diagnose of the internal organs.

2. Geometry of the problem and main equations

Fig. 1 shows the geometry of the problem. The paper deals with a three-layer film consisting of three magnetic layers. The thickness of the layers was in the range from 0.2 to 0.6 μm. Mark the film layers with the letters p, d, r. The thickness of the layers by the same letters will be denoted. The center of the coordinate system is selected in the center of the d layer. The DC magnetic field ***H*** was directed perpendicular to the film plane along the z-axis. The alternating magnetic field ***h*** was directed in plane of the structure and has a circular polarization.

The total thickness of the film is equal to half of the elastic wavelength. Elastic frequency is close to ferromagnetic resonance frequency (FMR). We assume that all layers have the same elastic properties. The film's layers have a cubic crystallographic symmetry; the crystallographic plane (100) coincides with the plane of the plate. The total energy density of the film's layers U in the field $\vec{H}_{ext} = \{h_x; h_y; H\}$ as the sum of the energy density of each layer is considered. The energy density of each layer includes the density of magnetic, magnetoelastic, elastic energy and anisotropy energy. We consider the interaction of the standing on the film thickness shear elastic waves with magnetic oscillations only. The energy density will have form [3]:

$$U = -M_0 h_x m_x - M_0 h_y m_y - M_0 H m_z + 2\pi M_0^2 m_z^2 + 2B_2(m_z m_y u_{zy} + m_x m_z u_{xz}) + 2C_{44}(u_{xy}^2 + u_{yz}^2 + u_{zx}^2) + K_1(m_x^2 m_y^2 + m_y^2 m_z^2 + m_z^2 m_x^2) + K_2 m_x^2 m_y^2 m_z^2 \quad (1)$$

where M_0 - saturation magnetization, m_i - components of unit magnetization vector, B_2 – magnetoelastic constant, C_{44} - elastic constant, u_{ik} – displacements tensor components, K_1, K_2 – first and second cubic anisotropy constants.

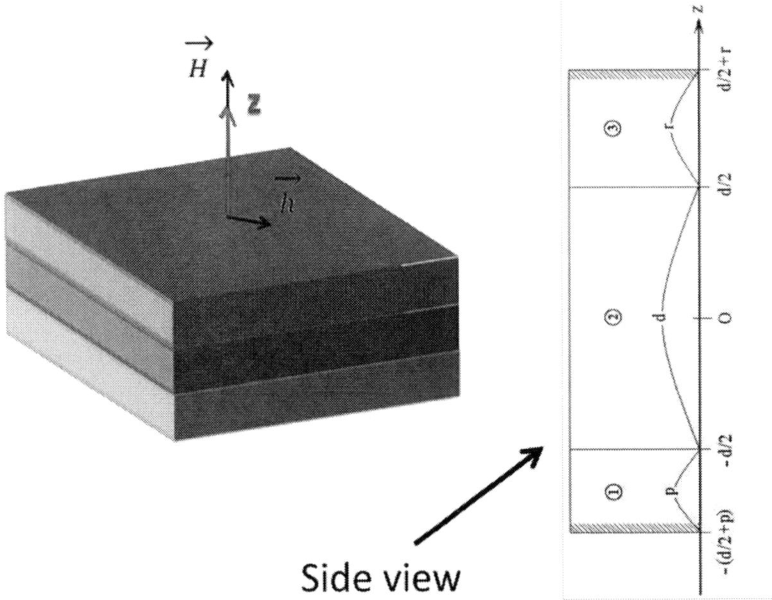

Figure 1. Geometry of the problem.

We divide the solution for elastic displacement into 2 parts:

$$u_x(z,t) = U_x(z,t) + v_x(z,t), \tag{2}$$

where U_x - the inhomogeneous part of the solution, for this part following condition is valid: $\partial^2 U/\partial z^2=0$ in all thickness and boundaries, $v_x(z,t)$ – the part satisfies the homogeneous boundary condition ($\partial v/\partial z=0$ on the external boundaries). The equality of the v functions and its derivations $\partial v/\partial z$ is on the internal boundaries between layers. We decompose the homogeneous part by eigenfunctions of the boundary value problem over the internal layer thickness (for example, p layer) in order to take off the coordinate dependence. As the result, we obtain:

$$v_{px}(z,t) = v_{0px}(t) + \sum_{n=1}^{\infty} v_{pn}(t)\cdot\cos\left(\frac{\pi\cdot n}{p}\cdot z + \frac{\pi\cdot n}{p}\cdot\frac{d}{2}\right) \tag{3}$$

We substitute $v_{1px}(t)$, $v_{0px}(t)$ values expressed by equation (3) to magnetoelastic equations and use equation for U. We can obtain following differential equations:

$$\frac{d^2 v_{1x}}{dt^2} + 2\beta\frac{dv_{1x}}{dt} + \frac{c_{44}}{\rho}\cdot\frac{\pi^2}{p^2}v_{1x} = -\frac{4p}{\pi^2}\cdot\left(\frac{d^2}{dt^2} + 2\beta\frac{d}{dt}\right)\cdot\left(-\frac{B_{p2}}{c_{44}}m_{px}m_{pz}\right) \tag{4}$$

$$\left(\frac{d^2}{dt^2} + 2\beta\frac{d}{dt}\right)\cdot\left(v_{0px}(t) + \frac{B_{p2}}{c_{44}}m_{px}m_{pz} + \frac{B_{d2}}{c_{44}}m_{dx}m_{dz}\right) = 0 \ . \tag{5}$$

We can find similar equations for the y-components by the changing x to y index. The similar mathematical manipulations we repeat for the other two layers. Thus, we obtain the system of ordinary differential equations, including the Landau-Lifshitz-Gilbert equation, equations of type (4), (5) for x

IOP Conf. Series: Materials Science and Engineering **613** (2019) 012043 doi:10.1088/1757-899X/613/1/012043

and y elastic displacements components for all three layers. Coupling between the layers in our model was carried out by the zero terms Fourier decomposition.

3. Results and discussions

The Runge-Kutta 4-5 orders method to solve the system differential equations was used. The frequency of the alternating field corresponded to the resonant frequency of the elastic oscillations of the film. The main parameters of the film material corresponded to the parameters of yttrium-iron garnet crystals. The time dependences of the magnetization and the elastic displacements for various frequencies were plotted. The resonance curves for the cases of linear and nonlinear oscillations were plotted (Fig. 2).

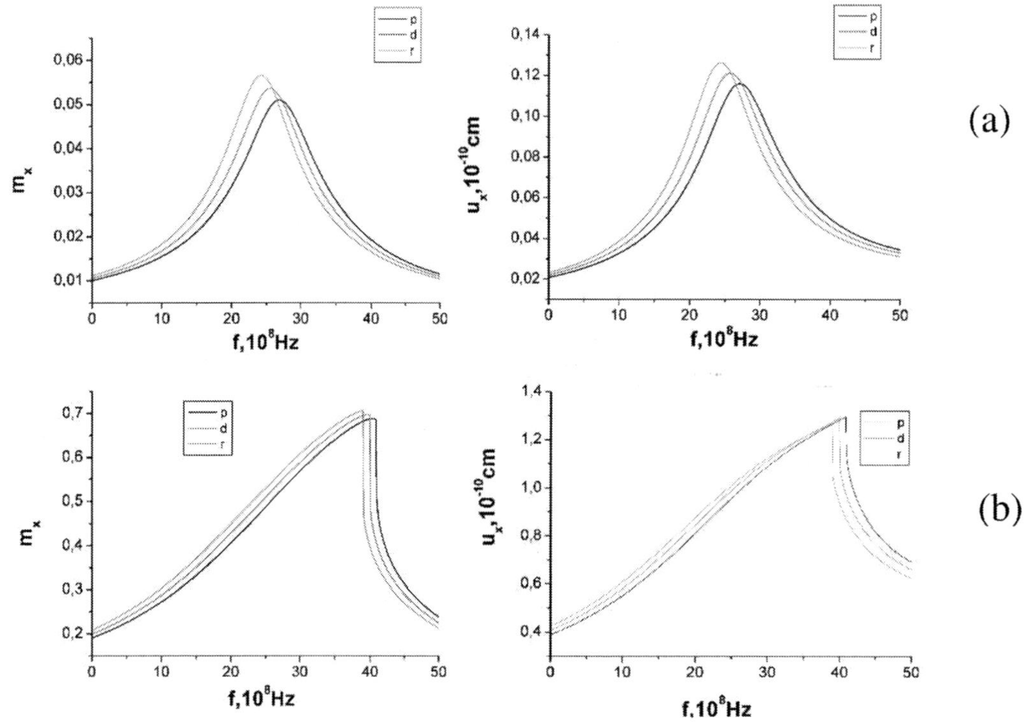

Figure 2. Frequency curves in the linear (h=1Oe) (a) and nonlinear (h=200 Oe) (b) cases. H_0=2750 Oe; $4\pi M_{0i}$ (in Gs) =1700 (i=p), 1750 (i=d), 1800 (i=r); B_2=13.92·10^6 erg cm^{-3}.

The maxima of the amplitudes of the magnetic and elastic oscillations are observed at the alternating field frequency close to the FMR frequency. The slope of the resonance curves is observed at large amplitudes of the alternating field. This slope is for the nonlinear case. The interesting dynamics is observed for the condition when the external constant field is close to the demagnetization field H~4πM0. In this case, the second-order magnetization precession modes were revealed. These regimes were previously observed for one ferrite layer in [4-5]. Due to the difference of the gyroscopic forces that are exerted on the magnetization vector at the positions of the maximum and minimum displacements relative to the axis z, the equilibrium position of the magnetization vector is shifted and average in time magnetization vector is involve in the precession along a large circle, so that the precession of equilibrium position takes place. Fig. 3 shows precession portraits of magnetization in the film layers with very different values of the magnetoelastic constants of the layers. The tendency of p layer magnetization to follow position of a new easy axis, different from the normal to the film plane is observed for Fig. 3 (a). This phenomenon occurs in a material with large magnetoelastic constants [3]. The precession regimes of the equilibrium position of the magnetization vector without the center coverage (in layer d and r) and with the center coverage (in layer p) are found. It is possible to distinguish the maximum amplitudes of the oscillations in layers at low frequencies (about 0.5 GHz) and at the high frequencies (about 25 GHz) taking into account these two oscillation modes.

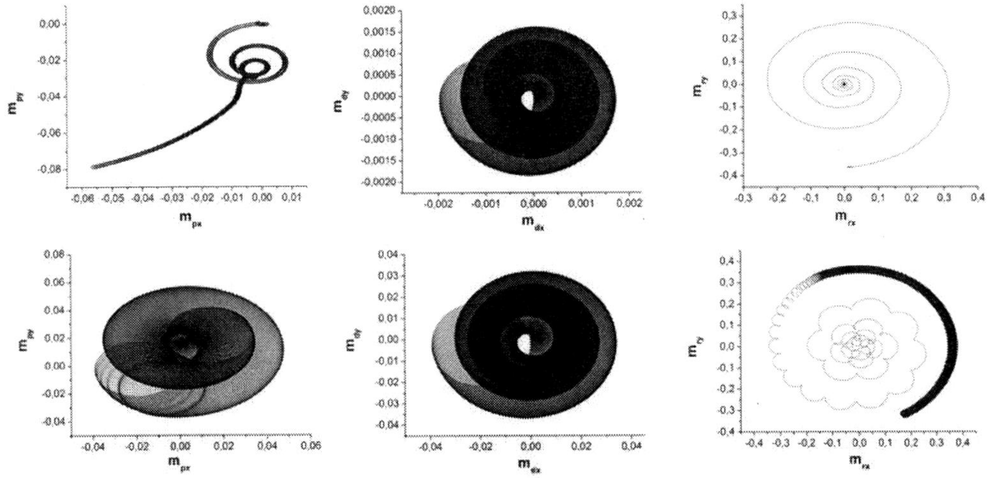

Figure 3. Precession portraits of the magnetization in different layers of the three-layer structure. The amplitude of the alternating magnetic field h=1 Oe (a), 20 Oe (b). $4\pi M_0$ - H= 5 Oe. H=1750 Oe; $4\pi M_0$=1755 Gs; value of the magnetic dissipation parameter α=0.1, B_{p2}=13.92·10^5 erg cm^{-3}, B_{d2}=13.92·10^6 erg cm^{-3}, B_{r2} =13.92·10^7 erg cm^{-3}.

4. Conclusions

This paper solves the problem of magnetic and elastic dynamics in a three-layer film in the most general nonlinear case. The generator of hypersonic vibrations can be constructed based on the results of studies of the dynamics of the structure. We can vary amplification coefficient of the elastic oscillations in the wide frequency range due to the presence of the magnetic and magnetoelastic coupling between the layers. The materials of the three-layer films can be metal-dielectric composite [6-7] or complex materials with extreme dynamic strength [8].

Acknowledgement

The authors would like to acknowledge to RFBR (grants 17-02-01138, 17-57-150001).

References

[1] J Janusonis, C L Chang, T Jansma, A Gatilova, V S Vlasov, A M Lomonosov, V V Temnov and R Tobey 2016 *Phys. Rev. B* **94** 024415 https://doi.org/10.1103/PhysRevB.94.024415

[2] C L Chang, A M Lomonosov, J Janusonis, V S Vlasov, V V Temnov, and R I Tobey 2017 *Phys. Rev. B* **95** 060409 https://doi.org/10.1103/PhysRevB.95.060409

[3] V S Vlasov, L N Kotov, V G Shavrov, V I Shcheglov 2010 *J. Com. Tech. and Electr.* **55** 645 https://doi.org/10.1134/S1064226910060069

[4] V S Vlasov, L N Kotov, V G Shavrov, V I Shcheglov 2011 *J. Com. Tech. and Electr.* **56** 1117 https://doi.org/10.1134/S1064226911080122

[5] V S Vlasov, M S Kirushev, L N Kotov, V G Shavrov, V I Shcheglov 2013 *J. Com. Tech. and Electr.* **58** 847 https://doi.org/10.1134/S1064226913080081

[6] L N Kotov, V S Vlasov, V A Ustyugov, V K Turkov, M Yu Dianov , Yu E Kalinin, A V Sitnikov , E A Golubev 2017 *IOP Conf. Ser.: Mater. Sci. Eng.* **175** 012021 https://doi.org/10.1088/1757-899X/175/1/012021

[7] L Kotov, V Ustyugov, V Vlasov, V Turkov, M Lasek, Yu Kalinin, A Sitnikov, E Golubev 2016 *IOP Conf. Ser.: Mater. Sci. Eng.* **123** 012009 https://doi.org/10.1088/1757-899X/123/1/012009

[8] L A Gömze and L N Gömze 2017 *IOP Conf. Ser.: Mater. Sci. Eng.* **175** 012001 https://doi.org/10.1088/1757-899X/175/1/012001

IOP Conf. Series: Materials Science and Engineering **613** (2019) 012044 doi:10.1088/1757-899X/613/1/012044

Amorphous iron formation following occupancy of iron atoms in vacant sites of epsilon-iron that transformed from alpha-iron during rubbing in a low vacuum

H Furuichi[1]

Furuichi Laboratory, 2878-1, Wadamachi, kofu 400-0001, Japan

Email: h-furuichi@msi.biglobe.ne.jp

Abstract. The main purpose of this paper is to present two TEM diffraction patterns taken from a layer (more than 100 μm thick) of piled-up adhered wear debris on a rubbed surface. The unidirectional rubbing was made in a low vacuum, regularly separating the slider from the specimen. The specimen and the slider were α-iron. One of the diffraction patterns consists of large spots and halo rings and the other consists of small faint spots and halo rings. The latter seems to be the state just before the perfect amorphization. All the spots in the diffraction patterns correspond to the structure formed due to the occupancy of iron atoms in O and T sites of ε-iron. Neither this way of amorphization nor the occupied structure mentioned above has not been reported. Both the amorphous state and the occupied structure remain at ambient conditions. The mechanism of high-pressure generation that transformed α-iron to ε-iron and the generation of high tensile stress that breaks the atomic bonds in wear debris are presented. The latter is considered to have enabled the occupancy.

1. Introduction

This article presents the transformation from α-iron to ε-iron that exists at 13 GPa and above [1], the occupancy of iron atoms in the T and O sites of the ε-iron structure [2] and the formation of amorphous iron following it, during unidirectional rubbing in a low vacuum, regularly separating the slider from the specimen. Hereafter, the iron after the occupancy and the unidirectional rubbing mentioned above will be called the occupied iron and unidirectional discontinuous rubbing. Neither the occupied iron nor this way of amorphous-iron formation has not been reported.

The mechanism of the high-pressure generation that transformed α-iron to ε-iron and that of the generation of high tensile stress that brakes the atomic bond are also presented. The latter enabled the occupancy.

The properties of the occupied iron and the amorphous iron mentioned above should be different from those of α-iron. The rubbing is inexpensive. Therefore, we can expect their practical application.

2. Material and methods

TEM diffraction patterns were taken from the layer of piled-up adhered wear debris generated by unidirectional discontinuous rubbing in a vacuum (7-13 mPa) between α-iron slider and α-iron specimen without lubricant at room temperature with the apparatus shown in figure 1.

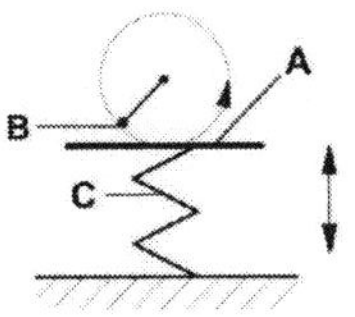

Figure 1. Schematic diagram of rubbing apparatus: A, specimen; B, slider; C, spring.

In this case, it was necessary to take diffraction patterns from only one of each pieces of debris to avoid the overlapping of the diffraction patterns. This is the reason why I took TEM diffraction patterns because an electron beam can be focused by the electron lenses, and, therefore, it is possible to irradiate a narrow area. The used iron was commercial pure polycrystal one (C<0.003%,Si< 0.005%,Mn<0.005%, P<0.005%, S<0.0003%, Al<0.005%, in mass). The specimens and the sliders were annealed and electropolished before rubbing. The apparent maximum pressure during the sweep of the slider, the sliding speed, the frequency and the sliding distance per cycle were 7 MPa, 32mms-1, 0.19 Hz and 20.5mm, respectively. The diffraction patterns were taken from the area where the apparent pressure was highest, after 2.5 x 10^4 traversals.

Even on extremely smoothly finished surfaces, there are many minute asperities [3]. When the slider contacts with the specimen, asperities on the specimen or those on the slider contact with the friction surface of the slider or of the slider and high pressure is generated at the contact areas; these contact areas have been called true contact areas and the generated high-pressure true contact pressure [4]. In the case of unidirectional discontinuous rubbing, when the slider leaves the specimen, the temperature of the true contact areas drops, and the asperities recover their hardness. This enables the repetition of the shearing at high pressure. After numerous repetitions of the rubbing, much debris piles up on the specimen and adheres to other debris to form a layer, whose thickness reaches more than 100 μm [5]. The thinning of the specimens to take TEM diffractions was made by fine emery polishing followed by electropolishing, only from the opposite side of the rubbed surface. Some TEM diffraction patterns had to be taken without a beam stopper, because it was impossible to make wide thin TEM-specimens owing to the severe surface roughness. After thinning, the specimens were dipped into extremely pure cyclohexane for at least three months at room temperature to know the survivability of the structure and state generated by the rubbing; cyclohexane prevents oxidation from iron.

3. Structure analysis and discussion

Figure 2 shows TEM diffraction patterns taken from the layer of piled-up adhered wear debris.

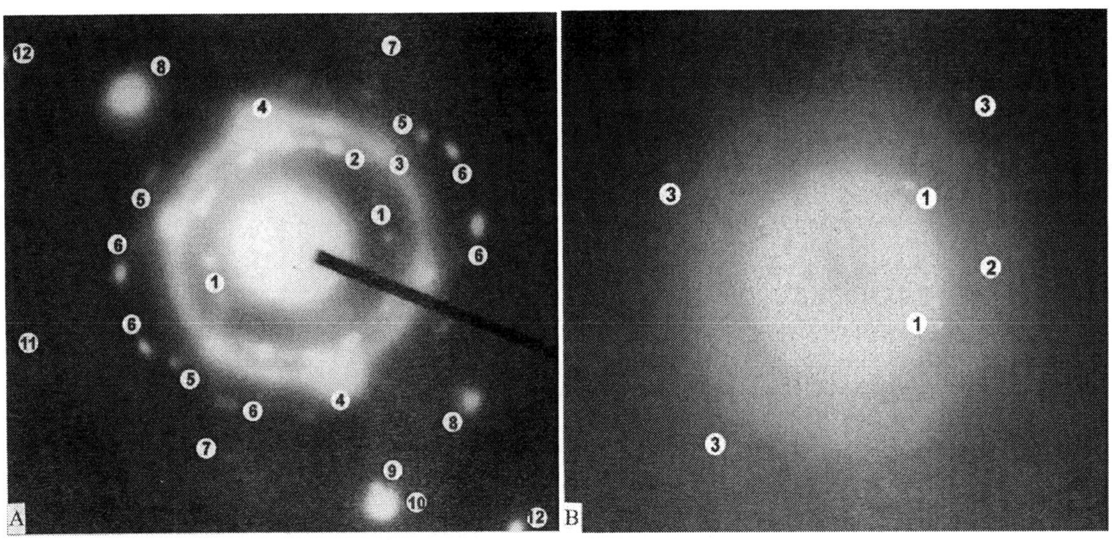

Figure 2. TEM diffraction patterns: A, less damaged area; B, more damaged area.

IOP Conf. Series: Materials Science and Engineering **613** (2019) 012044 doi:10.1088/1757-899X/613/1/012044
As demonstrated in Table 1, figure 2 does not correspond to α-iron [6].

Table 1. Data of spots in figure 2.

Spot number		Interplanar spacing (pm)	Reciprocal of fractional intercept	Intensity	
A	B			hcp	After occupancy
1		177	2 x (2 0 8/3)		
2		155	4 x (2 0 10) *		
3	1	122.5	(1 1 0)	$4f^2$	$64f^2$
4		102	(3/2 0 8/3)		$31f^2$
5		96	(3/2 0 3)		$23f^2$
6	2	86	(2 0 8/3)		$25f^2$
7		75	2 x (2 0 10) *		
8		71	(3 0 0)	$4f^2$	$64f^2$
9		68	(2 1 3)	$3f^2$	$28f^2$
10	3	61	(2 2 0)	$4f^2$	$64f^2$
11		54	(3 1 3)	$3f^2$	$32f^2$
12		46	(4 1 0)	$4f^2$	$64f^2$

*Intensity of (2 0 10) is $23f^2$.

Some spots correspond to ε-iron that exist at 13GPa and above [1]. The other spots correspond to the planes whose reciprocals of fractional intercept with a, b and c axes contain at least one nonintegral.

There are vacant sites in the ε-iron structure, i.e. O sites: 2/3, 1/3, 1/4; 2/3, 1/3, 3/4 and T sites: 1/3, 2/3, 1/8; 1/3, 2/3, 7/8; 0,0, 5/8; 0,0, 3/8 [2]. Comparison between these vacant sites and the nonintegral plane indices in Table 1 leads to the idea that the diffraction patterns were taken from the structure after the occupancy of iron atoms in the vacant sites of ε-iron. Occupancy of titanium atoms in the vacant sites of α-titanium by shearing at high pressure has been reported [7].

At the spot 1 in the pattern A, no spot ought to exist. There is no choice, therefore, but to think that the spot 1 is the superlattice of the spot 5. It has been reported that superlattice form in sheared iron up to 14 GPa [8]. For the structure mentioned above to occur, high pressure of 13 GPa and above that enable the transformation from α-iron to ε-iron must occur [1] and iron atoms must enter the inside of ε-iron structure. Generation of very high pressure by unidirectional discontinuous rubbing have been reported [9,10].

When the asperities on the slider or on the specimen contact with the other friction surface, they adhere to it and are pressed downward and laterally pulled with the movement of the slider and break to become wear debris. This mechanism has been called adhesive wear and has been accepted among tribologists [11]. For the high pressure mentioned above to be generated, the adhered asperities must be tensioned or compressed without yielding, i.e., without the movement of dislocations. The Young's modulus (E) of hcp iron in compression increases with the pressure [12]. Therefore. its smallest value appears just after the beginning of compression, i.e., fairly higher than 600 GPa [12]. High pressure 13 GPa is generated when the adhered asperities elastically deform a little higher than 2 %.

Without the movement of dislocations, the breaking stress of the adhered asperities is almost equal to the ideal strength, i.e., E/10 [13]. From the Hooke's law, the elongation of the interatomic distances just before the breaking of wear debris is about 10 %. Assuming that atom is a ball of radius r, the radius of ball that enter O sites and T sites before applying stress are 0.414r and 0.225r, respectively [2]. Judging from the facts mentioned above, it is impossible for iron atoms to occupy in the vacant sites of ε-iron, at this stage.

TEM diffraction patterns taken from unidirectionally discontinuously rubbed ceramics revealed a structure that comprises a part of the structure of compounds consisting of composing elements, some of them exist at high pressure [14,15], There was a case, however, where one of the composing elements of ceramics became an amorphous metal, besides the formation of the structure mentioned above [15]. This indicates that plural structures that comprise a part of the structure of compounds consisting of composing elements can exist. Without diffusion, it is impossible for the structure mentioned above to bond again.

Diffusion coefficients of ceramics are much lower than those of metals [16]. It can be considered, therefore, that the same phenomenon occurs with metals. Epsilon iron is composed only of iron atoms. If ε-iron breaks, diffused iron atoms occupy the widened vacant sites and as in the case of ceramics, broken atomic bonds were also bonded by diffused iron atoms. This process gives birth to the occupancy. There are many asperities on the friction surfaces. Wear debris repeat adhering to the slider and exfoliating from it, many times. Accordingly, a large number of occupied sites are generated. The halo rings in the diffraction patterns indicate the amorphous state formation.

This is as much explanation as possible, because no detailed elucidation of the mechanism of adhesive wear has been made [17].

The transformation from ε-iron to α-iron is martensitic [18]. Martensitic transformation is caused by a shift in which the relative position of atoms does not change. Therefore, reverse transformation can occur. However, in the case of this study, the reverse transformation cannot occur because the relative position of atoms has changed due to the occupancy of iron atoms in the vacant sites of ε-iron.

The properties of the occupied iron and the amorphous iron ought to be different from those of α-iron.

Unidirectional discontinuous rubbing is inexpensive. Therefore, we can expect to use them for practical purposes.

The occupancy causes the volume decrease near friction surfaces and increases the distance between two sliding surfaces, leading to many machines unusable.

The amorphous iron mentioned above is the extreme state of mechanically damaged metal structures.

4. Conclusions

TEM diffraction patterns were taken from the layer of piled-up adhered wear debris (more than 100 μm thick) generated by unidirectional discontinuous rubbing between α-iron specimen and α-iron slider in a low vacuum. They revealed a structure in which iron atoms occupy the vacant sites of ε-iron and amorphous iron. Both remain at ambient conditions. The rubbing is inexpensive and easy. From the facts mentioned above, it is expected that they are put into practical use.

High-pressure generation that transformed α-iron to ε-iron during rubbing in a low vacuum is also presented.

This article provides useful information for tribologists.

The amorphous iron formed by rubbing is the extreme state of mechanically damaged metal.

Acknowledgements

I would like to express my heart-felt thanks to Professor Emeritus K. Kato of Tohoku University, Professor Emeritus E. Ito of Okayama University, Dr. S. Asano (retired) of Nippon Steel Cooperation, Dr. C. Morita (retired) and Professor S. Arai of Nagoya University and Mr. S. Shirai, Director of ABT (retired) for their advices and help.

References

[1] Takahashi T and Basset A B 1964 *Science* **145** 483
[2] Greedan J E 2017 *Handbook of Solid State Chemistry*, ed Dronskoski R, Kikkawa S and Stein A (Weinheim, Germany: Wiley-VCH Verlag) 165
[3] Prakash A and Struik K 1981 *Precis.Eng.* **3** 215 https://doi.org/10.1016/0141-6359(81)90095-7
[4] Batchelon A W, Lam L N and Chandrasekaran M 2011 *Materials degradation and its control by surface engineering, third ed.* (Singapore World Scientific Publishing) 14

[5] Furuichi H, Nakamura S, Xuan JS and Yamazaki K 1995 *wear* **184** 99
[6] Swanson E, Fuyat R K and Ugrnic G M 1955 *Natl. Bur. Stand. (USA) Circl.* 539. **4** (Maryland, USA: National Bureau of Standard) 3
[7] Furuichi H and Kanno Y 2015 *Mater. Lett.* **161** 511 http://dx.doi.org/10.1016/j.matlet.2015.09.025
[8] Furuichi H, Fujii N, Kanno Y, Ito E, Watanabe S and Tanaka H 2004 *High Press. Res.* **24** 385 https://doi.org/10.1080/0895795041233132834
[9] Furuichi H and Matsuura K 1996 *J. Mater. Sci.* **31** 3201
[10] Furuichi H and Matsuura K 1997 *J. Mater. Sci. Lett.* **16** 1323
[11] Kato K and Adachi K 2000 *Modern Tribology Handbook* vol 1, ed B Bhushan (BocaRaton, USA: CRC Press.) 277
[12] Neuman G S, Stixrude L and Cohen R E 1999 *Phys. Rev. B* **60** 791
[13] Knott J 2015 *Phil. Trans. R. Soc.* **A 377** 19 https://doi.org/10.1098/rsta 2014.0126
[14] Furuichi H and Matsuura K 1996 *J. Mater. Sci.* **31** 3201
[15] Furuichi H, Fujii N and Oishi T 2000 *J. Mater. Sci.* **35** 561
[16] Lesage B 1994 *J. Phys. III France* **4** 1833
[17] Aghababaei R, Warner D H and Molinan J-F 2016 *Nat. Commun.* **7** 11816 https://doi.org/10.1038/ncomms11816

Synthesis and Powder Properties of Cerium Phosphate White Pigments

Hiroaki ONODA and Haruka MURAKI

Department of Informatics and Environmental Sciences, Kyoto Prefectural University, 1-5, Shimogamo Nakaragi-cyo, Sakyo-ku, Kyoto 606-8522, Japan

E-mail; h-onoda@kpu.ac.jp

Abstract. Catalytically active cerium dioxide is used as light reflection materials for cosmetic applications. Herein, as novel white pigments for use in cosmetics, cerium phosphates were prepared under various conditions. The chemical composition, powder properties, photocatalytic and oxidation catalytic activities, color phase, and smoothness of the cerium phosphates were evaluated. Samples prepared from cerium (+III) nitrate solution indicated XRD pattern of cerium phosphate hydrate. All samples prepared in this work had large particles and no specified shape. The obtained cerium phosphates exhibit little photocatalytic activity and no oxidation catalytic activity. These samples exhibit a little low reflectance within the range of visible light and 71-78 in L* value.

1. Introduction

Titanium dioxide and zinc oxide are used as white pigments for cosmetic applications [1]. These oxides are known to be photocatalytically active; therefore, cosmetics containing these pigments cause decomposition of a certain amount of sebum on the skin under ultraviolet radiation in sunlight [2,3]. In our previous works, titanium and zinc phosphates without photocatalytic activity were prepared as novel white pigments for cosmetics [4,5]. These compounds have a possibility to replace titanium dioxide and zinc oxide with photocatalytic activity. Cerium dioxide is also used as light reflection materials for cosmetic applications [6,7]. This oxide is known to be oxidation catalysts; therefore, cosmetics containing this pigment too much cause oxidation of the skin under. Therefore, the substitute for cerium oxide is required to prepare the cosmetics without oxidation activity.

Phosphates have been used as ceramic materials, catalysts, fluorescent materials, dielectric substances, metal-surface treatments, detergents, food additives, in fuel cells, as pigments, and so on [8-11]. Phosphate materials are recognized to have high biocompatibility and are therefore expected to be useful as white pigments in cosmetics [12,13]. White pigments that are not oxidation active are required for skin protection during cosmetic use. Furthermore, the moisture retention of the materials used in cosmetics is also important for preventing dry skin. Because phosphates were hydrophilic materials, phosphate pigments are suitable to use for cosmetics from this viewpoint.

The particle shape and the size distribution of phosphates are important parameters for cosmetic pigment application [14]. Spherical, homogenous particles are expected to spread well on the skin. However, overly small particles are unsuitable because the pigments may enter the pores of the skin. Generally, pigments with sub-micrometer dimensions are used in cosmetic applications.

In this study, to obtain a novel white pigment without the catalytic activity, cerium phosphate is prepared under various conditions. The respective chemical compositions, powder properties, photocatalytic and oxidation catalytic activities, color phases, and smoothness of the obtained materials were studied for application of the developed pigments in cosmetics.

2. Experimental

Cerium nitrate solution (0.5 mol•L^{-1}) was mixed with phosphoric acid (0.5 mol•L^{-1}) in a molar ratio of P/Ce(+III) = 1/1 and 2/1 at room temperature for more than 1 hour. The mixed solutions were treated with ultrasound for 1 hour. The precipitates were then filtered off, washed with water, and dried at room temperature over 3 days [5]. For comparison, samples were also prepared without ultrasonic treatment. Di-ammonium cerium nitrate solution (0.5 mol•L^{-1}) was also mixed with phosphoric acid in a molar ratio of P/Ce(+IV) = 4/3, and then treated with the above process. All chemicals were of commercial purity and were acquired from Wako Chemical Industries Ltd. (Osaka Japan) and used without further purification.

The crystalline phase compositions of these materials were analyzed using X-ray diffraction (XRD; MiniFlex; Rigaku Corp., Akishima, Japan) with monochromated Cu-Kα radiation. The IR spectra of materials were recorded on a HORIBA FT-IR 720 (Horiba Ltd., Kyoto, Japan) using the KBr disk method.

The shapes and sizes of the particles comprising the precipitates were estimated from the scanning electron microscopy (SEM; instrument: JSM-5510LV; JEOL Ltd., Akishima, Japan) images and particle size distributions. The particle size distributions of these materials were measured using a centrifugal precipitation particle-size distribution analysis system (SA-CP3L, Shimadzu Corp., Kyoto, Japan).

The cosmetic properties of the developed precipitates were estimated by analysis of the color phase, photocatalytic and oxidation catalytic activities, and smoothness. The color of the phosphate pigments was estimated using ultraviolet-visible (UV-Vis) reflectance spectra acquired with a UV2100, Shimadzu Corporation instrument (Kyoto, Japan) (reference compound: $BaSO_4$). The color of materials was also estimated with a TES135 plus color analyzer (TES Electrical Electronic Corp, Taipei, Taiwan). In this estimation, the L* value means the whiteness of powder, in which 100 is white, on the opposite site, 0 is black. The a* value means the redness of materials, with positive (maximum; +60) and negative (-60) values are corresponding with red and green, respectively. The b* value indicates the yellowish, in which positive (maximum; +60) and negative (-60) values are correspond with yellow and blue, respectively.

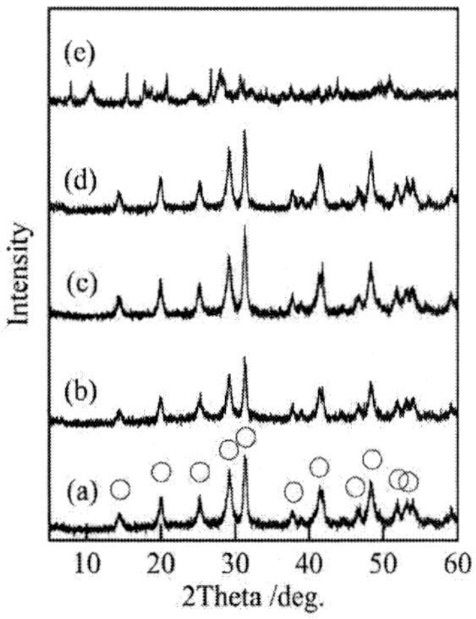

Fig. 1 XRD patterns of samples prepared in various conditions, (a) P/Ce(III) = 1/1, ultrasonic; 0 h, (b) P/Ce(III) = 1/1, 1 h, (c) P/Ce(III) = 2/1, 0 h, (d) P/Ce(III) = 2/1, 1 h, and (e) P/Ce(IV) = 4/3, 0 h, ○; $CePO_4$•nH_2O.

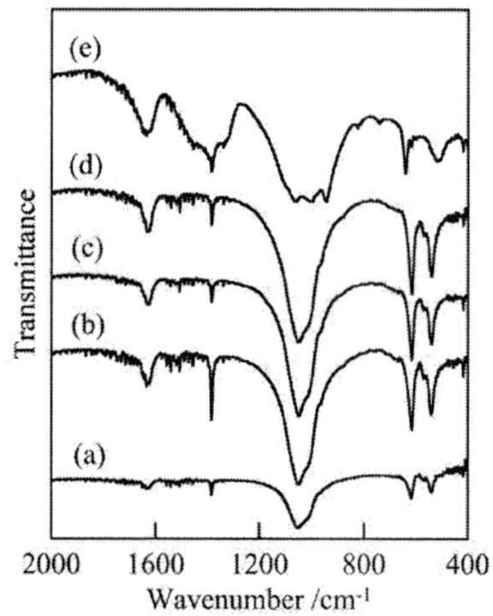

Fig. 2 IR spectra of samples prepared in various conditions, (a) P/Ce(III) = 1/1, ultrasonic; 0 h, (b) P/Ce(III) = 1/1, 1 h, (c) P/Ce(III) = 2/1, 0 h, (d) P/Ce(III) = 2/1, 1 h, and (e) P/Ce(IV) = 4/3, 0 h.

The photocatalytic activity of the samples was assessed from the decomposition of methylene blue using 365 nm irradiation [11–13]. A 0.01 g portion of the sample was placed in 4 mL of methylene blue solution (1.0×10^{-5} mol•L^{-1}) and this solution was then irradiated. The decrease in the absorption at about 660 nm was monitored over the course of 120 min. Furthermore, the oxidation catalytic activity of samples was estimated with ascorbic acid solution. A 0.05 g of sample was put into 2 mL of 0.01 mol/L of ascorbic acid solution. The color of solution changes to brown by oxidation catalyst.

As the smoothness of sample powders, the static friction resistances were estimated on artificial leather with Portable Friction Meter 94i-II (SHINTO Scientific Co., Ltd., Tokyo, Japan). The sample powders were spread on the leather, and then a sensor was run over the powders.

3. Results and Discussion

3.1. Chemical composition and powder properties of cerium phosphates

Figure 1 shows the XRD patterns of the samples prepared under various conditions. Samples prepared from cerium (+III) nitrate solution indicated the peak pattern of cerium phosphate hydrate (Fig.1(a)-(d)). On the other hand, sample prepared from di-ammonium cerium (+IV) nitrate solution had unknown peaks (Fig. 1(e)). Figure 2 shows IR spectra of the samples prepared under various conditions. In the same manner with XRD patterns, sample prepared from di-ammonium cerium (+IV) nitrate solution indicated the different IR spectrum with other samples. The absorption peak at 1630 cm^{-1} was attributed to the water of crystallization. The peaks at 1380 cm^{-1} were attributed to nitrate anion. The peaks at 1050, 620, and 540 cm^{-1} were attributed to phosphate anions [14].

Spherical morphology of the particles is desirable for cosmetic applications. Figure 3 shows representative SEM images of the samples prepared in various conditions.

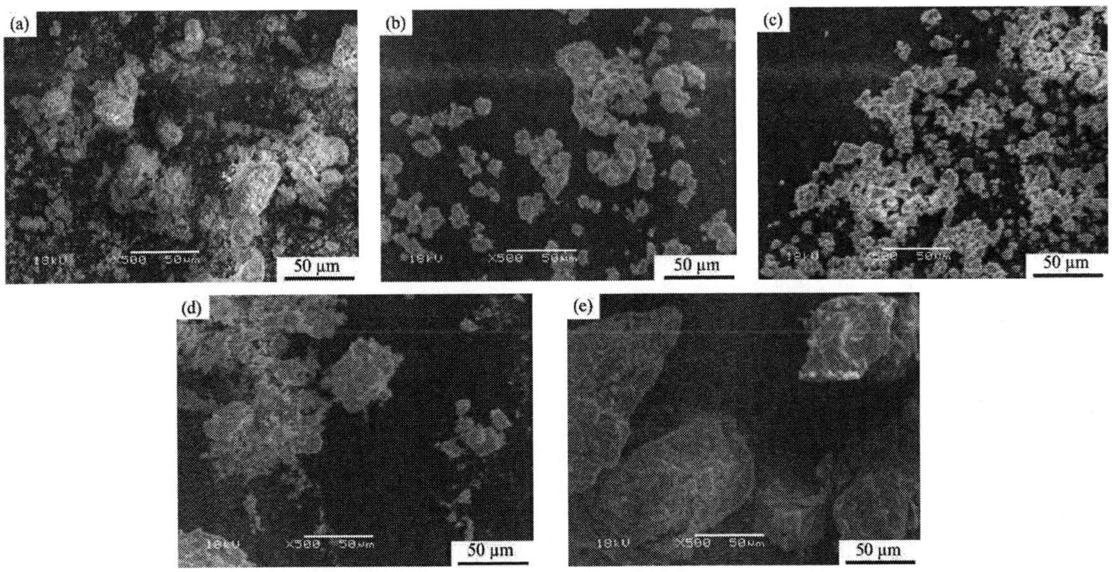

Fig. 3 SEM images of samples prepared in various conditions, (a) P/Ce(III) = 1/1, ultrasonic; 0 h, (b) P/Ce(III) = 1/1, 1 h, (c) P/Ce(III) = 2/1, 0 h, (d) P/Ce(III) = 2/1, 1 h, and (e) P/Ce(IV) = 4/3, 0 h.

All samples prepared in this work had large particles and no specified shape. Small and homogeneous particles are suitable for cosmetic applications. However, overly small particles have a major shortcoming in that they enter the pores of the skin [3]. Generally, pigments with sub-micrometer dimensions are used in cosmetics. The standard size of white pigment particles used in cosmetics is difficult to determine because the size of the pores in skin is affected by factors such as age, gender, and climate. Furthermore, overly large particles are inappropriate for cosmetic purposes owing to cracking of their coating on the skin. It is therefore important to control the particle sizes of the pigment. Figure

4 shows the particle size distribution of the samples prepared in various conditions. All samples prepared in this study had a large amount of particles with dimensions of over 1 μm. Sample prepared with ultrasonic treatment indicated higher ratio at 15 μm than those prepared without. Ultrasonic treatment produced the aggregation of phosphate particles.

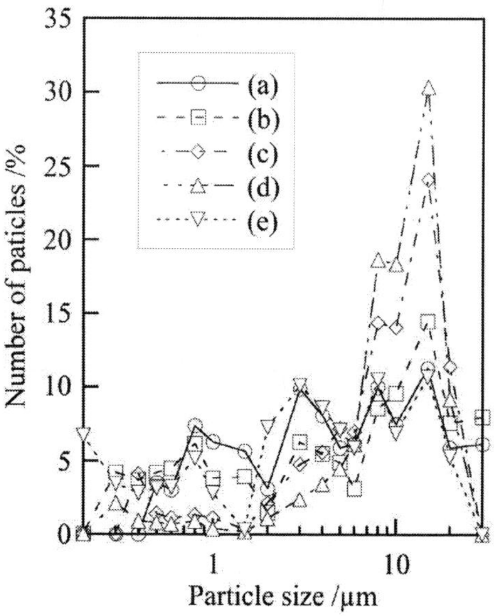

Fig. 4 Particle size distribution of samples prepared in various conditions, (a) P/Ce(III) = 1/1, ultrasonic; 0 h, (b) P/Ce(III) = 1/1, 1 h, (c) P/Ce(III) = 2/1, 0 h, (d) P/Ce(III) = 2/1, 1 h, and (e) P/Ce(IV) = 4/3, 0 h.

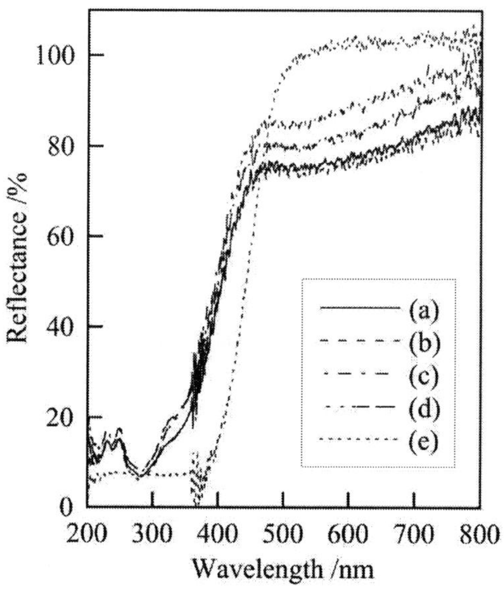

Fig. 5 UV-Vis reflectance of samples prepared in various conditions, (a) P/Ce(III) = 1/1, ultrasonic; 0 h, (b) P/Ce(III) = 1/1, 1 h, (c) P/Ce(III) = 2/1, 0 h, (d) P/Ce(III) = 2/1, 1 h, and (e) P/Ce(IV) = 4/3, 0 h.

3.2. Cosmetic properties of cerium phosphates

Figure 5 shows the UV-Vis reflectance spectra of the cerium phosphates prepared in various conditions. Sample prepared from di-ammonium cerium (+IV) nitrate solution indicated higher reflectance at 480-800 nm than those from cerium (+III) nitrate solution. The reference became lower dramatically at 500 nm in sample prepared from di-ammonium cerium (+IV) nitrate solution. Therefore, this sample is yellow powder. On the other hand, others are a little dark white powder.

The color of sample powder was also estimated by L*a*b* color space. Table 1 shows the L*a*b* values of samples prepared in various conditions. Samples prepared from cerium (+III) solution indicated from 71 to 78 in L* values, on the other hand, those from Ce(+IV) solution indicated over 80. Sample prepared from Ce(+IV) solution had higher values in a* and b* ones than others. These results were corresponding with the powder color and UV-Vis reflectance spectra of samples.

Figure 6 summarizes the respective photocatalytic activities of the samples prepared in various conditions. Zinc oxide, which is widely used as a white pigment in cosmetics, was evaluated for comparison with cerium phosphates [1]. Methylene blue decomposed under UV irradiation in the presence of zinc oxide (Fig. 6(b)). Samples prepared in this work had little photocatalytic activity in spite of the valence of cerium and the phosphorus / cerium ratios (Fig. 6(d)-(h)).

Figure 7 shows photograph of the oxidation catalytic estimation. Cerium oxide changed the color of ascorbic acid solution to brown (Fig. 7(a)). Samples prepared from cerium (+III) solution produced white dispersed solutions, on the other hand, sample prepared from cerium (+IV) solution was precipitated in colorless transparent solution. Samples prepared in this work had no oxidation catalytic activity.

As described above, pigments with a high level of smoothness spread well across the skin, and powder smoothness is another important factor for cosmetic applications [4]. As the powder smoothness, the static friction resistance of the samples prepared in various conditions was shown in Table 1. This value of cerium dioxide was 0.398. Samples prepared from cerium (+III) nitrate solution indicated from 0.32 to 0.38. The P/Ce ratio and ultrasonic treatment had less influence on the static friction resistance.

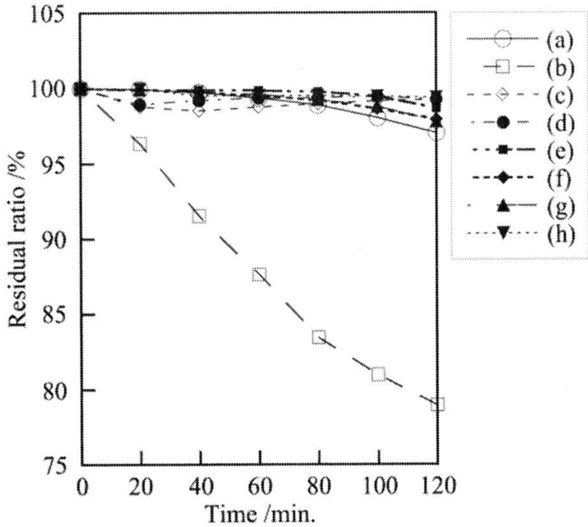

Fig. 6 Photocatalytic activity of samples prepared in various conditions, (a) blank, (b) ZnO, (c) CeO_2, (d) P/Ce(III) = 1/1, ultrasonic; 0 h, (e) P/Ce(III) = 1/1, 1 h, (f) P/Ce(III) = 2/1, 0 h, (g) P/Ce(III) = 2/1, 1 h, and (h) P/Ce(IV) = 4/3, 0 h.

(a) (b) (c) (d) (e) (f)

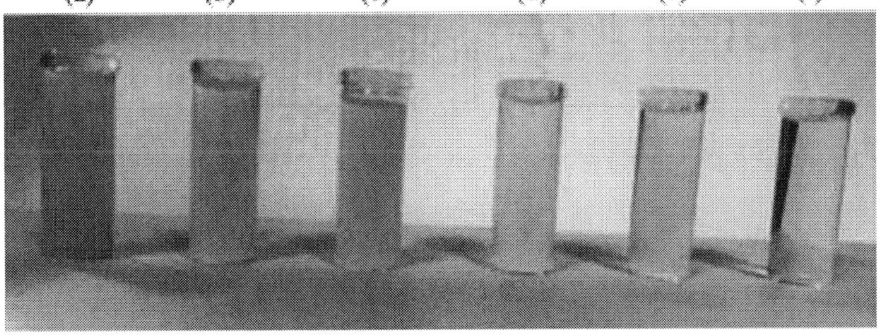

Fig. 7 Oxidation catalytic activity of samples prepared in various conditions, (a) CeO_2, (b) P/Ce(III) = 1/1, ultrasonic; 0 h, (c) P/Ce(III) = 1/1, 1 h, (d) P/Ce(III) = 2/1, 0 h, (e) P/Ce(III) = 2/1, 1 h, and (f) P/Ce(IV) = 4/3, 0 h.

Table 1 L*a*b* values and static friction resistances of samples prepared in various conditions

P/Ce	Ultrasonic /h	Ce	L*	a*	b*	Resistance /-
1/1	0	+III	77.53	1.48	-4.59	0.375
1/1	1	+III	71.53	1.75	0.17	0.321
2/1	0	+III	76.64	2.37	-1.89	0.354
2/1	1	+III	75.91	2.19	-1.10	0.375
4/3	0	+IV	80.82	6.38	5.60	0.272

4. Conclusion

Various cerium phosphates were prepared with and without ultrasonic treatment. Samples prepared from cerium (+III) nitrate solution indicated XRD pattern of cerium phosphate hydrate. All samples prepared in this work had large particles and no specified shape. The obtained cerium phosphates prepared in this work exhibit little photocatalytic activity and no oxidation catalytic activity. These samples prepared from cerium (+III) solution exhibit a little low reflectance within the range of visible light and 71-78 in L* value. The obtained samples indicated enough low static friction resistance.

References

[1] Monteiro-Riviere N A, Wiench K, Landsiedel R, et al 2011 *Toxicol. Sci.* **123** (1) 264 http://doi.org/10.1093/toxsci/kfr148
[2] Kockler J, Oelgemöller M, Robertson S, Glass BD 2014 *Cosmetics* **1** 128 http://doi.org/10.3390/cosmetics1020128
[3] Girigoswami K, Viswanathan M, Murugesan R, Girigoswami A 2015 *Mater. Sci. Eng. C* **56** (1) 501. http://doi.org/10.1016/j.msec.2015.07017
[4] Onoda H, Yamaguchi T 2012 *J. Mater. Chem.* **22** (37) 19826 http://doi.org/10.1039/c2jm33952h
[5] Onoda H, Haruki M, Toyama T 2014 *Ceram. Intern.* **40** (2) 3433 http://dx.doi.org/10.1016/j.ceramint.2013.9.088
[6] Davo-Quinonero A, Navlani-Garcia M, Lozano-Castello D, Bueno-Lopez A, Anderson J A 2016 *ACS catalysis* **6** 1723 http://doi.org/10.1021/acscatal.5b02741
[7] Lafaye G, Barbier Jr. J, Duprez D 2015 *Catalysis Today* **253** 89 http://dx.doi.org/10.1016/j.cattod.2015.01.037
[8] Jain A, Shore A M, Jonnalagadda S C, Ramanujachary K V, Mugweru A 2015 *Appl. Catal.* **489** 72 http://dx.doi.org/10.1016/j.apcata.2014.10.020
[9] Cai Y, Feng S, Ming C, Ren X, Qin Y, An L 2016 *Resul. Phys.* **6** 826 http://dx.doi.org/10.1016/j.rinp.2016.09.009
[10] Rao G V, Shashikala H D 2014 *J. Non-cryst. Solids* **402** (15) 204 http://doi.org/10.1016/j.j.noncrysol.2014.06007
[11] Lee J S, Murphy W L 2013 *Adv. Mater.* **25** (8) 1173 http://doi.org/10.1002/adma.201203370
[12] Hiromoto S, Inoue M, Taguchi T, Yamane M, Ohtsu N 2015 *Acta Biomater.* **11** (1) 520 http://doi.org/10.1016/j.actbio.2014.09.026
[13] Radovanovic Z, Jokic B, Veijovic D, Dimitrijevic S, Kojic V, Petrovic R, Janackovic D 2014 *App. Surf. Sci.* **307** (15) 513 http://doi.org/10.1016/j.apsusc.2014.04.066
[14] Lopes-Heras I, Madrid Y, Camara C 2014 *Talanta* **124** (15) 71 http://doi.org/10.1016/j.talanta.2014.02.029

IOP Conf. Series: Materials Science and Engineering **613** (2019) 012046 doi:10.1088/1757-899X/613/1/012046

Effects of weathering grade of metavolcanic aggregates on mortars

M Abdelfattah*, I Kocserha, R Géber

Institute of Ceramic and Polymer Engineering
University of Miskolc, Miskolc, Hungary

*E-mail: madatow@yahoo.com

Abstract. The aim of this work is to study the effect of weathering grade on physical, mechanical and microstructure properties of metavolcanic aggregates and on mortars containing aggregates. Samples were collected from several locations in the Central and Southern Eastern desert in Egypt and they were characterised by XRF, polarized light, and SEM microscopy. Beside compressive strength, bulk density, apparent porosity and water absorption, modified weathering potential index (MWPI) was also determined based on the chemical composition of the rock aggregates. Seven mortars with metavolcanic aggregates were mixed and specimens were formed. Physical and mechanical properties of specimens were measured according to standards. Results showed that MWPI had the good relation with geotechnical properties of metavolcanic aggregates. Physical and mechanical properties of metavolcanic aggregates and concretes depended on weathering grade of rocks and on ratio of weathering minerals in aggregates.

1. Introduction

Weathering of both soils and rocks is one of the most important problems, which the engineering geologist has to concern [1]. Weathering takes place in all environments, but it is most intense in hot, wet climates, where weathering may be expected to extend to great depths. Physical weathering led to open up of discontinuities by rock fractures. On the other hand, the chemical changes of minerals were the most crucial key for chemical weathering. Physical and chemical weathering greatly affected engineering structures found at or near the earth's surface [2,3,4]. Weathering degree of rock can be measured by many qualitative and quantitative prediction models. Qualitative models are based on observation and descriptions. Results of these models are the colour of rock, weathering degree of minerals and observational description of physical weathering, like exfoliation [5,6,7]. Quantitative approaches result consistent and objective rock weathering degree [6]. Rock weathering has several states which affects the aggregate characteristics. Therefore, this research is aimed to study the effect of weathering grade on physical and mechanical properties of metavolcanic aggregates and also on mortars based on aggregates.

2. Materials and methods

Metavolcanic samples were collected from different locations in the Southern and Central Eastern Desert, Egypt. Uniaxial compressive strength; bulk density; apparent porosity and water absorption were measured according to ASTM C-170 (2006), ASTM C-20 (2005), ASTM C-33 (1986) and ASTM C-127 (2006) [9,10,11,12], respectively. Chemical composition of the metavolcanic samples was analyzed by XRF (Axios, Sequential WD_XRF Spectrometer, PANalytical 2005 The microstructure of the samples was investigated by the use of scanning electron microscopy (JSM 6300) and transmitted light microscopy (TLM - B-150POL-MALC LM). In order to describe the effect of weathering on the mechanical and physical properties of metavolcanic aggregates, based on data originated from result of XRF, weathering indices were calculated according to the following equations:

MWPI (weathering potential index) = [K_2O + Na_2O +CaO + MgO) 100 / SiO_2 + Al_2O_3 + Fe_2O_3 + CaO + MgO + K_2O + Na_2O], CIA (chemical index of alteration) = Al_2O_3 *100 / (Al_2O_3+ CaO + Na_2O+K_2O),

CIW (chemical index of weathering) = ((Al_2O_3 + K_2O) / (Al_2O_3+ CaO + Na_2O)) *100.

The chemical weathering indices are definite equations used to quantify the degree of depletion of mobile components relative to immobile components [13,14].

Seven different mortars were prepared for tests. Three components, cement (CEM I 42.5N); fine and coarse aggregates were added to mixes in the ratio of 1:2:4. Mixes were devoid of sand fraction. Water-cement ratio was set to 0.5. Each mix was thoroughly blended to ensure homogeneous admixture. Physical and mechanical tests of samples from metavolcanic mortars were determined on cubic samples (50x50x50mm) according to the ASTM C109/C109M-16a [15]. Three cubes samples were produced for each mix. Curing time of samples was 28 days under water.

3. Results and discussion

3.1. Characteristics of metavolcanic aggregates

Geotechnical characteristics of the metavolcanic aggregates were summarized in Table 1. Results show that the values of MWPI increase as the weathering grade decrease and density increase. This is similar to findings from Jayawardena and Izawa [16] and Gupta and Rao [17]. Alternatively, CIA and CIW values show a continuous increase with increasing degree of weathering. Weathering index MWPI had good relation with geotechnical properties of metavolcanic aggregates (Figure. 1). Compressive strength, bulk density and specific gravity were a direct relation with the MWPI (Figures. 1a, 1b, respectively). On the contrary, apparent porosity had negative relation with the MWPI (Figure 1b).

Table 1. Weathering indices, physical and mechanical properties of the aggregates and mortars.

		Mechanical and physical properties of aggregates				Mechanical and physical properties of mortars		Weathering indices of aggregates		
Rock type	Rock unit	UCS, (MPa)	BD, (g/cm3)	AP, (%)	SG,	UCS, (MPA)	BD, (g/cm³)	(MWPI)	(CIA)	(CIW)
Metabasalt	AMB1	140	3.05	0.46	3.4	32.51	2.48	24.22	50.76	52.25
	ESMB2	104	2.80	0.26	2.89	21.01	2.27	19.83	57.25	60.41
	SMB3	87.76	2.7	0.48	2.78	32.01	2.34	19.11	60.93	62.36
	SMB4	40.21	2.9	1.14	3	32	2.39	20.25	55.85	64.49
Metadacite	UKMD5	40	2.5	1.80	2.6	10.71	2.07	19.63	54.94	60.35
Meta-andesite	SMA6	31.26	2.49	2.04	2.67	7	1.9	12.76	63.01	76.91
Basaltic meta-andesite	ESMBA7	102.55	2.87	0.76	2.9	24.06	2.29	21.34	57.0	57.58

UCS: uniaxial compressive strength; BD: bulk density; AP: apparent porosity; SG: specific gravity; modified weathering potential index: MWPI; chemical index of alteration: CIA; chemical index of weathering: CIW.

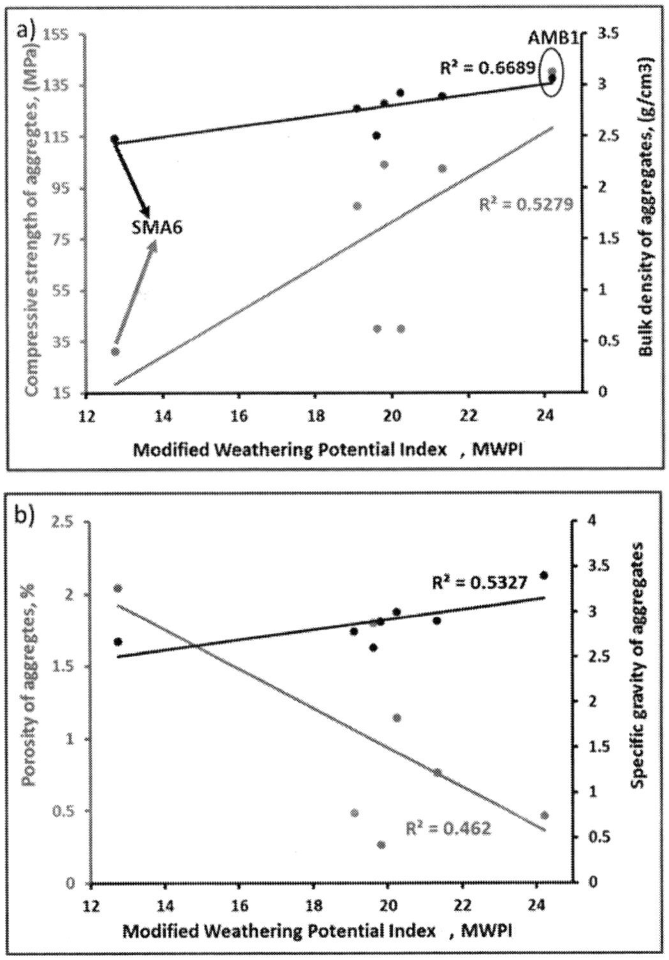

Figure 1: a) Relations between MWP index and compressive strength - bulk density b) Relations between MWP index specific gravity - apparent porosity

Mineralogical, weathering indices, physical and mechanical characteristics of the AMB1 were confirmed by its blocky and massive nature and its compact microstructure. Studying the microstructure of AMB1 by TLM, beside Ca-plagioclase, augite, olivine, actinolite, it contained only a small amount of weathering contents like chlorite and epidote (Figure 2 a). Interpretation of results focused only on the highest and the lowest values of all properties. Density and compressive strength of AMB1 (metabasalt) were the highest values, whilst the apparent porosity value of it was the lowest one (3.05 g/cm3, 140 MPa and 0.46 %, respectively) compared to the other tested aggregates. The MWPI value for AMB1 was the highest 24.22, whilst the CIA and CIW values are the lowest, 50.76 and 52.25 respectively, compared with the other metavolcanic samples (Table 1). Therefore, the AMB1 sample seemed to the most suitable metavolcanic aggregate in concrete. On the other hand, SMA6 consisted of highly fractured and weathered minerals (Figure. 2 b). It mainly composed of plagioclase, hornblende chlorite, quartz and fractured calcite (Figure.2b). The plagioclase in meta-andesite sample (SMA6) was highly-altered into sericite. Density and compressive strength of SMA6 were the lowest values, whilst apparent porosity was the highest value (2.49 g/cm^3, 31.26 MPa and 2.04 %, respectively). SMA6 has the lowest MWPI value of 12.76, but values of CIA and CIW were the highest 63.01 and 76.91, respectively. Therefore, the SMA6 sample is the unsuitable metavolcanic aggregates for concrete.

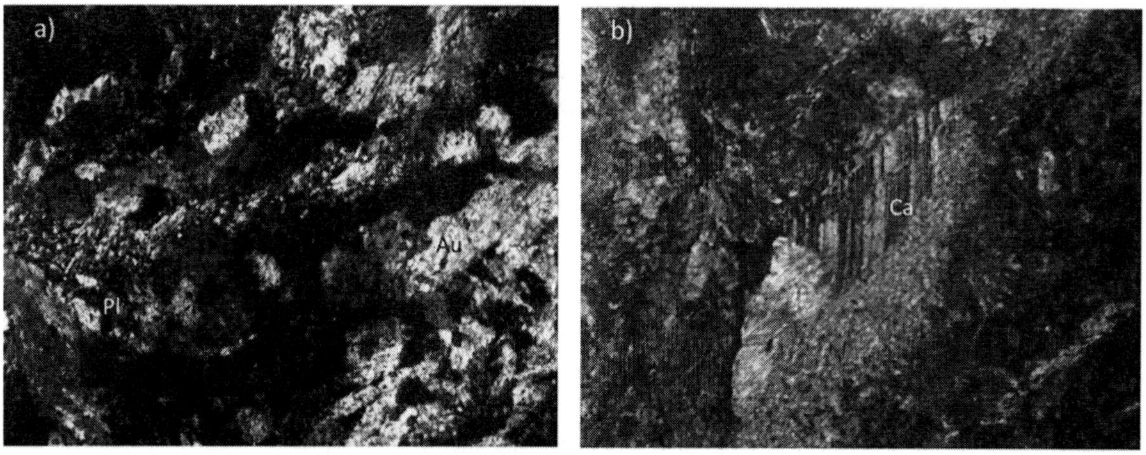

Figure 2: a) Augite minerals with compacted structure of AMB1 (cross nicole (XN)).
b) Photomicrograph showing highly fractures veinlets of calcite SMA6 (cross nicole (XN)).
Au: augite, Pl: plagioclase, Ca: calcite

3.2. Characteristics of metavolcanic mortars

According to ASTM C109/C109M-16a (2016) standard, seven mortars containing metavolcanic aggregates were mixed, samples were formed, cured and measured. Table 1 includes the average results for each tested mortars (three samples per mortar). Physical and mechanical characteristics of aggregates affected the physical and mechanical properties of its mortar. Therefore, density and compressive strength of aggregates were in direct relation with the compressive strength of mortars (Fig. 3). Compressive strength and bulk density of AMB1 mortar were the highest values (32.51 MPa, 2.48 g/cm^3). On the other hand, the compressive strength and bulk density of SMA6 mortar sample were the lowest values (7 MPa, 1.9 g/cm^3) (Fig. 3).

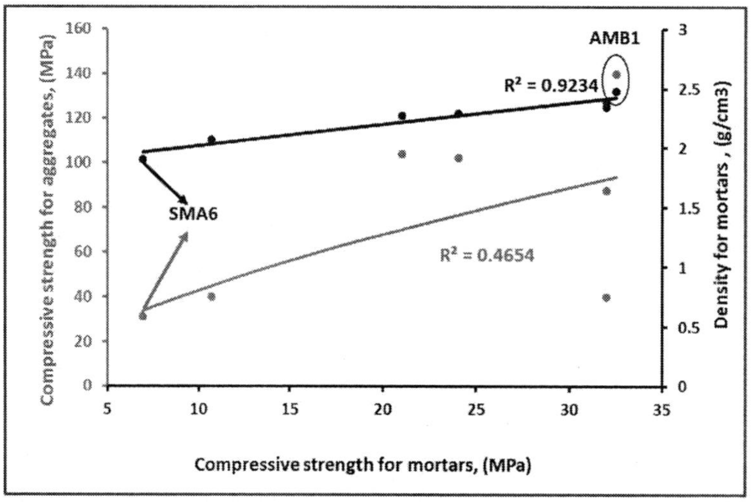

Figure 3: Relations between compressive strength of mortars and compressive strength-bulk density of aggregates.

3.3. SEM micrographs

Figure 4a illustrates the SEM micrograph of AMB1 mortar. Structure was highly compacted, no pore observed on analysed surface. It can be seen interlocking phases of aggregates and cement. AMB1 contained cementitious compounds such as calcium hydroxide ($CaOH_2$) and calcium silicate hydrate (CSH). On the contrary, the SMA6 aggregate was extensively weathered, where plagioclase and hornblende crystals are highly altered into sericite. In addition, SMA6 sample was characterized by fractured calcite veinlets indicating existence of micro-pores (Figure 2b). Therefore, SMA6 mortar was showed uncompact microstructure with large pores with cotton shape CSH phases. Latter indicated on Figure 4b with white arrows.

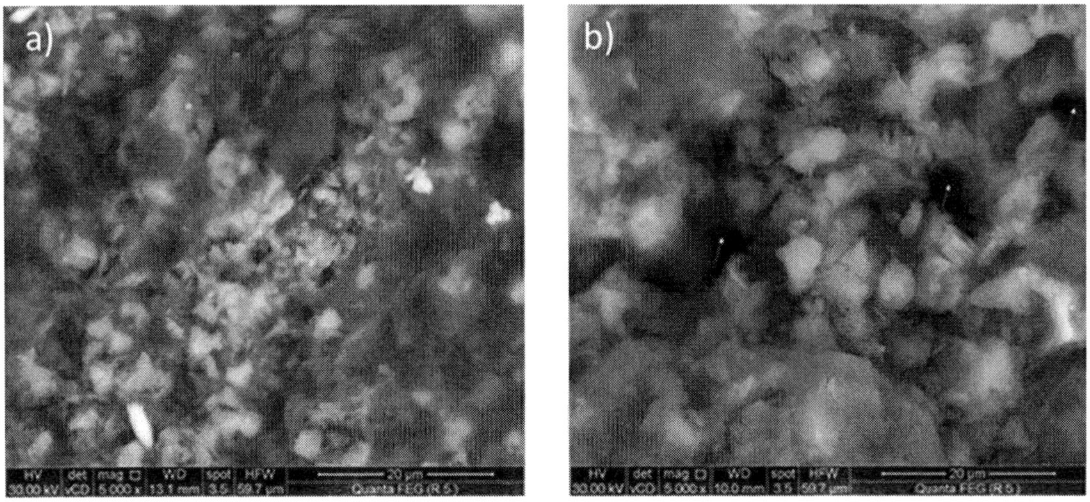

Figure 4: (a. SEM micrograph of AMB1 mortar; (b SEM micrograph of SMA6 mortar.

4. Conclusions

The weathering grade is the vital key for effecting on metavolcanic aggregates, which can be used in concrete. The modified weathering potential index was calculated based on the chemical composition of the rock aggregates. This index has good relation with geotechnical properties of metavolcanic aggregates. Compressive strength, bulk density and specific gravity were in direct relation with the MWPI. On the contrary, apparent porosity indicated a negative relation with the MWPI.

MWPI proves that the SMA6 sample is the most weathered among all samples. Therefore, physical and mechanical properties of its mortar were very low.

On the other hand, AMB1 metabasalt had the highest compressive strength and bulk density and the lowest porosity among of tested aggregates hence mortars based on AMB1 had also good properties. Highest values of compressive strength and density of AMB1 metabasalt aggregate and mortar were explained by its compact microstructure. In addition, the development of the cementitious phases contributes to the acceptable values of apparent porosity of AMB1 mortar. Application of AMB1 metabasalt is recommended as aggregates in concrete.

5. Acknowledgement

The described study was carried out as part of the EFOP-3.6.1-16-2016-00011 "Younger and Renewing University – Innovative Knowledge City – institutional development of the University of Miskolc aiming at intelligent specialisation" project implemented in the framework of the Szechenyi 2020 program. The realization of this project is supported by the European Union, co-financed by the European Social Fund.

6. References

[1] G Shrivastava 2014 *Journal of Mechanical and Civil Engineer* 18 e-ISSN: 2278-1684, p-ISSN: 2320-334X

[2] F Arıkan, R Ulusay, N Aydın 2007 *Bulletin Engineering and Geology Environment* **66** 415 https://doi.org/10.1007/s10064-007-0087-0

[3] M. Abed – R. Nemes 2018: *Építőanyag-JSBCM* **70** (5) 152 https://doi.org/10.14382/epitoanyag-jsbcm.2018.28

[4] B. Benabed 2018: *Építőanyag-JSBCM* **70** (5) 155 https://doi.org/10.14382/epitoanyag-jsbcm.2018.29

[5] T Y Irfan, G E Powell 1985 *Bulletin of the International Association* **32** 67 https://doi.org/10.1007/BF02594767

[6] C Gokceoglua, K Zorlub, S Ceryanc, H A Nefeslioglud 2009 *Material Characterization* **60** 1217 https://doi.org/10.1016/j.matchar.2009.06.006

[7] P Guan, N Cww, M Sun, W Tang 2001 *Engineering Geology* **59** 147 https://doi.org/10.1016/S0013-7952(00)00071-5

[8] C Gokceoglu, H Aksoy 2000 *Engineering Geology* **58** 1 https://doi.org/10.1016/S0013-7952(00)00032-6

[9] ASTM C170/C170M-17 2017 Standard Test Method for Compressive Strength of Dimension Stone *ASTM International, West Conshohocken, PA* www.astm.org https://.doi.org/10.1520/C0170_C0170M-17

[10] ASTM C20-00 2015 Standard Test Methods for Apparent Porosity, Water Absorption, Apparent Specific Gravity, and Bulk Density of Burned Refractory Brick and Shapes by Boiling Water *ASTM International West Conshohocken PA* www.astm.org https://.doi.org/10.1520/C0020-00R15

[11] ASTM C33/C33M-18 2018 Standard Specification for Concrete Aggregates, *ASTM International, West Conshohocken, PA* www.astm.org https://.doi.org/10.1520/C0033_C0033M-18

[12] ASTM C127-15 2015 Standard Test Method for Relative Density (Specific Gravity) and Absorption of Coarse Aggregate *ASTM International West Conshohocken*, PA www.astm.org https://.doi.org/10.1520/C0127-15

[13] L Harnois 1988 *Geology* **55** 319 https://doi.org/10.1016/0037-0738(88)90137-6

[14] W Nadlonek, I Bojakowska, 2018 *Applied Ecology and Environmental Research* **16** (3) 2453 http://dx.doi.org/10.15666/aeer/1603_24532473

[15] ASTM C109/C109M-16a 2016 Standard Test Method for Compressive Strength of Hydraulic Cement Mortars (Using 2-in. or [50-mm] Cube Specimens) *ASTM International West Conshohocken PA* www.astm.org https://.doi.org/10.1520/C0109_C0109M-16A

[16] U Jayawardena, S De, E Izawa 1994 *Bulletin International Association of Engineering Geology* **49** 55 https://doi.org/10.1007/BF02595001

[17] A S Gupta, K S Rao 2001 *Bulletin of Engineering Geology and the Environment* **60** 201 https://doi.org/10.1007/s100640100113

IOP Conf. Series: Materials Science and Engineering **613** (2019) 012047 doi:10.1088/1757-899X/613/1/012047

Effects of Pumice Powder on Silicone Exterior Coating Material

Mehtap Demir[1], Taha Oğuz[1]

[1]Faculty of Engineering, Department of Metallurgy and Material Engineering, Adiyaman University, Turkey

mehtapdemir@adiyaman.edu.tr

Abstract. Pumice has been used mainly in structural applications such as light-weight concretes, cements, and filters. Despite being used mostly in these areas of construction industry, pumice has very few applications in other sectors. Especially in the present investigation, it has been aimed to spread the usage of pumice in different sectors by analysing the particle size of the material. In cooperation with the factory and the processes of production of building chemicals was observed and it was investigated the characterization of the material produced by using FE-SEM. So that the microstructure images and Energy Dispersive X-ray Spectroscopy (EDS) composition analyses were obtained. It was observed that there has been a negligible amount of reaction between the Silicon and the pumice particles. In similar studies, when the doping is carried out at high temperatures, unwanted reactions (insulation, mechanical strength, etc.) occur between the particle and the base material and this affects the physical properties of the material in a negative way. In this study, due to the doping is in room conditions, there is little reaction between the particle and Silicon.

1. Introduction

Pumice is a silica-base porous aggregate which forms naturally during volcanic eruption [1,6,7]. The pressure drops due to the rapid expulsion of magma into the atmosphere causes there leave of dissolved gases which forms magma froth. The falling magma froth turns to a porous ceramic after solidification [2]. Pumice is mainly surface-mined in large quantities and has been used in structural applications such as light-weight concretes [3], cements [4,8], and filters [5]. Despite being used mostly its applications of construction industry, pumice has very few applications in other sectors. Especially in the present investigation, it has been aimed to spread the usage of pumice in different sectors by analysing the particle size of the material.

2. Material and Methods

2.1. Fabrication process

Pumice is extracted in surface mining and hence comes with some impurities [1]. The as-received pumice powder, provided by Novelart Construction Chemicals & Exterior Insulation Accessories Company, was put on precision weighing before starting sieve analysis. The mixture as-weighed 342 g and scaled 3 mm was micronized using the sieve shown in Figure 1. Then the micronized pumice particles are doped to the pre-prepared Silicon in the specified amount to distribute the pumice particles in a homogeneous manner with the help of the mixer rotating at a specific time interval and for a certain period. The hydraulic mixer shown in Figure 1 was used to enable the doping to take place

IOP Conf. Series: Materials Science and Engineering **613** (2019) 012047 doi:10.1088/1757-899X/613/1/012047

homogeneously. The mixing period of the mixer was set to 1600 rev/min before the doping was made (1600 revolution in 1 minute). These samples are naturally solidified under atmospheric conditions without any heat treatment. The absence of extra processing for maximum wetting between pumice particles and Silicone makes this study exceptional.

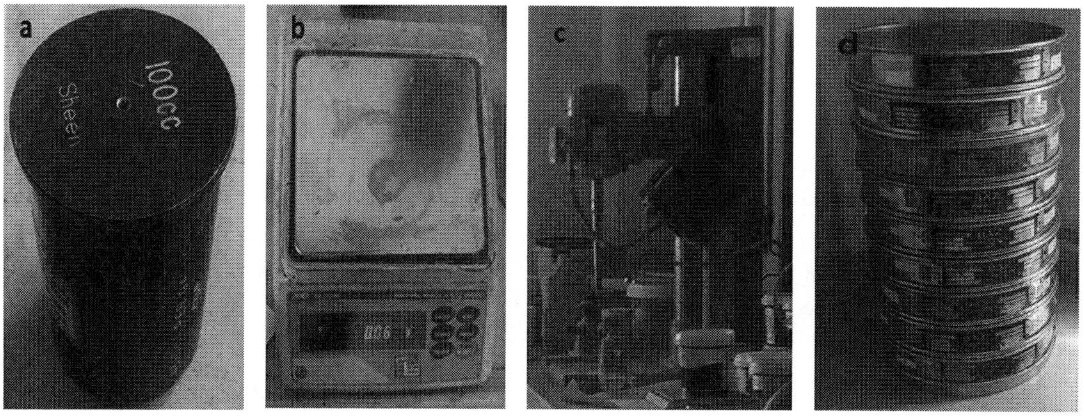

Figure 1. The devices used in the preparation of the micronized pumice powder (a) Density container (b) Precision weighing (c) Hydraulic mixer (d) Sieve in different scales

2.2. Characterisation of samples

Pumice is a porous glassy volcanic rock, which is formed by cavities, spongy, volcanic events and resistant to physical and chemical factors [7]. During its formation, it contains numerous porous ranging from macro scale to micro scale due to sudden cooling and sudden abrasion of the gases in the structure. As the porous are generally (especially microporous) unconnected hollow, heat and sound insulation is very high.

As a result of volcanic events, pumice is two types; acidic and basic. The most common is acidic pumice. Acidic pumice is white & dirty white color. According to the Mohs Scale; its hardness 5 ~ 6 and its density 0.5 ~ 1 gr / cm^3. Basic pumice is brown & black and heavier. According to the Mohs Scale; its hardness 5 ~ 6 and its density 1 ~ 2 gr/cm^3[9].

500 microns mixed pumice; black and white pumice is shown in Figure 2. And also, the smallest pumice particles obtained in the experiments are seen as 90 microns.

Figure 2. Obtained micronized pumice powder samples; 0.5 mm and 0.09 mm

IOP Conf. Series: Materials Science and Engineering **613** (2019) 012047 doi:10.1088/1757-899X/613/1/012047

3. Results and Analysis

3.1. Chemical composition of pumice

The pumice powder was micronized at 90, 250, 500, 700 microns scales by sieve analysis. The pumice density was obtained about 0.5-1 g / cm^3. Microstructure images have been analyzed by Scanning Electron Microscopy (SEM), and composition analysis by Energy Dispersive X-ray Spectroscopy (EDS). In the measurements, Cukurova University Central Research Laboratory, FEI Quanta 650 model Field Emission Scanning Electron Microscope (FE-SEM) and BSD detector have been used.

In the SEM analyzes carried out on the samples, it has been observed that there has been a negligible amount of reaction between the Silicon and the pumice particles. In Figure 3, SEM images of 700 microns black pumice powder and 700 microns white pumice powder have been given. In figure 4, EDS results of these pumice powders have been seen. So that the element content of the sample have been determined. It is clear that, as the Silicon ratio increases, the pumice stone's colour is close to white. Each table showing the elemental composition of pumice particles is given in the relevant micrograph. In Figure 5, we see EDS results of silicone exterior coating material when pumice doped or not doped. Due to the porous and spongy structure of the pumice, it shows the maximum wettability between pumice particles and silicon. The wetness of the new material between the particle and the silicone is proved to be good. The advantage of this is that the external cracks that may occur in buildings are eliminated due to the flexibility of pumice doped silicone. This is an advantageous situation in new buildings. In Figure 6 and Figure 7, EDS results of silicone exterior coating material when pumice doped or not doped have been given. When we look silicone rates, we see, while the silicon is heavy in pure form, the new material obtained with pumice doped silicon is lighter. This is due to the porous structure in the pumice.

3.2. Physical properties of pumice

Generally, the physical properties of pumice are caused by ceramic based elements in the composition. The SiO_2 content of the pumice gives the stone an abrasion characteristic. Al_2O_3 compound, which is also found in the content of pumice, gives a high resistance to fire and heat. The high amount of porous in the pumice gives very low proportions of specific gravity and high insulating properties.

Figure 3. SEM images of 700 microns black pumice powder and 700 microns white pumice powder

IOP Conf. Series: Materials Science and Engineering **613** (2019) 012047 doi:10.1088/1757-899X/613/1/012047

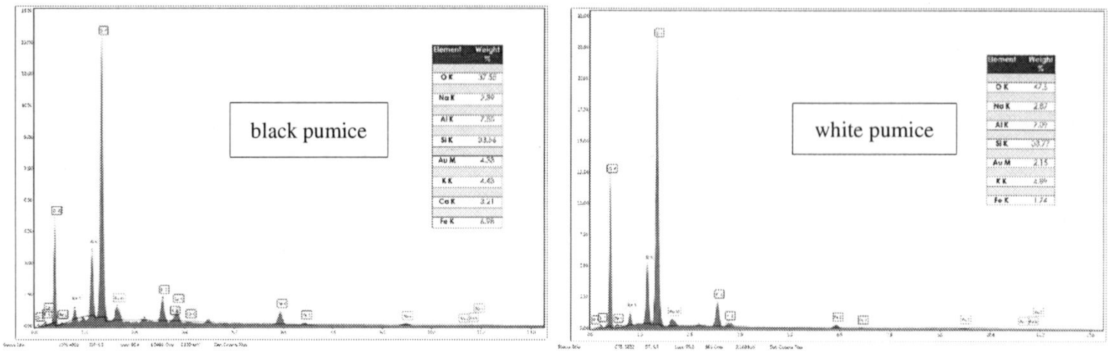

Figure 4. EDS results of 700 microns black pumice powder and 700 microns white pumice powder

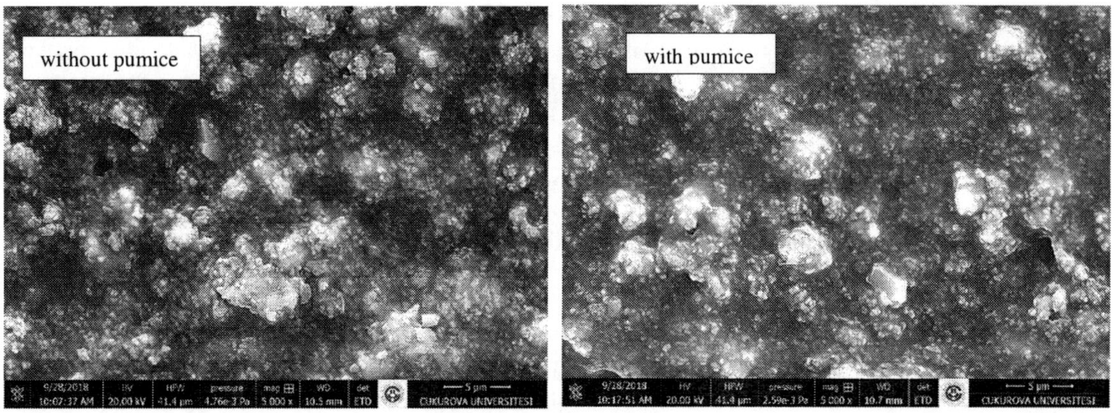

Figure 5. SEM images of silicone exterior coating material when pumice doped or not doped.

4. Conclusions

Doping was made in laboratory environment at room temperature. Making the doped at room temperature is one of the aspects that make this study privileged because of at room temperature will minimize both energy savings and undesirable reactions.

As a result of microstructure observations in present investigate, we see that the pumice structure has numerous porous structures ranging from a large number of macro scale to micro scale. It is very light due to its porous, so it is highly preferred in works without high weight. There are further experiments are required to find the density evulations and high insulating properties of pumice doped silicone exterior coating material.

Acknowledgements

The authors thank to Prof. Dr. Amirullah M. Mamedov for valuable comments and suggestions to improve the quality of this study. Also they thank to Novelart Construction Chemicals & Exterior Insulation Accessories Company, and its chemical engineer Metin Gökalp, and Kadir Deniz Takcı who works in Cukurova University Central Research Laboratory (Cumerlab), for helping on preparation of samples by using SEM device. This study was supported by Adıyaman University Scientific Research Project Unit within the scope of MÜFYL / 2018-0004 project.

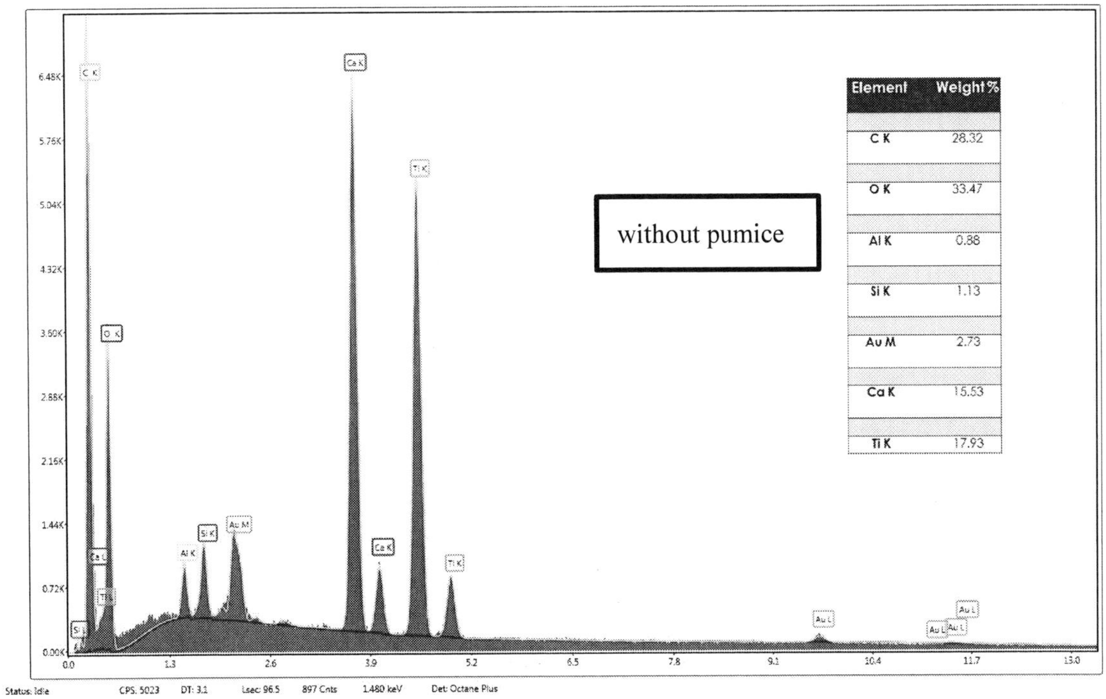

Figure 6. EDS results of silicone exterior coating material when pumice not doped.

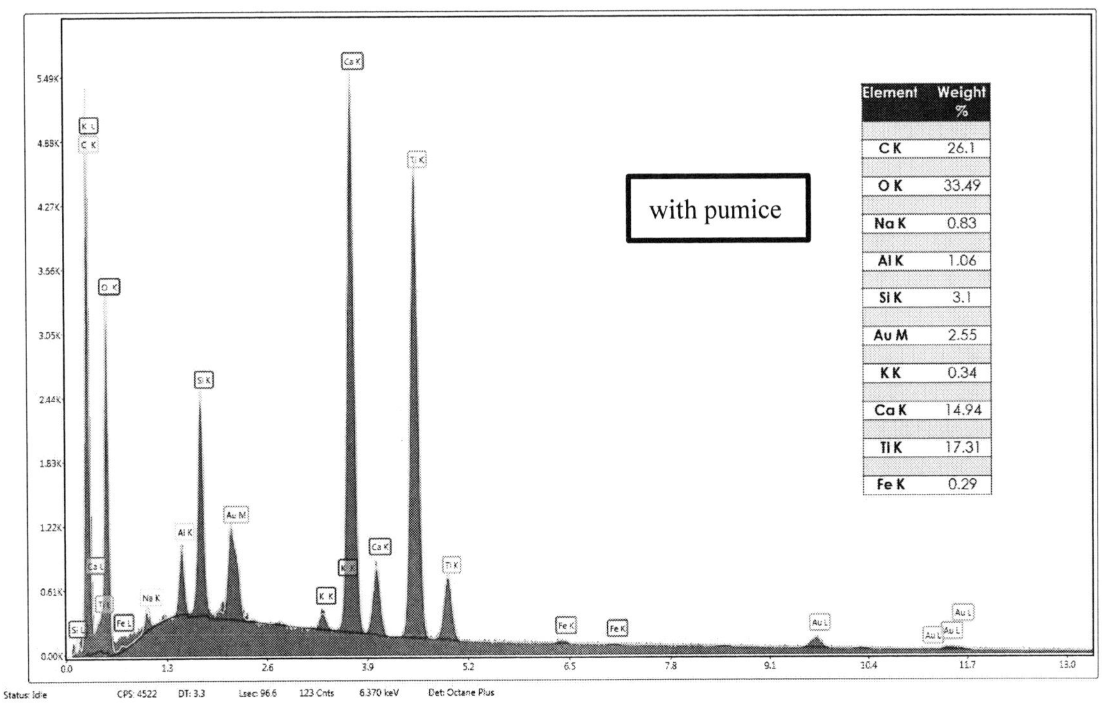

Figure 7. EDS results of silicone exterior coating material when pumice do

IOP Conf. Series: Materials Science and Engineering **613** (2019) 012047 doi:10.1088/1757-899X/613/1/012047

References

[1] M Taherishargh, I V Belova, G E Murch, T Fiedler 2015 *Materials Science & Engineering A* **635** 102 https://doi.org/10.1016/j.msea.2015.03.061
[2] C A Fleischer, M Zupan 2010 *J Compos. Mater.* **44** (23) 2679 https://doi.org/10.1177/0021998310369575
[3] K M A Hossain, S Ahmed, M Lachemi 2011 *Constr. Build. Mater.* **25** (3) 1186 https://doi.org/10.1016/j.conbuildmat.2010.09.036
[4] K M A Hossain 2003 *Cem. Concr. Res.* **33** (10) 1601 https://doi.org/10.1016/j.conbuildmat.2010.09.036
[5] U Kelm,V Sanhueza, C Guzman 2003 *Appl. Clay Sci.* **24** (1–2) 35 https://doi.org/10.1016/j.clay.2003.07.004
[6] G Betis, P Pantazopoulou, J Louvaris, E Phedros 1995 *International Symposium on Structural Lightweight Aggregate Concrete Sandefjord Norway* 421
[7] W P Bolen 2000 *U S Geological Survey Minerals Yearbook-1999 Virginia USA* 60.1 - 60.2
[8] M K A Hossain 2004 *Cement and Concrete Research* **34** (2) 283 https://doi.org/10.1016/j.cemconres.2003.08.004
[9] K Onouea, H Tamai, H Susenoc 2015 *Construction and Building Materials* **83** 261 https://doi.org/10.1016/j.conbuildmat.2015.03.019

IOP Conf. Series: Materials Science and Engineering **613** (2019) 012048 doi:10.1088/1757-899X/613/1/012048

Porous geopolymers: processing routes and properties

Tomáš Kovářík[1] and Jiří Hájek[1]

[1]New Technologies - Research Centre, University of West Bohemia (Czech Republic)

E-mail: toko@ntc.zcu.cz

Abstract. Cellular ceramics constitute a specific class of materials containing a high level of porosity, greater than 60 vol%, which are characterized by the presence of three-dimensionally arranged open and/or closed cells. Because of their structure, cellular ceramics exhibit a unique combination of properties such as low density, low thermal conductivity, low thermal mass, high permeability, high thermal shock resistance and high specific surface area, making them essential for various engineering application. Nowadays, porous geopolymers have been the focus of promising research in the field of porous inorganic materials because of their unique combination of good thermal stability and excellent mechanical properties. An interesting technological feature is that their solidification kinetics is easily adjustable and thanks to ceramic-like structure they have significant structural stability at elevated temperatures. The processing methods used for the fabrication of porous geopolymers can be divided into various approaches. Many methods have been explored to synthesize reproducible porous sponges or foams from geopolymer systems, such as gaseous method, rapid solidification, foaming, freeze-casting and/or combination of them. Concerning the macro/micro-structure of cells, the processing can be strongly influenced by various characteristics as a chemical composition, rheological behaviour of the slurries, kinetics of pore formation and hardening behaviour. This study presents methods of processing and manufacturing approaches with respect to types of porous materials and experimental results in this field.

1. Introduction

Porous inorganic materials have been attracting significant attention in widespread application fields such as catalysis supports, electrolyte carriers, membranes for separation and purification, insulators, filters, sensor carriers and biomedical scaffolds. Mainly, for these applications ceramic materials are suitable due to their exceptional combination of properties such as excellent chemical resistance, high melting point and thermal conductivity, thermal shock resistance, low thermal expansion coefficient, controlled permeability, high surface area, low density and high specific mechanical strength [1].

Extensive research efforts have been directed in development of porous ceramics because of possessing pivotal microstructural properties such as stable and uniform porous structure, high surface area, tunable pore size and well-defined surface properties. The role of porous structure in ceramic materials for different application is not identical and porous ceramics have several main types of structure, which mostly depend on the pore-formation method. Porous materials are classified into three classes depending on the pore diameter, d: macroporous ($d > 50$ nm), mesoporous ($50 \text{ nm} > d > 2$ nm), and microporous ($d < 2$ nm), according to the nomenclature of IUPAC (International Union of Pure and Applied Chemistry). Moreover, porous materials can be described through pore type, morphology and orientation. Besides the classification to open (interconnected) and closed (isolated) porosity the shape of the pores can be design as honey-comb, lotus-type, reticulated or bubble-like foamed materials and/or combination of these [2, 3].

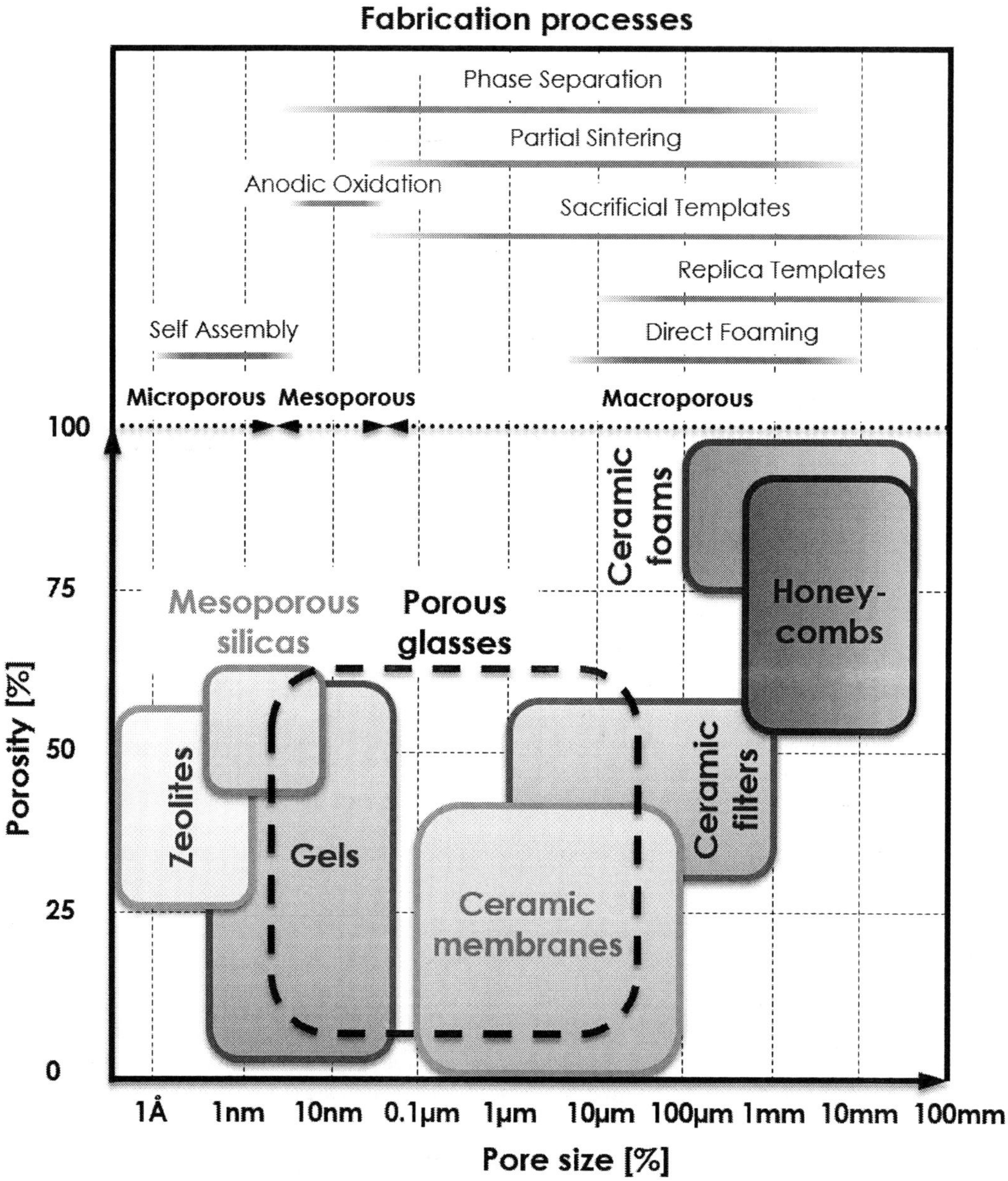

Figure 1. Pore size and porosity of typical porous materials and fabrication processes.

There are numerous pore-formation methods, which differ in principle: a) impregnation of a polymer cellular matrix with a ceramic suspension; b) templating of natural porous materials; c) introduction of foam-forming additives producing pores into an initial mixture; d) mechanical incorporation of air (gases) into ceramic suspension; e) swelling of the ceramic mixtures; f) introduction of filler grains with their own porosity into the ceramic mixture; g) mechanical pultrusion/injection or additive manufacturing of ceramic slurries; h) electrochemical treatment/deposition and varied modifications of these. Subsequently, many manufacturing techniques are applied to obtain preferred porous structure of ceramic mixtures such a freezing, drying, evaporation, dissolution and thermal treatment [4, 5].

Figure 1 shows schematic illustration of typical porous ceramic/ceramic-like materials depending on porosity and pore size, supplemented by fabrication processes. Micro-macro porous ceramics of vast chemical compositions have been prepared using these fabrication routes, based on Al_2O_3, SiO_2, ZrO_2, TiO_2, MgO, Cr_2O_3, Fe_2O_3 and their binary/ternary compounds or more complex compositions. Besides, carbides, nitrides, borides, silicides and combination of these were intensively studied. Also, natural clays and minerals, phosphate-based compounds or compositions with exotic elements were extensively prepared and tested [6].

Comparing to traditional ceramic materials, geopolymers are a relatively new group of materials that have been extensively studied for over three decades [7]. In general, geopolymers can be prepared by dissolution of natural or synthetic aluminosilicate raw materials (clays, slags, fly ashes, glasses) under strong alkaline conditions. The alkalination has the effect of depolymerization of silicates/aluminates. The structural reorganization results in polycondensation followed by formation of rigid aluminosilicate framework [8]. Over the past years, the geopolymers showed a variety of interesting properties and characteristics, such as excellent thermal stability, mechanical strength, acid resistance, low shrinkage, low thermal conductivity, adjustable setting at relatively low production costs. Their physical properties make them a viable alternative for many conventional binders and their synthesis at low temperature is an important benefit in industrial scale. Also, the chemical composition allows transformation of predominantly amorphous structure into various crystalline forms due to the applied heat treatment [9-12].

In recent years, porous geopolymers have gained a lot of attention because of unique properties associated with thermal and chemical resistance. Moreover, significant material and technological advantages are ceramic-like structure with high melting point and easy control of solidification kinetics [13]. Many methods have been explored to synthesize reproducible porous geopolymer sponges, foams, graded and/or hollow structures. In general, processing techniques can be divided into several methods such as: a) Direct foaming, b) Sacrificial template, c) Emulsion templating, d) Additive manufacturing, e) Replica templating and combination of these. Detailed processing features of each approach have been provided by excellent review article published elsewhere [14].

2. Processing techniques

2.1. Direct foaming

In the direct foaming technique, fresh geopolymer foam is produced by incorporating of air or gas products into a wet geopolymer slurry which is subsequently cured and solidified under slightly increase temperature (40-80°C). Gaseous products are usually formed by the dissolution of chemical additives (blowing agents) under alkaline conditions of geopolymer slurries. Examples of these additives may be aluminum or silicon metal powders or silicon-containing compounds [15-17]. Also, the most commonly used solution-based additive is hydrogen peroxide or other compounds based on hypochlorites and perborates [18,19]. It should be noted that blowing agents generate porosity through gas bubbles (mostly O_2 and H_2) under unstable conditions (high surface tension, coalescence, drainage). For these reasons, the resultant porous structure is formed by heterogeneous pores of varying size and distribution. In order to prevent or reduce such a structural phenomenon a number of stabilizing agents can be added to the suspension (surfactants, particles, fibers). **Figure 2** shows schematic fabrication route of direct foaming incl. applied additives and key properties. A wide variety of material and processing properties (chemistry, blowing agents, surfactants, rheology, solidification conditions) leads to the formation of porosity which is ranging from closed inhomogeneous cell size distribution to cellular morphology with open-cell structure and interconnected pores [20]. In summary, the experimental results reveal that using of direct foaming technique in geopolymer mixtures generates both open and closed-cell structures with a wide range of pore size (5 – 3000 μm) and total porosity (30 – 85 vol%).

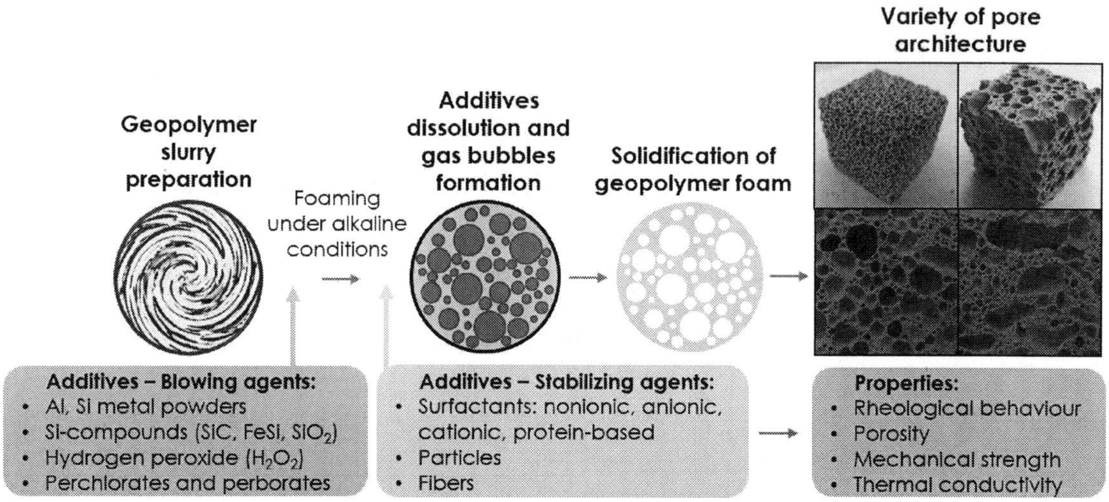

Figure 2. Schematic fabrication route of direct foaming method.

2.2. Sacrificial template

The sacrificial template method is one of the fabrication techniques where porous geopolymer is formed by mixing appropriate amounts of sacrificial template as a pore forming additive with a geopolymer slurry. This method leads to formation of porous material displaying a negative replica of the original sacrificial template. The porosity can be generated by extraction of template from the biphasic material through thermal decomposition by applying of thermal treatment and/or melting, dissolution and sublimation. The way that the sacrificial template is extracted from the biphasic material depends primarily on the type of pore former (e.g. natural/synthetic organics, liquids, salts).

The experimental results reveal successful synthesis of lamellar macro-porous potassium-based geopolymer by ice-templating via freeze-casting procedure. Ice-templated hierarchical structure had 53–83% total porosity. A broad mesopore distribution was detected between 4 and 100 nm with maxima from 5 to 7 nm, while macropores in the range from 1 to 100 μm [21]. Different amounts of water (20, 50, 70 vol%) were mixed with the geopolymer slurry to induce lamellar ice growth by unidirectional freezing. A lower water content (30%) and curing at 80°C after maturation at room temperature was conducive to a narrow lamellar pore width distribution in the range from 30 to 130 μm [22]. **Figure 3** shows schematic fabrication route of sacrificial template method with an example of polymer filler and water/ice-based templating.

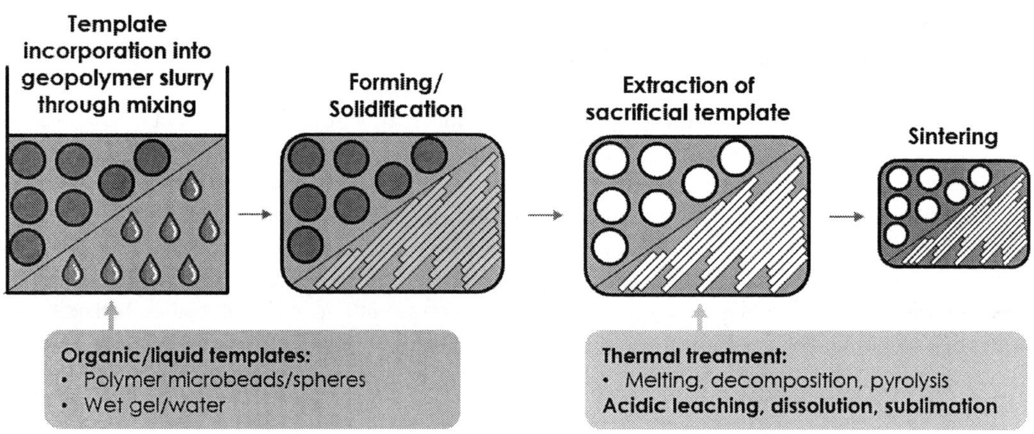

Figure 3. Schematic fabrication route of sacrificial template method.

2.3. Additive manufacturing

In the past few years, additive manufacturing (3D printing) has received considerable attention from a number of grounded reasons. This technology brings benefits to the creation of materials with a complex geometry and high precision, maximum material savings and flexibility in design. A wide range of material are currently used in 3D printing include metals, polymers, ceramics and inorganic pastes.

Recently, highly porous ceramic lattices (porosity up to ~ 71 vol%) were fabricated with ~ 0.8 mm struts based on metakaolin and sodium silicate solution geopolymer mixtures. Rheological properties of geopolymer pastes were optimized to possess Bingham shear thinning behavior and graded lattice with increasing spacing between subsequent struts was designed [23].

Also, innovative methodology of powder-based 3D printing process was developed to fabrication of slag-based geopolymer. Cubic and plate structures with complex geometries were manufactured by depositing binder liquid selectively into to powder bed. The cubic structure revealed apparent porosity ~ 57% and uniaxial compressive strength ~ 0.9 MPa. A post-processing method (1-day alkaline solution immersion) reduced the anisotropy of the printed material and increased the strengths up to ~ 10 MPa [24].

Moreover, additive manufacturing of geopolymer based on fly ash, granulated blast-furnace slag, silica fume and potassium hydroxide/silicate solution was introduced. Comparing to casted, printed sample exhibited similar or higher compressive strength when load was applied in a plane perpendicular to the layers ranging from 30 to 40 MPa after 28 days [25].

Besides, extrusion based 3D printing graphene oxide/geopolymer nanocomposite was reported. The nano-graphene oxide in the Na-based geopolymer system resulted in proper rheological properties for extrusion, the higher compressive strength of about 30 MPa and increased electrical conductivity [26].

2.4. Replica templating

In the replica technique, a highly porous polymeric sponge is initially soaked into a ceramic suspension until the internal pores are filled in with ceramic material. During the next stage, the suspension has to be sufficiently fluid to be partially removed by applied pressure but viscous enough to avoid dripping. Hence, ceramic suspensions exhibiting shear-thinning behavior are needed to efficiently coat the polymeric template. The ceramic-coated polymeric template is subsequently dried and heat treated to allow the gradual decomposition and diffusion of the polymeric material. Densification by sintering at desired temperature by specific heat treatment regime usually follows.

This procedure has been successfully applied for synthesis of leucite-based ceramic foam derived from potassium-based geopolymer precursor. Compared with traditional ceramics, the fabrication process has been greatly time reduced by accelerating the solidification kinetics of geopolymer mixture through heat treatment at 60 °C for 30 min in a closed container. Compressive strength reached ~ 0.85 MPa after sintering at 1300 °C, with an open porosity of ~ 79 vol%. The entire structural grid is presented by interconnected network with cellular structure where pore size ranges from 1 to 2 mm. It was demonstrated that the predominantly amorphous geopolymer matrix can be effectively reinforced by fine ceramic filler for the applied heat treatment [27]. Also, different types of ceramic fillers (alumina, cordierite) were tested and demonstrated a positive influence on the thermal resistance of the geopolymer composite during the heat treatment [28].

Figure 4 shows schematic fabrication route of replica template method for geopolymer mixtures with an important processing parametres in the sense of step by step process.

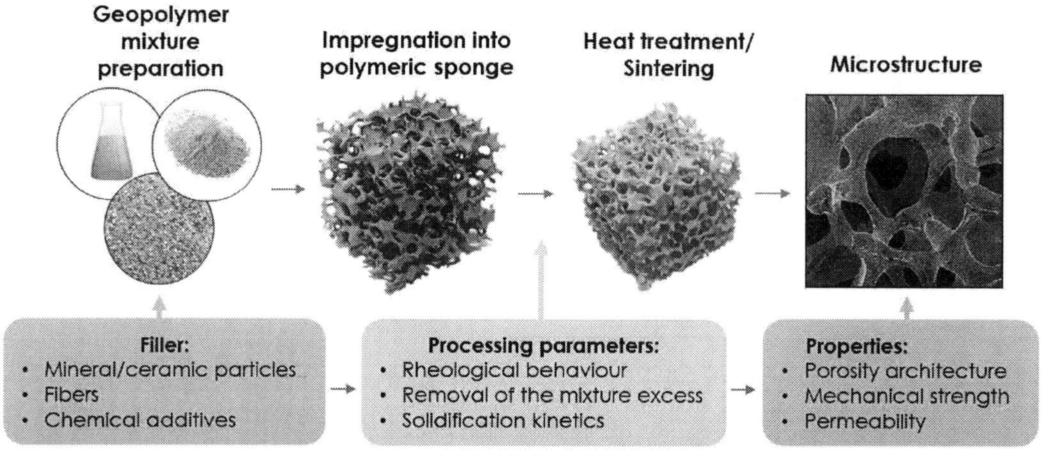

Figure 4. Schematic fabrication route of replica template method for geopolymer mixtures.

3. Conclusion

Various processing routes using direct foaming, replica, sacrificial template or additive manufacturing methods are nowadays available for the production of meso/macroporous ceramics. The use of this knowledge has led to the successful application of these techniques in the preparation of many types of porous geopolymers. The combination of these techniques or the modification of the manufacturing process may lead to the preparation of non-traditional porous materials with unique properties. The diversity of the chemical composition of geopolymer materials, including the conditions of preparation, implies the need for innovative processes. The resultant porous structures can therefore be an improved alternative to traditional ceramic materials in certain applications.

Acknowledgements

The result was developed within the CENTEM project, reg. no. CZ.1.05/2.1.00/03.0088, cofunded by the ERDF as part of the Ministry of Education, Youth and Sports OP RDI programme and, in the follow-up sustainability stage, supported through CENTEM PLUS (LO1402) by financial means from the Ministry of Education, Youth and Sports under the National Sustainability Programme I.

References

[1] Scheffler M, Colombo P 2005 Cellular Ceramics: Structure, Manufacturing, Properties and Applications *Wiley-VCH Verlag GmbH & Co. KGaA* ISBN: 978-3-527-31320-4

[2] Ohji T 2013 Handbook of Advanced Ceramics: Materials, Applications, Processing, and Properties (Porous Ceramic Materials, Chapter 11.2.2) *Elsevier* ISBN: 9780123854704

[3] Liu P, Chen G-F 2014 Porous Materials, Processing and Applications (General Introduction to Porous Materials, Chapter 1) 1st Edition *Elsevier* ISBN: 9780124078376

[4] Okada K, Isobe T, Katsuma K, Kameshima Y, Nakajima A, MacKenzie K J D 2011 *Sci. Technol. Adv. Mater.* **12** 11 https://doi.org/10.1088/1468-6996/12/6/064701

[5] Sarkar N, Kim IJ 2015 *Advanced Ceramic Processing, (Porous ceramics*, Chapter 3) 1st Edition *IntechOpen Limited* https://doi.org./10.5772/61047

[6] Studart A R, Gonzenbach U T, Tervoort E, Gauckler L J 2006 *J. Am. Ceram. Soc.* **89** (6) 1771 https://doi.org/10.1111/j.1551-2916.2006.01044.x

[7] Davidovits J 2015 *Geopolymer Chemistry and Applications*, 4th Edition (Saint-Quentin: Institut Geopolymere) ISBN: 9782951482098

[8] Shi C, Feznández-Jiménez A, Palomo A 2011 *Cem. Concr. Res.* **41** 750 https://doi.org/10.1016/j.cemconres.2011.03.016

[9] Duxson P, Provis J L, Lukey G C, Mallicoat SW, Kriven W M, Van Deventer J S J 2005 *Colloids Surf. A* **269** 47 https://doi.org/10.1016/j.colsurfa.2005.06.060
[10] Bakharev T 2005 *Cem. Concr. Res.* **35** 658 https://doi.org/10.1016/j.cemconres.2004.06.005
[11] Kovářík T, Rieger D, Kadlec J, Křenek T, Kullová L, Pola M, Bělský P, Franče P, Říha J 2017 *Constr. Build. Mater.* **143** 599 https://doi.org/10.1016/j.conbuildmat.2017.03.134
[12] Kamseu E, Djangang Ch, Veronesi P, Fernanda A, Melo U Ch, Sglavo V M, Leonelli C 2015 *Mater. Design* **88** 336 https://doi.org/10.1016/j.matdes.2015.08.151
[13] Khale D, Chaudhary R 2007 *J. Mater. Sci.* **42** (3) 729 https://doi.org/10.1007/s10853-006-0401-4
[14] Bai Ch, Colombo P 2018 *Ceram. Inter.* **44** 16103 https://doi.org/10.1016/j.ceramint.2018.05.219
[15] Sanjayan J G, Nazari A, Chen L, Nguyen G H 2015 *Constr. Build. Mater.* **79** 236 https://doi.org/10.1016/j.conbuildmat.2015.01.043
[16] Medri V, Papa E, Dedecek J, Jirglova H, Benito P, Vaccari A, Landi E 2013 *Ceram. Int.* **39** (7) 7657 https://doi.org/10.1016/j.ceramint.2013.02.104
[17] Prud'homme E, Joussein E, Rossignol S 2015 *Eur. Phys. J. Spec. Top.* **224** (9) 1725 https://doi.org/10.1140/epjst/e2015-02494-7
[18] Bai C, Colombo P 2017 *Ceram. Int.* **43** (2) 2267 https://doi.org/10.1016/j.ceramint.2016.10.205
[19] Novais R M, Buruberri L H, Seabra M P, Bajare D, Labrincha J A 2016 *J. Clean. Prod.* **124** 395 https://doi.org/10.1016/j.jclepro.2016.02.114
[20] Verdolotti L, Liguori B, Capasso I, Errico A, Caputo D, Lavorgna M, Iannace S 2015 *J. Mater. Sci.* **50** (6) 2459 https://doi.org/10.1007/s10853-014-8801-3
[21] Papa E, Medri V, Benito P, Vaccari A, Bugani S, Jaroszewicz J, Swieszkowski W, Landi E 2015 *Mic. Mes. Mater.* **215** 206 https://doi.org/10.1016/j.micromeso.2015.05.043
[22] Papa E, Medri V, Benito P, Vaccari A, Bugani S, Jaroszewicz J, Swieszkowski W, Landi E 2016 *RSC Adv.* **6** (29) 24635https://doi.org/10.1039/c6ra02232d
[23] Franchin G, Scanferla P, Zeffiro L, Elsayed H, Baliello A, Giacomello G, Pasetto M, Colombo P 2017 *J. Eur. Ceram. Soc.* **37** (6) 2481 https://doi.org/10.1016/j.jeurceramsoc.2017.01.030
[24] Xia M, Sanjayan J 2016 *Mater. Des.* **110** 382 https://doi.org/10.1016/j.matdes.2016.07.136
[25] Panda B, Paul S Ch, Hui L J, Tay Y W D, Tan M J 2017 *J. Clean. Prod.* **167** 281 https://doi.org/10.1016/j.jclepro.2017.08.165
[26] Zhong J, Zhou GX, He P G, Yang ZH, Jia D Ch 2017 *Carbon* **117** 421 https://doi.org/10.1016/j.carbon.2017.02.102
[27] Kovářík T, Křenek T, Rieger D, Pola M, Říha J, Svoboda M, Beneš J, Šutta P, Bělský P, Kadlec J 2017 *Mater. Lett.* **209** 497 https://doi.org/10.1016/j.matlet.2017.08.081
[28] Kovářík T, Křenek T, Pola M, Rieger D, Kadlec J, Franče P 2017 *IOP Conf. Ser.: Mater. Sci. Eng.* **175** 012044 https://doi:10.1088/1757-899X/175/1/012044

IOP Conf. Series: Materials Science and Engineering **613** (2019) 012049 doi:10.1088/1757-899X/613/1/012049

Detection technique of the optical and thermoradiative characteristics with compensation effect of reflection and transmittance indicatrices for the semitransparent materials with high subsurface scattering

Vladimir G. Merzlikin[1, 2] Andrey V. Bystrov[2] and Sergey V. Smirnov[3]

[1]Department of Technology and Equipment of Mechanical Engineering, Moscow Polytechnic Un, Bol'shaya Semenovskaya, 38, Moscow 107023, Russia
[2]Department of Industrial Economics, Plekhanov Russian Un of Economics, Stremyanny, 36, Moscow 117997, Russia
[3]Department of Mechanical and Instrumental Engineering, Peoples' Friendship Un of Russia (RUDN University), Miklukho-Maklaya, 6, Moscow, 117198, Russia

E-mail: MerzlikinV@mail.ru, bistrov-sun@mail.ru and smirnof61@mail.ru

Abstract. The experimental set-up is developed for detection of reflection and transparence coefficients of semitransparent materials within visible and IR radiation diapasons. The peculiarity of the set-up is using the spectral photometric integrating spheres for effect compensation of reflection indicatrices of tested materials with fixed wavelengths 0.63 μm 1.15 μm (near IR), 2.79 μm and 3.39 μm (middle infrared) of probing laser radiation. In the first case the integrating sphere is a photometric ball which made from porous fluoroplastic with high reflectivity up to 97-99%. For the second range an aluminum sphere (sand blasting) was used. The application of the photometric sphere reduced the measurement error, connected with different sensitivity of photosensors to beam slope angle of detected radiation reflected (transmitted) for flat material samples with different indicatrices of reflection (transmission). The proposed method of experimental measurements of spectral photometric characteristics of reflection and transparent improve the accuracy of theoretical estimations of the optical (thermoradiation) parameters (absorption and scattering indexes), as well as the functions of a thermal radiation heat source for the calculation of temperature fields in semitransparent materials. For the first time, a computational and experimental estimation of scattering and absorption indexes (with an accuracy not less than 1%) were carried out for semitransparent materials with high scattering (porous ceramics based on stabilized zirconium dioxide) in the middle infrared range of wavelengths which is most problematic for optical measurements.

1. Introduction

In traditional studies of complex heat transfer and radiant heat transfer for the heat protection elements of the combustion chambers walls of diesel engines and turbines, blades of aerospace turbine, traps of overheated coolants of nuclear reactor, usually it is necessary to consider the effect of the integral heat flux using the contribution of the penetrating radiant component in the short wavelength range ~ 1-3 μm in the heating process of heat-insulating (heat-shielding) coatings.

Applicate ceramic insulating coatings (based on oxides of silicon, aluminum, zirconium) are partially transparent to the specified radiant exposure. But, as a rule, this effect is not taken into account [1, 2], although the use of these coatings becomes relevant, based the trends of the conceptual development of similar heat protection in mentioned above fields of science and technology [3, 4].

Due to the lack of optical heat exchange models the study of radiation-convective action was limited by analysis of effective heat transfer using only the exposed surface without analysis of subsurface overheating caused by penetrating radiation [1, 2].

Thus, the correct evaluation of the radiation and hence temperature fields was absent. Instead, there is the use of numerous empirical formulas or incorrect physical models (for example, for non-stationary heat transfer coefficient according to the Woschni formulas in combustion engines of ICE [1, 2, 4]), obtaining of which requires time-consuming laboratory measurements and costly full-scale tests of model and industry heat-resistant samples.

Therefore, a preliminary experimental evaluation of high-precision optical characteristics of materials and coatings will reduce the number of bench tests and predict the most appropriate selection of materials for elements of heat protection [3, 4].

2. Optical set up for experimental measurements of reflection and transparence

To evaluate the optical characteristics of a semitransparent material and coating optical set up for experimental measurements of reflection and transparence coefficients (using a photometric ball, additionally introduced into the optical layout) has been developed [5]. These data allow us to estimate the absorption and scattering indices that determine the function of the heat source in solving different problems of radiative heat transfer and produced temperature fields [3, 4].

Measurements of the reflection coefficients of experimental model samples of semitransparent heat-insulating materials were carried out on the setup shown in Figure 1*a*.

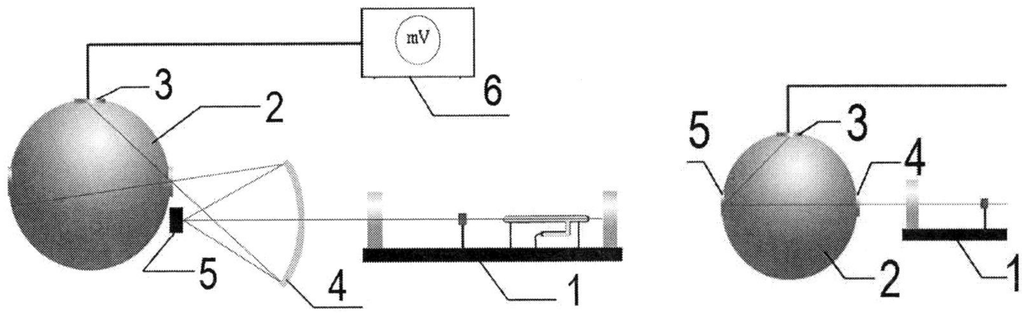

a *b*

Figure 1. The optical layout of the test set up for measuring the reflection coefficients (*a*) and transparence (*b*) of samples of semitransparent materials respectively: in the 5*a* and 4*b* positions (probe laser - 1, integrating sphere - 2, photodetector - 3, reflective hemisphere with a hole for probe radiation - 4, the opaque reference sample – 5*b*, nanomicro voltmeter - 6)

The following radiation sources were used: tricolor helium-neon laser "ЛГ-126" (here and below Russian abbreviation is used) with working laser wavelength of 0.63 μm, 1.15 μm, 3.39 μm); solid-state laser "ЛТ-2" with a wavelength of 1.06 μm; solid-state laser "УЛТ-3" with a wavelength of 2.79 μm.

The probing radiation reached the test sample, after passing through the central hole of the mirror hemisphere, then this beam was diffusely reflected from the sample and was directed (by a focusing mirror) into a photometric ball, equipped with a photodiode "ФД-24К" (3) for measurements at wavelengths of 0.63, 1.06, and 1.15 μm (visible and near infrared range) or the photoresistor "ФР-188" for measurements at wavelengths of 2.79 μm and 3.39 μm (middle range).

The focusing mirror was a glass spherical segment (4*a*) with a radius of 0.2 m and a sphere section size of 0.4 m. Aluminum was sprayed onto the working surface of the mirror, having a limiting reflection coefficient ~ 90% in the visible, near and middle IR regions of the spectrum.

At the investigation in a darkened room (in the absence of extraneous light sources) the ratio of recorded measuring signals and noise one did not exceed 1%.

The use of a photometric ball (integrating sphere) made it possible to eliminate the measurement error, connected with different sensitivity of photo detectors for beam slope angle of the incident radiation (i.e. various indicatrices) [4, 6], that allows to increase accuracy of measurements of optical and thermo-radiation characteristics and improve the evaluation of the function of thermal radiation heat source to calculate temperature fields in the exposed materials and coatings.

In addition, a simplified layout (shown in Figure 1*b*), allowing to measure transparence was used. In this case, when measuring the reflection coefficients in the 5*a* position, the reference standard sample and the tested one were alternately placed.

When measuring the transmittance of the researched sample, it was located in position 4*b*, but the reference standard was located in the 5*b* position.

Radiation attenuation was carried out using light filters or rotating disk attenuators. The measurement accuracy at using a photodiode "ФД-24К" and photoresistance "ФР-188" accordingly was not worse than 2% and 0.5%.

As reference samples (standards) quartz ceramics (made from high purity synthetic powder with Russian brand "ОСЧ-14") and aluminum with brand "АМц" (sandblasted).

3. Numerical evaluation of the absorption and scattering parameters

Absorption and scattering rates were calculated according to the methodology (developed by the authors), in the two-flux approximation of the solution of the radiative transfer equation based on the measurement of the reflection coefficients *r (H)* and transmittance ones *τ (H)* for samples of different thickness *H* in the form of compressed powder tablets of stabilized zirconia ZrO_2+8%Y_2O_3 (Table 1) [4].

The values of unknown spectral absorption κ and scattering σ indices may be calculated from experimental spectrophotometry measurements of the reflectance and the transparence of flat sample with different thickness *H*:

$$\kappa = \frac{b_{rad} \cdot (1-A)}{1+A}, \sigma = \frac{b_{rad}^2 - \kappa^2}{2\kappa}, b_{rad} = \frac{1}{H} \ln \frac{2U}{V - \sqrt{V^2 - 4U^2}}, r(H \to \infty) = \frac{Y-U}{UY-1}, \quad (1)$$

where $U = r(H) + \tau(H)$; $V = [r(H)/\tau(H)](1-U^2) + U^2 + 1$; $Y = \exp(-b_{rad}H)$; b_{rad} - extinction index; $A = r(H \to \infty)$ - albedo (reflection coefficient for semi-infinite substance layer).

Table 1. Spectral values of experimentally measured reflection coefficients *r(H)* and transparence *τ (H)* and calculated parameters: $r(H \to \infty)$ - albedo; absorption κ and scattering σ indices for samples of pressed powders (ZrO_2 + 8% Y_2O_3) with thickness *H* = 7 mm

λ, μm	r(H), %	τ(H), %	r(H→∞), %	κ, 1/m	σ, 1/m
0.63	97.0	2.6	98.5	0.6	5080
1.15	92.0	4.3	93.2	6	2430
2.79	91.0	3.7	92.5	9	2470
3.39	89.0	3.3	89.5	14	2400

There is a significant value of the reflection coefficient of model ceramic samples up to a wavelength of 3.39 μm.

Low absorption materials (for example, samples of corundum or quartz) have high reflectivity values too. But only zirconia has a transparency band up to ~3.39 μm unlike samples of other oxides whose transparency bands are limited by 2-3 μm.

4. Conclusion

Experimental set up for measuring reflection and transparence coefficients of materials and coatings in the visible and infrared ranges has been developed. Using a photometric ball in an optical set up for recording reflection and transparent coefficients allowed to eliminate the measurement error, connected with different sensitivity of photo detectors with respect to the incident (on its surface) radiation, whose diffuse-directed indicatrix is indefinite. This technique improves measurement accuracy optical (thermoradiation) characteristics and estimates of the function of thermal radiation heat source for solving different problems of radiation - conductive heat transfer.

First, the optical parameters are estimated for technological samples of semitransparent coatings based on zirconium dioxide in the middle infrared: absorption ~ 14 1/m and scattering ~ 2400 1/m indexes with a reflection coefficient of a semi-infinite layer (albedo) ~ 89.5%.

Performed experimental measurements in the visible and near IR diapasons corresponds to the data reported by other authors [3, 7].

Presented method of high-precision optical recording is important for selection the type of semitransparent Thermal Barrier Coatings [1-3, 7] with correctly certified parameters to study specific radiant heat transfer in heat-stressed elements in the IR range for heat and power plants, engines (combustion chambers, furnaces) and vehicles.

References

[1] M Ciniviz, M S Salman, E Canl, H Köse, Ö Solmaz 2012 *Ceramic Coatings – App. in Engineering* 195 https://doi.org/10.5772/29993

[2] M A Azadi, M Baloo, G H Farrachi, S M Mirsalim 2013 *Int. J. of Autom. Eng* **3** (1) 305 http://www.iust.ac.ir/ijae/search.php?sid=1&slc_lang=en&auth=Farrahi

[3] L Wang, J I Eldridge, S M Guo 2013 *Acta Materialia* **64** 402 https://doi.org/10.1016/j.actamat.2013.10.053

[4] V Merzlikin 2007 *Proc. of8th Int. Conference on Engines for Automobile - ICE2007 (Capri, Napoly)* 24-0066 https://www.sae.org/technical/papers/2007-24-0066

[5] Integrating Sphere Radiometry and Photometry (Advancing the Technology of Light) 2017 *Technical Guide Labsphere* 22 https://www.labsphere.com/site/assets /files/2550/radiometry_and_photometry_tech_guide-1.pdf

[6] L Hanssen, S Kaplan, R Datla 2015 *Infrared Optical Properties of Materials* 6 https://pdfs.semanticscholar.org/f278/3f28e85b22aa2f8bdfa3b1dc7ab342d81479.pdf

[7] L Dombrovsky, B Rousseau, P Echegut, D Baillis 2011 *Journal of the American Ceramic Society* **94** (12) 4310 https://doi.org/10.1111/j.1551-2916.2011.04655.x

IOP Conf. Series: Materials Science and Engineering **613** (2019) 012050 doi:10.1088/1757-899X/613/1/012050

The effect of short carbon fibers on rheological behaviour and mechanical properties of metakaolin-slag geopolymer binder

David Rieger[1], Michal Pola[1], Petr Franče[1], Jaroslav Kadlec[1], Tomáš Kovářík[1] and Pavel Kopačka[2]

[1] New Technologies - Research Centre, University of West Bohemia, Univerzitni 8, 306 14 Pilsen, Czech Republic

[2] Diamorph Hob CerTec s.r.o., Tovární 739, 330 12 Horní Bříza, Czech Republic

Email: DavidPJ@ntc.zcu.cz

Abstract. This study investigates the effects of short carbon fibers addition on rheological behaviour of geopolymer binder and final mechanical strength of prepared composite. The pure geopolymer binder and variously carbon fiber reinforced composites were tested. The pure geopolymer binder was synthesized by alkalination of calcined claystone powder and milled blast furnace slag by potassium silicate solution. The binders reinforced by carbon fibers were prepared under the same conditions. Mixtures were accordingly reinforced by 20 wt% of short and milled fibers, respectively. The effect of fibers length (3 mm and $\leq$ 0.5 mm) was studied with regard to rheological behaviour of prepared mixtures and mechanical properties of final composites. The rheological properties were determined in accordance to flow properties by rotational rheometry. The final properties were determined by measurements of flexural strength after 3, 28 and 280 days and by means of optical and scanning electron microscopy. The results indicate that milled carbon fibers had a slight effect on flow properties compared to pure binder. In contrast, 3 mm long carbon fibers significantly increased viscosity and noticeably decreased workability by interlocking mechanism of individual fibers. Moreover, it was proved that addition of milled fibers had a low reinforcement effect on composite samples and led only to slight improvement of stress-strain behaviour. On the contrary, the reinforcement by 3 mm carbon fibers improved the stress-strain properties and significantly increase ultimate flexural strength (approximately 5 times) compared to pure geopolymer binder in all studied time intervals.

1. Introduction

Inorganic polymer cements can be synthesized by alkali-activation of a variety of materials including thermally activated clays and coal fly ashes to produce material with mechanical and thermal properties suitable for wide range of industrial applications [1, 2]. Davidovits entitled this type of materials as "geopolymers" because of polymer like chemical structure and introduced pioneering work on alkali-activated binders based on calcined clays [3] and produce sufficient mechanical properties [4]. Geopolymers are products of chemical reaction between aluminosilicate material and liquid alkaline environment where chemical cleavage of the Si-O and Al-O bonds in parent material leads to saturation of liquid solution and subsequent polycondensation of amorphous aluminosilicate matrix [5-7]. In some cases slag is used as an additive for improvement of setting time and mechanical strength [8, 9].

However, by ceramic-like nature have geopolymeric binders brittle character of fracture [10]. Using of fibers can change the character of fracture into quasi-brittle and prevent the crack propagation in material. Wide range of fibers is used for this purpose from organic natural and synthetic materials [10-12], mineral materials as basalt, glass, zirconia and others to high performance carbon-based materials [13-16]. Fiber reinforcement of inorganic binders is mainly used in form of short fibers for casting processes or as woven and non-woven composites prepared by impregnation.

This experimental research focuses on rheological and mechanical properties of geopolymeric matrix reinforced by carbon fibers of different length. The binder comprises of calcined clay stone powder and milled blast furnace slag. Carbon fibers in length 3 mm and ≤ 0.5 mm were applied in amount of 20 wt% related to powder part and the effect on flow properties and final mechanical strength was evaluated.

2. Materials

The mixture of calcined claystone powder and milled blast furnace slag (CCS) was used as a binder and the morphology can be seen in Figure 1. The slag part of binder is represented mainly by bigger particles, consisting of amorphous glassy material with minor accessory minerals merwinite and akermenite. Calcined claystone part consists almost completely of reactive metakaolin with trace amount of accessory minerals quartz and mullite. The granularity, with characteristic volumetric values d10, d50 and d90 of 0.37, 5.03 and 11.37 μm respectively, was determined by laser light scattering technique. For alkaline activation of geopolymerization reaction an aqueous solution of potassium silicate with silicate module 1.61 was used. These materials were supplied by České lupkové závody a. s, Czech Republic and their chemical composition is presented in Table 1. The reactivity of this binder was studied in details in references [16, 17].

The short carbon fibers with initial length of 3 mm and diameter of 14 μm, provided by GRM Systems s.r.o., Czech Republic, were used and marked as "sCF". The milling process of fibers was conducted in laboratory vibrational mill for 30 seconds and resulted fibers with final length below 0.5 mm were marked as "mCF". Images of used carbon fibers, before and after milling process, can be seen in Figure 2 and 3 respectively.

Table 1. Chemical composition of raw materials [mass %]

Composition	SiO_2	Al_2O_3	CaO	MgO	TiO_2	Fe_2O_3	K_2O	LoI	Others
CCS	46.37	28.72	15.46	3.40	2.01	0.93	0.56	1.92	0.63
Activator	17.60	-	-	-	-	-	17.13	65.27	-

Figure 1. SEM image of CCS binder.

Figure 2. SEM image of short carbon fibers.

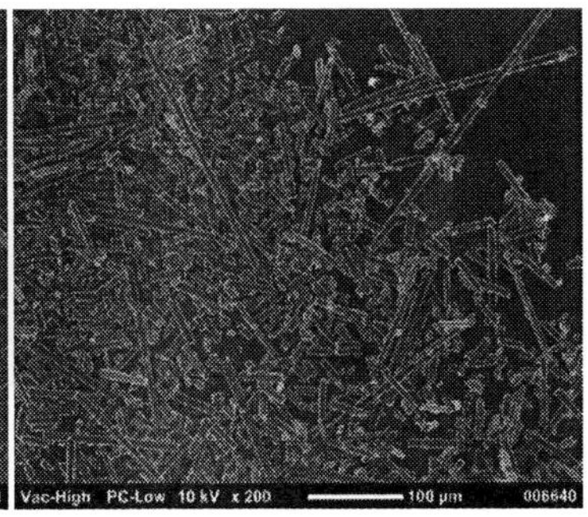

Figure 3. SEM image of carbon fibers after milling process.

3. Procedures and results

3.1. Preparation of binder and composites

Binder pastes were prepared by mixing of 100 weight parts of binder powder together with 100 weight parts of activator solution. Subsequently, 20 wt% of fibers was added into the paste and mixed together

IOP Conf. Series: Materials Science and Engineering **613** (2019) 012050 doi:10.1088/1757-899X/613/1/012050

for 5 minutes in laboratory vacuum mixer. Pastes were casted into moulds with dimensions of 120x20x20 mm. After 2 days were specimens unmoulded and aged for defined time in polypropylene bags.

3.2. Flow properties

The rheological properties were determined in accordance to flow properties by rotational rheometry on the rheometer TA Instruments Ares G2 in plane-plate geometry of 40 mm in diameter. Measurements were performed after 15 minutes from beginning of mixing. The difference in stress behaviour according to shear rate can be clearly seen in Figure 4 and the sCF highly increases the viscosity and reduces workability.

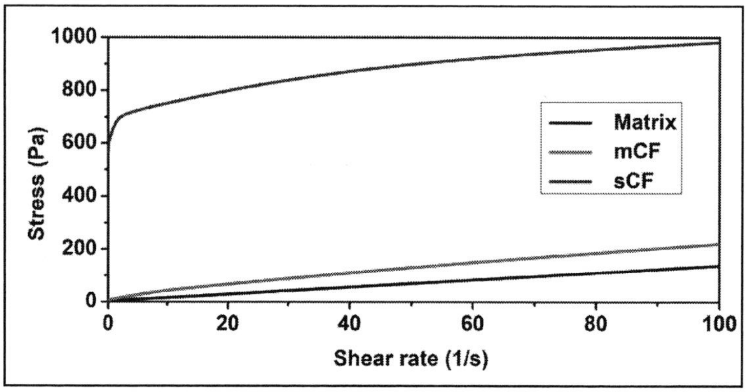

Figure 4. Stress-shear rate behavior of pure matrix and matrix with short and milled carbon fibers.

3.3. Flexural properties and microstructure

The flexural strengths of individual compositions were measured after 3, 28 and 280 days respectively on 20x20x120 mm samples. How it can be seen in Figure 5 that the effect of carbon fibers is remarkable in both cases and the flexural strength is affected by the length of fibers.

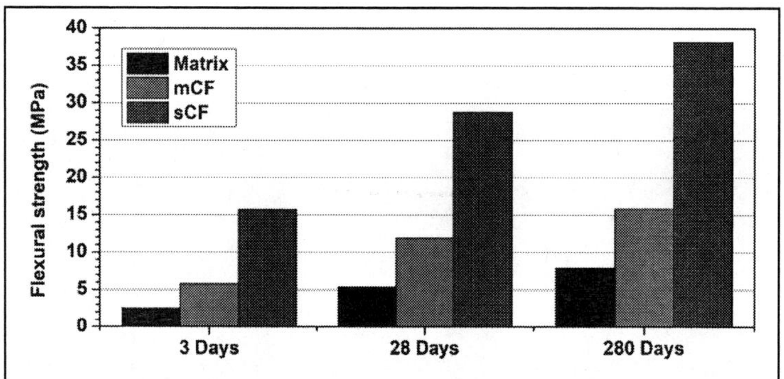

Figure 5. Flexural strength of individual compositions measured in time intervals of 3, 28 and 280 days.

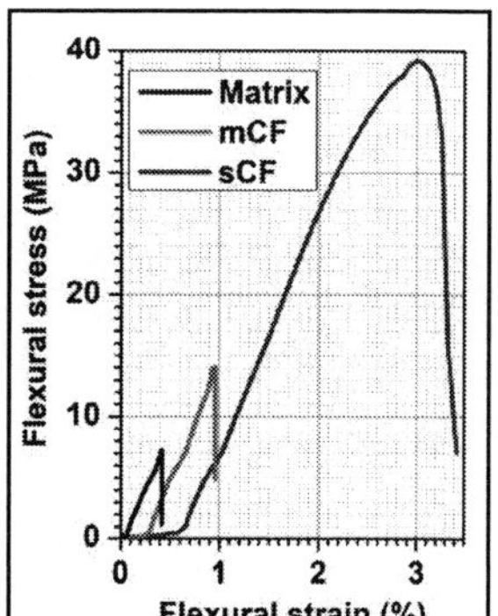

Figure 6. Stress-strain behaviour of individual compositions after 280 days.

Addition of mCF remarkably increases flexural strength in contrast to pure Matrix. From stress-strain behaviour can be concluded that in the case of mCF this effect is mainly by bridging of initial short cracks as it can be seen on stress-strain curve in Figure 6. When the ultimate flexural stress is reached the stress sharply decreases without bridging effect and in this case the mCF behaves more like the particle than fiber filler. The effect of sCF is different in stress-strain behaviour and further increase of ultimate stress can be seen. Moreover, effect of longer fibers and the reinforcement by the pull-out mechanism can be assumed from residual stress. This assumption is further supported by analysis of microstructure.

The microstructure of fracture surfaces was evaluated by scanning electron microscopy and by optical microscopy in bright field (BF). The SEM images are presented in Figures 7-9. The Figure 7 represents the structure of pure geopolymer binder. In Figure 8 the presence of carbon fibers with length of tens of micrometers can be clearly seen with residual imprints in geopolymer matrix. The observed length corresponds to stress-strain behaviour and explains the negligible bridging effect. In the case of sCF presented in Figure 9 the short carbon fibers can be seen going across the field of view without breakage what confirms the pull-out mechanism of reinforcement. From the microstructure of imprints in mCF and sCF can be concluded that the wetting of carbon fibers by this geopolymer binder is sufficient. This situation is further documented in Figures 10-12 taken by optical microscopy in BF where the contrast between light geopolymer matrix and dark carbon fibers is better and more intensive.

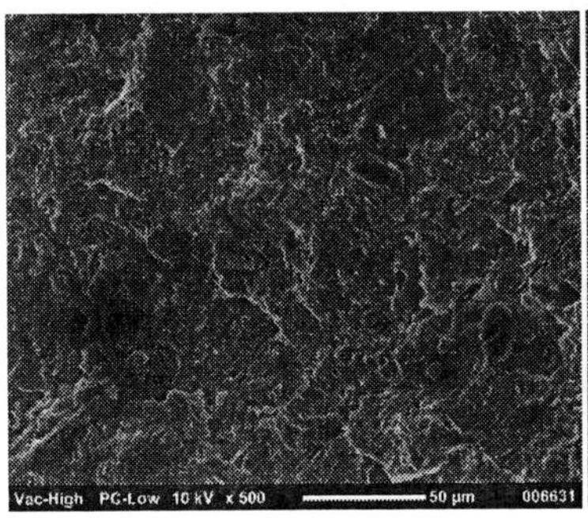

Figure 7. SEM image of pure Matrix.

Figure 8. SEM image of matrix with mCF.

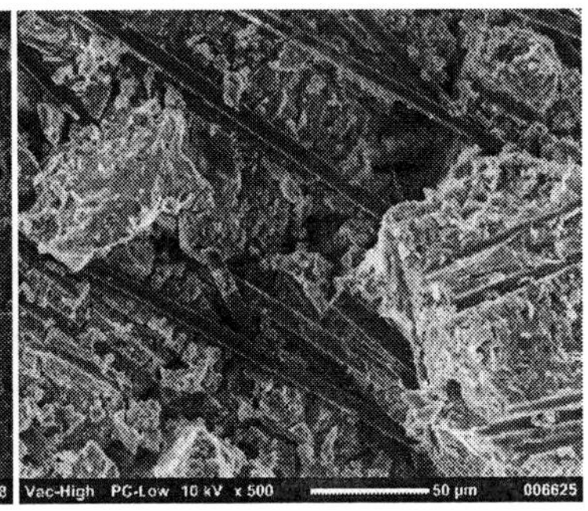

Figure 9. SEM image of matrix with sCF.

Figure 10. Image of Matrix in BF.

Figure 11. Image of matrix with mCF in BF.

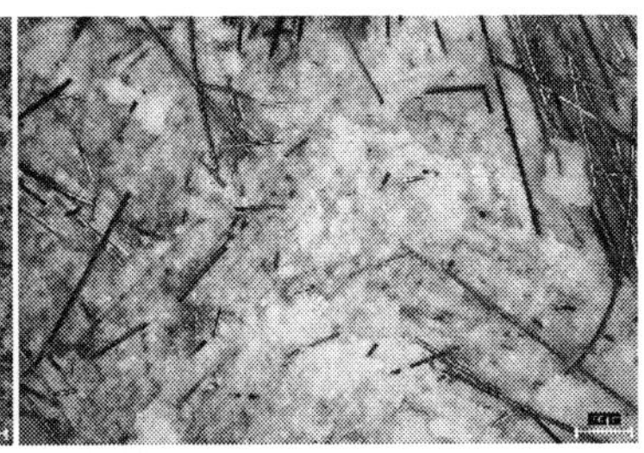

Figure 12. Image of matrix with sCF in BF.

4. Conclusions

In this study the effect of carbon fiber addition on flow properties of prepared suspensions and stress-strain behavior was investigated. It can be concluded that the length of carbon fibers play important role in flexural strength, flow properties and workability. The addition of carbon fibers raises both the ultimate flexural strength and viscosity. Both these effects are more significant with increasing of fibers length. Addition of sCF produces good composite material with residual flexural strength by bridging effect of fibers and with very good flexural strength of 15.7, 28.8 and 38.1 MPa after 3, 28 and 280 day respectively. It is in contrast to values of 5.8, 11.9 and 15.8 MPa for mCF composite and 2.6, 5.4 and 7.2 MPa in the case of pure Matrix. On the other hand, these remarkable properties are on the cost of decreased fluidity and workability. From this study can be moreover concluded that the wetting of carbon fibers by studied geopolymer binder is good and sufficient and this effect can be assigned to surface properties of used potassium silicate solution.

Acknowledgments

The result was developed within the CENTEM project, reg. no. CZ.1.05/2.1.00/03.0088, cofounded by the ERDF as part of the Ministry of Education, Youth and Sports OP RDI programme and, in the follow-up sustainability stage, supported through CENTEM PLUS (LO1402) under the NSP I.

References

[1] Duxson P, Provis J L, Lukey G C and Van Deventer J S J 2007 *Cem Concr Res* **37** 1590 https://doi.org/10.1016/j.cemconres.2007.08.018
[2] Rieger D, Kullová L, Čekalová M, Novotný P, Pola M 2017 *J Phys Conf Ser* **790** 012030 https://doi.org/10.1088/1742-6596/790/1/012030
[3] Davidovits J 1991 *J Therm Anal* **37** 1633–56, https://doi.org/10.1007/BF01912193
[4] Rieger D, Pola M, Franče P, Kadlec J, Kovářík T 2018 *J Phys Conf Ser* **1045** 012037 https://doi.org/10.1088/1742-6596/1045/1/012037
[5] Kouassi S S, Tognonvi M T, Soro J and Rossignol S 2011 *J Non-Cryst Solids* **357** 3013 https://doi.org/10.1016/j.jnoncrysol.2011.04.006
[6] Shi C, Fernández-Jiménez A, Palomo A 2011 *Cem Concr Res* **41** 750 https://doi.org/10.1016/j.cemconres.2011.03.016
[7] Šesták J and Foller B 2012 *J Term Anal Calorim* **109** 1 https://doi.org/10.1007/s10973-011-1625-3
[8] Yip Ch K, Lukey G C, Provis J L, Van Deventer J S J 2008 *Cem Concr Res* **38** 554 https://doi.org/10.1016/j.cemconres.2007.11.001
[9] Kadlec J, Rieger D, Kovářik T, Novotný P, Franče P, Pola M 2017 *IOP Conf Ser Mater Sci Eng* **175** 012052 https://doi.org/10.1088/1757-899X/175/1/012052
[10] Toutanji H, Xu B, Gilbert J, Lavin T 2010 *Const Build Mater* **24** 1 https://doi.org/10.1016/j.conbuildmat.2009.08.023
[11] Yan L, Kasal B, Huang L 2016 *Composites Part B* **92** 94 https://doi.org/10.1016/j.compositesb.2016.02.002
[12] Alzeer M, MacKenzie K 2013 *Appl Clay Sci* **75–76** 148 https://doi.org/10.1016/j.clay.2013.03.010
[13] Peigang He SY, Jia D, Yang Z, Duan X, Wang S, Zhou Y 2016 *Ceram Int* **42** 7837 https://doi.org/10.1016/j.ceramint.2016.01.197
[14] Rieger D, Kadlec J, Pola M, Kovářik T, Franče P 2017 *IOP Conf Ser Mater Sci Eng* **175** 012054 https://doi.org/10.1088/1757-899X/175/1/012054
[15] Pernica D, Reis P N B, Ferreira J A M et al. 2010 *J Mater Sci* **45** 744 https://doi.org/10.1007/s10853-009-3994-6
[16] Rieger D, Kovářík T, Pola M, Franče P, Kadlec J 2018 *J Phys Conf Ser* **1045** 012036 https://doi.org/10.1088/1742-6596/1045/1/012036
[17] Rieger D, Kullová L, Čekalová M, Kovářík T 2017 *J Phys Conf Ser* **790** 012029 https://doi.org/10.1088/1742-6596/790/1/012029

IOP Conf. Series: Materials Science and Engineering **613** (2019) 012051 doi:10.1088/1757-899X/613/1/012051

Plasticity of Red Mud and Clay Mixtures

A Hamza[1], I Kocserha[2*], R Géber[3], A Buzimov[4]

[1,2,3] Institute of Ceramic and Polymer Engineering,
University of Miskolc, Miskolc, Hungary
[4] Institute of Strength Physics and Materials Science SB RAS, Tomsk, Russia

*E-mail: istvan.kocserha@uni-miskolc.hu

Abstract. Recent study is focusing on the plasticity properties of clay and red mud (RM) mixtures. Two different clays were examined with (carbonate clay, CC) and without earth alkali carbonate (carbonate free clay, CFC) content. Red mud with 32 wt% moisture content was added for dried and powdered clays. Maximum applicable red mud content (30 wt%) was determined by mixing experiment. The mixtures were prepared by pan mill and samples of 33 mm in diameter and 40 mm in height were pressed. Measurements were obtained with 20, 22, 24 and 26 wt% moisture content in each mixture. Two methods were used for measurement and characterization of the plasticity of clay body. Beside determination of Pfefferkorn plasticity index, deformation curves of different samples were also recorded and analysed. Results indicated that red mud affected the forming behaviour of clays. Addition of red mud may rise the energy consumption of extrusion process.

1. Introduction

Red mud (RM) is highly alkaline by-product of Bayer-process. As a huge volume can be found worldwide of this material, utilization of it generated many research works. Based on lots of scientific studies on several fields of application, RM may be considered as secondary raw material. It seems to be useable in several branch in ceramic and silicate industry, like cement production, glass ceramics, ceramic glazes, concrete and mortar production or geopolimerization. Among them the heavy-clay ceramic industry seems to be one of the possible fields, where huge volume of RM may be consumed [1-17]. The re-use of RM is possible not only in powdered form [5-15], but it can be added to clay in plastic consistency, as well [18] In latter, there is a restriction in water content of red mud. If it is too high, red mud became very sticky and its dosage became impossible. In deposits, red mud usually has high water content, however its surface level could partly get dry. Dewatering technique are required, like filter press [19] in order to decrease water content to 32-35 wt%. Mixing red mud into clay is only possible under this water content value [18]. For extrusion technique in heavy clay industry, clay mixtures should have sufficient plasticity. Clay minerals, additive materials, water content, particle size distribution effects on the plasticity of a clay or a clay mixture – generally on clay-water system [20,21]. Several practical tests have been developed to determine and describe plastic behaviour with certain figures or indices like Pfefferkorn, Atterberg or cone peneteration methods. Beside them, more precise, but time-consuming techniques, like capillary rheometry are also used, which resulted the total flow curve of a plastic material [22,23]. As strength, porosity or wear resistance would be the main behaviour of ceramic materials [24-28], former studies focused mainly on the analysis of products based on raw materials incorporated with red mud [5-15]. The effect of red mud addition on the plasticity of brick clay mixtures is not fully known [29]. As red mud has paste consistency and it consist of mainly fine particles, it would be expected that it behaves like plastic body. However, it contains a small amount of clay minerals, so the flow behaviour and the amount of adsorbed water are caused by the high surface area and not the ions as it is in case of the clays and silts [30,31]. This would be the base for that a rheological analysis of red mud indicated its thixotropic behaviour, so its viscosity changes during forming. Adding RM to clay, plasticity changes, which affects the operation and energy demand of shaping technology. The aim of this study is to examine and describe the effects of RM on the plasticity with the help of the two different measuring method.

2. Materials and Methods

2.1 Raw materials

For the purpose of examining clay and red mud mixtures, three clays were obtained from two Hungarian brickworks. Two of them consist considerable weight percentage of carbonate, while third clay is carbonate free. The clays were named as follows: Grey clay – carbonate clay: GCC; Yellow clay – carbonate clay: YCC and the carbonate free clay: CFC. The red mud (RM) was obtained from the upper surface of an old Hungarian dump. Before the preparation of the mixtures, mineral composition of the raw materials was analysed by X-ray Powder Diffraction (Rigaku Miniflex II, Cu Kα, 2θ range from 3 to 90°). The quantitative results were calculated by Rietveld full-profile refinement analysis. The XRD patterns and the results are presented in Figure 1. and Table 1. respectively.

Table 1. Mineral composition of raw materials

Raw.mat.	Q	I	M	A	SM	G	K	M	C	D	CH	AM	WC
GCC	43.2	13.7	5.3	6.7	4.5	1.1	0.3	6.9	6.3	11.1	0.8	8	12.7
YCC	29.7	25.1	6.8	3.2	7.1	0.5	6.1	1.9	13.4	3.6	2.7	0	14.3
CFC	41.1	24.2	5.0	5.5	6.1	3.3	6.7	4.4	-	-	2.3	10	12.5
	Q	GI	C	H	CA	G	KI						
RM	1.0	7.2	5.5	44.2	32.9	4.5	4.7						

(Q:quartz; I:illite; M:muscovite; SM. smectite; A: albite; G: goethite; K:kaolinite; M:microcline; C:calcite; D:dolomite; CH: chabazite; AM:amorphous content; WC: water content; H:hematite; GI: gibbsite; CA: cancrinite; KI: kimezeyite)

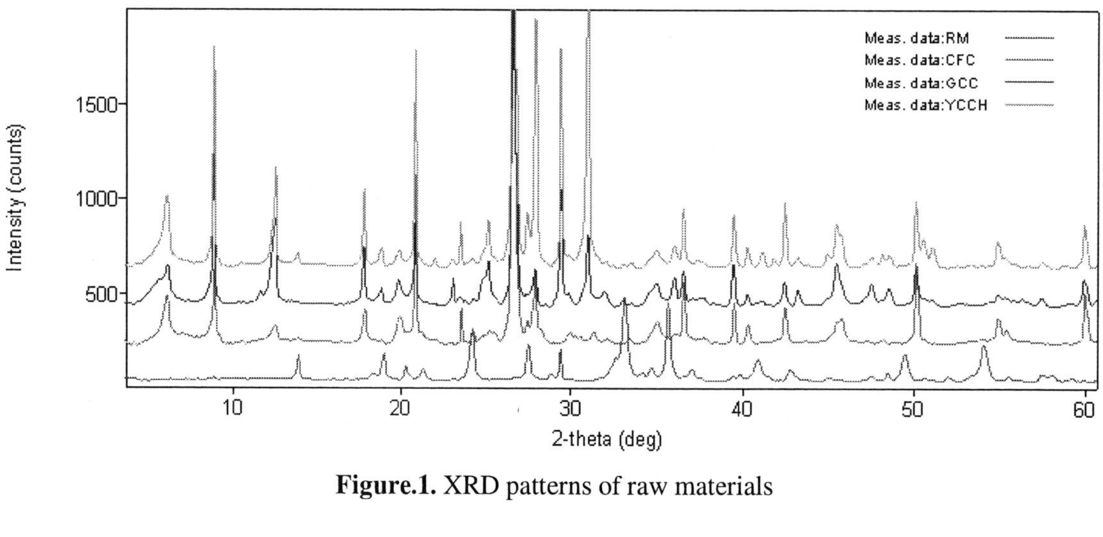

Figure.1. XRD patterns of raw materials

2.2 Mixing and sample preparation

An industrial clay mixture was prepared by mixing GCC and YCC and used as carbonate clays (CC). RM with 32 wt% water content was added for dried and powdered clays. Maximum applicable RM content (30 wt%) was determined by mixing experiment [18]. After that mixtures were prepared by pan mill, cylindrical samples of 33 mm in diameter and 40 mm in height were manually pressed for both measurement method. The moisture contents of mixtures were set to between 20 and 28 wt% in order to get more point to distinguish from each other. Table 2. indicated the composition of mixtures. The mixtures were marked as follows: code for clay (CC or CFC) + RM content - e.g. CC10 means carbonate clay with 10wt% RM content.

Table 2. Composition of prepared mixtures in wt%

Carbonate clay (CC)/RM	100/0	90/10	80/20	70/30
Carbonate free clay (CFC)/ RM	100/0	90/10	80/20	70/30

2.3 Testing methods for plasticity

In order to describe the effects of red mud on the clays, numerical values of plasticity (or formability) were determined by applying two methods. For both techniques, three samples were used and average value were represented on figures as a results. Overall 48 samples were prepared for each test, separately.

Pfefferkorn method

Pfefferkorn test was the first method for measuring of plasticity. This is a simple practical method and it is based on the principle of impact deformation. Pfefferkorn's plasticity number or index shows the percentage water content, at which the sample is pressed to 1/307 ratio. A defined sample with a diameter of 33 mm and an initial height of 40 mm, is deformed by a freefalling plate with a mass of 1.192 kg. The initial height is related to the impact deformation, the result is the ratio of deformation. The ratios of deformations or the impact deformation heights (H_0, initial height; H_f, final height) are plotted against the water content. [20,21]

Compression test

Compression test was also applied on cylindrical samples to evaluate the plasticity of clay and RM mixtures. Samples were compressed with deformation rate of 60 mm/min on an Instron 5570 Universal Material Tester and stress-deformation curves were recorded (96 curves). Similar to metals, yield stress was obtained from the curves according to Figure. 2. Stress value was the basis of comparison. [20]

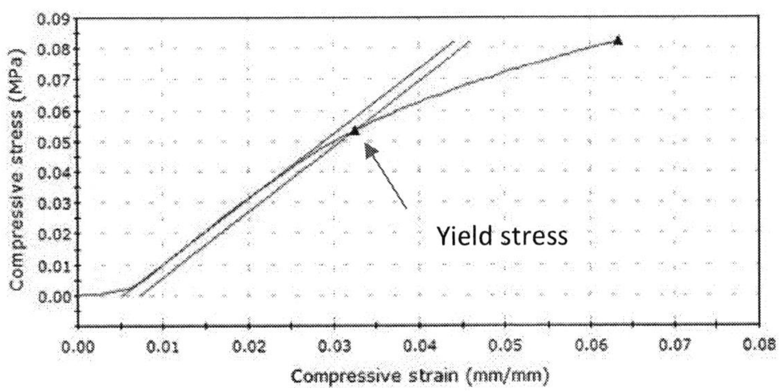

Figure.2. Determination of yield stress from stress-strain curve

3. Results and discussion

3.1. Pfefferkorn plasticity

Plasticity curves according to Pfefferkorn can be seen on Figure.3 and Figure 5. for CC and CFC clay and RM mixtures. Figure. 4 and Figure 6 show an example for deformation of samples with different water content. Plasticity indices and determination coefficients obtained from curves were summarized in Table 2.

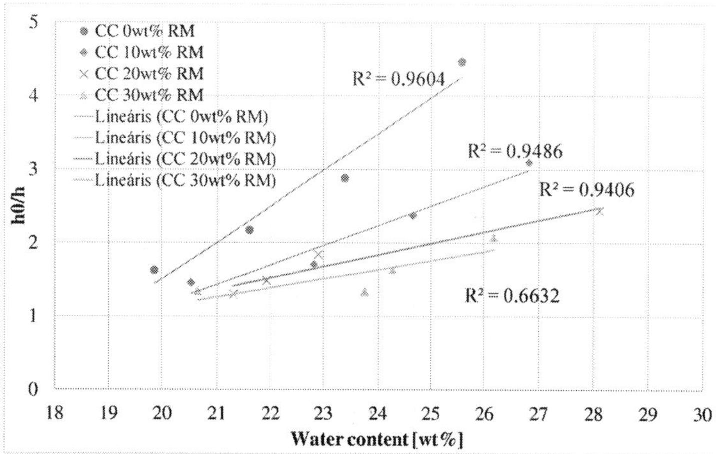

Figure 3. Pfefferkorn plasticity curves for CC mixes

Figure 4. CC mix samples with 30 wt% RM content

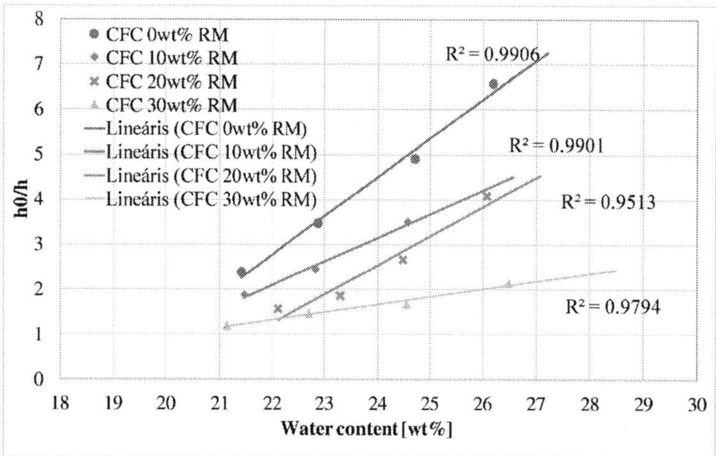

Figure5. Pfefferkorn plasticity curves for CFC mixes

Figure 6. CFC samples with 30 wt% RM content

Table 3. Pfefferkorn plasticity index for CC and CFC clay mixtures

Mixture	RM [wt%]	R^2	m	b	Pfefferkorn index	Mixture	R^2	m	b	Pfefferkorn index
CC0	0	0.96	1.94	17.19	23.49	CFC0	0.99	1.15	18.84	22.57
CC10	10	0.95	3.54	16.06	27.57	CFC10	0.99	1.89	18.05	24.19
CC20	20	0.94	5.94	13.05	32.36	CFC20	0.95	1.47	20.27	25.04
CC30	30	0.66	5.32	15.23	32.50	CFC30	0.98	5.70	14.53	33.05

The steeper the curve, the "shorter" the body, i.e. the more intensely the body will react to variations of the water content [21]. Regarding plasticity index of clays without RM, CFC had higher plasticity, than CC i. e. less water was enough to reach to 30% contraction. Adding RM to clay, plasticity index increased for both clays. Increasing the amount of RM from 0% to 30% resulted an increase in index value by 18% for CC and by 32% for CFC mixtures. Results show that the use of RM as clay replacement material made clay mixtures harder, energy demand for processing would be higher.

3.2. Compression test and yield stress determination

Compression test was performed on all mixtures with different RM and water content. Figure 7. and Figure 8. present two typical stress-displacement curves for CC and CFC with 22 w% water content. According to Figure. 2., yield stress values obtained from curves and their average values were illustrated on Figure 9. and Figure 10. as function of RM content. Numerical values were summarized in Table 4.

Analyzing Figure 7. and Figure 8, it can be seen, that higher RM content means steeper curve, so clay-RM mixtures become stiffer. Forming of mixtures with RM require more force and higher energy consumption. This tendency was the same for all CC and CFC compounds.

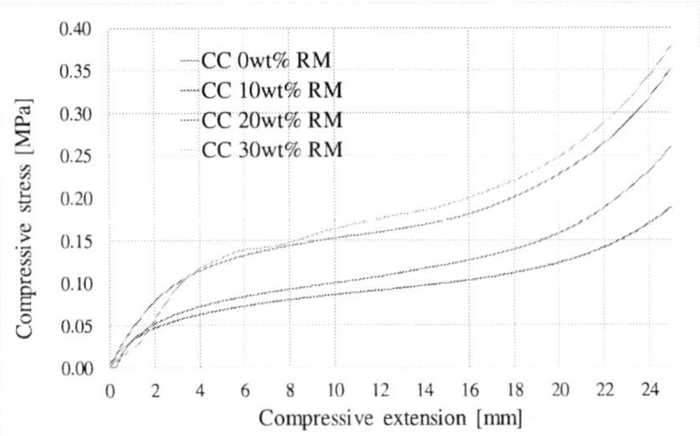

Figure 7. Stress-deformation curves of CC and RM mixes (22 wt% water content)

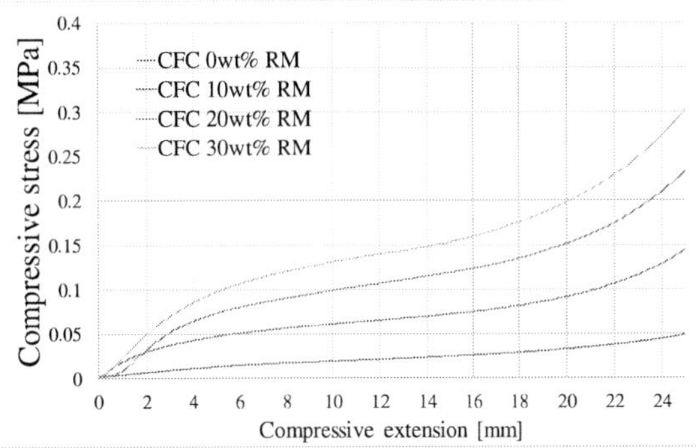

Figure 8. Stress-deformation curves of CFC and RM mixes (22 wt% water content)

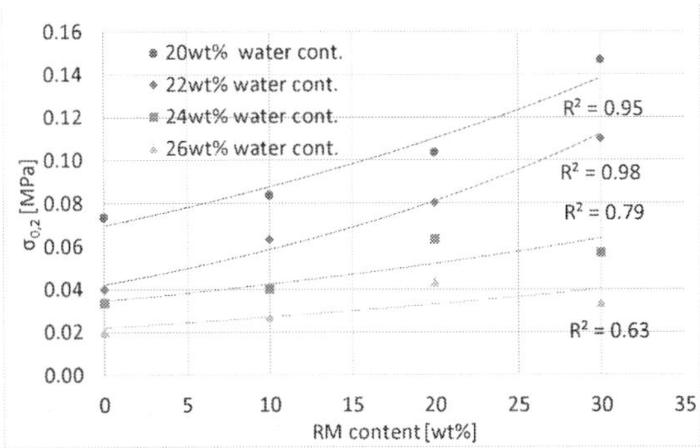

Figure 9. Yield stress as function of RM content for CC

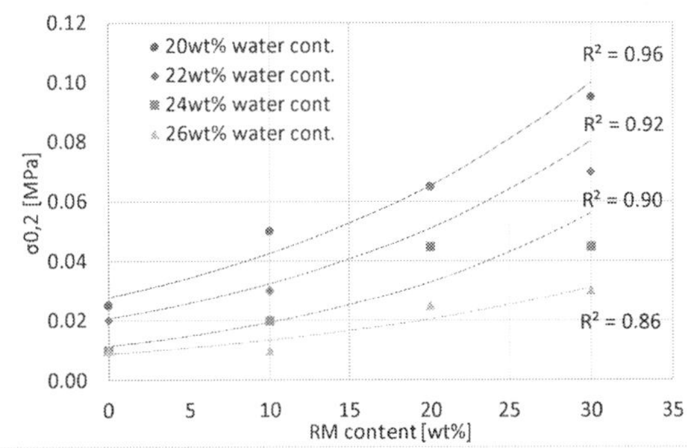

Figure 10. Yield stress as function of RM content for CFC

Table 4. Yield stresses of clay-RM compounds

	Mixture name							
Water content	CC0	CC10	CC20	CC30	CFC0	CFC10	CFC20	CFC30
[wt%]	Yield stress $\sigma_{p0,2}$ [MPa]							
20	0,073	0,083	0,103	0,147	0,025	0,050	0,065	0,095
22	0,040	0,063	0,080	0,110	0,020	0,030	0,065	0,070
24	0,033	0,040	0,063	0,057	0,010	0,020	0,045	0,045
26	0,020	0,027	0,043	0,033	0,010	0,010	0,025	0,030

4.Conclusion

Plasticity of mixtures of carbonate and carbonate free clays with RM were analyzed by Pfefferkorn and compression test methods. Results obtained from both methods proved that RM decreases the plasticity of clay mixtures. Same water content, but lower plasticity means higher forming force and higher energy consumption in processing and forming technology. According to these and former [18] results, the usage of RM in heavy clay industry is possible without expensive drying and grinding of RM up to 30 wt%. However, in heavy clay industry if clay as mined is too 'fat', opening agents (usually natural sand) are added to the mixture in order to decrease the plasticity. Taking into account environmental issues, beside incorporating in clay, primer raw materials, like sand can be replaced with RM.

5. Acknowledgement

The described article was carried out as part of the Sustainable Raw Material Management Thematic Network – RING 2017, EFOP-3.6.2-16-2017-00010 project in the framework of the Széchenyi2020 Program. The realization of this project is supported by the European Union, co-financed by the European Social Fund

6. References

[1] Y Pontikes, P Nikolopoulos and G N Angelopoulos 2007 *Journal of the European Ceramic Society* **27** 1645https://doi.org/10.1016/j.jeurceramsoc.2006.05.067

[2] V M Sglavo, S Maurina, A Conci, A Salviati, G Carturan and G Cocco 1999 *Journal of the European Ceramic Society* **20** 245 https://doi.org/10.1016/S0955-2219(99)00156-9

[3] L Pérez-Villarejo, F A Corpas-Iglesias, S Martínez-Martínez, R Artiaga, J Pascual-Cosp 2012 *Construction and Building Materials* **35** 656 https://doi.org/10.1016/j.conbuildmat.2012.04.133

[4] R K Paramguru, P C Rath and V N Misra 2005 *Mineral Processing & Extractive Metall. Rev.* **26** 1 https://doi.org/10.1080/08827500490477603

[5] W Liu, J Yang and B Xiao 2009 *Int. J. Miner. Process.* **93** 220 https://doi.org/10.1016/j.minpro.2009.08.005

[6] C Klauber, M Gräfe and G Power 2011 *Hydrometallurgy* 108 11 https://doi.org/10.1016/j.hydromet.2011.02.007

[7] H He, Q Yue, Y Su, B Gao, Y Gao, J Wang and H Yu 2011 *Journal of Hazardous Materials* **213-214** 53 https://doi.org/10.1016/j.jhazmat.2011.11.095

[8] H He, Q Yue, Y Qi, B Gao, Y Zhao, H Yu, J Li, Q Li and Y Wang 2012 *Applied Clay Science* **70** 67 https://doi.org/10.1016/j.clay.2012.09.022

[9] S Pişkin, A K Figen, E Özkan and Ü Özçay 2013 *IFAC Proceedings* **46** 484 https://doi.org/10.3182/20130825-4-US-2038.00010

[10] M P Babisk, T P Altoé, H J O Lopes, U S Prado, M C B Gadioli, S N Monteiro and C M F Vieira 2014 *Materials, Science Forum* ***798-799*** 509 https://doi.org/10.4028/www.scientific.net/MSF.798-799.509

[11] S Liu, X Guan, S Zhang, C Xu, H Li and J Zhang 2016 *Materials Letters* **191** 222 https://doi.org/10.1016/j.matlet.2016.12.028

[12] C Scribot, W Maherzi, M Benzerzour, Y Mamindy-Pajany and N E Abriak 2018 *Construction and Building Materials* **163** 21 https://doi.org/10.1016/j.conbuildmat.2017.12.092

[13] P E Tsakiridis, S Agatzini-Leonardou, and P Oustadakis 2004 *Journal of Hazardous Materials* **B116** 103 https://doi.org/10.1016/j.jhazmat.2004.08.002

[14] L Senff, D Hotza, J A Labrincha 2016 *Construction and Building Materials* **25** 163 https://doi.org/10.1016/j.conbuildmat.2010.06.043

[15] R Liu, Chi Poon 2016 *Journal of Cleaner Production* **112** 384 https://doi.org/10.1016/j.jclepro.2015.09.049

[16] Z Sas, J Somlai, G Szeiler and T Kovács 2015 *Journal of Radioanalytical and Nuclear Chemistry* **306** 271 https://doi.org/10.1007/s10967-015-3966-z

[17] Z Sas, J Szántó, J Kovács, J Somlai and T Kovács 2015 *Journal of Environmental Radioactivity* **148** 27 https://doi.org/10.1016/j.jenvrad.2015.06.002

[18] I Kocserha, A Hamza, R Géber 2018, *IOP Conference Series: Materials Science and engineering* **426** 012026 https://doi.org/10.1088/1757-899X/426/1/012026

[19] S Arslan, G K Demir, B Celikel, M Baygul, C E Suarez 2012 *Light Metals* 81 https://doi.org/10.1007/978-3-319-48179-1_15

[20] L A Gömze, S N Kulkov, E Kurovics, A S Buyakov, S P Buyakova, A Y Buzimov, R Géber, M V Grigoriev, I Kocserha, A S Kulkov, T Yu. Sablina, N L Savchenko, I N Sevostyanova, A Simon 2018 *Építőanyag*-JSBCM **70** (1) 8 https://doi.org/10.14382/epitoanyag-jsbcm.2018.2

[21] L A Gömze, S N Kulkov, E Kurovics, A S Buyakov, A Y Buzimov, M V Grigoriev, B I Kanev, T V Kolmakova, R V Levkova, R V Levkov, S A Sitkevich 2018 *Építőanyag*-JSBCM **70** (1) 13 https://doi.org/10.14382/epitoanyag-jsbcm.2018.3

[22] F A Andrade, H A Al-Qureshi, D Hotza 2010 *Applied Clay Science* **51** 1 https://doi.org/10.1016/j.clay.2010.10.028

[23] K Göhlert, M Uebel 2007. Rheology of Ceramic Bodies. In: Haendle, F., et al. (Ed.), *Extrusion in Ceramics.* Springer 381 https://doi.org/10.1007/978-3-540-27102-4_21

[24] E S Kalatur, S P. Buyakova, S N. Kulkov, I Gotman, I Kocserha 2014 *Építőanyag - JSBCM* **66** (2) 31 http://dx.doi.org/10.14382/epitoanyag-jsbcm.2014.6

[25] N Savchenko, I Sevostyanova, T Sablina, L Gömze, S Kulkov 2014 *AIP Conference Proceedings* **1623** (1) 547 http://doi.org/10.1063/1.4899003

[26] S N Kulkov, S P Buyakova, L A Gömze 2014 *Építőanyag* - JSBCM **66** (1) 2 http://dx.doi.org/10.14382/epitoanyag-jsbcm.2014.1

[27] T Sadowski, L Marsavina 2011 *Computational Materials Science* **50** (4) 1336 https://doi.org/10.1016/j.commatsci.2010.04.011

[28] I Kocserha and L A Gömze 2010 *Applied Clay Science* **48** (3) 425 https://doi.org/10.1016/j.clay.2010.01.017

[29] Y Pontikes, G N Angelopoulos, U Kim, H Le, W Carty 2006 *Advances in Science and Technology* **45** 2240 https://doi.org/10.4028/www.scientific.net/AST.45.2240

[30] G Nagy 2014, *Pollack Periodica* **9** (2) 131 https://doi.org/10.1556/Pollack.9.2014.2.13

[31] O Kotova, Clay Minerals 2013 *IOP Conference Series: Materials Science and Engineering* **47** 012037 https://doi.org/10.1088/1757-899X/47/1/012037

IOP Conf. Series: Materials Science and Engineering **613** (2019) 012052 doi:10.1088/1757-899X/613/1/012052

Damage of thin-walled box beams with omega cross-section joined by adhesive layers subjected to bending deformation process

T Sadowski[1)*)], **M.Nowicki**[1)]

[1)]Department of Solid Mechanics, Lublin University of Technology, 20-618 Lublin, Poland

[*)]t.sadowski@pollub.pl

Abstract. The subject of this study is aluminum box beams of omega cross-section, consisting of an omega profile and connected to the metal by the adhesive interface of the area 120 cm^2. In the laboratory tests were used specimens made from 7075 aluminum with dimensions 300 x 84 x 36 mm. To connect the components of the specimens an adhesive layer Loctite 9545 was used.

The samples were subjected to 3-point bending (3-PB). Laboratory tests were conducted using a MTS (25kN) and a Digital Image Correlation (DIC) system, in order to accurately determine states of deformation of the box beams. With the load increase the profiles deform elastically at the beginning and then in the final stage of deformation process, a collapse of the side walls and local plastic deformation occured. This causes a sudden drop of the load capacity of the entire profile.

To complete the experimental analysis, the numerical model of the thin-walled box beam was created. The model allows for observation of stress and displacement fields variation in the bending element, as well as places of their concentration. Moreover, a detailed analysis of damage growth and the final destruction of the connecting elements associated with the load increase was performed. Numerical simulations were done using ABAQUS code. The FEM analysis results confirm correctness of the assumptions in the numerical model and allow for a more detailed description of the process decohesion adhesive layer and the damage of aluminium elements of the thin-walled box beams.

1. Problem formulation

Different connection types can be currently used in the joining of structural elements: purely mechanical, purely adhesive, e.g. [1-6] and hybrid joints, e.g. [7-14].

Effectiveness of the purely adhesive joints depends on geometry of connected structural elements, material applied and technology, e.g. [1–6, 15]. All the above include (e.g. [1-17]): (a) shape and dimensions of joined parts and lap areas, (b) surface preparation of lap areas, (c) thickness of the adhesive and adherends.

The strength of purely adhesive joints can be increased by means of several techniques, such as the shaping of geometric or material properties of the joint by: filleting of the adhesive layers, e.g. [18-23], introducing functional gradation of the adhesive properties, e.g. [24], varying the adherend shape via filleting or chamfering of the overlap edges of the adherends, e.g. [25-32], manufacturing of multistep-lap joints, e.g. [33].

IOP Conf. Series: Materials Science and Engineering **613** (2019) 012052 doi:10.1088/1757-899X/613/1/012052

Up to now, majority of research papers deals with reporting of different experiments and modeling of simple connections, like: single lap of double lap joints, etc. However, there exist more complex connections with the real 3-D shape, having thin-walled structures, e.g. [34].

Generally, the structural elements should have the smallest mass with the highest strength. In this context, the good examples are thin-walled metal parts, which are often used to form the support frames and stiffeners of many structures in different branches of modern engineering.

For the construction of thin-walled profiles a variety of solutions is used. They may be formed as single profiles through the extrusion process or can be joined by: welding, riveting or gluing. The last type of connecting elements is becoming increasingly popular, as cohesive connections are more and more common in technology.

In this study we present the new results of laboratory tests and the novel FEA model of laminated thin-walled aluminum profile with the omega cross section. The samples of the length 30 cm were made of 2 aluminum parts (an omega roll formed shape and a flat strip) connected with adhesive to estimate effectiveness of each joints, figure 1. The joined aluminum profiles by the adhesive Loctite 9545 were cured in high temperatures according to the manufacturer's requirements. The samples were further subjected to 3-PB deformation controlled process up to the final failure.

The paper removes lack of the FEA model in the literature for this kind of profile, e.g. [34]. The numerical model describes a cohesive behavior of the adhesive layer including damage process and the bending collapse mechanisms associated with a ductile damage model in the aluminium parts of the specimens. This allows for description of the longeron gradual degradation until the failure, including: (1) cracking inside of the aluminium parts and (2) delamination of the adhesive layers under increase of deformation.

2. Samples preparation

The thin-walled box beams (figure 1) with omega-cross sections consisted of 3 parts: (1) an omega profile made of aluminium 7075(T6), (2) a flat sheet made of aluminum and (3) 2 adhesive layers made Loctite Hysol 9514. Mechanical chracteristics of the aluminum 7075 are following: the Young's modulus - 69 GPa, the Poisson ratio - 0.327, the yield limit – 470 MPa, the ultimate strength – 600 MPa. Characteristics of the Loctite Hysol 9514 (according to [35]) are the following: the Young's modulus – 1.46 GPa, the tensile strength – 44 MPa, the lap shear strength – 45 MPa.

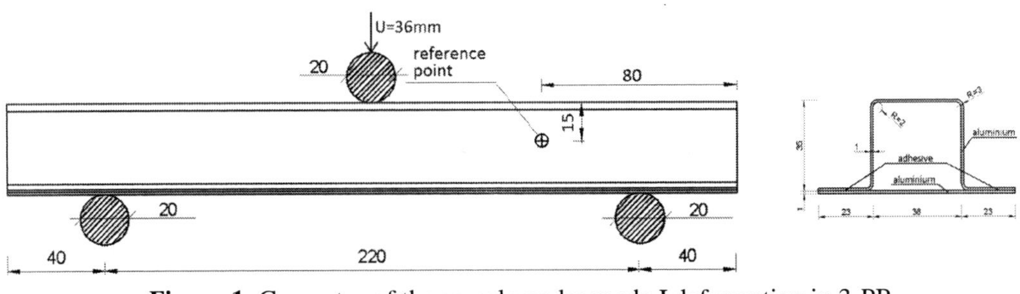

Figure 1. Geometry of the sample under mode I deformation in 3-PB.

3. Finite element numerical model

The numerical analysis of deformation processes were done with Abaqus Explicite. The material of the joined aluminum parts was described with a shell finite element (S4R) with plastic damage, while the bonding layers were described with surface-based cohesive contact.

Ductile Damage Model (DDM) of elasto-plastic materials for adherends was applied. Damage process initiates in the material as voids initiation and coalescence (e.g. [36 – 42]) at maximum stress.

The state of stress σ is the function of damage parameter D (e.g. [43-52]) and stress of the virgin material $\bar{\sigma}$ according to $\sigma = (1-D)\bar{\sigma}$, whereas the Young's modulus is described by $E = (1-D)E_0$.

E_0 is the initial elastic modulus. In the DDM the other parameter is used, i.e. G_f – a facture energy needed to open a unit area of crack in the aluminuim 7075.

To describe the mechanical response of the 2 adhesive layers (figure 1) the surface-based cohesive behaviour was applied. A negligibly small interface thickness was introduced between the two parts of the aluminium profiles and modelled using the traction-separation constitutive relation, e.g. [2, 3, 6, 53-58]. Damage initiation occurs when the maximum stress criterion is satisfied and crack propagation is related to the fracture energy G_c.

4. Experimental tests

The thin-walled box beams were subjected to 3-PB deformation controlled process under 2 modes with different placement of the cross-sections, figure 2. A speed of the cross-head was 0.036 mm/s and maximum deflection U_{max} = 36 mm. In order to obtain a precise monitoring of the whole deformation, the DIC system ARAMIS was used.

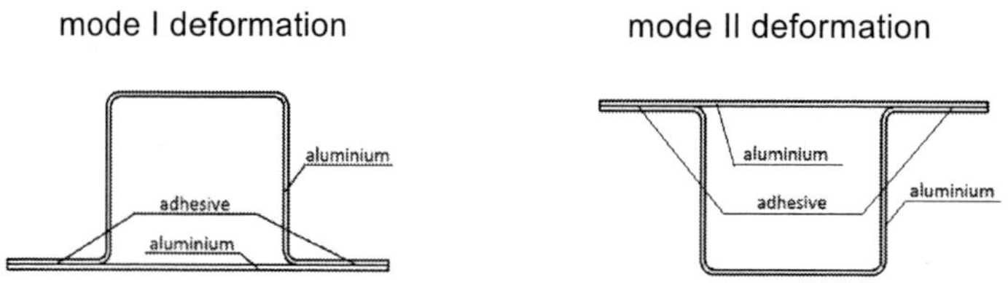

Figure 2. Placement of the cross-section in mode I and II during 3-PB deformation.

The results of experimental tests were shown in figure 3.

a) 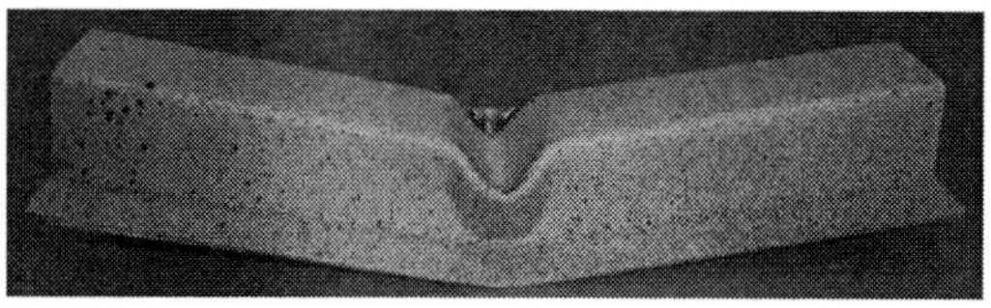 b) 

Figure 3. 3-PB modes of deformation for U_{max} = 36 mm: a) mode I, b) mode II.

After exceeding the yield strength in the aluminium one can observe breakdown of the side walls after initiation of the local plastic damage and further cracks propagation (related to G_f = 22.1 mJ) in the top part of the samples. The deformed shape of the damaged profile is characteristic for the bent thin-walled box beams – i.e. the plastic collapse mechanism, see [34, 59 - 61], whereas the sheet plate is elongated and bent. Delamination of samples depend on the mode of deformation. When the state of stress in deformation process of the adhesive layer reaches the critical value of the maximum stress (corresponding to G_c = 900 J/m^2, [62]) the bonding layer starts to damage in mode II. Finally, delamination between 2 parts of the samples takes place. In case of mode I this process do not take place.

a) 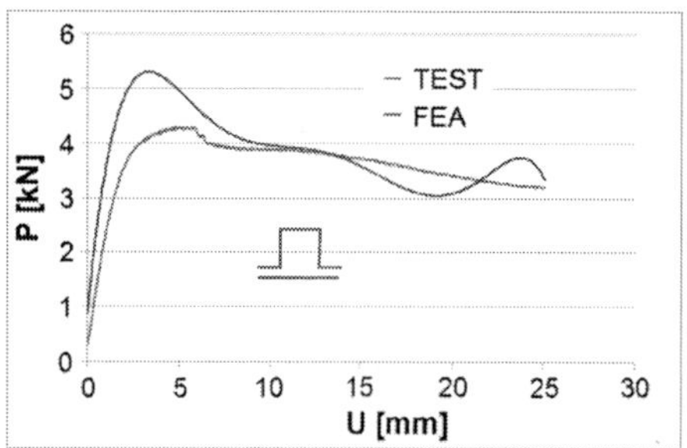

b) 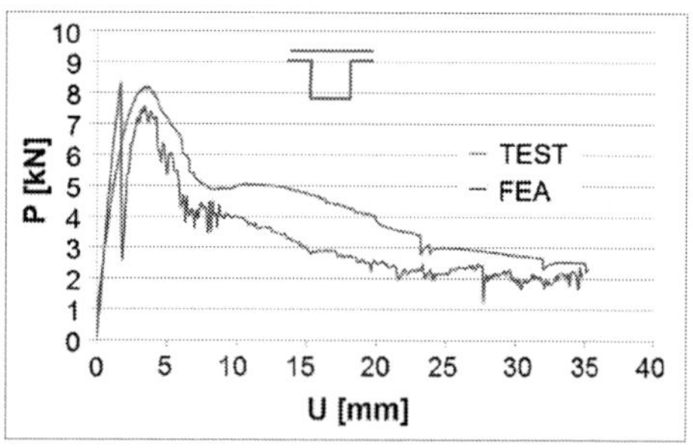

Figure 4. Force-displacement curves for: a) mode I and b) mode II deformations

One can notice that load capacity in mode II is approximately 2 times higher in comparison to mode I.

5. Numerical model

The FEA was performed using the Abaqus Explicite.

The joined metallic parts were modelled using 17 550 finite elements of the type S4R having 17969 nodes. For calculation we adopted assessed data from the experiments collected chapters 2 and 4. In particular the fracture energy for the considered material is equal to $G_f = 22.1$ mJ.

The adhesive layer was modelled as a surface-based cohesive behavior with damage. The adhesive material data were taken from [35], whereas the fracture energy parameters $G_c = 900$ J/m^2, [62].

The results of numerical analysis are shown in Fig. 5 for mode I and II deformation. One can notice that the deformed shape of the specimen in the end of deformation ($U_{max} = 36$ mm) is very similar to those obtained in experimental tests (figure 3). Figure 5 presents distribution of the reduced Huber – von Mises stress and the maximum values of the reduced stresses (red colour) were concentrated just under the loading point. Locally we have values equal to σ_{max}, i.e. damage initiation in the aluminum parts took place. One can also observe full degradation (plastic collapse) of the aliminium alloy – blue colour - i.e. visible discontinuities were created in the form of open spaces or cracks. This excellently corresponds to experimental results – Fig. 3. The adhesive layer is not subjected to delamination in the mode I deformation. However, in the mode II deformation a complete degradation of the bonding

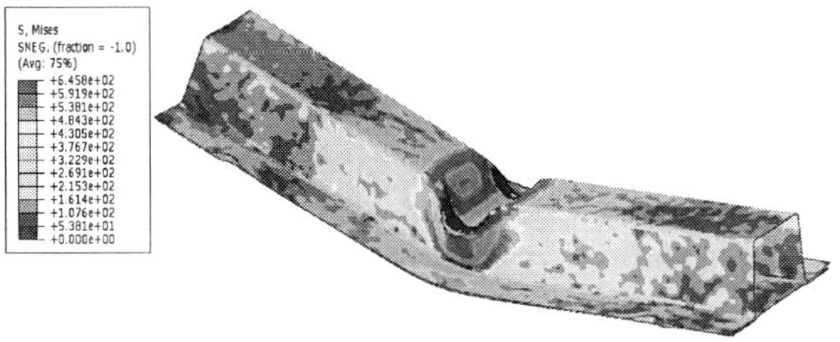

a)

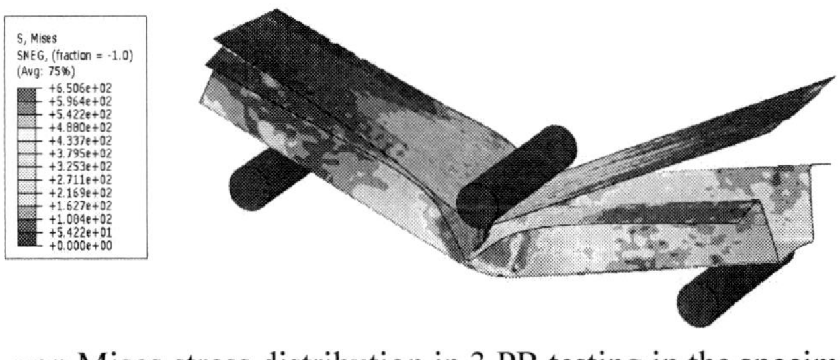

b)

Figure 5. Huber – von Mises stress distribution in 3 PB testing in the specimens bonded with Hysol 9514 after deflection $U_{max} =$ 36 mm for: a) mode I and b) mode II deformations

layer took place (figures 3b and 5b). Huber – von Mises stress concentration and plastic collapse occurs in the central cross section of the sample. After the destruction of cohesive layer and separation of the flat sheet from the omega profile, the load capacity decreases. Then the stiffness of the joined structural element decreases and it behaves as in [63].

Figure 4 presents comparison of all experimental and numerical force-deflection diagrams for the both modes deformation. All curves have similar shapes; however the load capacity is higher for the mode II deformation. The first part of diagrams are linear elastic and becomes slightly nonlinear up to the maximum force P_{max}. After exceeding this value the plastic collapse took place in the omega profile and initiation of the cohesive layer degradation is observed for mode II deformation, e.g. [34, 57, 58]. There is a sudden drops of the load capacity P and the further deformation of the joined structural element is associated with progressive: (1) cracking in the omega profile and (2) deterioration of the adhesive layer (in case of mode II). Discrepancy between experimental and the numerical analysis is caused by lack of the ideal geometrical shape of the samples and causes a local: (1) damage of the metallic profile or (2) degradation of the bonding layer.

6. Concluding remarks

The general conclusion from the above investigations is that newly formulated numerical model with good accuracy corresponds to experimental results obtained for the 2 modes of 3-PB deformation processes. The geometry after deformation and size of discontinuities after the maximum displacement U_{max} both in the experiments and numerical tests are very similar.

The particular conclusion can be formulated as follows:

- the higher maximal forces P_{max} are obtained for the second mode of deformation (the omega profile in the bottom position), figure 4,
- in mode I and II deformations (figures 3 and 5) the displacement distributions in the thin-walled aluminium adhesive joints are not uniform due to concentration of deformations in the central part of the samples and formation of the plastic hinge with local macroscopically visible cracks,
- just after the maximum force P_{max} the plastic collapse with cracking in the aluminium parts took place and we observed the sharp decline the load carrying force P of the joints to the residual value.
- the adhesive layers in the mode I deformation are not subjected to delamination. In the mode II deformation all specimens exhibit complete destruction of the adhesive layers at the maximum deflection $U_{max} = 36$ mm, i.e. degradation of the joint is much faster in this case.

Acknowledgment

This work was financially supported by the Ministry of Science and Higher Education (Poland) under statutory research project No S/20/2018.

References

[1] Adams R D, Comyn J and Wake W C 1997 Structural adhesives joints in engineering. *London, UK: Chapman and Hall*

[2] da Silva L F M and Ochsner A 2008 Modelling of adhesively bonded joints *Springer* https://doi.org/10.1007/978-3-540-79056-3

[3] da Silva L F M, Ochsner A and Adams R D 2011 Handbook of Adhesion Technology. *Springer* https://doi.org/10.1007/978-3-642-01169-6

[4] da Silva L F M, das Neves P J C, Adams R D and Spelt J K 2009 *Int. J. Adhes. & Adhes* **29** 319 https://doi.org/10.1016/j.ijadhadh.2008.06.005

[5] da Silva L F M, das Neves P J C, Adams R D and Spelt J K 2009 *Int. J. Adhes. & Adhes* **29** 331 https://doi.org/10.1016/j.ijadhadh.2008.06.007

[6] da Silva L F M, Raul D and Campilho S G 2011 Advances in Numerical Modelling of Adhesive Joints *Springer-Verlag Berlin Heidelberg* https://doi.org/10.1007/978-3-642-23608-2

[7] Sadowski T, Kneć M and Golewski P 2010 *Int. J. Adhesion and Adhesives* **30** 338 https://doi.org/10.1016/j.ijadhadh.2009.11.004

[8] Pirondi A and Moroni F 2011 *Advanced Structured Materials* **6** 109 https://doi.org/10.1007/8611_2010_36

[9] Sadowski T, Golewski P and Zarzeka-Raczkowska E 2011 *Comput. Mater. Sci.* **50** 1256 https://doi.org/10.1016/j.commatsci.2010.06.022

[10] Balawender T, Sadowski T and Golewski P 2011 *J. Adh. Sci. Technol.* **25** 2391 https://doi.org/10.1163/016942411X580108

[11] Balawender T, Sadowski T and Golewski P 2012 *Comput. Mater. Sci.* **64** 270 https://doi.org/10.1016/j.commatsci.2012.05.014

[12] Sadowski T and Golewski P 2013 *Arch. Metall. Mater.* **58** 579

[13] Sadowski T, Balawender T, Śliwa R, Golewski P and Kneć M 2013 *Arch. Metall. Mater.* **58**

163 https://doi.org/10.2478/v10172-012-0168-3
[14] Sadowski T, Kneć M and Golewski P 2014 *J. Adhesion* **90** 346 https://doi.org/10.1080/00218464.2013.766599
[15] Ozel A, Yazici B, Akpinar S, Aydin MD and Temiz S 2014 *Comp. Part B* **62** 167 https://doi.org/10.1016/j.compositesb.2014.03.001
[16] Gultekin K, Akpinar S and Ozel A 2014 *Comp. Part B* **60** 736 https://doi.org/10.1016/j.compositesb.2014.01.022
[17] da Silva L F M, Carbas R J C, Critchlow G W, Figueiredo M A V and Brown K 2009 *Int. J. Adhesion & Adhesives* **29** 621 https://doi.org/10.1016/j.ijadhadh.2009.02.012
[18] Lang TP and Mallick PK 1998 *Int. J. Adhesion & Adhesives* **18** 167 https://doi.org/10.1016/S0143-7496(97)00056-0
[19] Quaresimin M and Ricotta M 2006 *Comp. Sci. Technol.* **66** 176 https://doi.org/10.1016/j.compscitech.2005.04.026
[20] Kim S K, Yoo J S, Yi Y M and Kim CG. 2006 *Comp. Struct.* **72** 477 https://doi.org/10.1016/j.compstruct.2005.01.023
[21] Belingardi T, Goglio L and Tarditi A 2002 *Int. J. Adhesion & Adhesives* **22** 273 https://doi.org/10.1016/S0143-7496(02)00004-0
[22] Deng J and Lee M M K 2008 *Comp. Part B* **39** 731 https://doi.org/10.1016/j.compositesb.2007.05.004
[23] Akpinar S, Doru M O, Özel A, Aydin M D and Jahanpasand H G 2013 *Comp. Part B* **55** 55 https://doi.org/10.1016/j.compositesb.2013.05.056
[24] Stein N, Mardani H and Becker W 2016 *Int. J. Adhesion & Adhesives* **70** 117 https://doi.org/10.1016/j.ijadhadh.2016.06.001
[25] Adams R D, Atkins R W, Harris J A and Kinloch A J. 1986 *J. Adhesion* **20** 29
[26] Sancatar E and Nirantar P 2003 *J. Adhesion Sci. Technol.* **17** 655 https://doi.org/10.1163/156856103321340796
[27] Sancatar E and Simmons S R 2000 *J. Adhesion Sci. Technol.* **14** 1363 https://doi.org/10.1163/156856100742258
[28] Ganesh V K and Choo T S. 2002 *J. Comp. Mater.* **36** 1757 https://doi.org/10.1177/0021998302036014172
[29] da Silva L F M and Adams R D 2007 *Int. J. Adhesion & Adhesives* **27** 227 https://doi.org/10.1016/j.ijadhadh.2006.04.001
[30] You M, Yan Z M, Zheng X L and Yu H Z 2007 *Int. J. Adhesion & Adhesives* **28** 71 https://doi.org/10.1016/j.ijadhadh.2007.06.001
[31] Vallee T and Keller T 2006 *Comp. Part B* **37** 328 https://doi.org/10.1016/j.compositesb.2005.11.002
[32] Golewski P and Sadowski T 2017 *Int. J. Adhesion & Adhesives* **77** 174 https://doi.org/10.1016/j.ijadhadh.2017.04.007
[33] Akpinar S 2014 *Comp. Part B* **67** 170 https://doi.org/10.1016/j.compositesb.2014.06.023
[34] Belingardi G and Scattina A 2013 *Int. J. Adhesion & Adhesives* **40** 31 https://doi.org/10.1016/j.ijadhadh.2012.08.002
[35] Technical Data Sheet Hysol Hysol 9514, November 2012
[36] Hooputra H, Gese H, Dell H and Werner H 2004 *Int. J. Crashworthiness* **9** 449 https://doi.org/10.1533/ijcr.2004.0289
[37] Postek E and Sadowski T 2011 *Comp. Interf.* **18** 57 https://doi.org/10.1163/092764410X554049
[38] Birsan M, Sadowski T, Marsavina L, Pietras D. 2013 *Int. J. Solids Struct.* **50** 519 https://doi.org/10.1016/j.ijsolstr.2012.10.011
[39] Sadowski T, Kneć M and Golewski P 2014 *Comp. Struct.* **112** 66 https://doi.org/10.1016/j.compstruct.2014.01.008
[40] Postek E and Sadowski T 2018 *Comp. Struct.* **203** 498 https://doi.org/10.1016/j.compstruct.2017.10.092

IOP Conf. Series: Materials Science and Engineering **613** (2019) 012052 doi:10.1088/1757-899X/613/1/012052

[41] Postek E and Sadowski T 2018 *Comp. Struct.* **194** 494 https://doi.org/10.1016/j.compstruct.2018.04.014
[42] Postek E and Sadowski T 2018 *Int. J. Refaract. Metals Hard Mater.* **77** 6 https://doi.org/10.1016/j.ijrmhm.2018.07.009
[43] Lemaitre J and Chaboche J 1978 *J. Mecanique Applique* **2** 317 http://worldcat.org/issn/03990842
[44] Kachanov L M 1986 Introduction to Continuum Damage Mechanics. *Dordrecht: Martinun Nijhoff Publishers* https://doi.org/10.1007/978-94-017-1957-5
[45] Sadowski T and Marsavina L 2011 *Comput. Mater. Sci.* **50** 1336 https://doi.org/10.1016/j.commatsci.2010.04.011
[46] Sadowski T and Samborski S 2008 *Comput. Mater. Sci.* **43** 75 https://doi.org/10.1016/j.commatsci.2007.07.041
[47] Sadowski T, Marsavina L, Peride N and Craciun EM 2009 *Comput. Mater. Sci.* **46** 687 https://doi.org/10.1016/j.commatsci.2009.06.006
[48] Gajewski J and Sadowski T 2014 *Comput. Mater. Sci.* **82** 114 https://doi.org/10.1016/j.commatsci.2013.09.025
[49] Marsavina L, Linul E, Constantinescu D, Apostol D, Voiconi T and Sadowski T 2014 *Eng. Fract. Mech.* **129** 54 https://doi.org/10.1016/j.engfracmech.2013.12.006
[50] Sadowski T. and Neubrand A 2004 *Int. J. Fract.* **127** L135 https://doi.org/10.1023/B:FRAC.0000035087.34082.88
[51] Sadowski T and Bec J 2011 *Comput. Mat. Sci.* **50** 1269 https://doi.org/10.1016/j.commatsci.2010.04.014
[52] Sadowski T 2012 *Comput. Mat. Sci.* **64** 209 https://doi.org/10.1016/j.commatsci.2012.01.034
[53] Pirondi A and Moroni F 2009 *J. Adh. Sci. Technol.* **23** 1547 https://doi.org/10.1163/156856109X433063
[54] Sadowski T and Golewski P 2012 *Comput. Mater. Sci.* **52** 293 https://doi.org/10.1016/j.commatsci.2011.02.027
[55] Sadowski T and Golewski P 2012 *Comput. Mater. Sci.* **64** 285 https://doi.org/10.1016/j.commatsci.2012.02.048
[56] Dębski H and Sadowski T 2014 *Comput. Mat. Sci.* **83** 403 https://doi.org/10.1016/j.commatsci.2013.11.045
[57] Golewski G and Sadowski T 2014 *Construct. Building Mater.* **51** 207 https://doi.org/10.1016/j.conbuildmat.2013.10.044
[58] Ivanov I, Sadowski T and D.Pietras 2013 *Europ. Phys. J.- Special Topics* **222** 1587 https://doi.org/10.1140/epjst/e2013-01947-3
[59] Rusiński E, Kopczyński A and Czmochowski J 2004 *J. Mater. Proc. Technol.* **157** 405 https://doi.org/10.1016/j.jmatprotec.2004.09.063
[60] Kotełko M 2004 *Thin-Walled Struct.* **42** 153 https://doi.org/10.1016/S0263-8231(03)00055-7
[61] Kotelko M, Lim TH and Rhodes J 2000 *Thin-Walled Struct.* **38** 179 https://doi.org/10.1016/S0263-8231(00)00032-X
[62] Pirondi A, Fersini D, Perotti E and Moroni F 2007 *Atti del Congresso IGF* **19** 85
[63] Pastor M M and Roure F 2008 *Thin-Walled Struct.* **46** 476 https://doi.org/10.1016/j.tws.2007.10.013

IOP Conf. Series: Materials Science and Engineering **613** (2019) 012053 doi:10.1088/1757-899X/613/1/012053

Examination of the utilization of aluminum dross in road construction

Bella Udvardi[1], Róbert Géber[1], István Kocserha[1]

[1]Institute of Ceramics and Polymer Engineering, University of Miskolc, Hungary

E-mail: ubella07@gmail.com, robert.geber@uni-miskolc.hu, istvan.kocserha@uni-miskolc.hu

Abstract. Aluminum dross is a recyclable by-product of the aluminum manufacturing process. Since the use of aluminum alloys increased in large quantities, which cause a significant amount of aluminum dross produced, therefore, the need for alternative solutions are also increased. The goal of this study is to investigate the utilization possibility of dross as an alternative asphalt filler. During the research aluminum dross and limestone *(as reference material)*, which is well-known filler in asphalt technology, were investigated. Different tests were done on the samples: specific gravity, particle size distribution, BET specific surface area and Rigden voids were determined on the fillers. Scanning electron microscope was used to characterize the morphological features of fillers. To reveal the affinity of fillers to bitumen hydrophilic/hydrophobic properties and oil adsorption ability were also investigated. These examinations help to better to understand the interaction between fillers and binder material. With the use of fillers asphalt mixtures and specimens were also prepared for stiffness tests and asphalt cracking temperature tests. The results show a good correlation between dross and limestone fillers, therefore aluminum dross may be an alternative to partly substitute limestone filler.

1. Introduction

In this research work aluminum dross, which is a by-product of aluminum manufacturing process was investigated. Handling of this material causes a large problem not only in Hungary, but all over the world. Storaging in landfills is not a proper solution, therefore the industrial utilization of dross is environmentally and economically important. Road construction may be a potential area on utilizing, but it has been used in other areas [1].

Aluminum dross can be classified into two different types *(white and black dross)*, which are depend on the aluminum content. The white *(or non salt-containing)* dross contains more aluminum, than the black (*or salt containing*) dross [2, 3, 4, 5, 6, 7].

During the research limestone powder was used as a reference material, because of its current usage in asphalt technology. Different tests were made on the fillers in order to determine their physical properties: specific gravity, particle size distribution, BET specific surface area, Rigden voids, morphological test by SEM, and hydrophilic coefficient. By the use of fillers standard asphalt specimens were prepared, on which stiffness and asphalt cracking temperature were determined.

2. Experimental

2.1 Sample preparation

The as-received form of aluminum dross was a solid material, which needed to crush by hammer. Smaller pieces of dross was comminuted by planetary ball mill to. Limestone was in a bulk state as-received from the manufacturer. According to the standard *(ÚT 2-3.602 (e-UT 05.01.13):2008)* [8] the required particle size *(d<0.063 mm)* was prepared using standard sieves. Samples were dried for weight-constancy in a laboratory drying chamber.

2.2 Preparation of asphalt samples

Two types of standard asphalt samples (*Marshall specimens and beams*) were prepared for stiffness tests and low temperature cracking *(LTC)* tests, respectively. The following mineral aggregates and fillers were used in this study:

- fillers (*aluminum dross and limestone, d<0.063 mm*)
- dolomite 0/2,
- basalt 2/4,
- basalt 4/8,
- basalt 8/11.

The binder used in the mixtures was a standard 25/55-65 bitumen. A special additive material, called Viatop Premium was also used in the mixture. It is a natural fibre additive material, which provides fast and complete dispersion in the asphalt mix [9].

Marshall specimens were prepared *(Fig. 1.)* for stiffness test. Raw materials were dried on 180°C for 2 hours and then bitumen was heated up to 180°C. As bitumen is a viscoelastic material, and its rheological properties are important [10, 11], therefore it was necessary to reach this temperature. Raw materials and bitumen were mixed together by laboratory mixer. After mixing 1200 g of hot asphalt was poured into a preheated mold and was compacted by 50 blows in both sides.

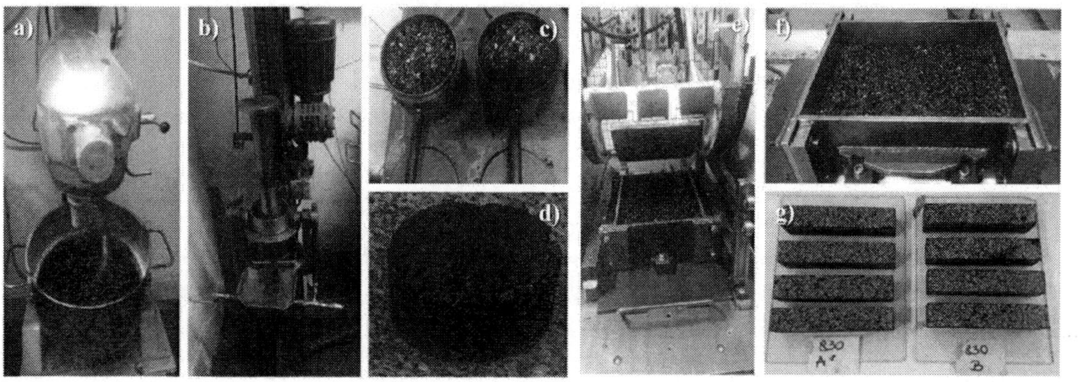

Figure 1. Preparing of Marshall specimens and asphalt beams
a) asphalt mixer, b) Marshall-compactor, c) compacted samples in the molds d) Marshall-specimen e) compactor apparatus f) asphalt specimen in mold g) cutted specimens for LTC test

Specimens for LTC tests were prepared using the following method: after mixing the hot asphalt mixture was poured into a steel mold and was compacted using a compactor apparatus. After cooling down, the required geometrical sizes of specimens were obtained by cutting.

2.3 Testing methods

Specific gravity of fillers was determined using a standard picnometer [12]. Particle size distribution was measured using Malvern Mastersizer X instrument using an air dispersion unit. For morphological tests Carl Zeiss EVO MA10 scanning electron microscope (*SEM*) was used. High resolution electron micrographs were taken in different magnifications for further observation. Specific surface area (*SSA*) of the powders (*by BET-method*) were determined by TriStar 3000 instrument (*manufactured by Micromeritics*). The method of this instrument based on physical absorption and capillary condensation, and it works in volumetric mode. The fine powder samples were prepared with the use of SMARTPREP instrument. Rigden Void was determined using a standard Rigden apparatus [13]. Hydrophilic coefficient of powders was determined by sedimentation method. 5 grams of fine materials were weighed and put in a 25 ml measuring cylinder. Parallel 10 ml of water and 10 ml of paraffin were poured into the cylinder. Powders and liquids were thoroughly mixed together then the cylinders were filled with more liquid and were closed by plugs. Volume of powders were determined after 72 hours of sedimentation.

3. Results and discussion

Table 1. summarizes the results of tests.

Table 1. The results of fillers

Parameter	Unit	Limestone	Aluminum dross
Specific gravity, ρ [7]	g/cm^3	2.82	2.90
Average particle size, d_{50}	μm	13.77	15.62
Specific surface area, SSA	m^2/g	1.55	1.26
Rigden Void, RV [8]	vol%	47.90	43.90
Hydrophilic coefficient, η	-	0.57	0.74

Values of the specific gravity shows that limestone has a lower density, than aluminum dross, but the difference is not significant. The reason of the difference is that dross contains elements with higher density *(for example Al, Mg)*. Particle size distribution is very important in asphalt technology, because the size and distribution of fine aggregates affects the stiffness of the asphalt pavement. Table 1. illustrates the average particle sizes of fillers. Based on it limestone has a 50% of the particles below 14 μm, while average particle size of dross is around 16 μm. During the research it was needed to determined the BET specific surface area, because the materials with high specific surface area are require bitumen surplus, which can reduce the strength of asphalt pavement. Specific surface area of the dross is lower, than the limestone. The reason of this is due to the particle size, because dross consists coarser particles. Rigden void test is used to evaluate the stiffening effect of fillers on asphalt mixtures. Table 1. clearly shows that limestone powder has a higher Rigden voids, than dross. The value of Rigden void depends on the particle size distribution, particles size, shape and specific gravity of filler. It was concluded, that those material, which has higher specific surface area, has a higher Rigden voids. According to the hydrophilic coefficient test both fillers are hydrophobic, therefore fillers can generate a good relation with the binder. During the research scanning electron micrographs were taken to observe the microstructure and the surface features of fillers. It is important that the raw material need convenient geometrical features, because it affects the function of mineral skeleton and bearing capacity of the asphalt pavement. Fig. 2. show the electronmicrographs, which were taken on the fillers.

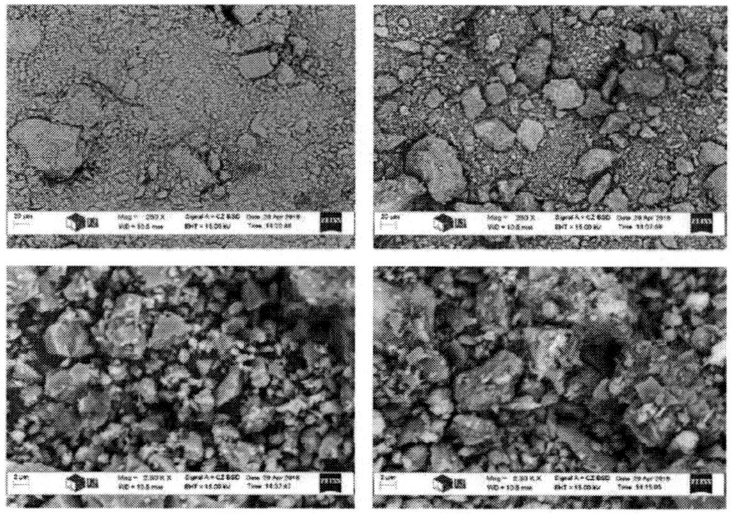

Figure 2. Morphological features of fillers

3.1 Indirect tensile test *(IT-CY)*

Research of asphalt technology focuses on the stiffness of asphalt as based on the sizing of the traffic load on asphalt pavements.

One of the most important properties of asphalt mixture is stiffness, which depends on the composition of mixture, temperature and loading time. The measurements were carried out according to the method described in MSZ EN 12697-26-2005 [10]. During the test one of the diameters of the sample is loaded, while transverse elongation of the specimen was measured. Stiffness modulus of the material is the function between the loading force and elongation [14, 15, 16, 17, 18]. During the research stiffness was determined at different temperatures *(5°C, 20°C and 50°C)*. Fig. 3. shows the results of stiffness tests.

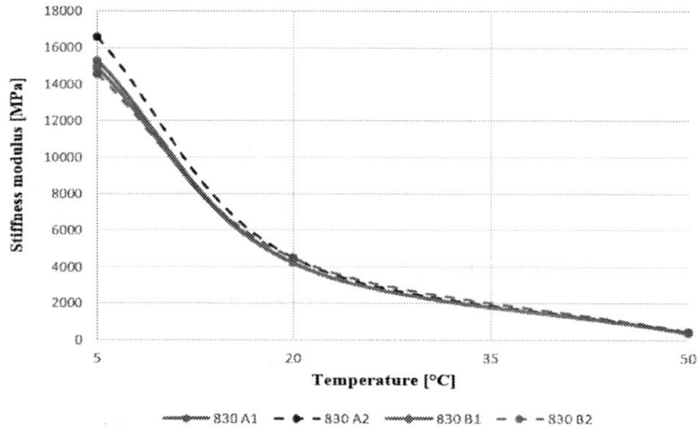

Figure 3. Result of stiffness test

Based on the results, it can be stated that with the increasing of temperature stiffness modulus are decreases. As the temperature rises, the difference between the two fillers is smaller.

3.2 Low temperature cracking *(LTC)*

In the late autumn, winter and early spring period, as a result of rapid cooling, thermal tensile stresses are generated in the asphalt pavement, whose values can achieve the tensile strength of the asphalt, so it can cause cracks in the asphalt layer. The cracks will appear first in the wearing course and as a results of water penetration and the reduction of the cross-section of the asphalt pavement it will spread into the lower layers of the asphalt. As a result, the bearing capacity of the asphalt pavement will reduced.

Both the cross and longitudinal cracks may appear on the wearing course, therefore mosaic-like ruptures and potholes will generate.

In this paper low temperature cracking values of asphalt specimens were determined. The formation of fracture is depend on the cooling rate. A uniform cooling rate was applied on beam specimens *(50x50x250 mm)*, which prevents the longitudinal shrinkage caused by the cooling. During the test, the longitudinal reduction was continuously measured, which always reinstate by the control pull, so it can prevent the shrinkage of the longitudinal dimension. Decrease of the temperature asphalt cracks were occured at a certain temperature, which is called low temperature cracking *(Fig. 4.)*[17, 18].

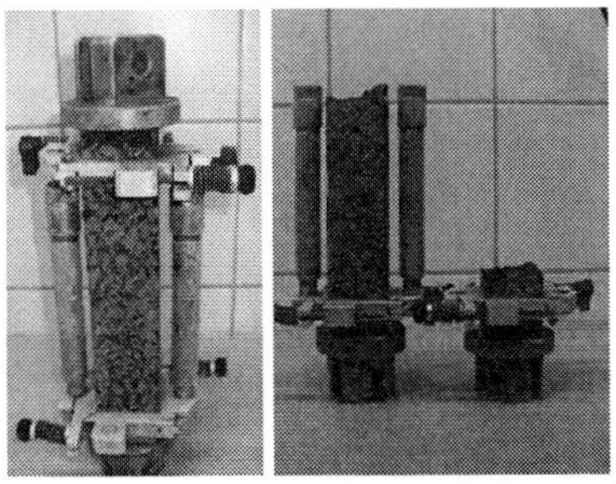

Figure 4. Cracking of the asphalt specimen

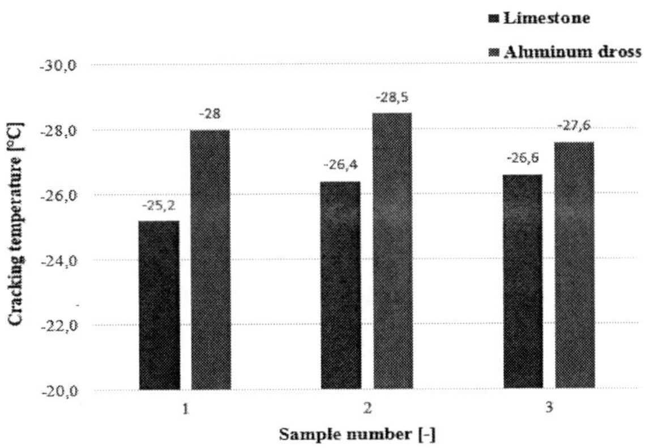

Figure 5. Results of LTC tests

The mixture, which contains limestone has an average cracking temperature at T=-26,1 °C, while the mixture contains aluminum dross has an average cracking temperature at T=-28°C. It can be stated that the mixture with aluminum dross has a better cracking property, therefore its behaviour in the lower temperature region is better, than the reference mixture.

4. Conclusion

Test results show that aluminum dross behaves very similar to limestone powder as a filler. The results of stiffness tests indicate, that stiffness of both asphalt mixtures decreases as temperature rises. Asphalt cracking temperature of the mixture made with aluminum dross is more resistant to low temperatures. It can be concluded, that dross may be a potential asphalt filler to *(partly)* substitute limestone powder. However, further investigations and more standard asphalt tests are needed to determine the exact amount of dross in asphalt mixtures.

Acknowledgement

The described work/article was carried out as part of the „Sustainable Raw Material Management Thematic Network – RING 2017", EFOP-3.6.2-16-2017-00010 project in the framework of the Széchenyi2020 Program. The realization of this project is supported by the European Union, co-financed by the European Social Fund.

IOP Conf. Series: Materials Science and Engineering **613** (2019) 012053 doi:10.1088/1757-899X/613/1/012053

Thanks to the employees of the Budapest University of Technology and Economics (*Department of Highway and Railway Engineering*) for their assistance in preparing asphalt specimens and measuring on the samples. The research results were presented in ic-cmtp5 Conference, in Miskolc-Lillafüred, October, 2018.

5. References

[1] H Mustafa, M Khater 2018 *Építőanyag-JSBCM* **70** (6) 186 https://doi.org/10.14382/epitoanyag-jsbcm.2018.33
[2] G J Kulik and J C Daley, 1990 *The Minerals, Metals & Materials Society-AIME* 427 https://pdfs.semanticscholar.org/c0fe/c7f460d79a5406c575a47032a3d4f024f455.pdf
[3] R D Peterson and L Newton, 2002 *TMS Annual Meeting* (Warrendale, Pennsylvania) 1029
[4] L S Carvalho, 1991 *III Seminário de Tecnologia da Indústria do alumínio* 169
[5] D S Gwinner, 1996 Environmental Issues in the Aluminum Reclamation Industry *ABAL (Associação Brasileira do Alumínio)* 40
[6] M S Reddy, D Neeraja, 2018 *Journal of Material Science* **43** 124 https://doi.org/10.1007/s12046-018-0866-2
[7] J Y Hwang, X Huang, Z Xu, 2006 *Journal of Minerals & Materials Characterization & Engineering* **5** (1) 47 https://doi.org/10.4236/jmmce.2006.51003
[8] ÚT 2-3.602 (e-UT 05.01.13):2008-*Minerals.* Aggregates for roads, airports and other public areas for asphalt mixtures and surface coating. Hungarian technical regulation for road construction. (in Hungarian)
[9] Cs Tóth, Z Soós 2015 *Építőanyag-JSBCM* **67** (4) 126 http://dx.doi.org/10.14382/epitoanyag-jsbcm.2015.20
[10] L A Gömze, L N Gömze, N S Kulkov, L I Shabalin, I Gotman, F Pedraza, G L Lecomte, T Mayorova, E Kurovics, A Hamza, 2015 *Építőanyag-JSBCM* **67** (4) 143 http://dx.doi.org/10.14382/epitoanyag-jsbcm.2015.24
[11] S Cho, Cs Tóth, Z Soós, 2018 *Építőanyag-JSBCM* **70** (6) 204 https://doi.org/10.14382/epitoanyag-jsbcm.2018.35
[12] MSZ EN 1097-7:2008-Tests for mechanical and physical properties of aggregates. Part 7: Determination of the particle density of filler. Pyknometer method. MSZ EN 1097-4:2000- Tests for mechanical and physical properties of aggregates. Part 4: Determination of the voids of dry compacted filler. *(Hungarian standard)*
[13] MSZ EN 12697-26:2005 – *Bituminous mixtures*. Test methods for hot mix asphalt. Part 26: Stiffness. *(Hungarian standard)*
[14] L Pethő, 2008 Influence of temperature distribution on the fatigue and mix design of asphalt pavement sturctures, Summary of Ph.D. thesis, *Budapest University of Technology and Economics* (in Hungarian)
[15] Cs Tóth 2013 *Journal of Road Affairs, Spring* **1** (in Hungarian) http://utugyilapok.hu/cikkek/hazai-aszfaltkeverekek-merevsegingadozasanak-tapasztalatai/
[16] K Török and I Pallós, 2015 *Journal of Road Affairs Spring* **5** (in Hungarian) http://utugyilapok.hu/cikkek/az-aszfaltok-teli-hidegviselkedeset-befolyasolo-anyagtulajdonsagok-laboratoriumi-vizsgalatai/
[17] Cs Tóth, 2010 Stiffness of asphalt mixtures depending on the load time, temperature and aggregates distribution, PhD dissertation, *Budapest University of Technology and Economics* (in Hungarian).
[18] K Almássy, 2010 Behaviour of asphalt roadway construction grids, PhD dissertation, *Budapest University of Technology and Economics* (in Hungarian)

IOP Conf. Series: Materials Science and Engineering **613** (2019) 012054 doi:10.1088/1757-899X/613/1/012054

Dependence of oxidation rate from structures in case of graphites and diamonds

J Lakatos[1], G Muránszky[1], P Rugóczky [1,2]

[1]Institute of Chemistry, Faculty of Material Science and Engineering, University of Miskolc, Miskolc-Egyetemváros, H-3515, Hungary
[2] Fraisa Hungária Kft, Sárospatak (Hungary)

E-mail: mtasotak@uni-miskolc.hu, gabor.muranszky@gmail.com, prugoczky@gmail.com

Abstract. The oxidation of graphite and diamonds in gas and solution phase was investigated to find how much the oxidation can indicate the structural differences. Natural and artificial high purity (spectral) graphite and natural and synthetic diamonds as super abrasive materials and nano diamond deposited to the surface of Mo road were tested and compared.

1. Introduction

The gas phase oxidation of graphite is well documented since it has great importance of the lifetime of this substance at the high temperature processes [1]. Contescu et al. [1] based on Clark et al. and Propp works (ref. 10-11 in [1]) stated that the graphite oxidation at low temperature depends on the material reactivity which in turn depends on the material's microstructure. Bom at al. [2] using thermogravimetric analysis (TGA) has demonstrated that multiwalled nanotubes (MWNTs) annealed at 2200 to 2800 °C are more air stable than as-produced MWNTs, diamond, graphite, and annealed diamond, since the thermal annealing removes the structural defects. The rate of oxidation in solution phase also was found has structural dependence [3, 4]. Milley [3] showed that the redox potential curves measured in coal-Cr(VI) system depends on the type of the tested coal. Investigation of Cr(VI) reduction capacity of different rank coals, Lakatos et al.[4] have found that the Cr(VI) reduction ability decreases in function of coal rank in the lignite to antracite sequence. Oxidative treatments often applied to functionalization of graphite [5-7] and diamonds [8,9]. These examples indicate that the rate of oxidation at both cases: gas and solution phases can provide structural information of the oxidised carbon materials. Based on this way we made attempts to get structural information from the oxidation data of different carbon structures: coals, chars, different graphites and diamonds.

2. Materials and methods

Oxidation test of the different carbons in air were carried out using thermogravimetric method at 5 °C/min heating rate at 10 l/h air flow, MOM Derivatograph D type TG analyser was used. The oxidation test in solution phase (Sulphuric acid-sodium-chromate solution) were carried out using batch methods. During the relatively long term oxidation (1-10 days) the samples was shaked up once a day. The Cr(VI) oxidation power highly depends on the pH of the solution which was set by the sulphuric acid concentration of the solution. Higher acid concentration will provide higher oxidation power as can be seen at the Fig. 1.

Two graphites one of them is artificial high purity spectral graphite and the other is a natural (source is Madagascar) and different diamond powders used as superabrassives sold by Lands [10], Dia Plus diamond powder [11] and chemical vapor deposited nano diamond powder collected from polished

Mo rod were tested. Some comparisons are given of these materials on Figs.2,3,4 using FTIR and Raman spectrums.

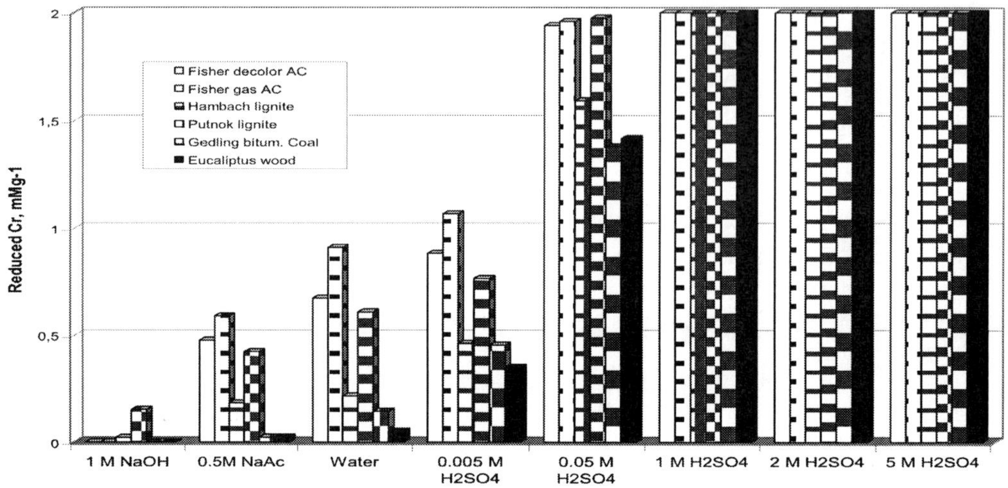

Figure 1. Oxidation of activated carbons (AC), lignites, bituminous coal and Eucaliptus wood saw dust by Cr(VI) at different pH. Solid Liqid ratio: 1:40, c_0(Cr(VI):50 mM , Contact time: 7 days

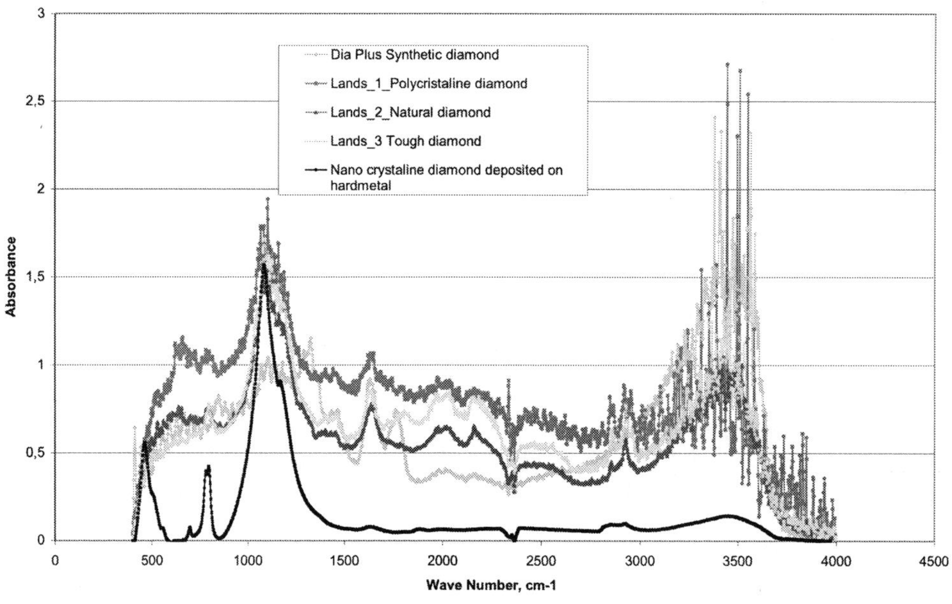

Figure 2. FTIR spectrums of tested diamonds. Brucker Vertex 70 transmission mode, 250 mg KBr/25 mg sample.

The FTIR spectrums indicate differences between the nano, the diaPlus and the Lands superabrassive diamonds. Only one characteristic peak can detect by Raman, but in case of the powders it superposed on a wide band. The typical sharp characteristic diamond peak could be detected only the large one crystal case. The Raman spectrums of the nano diamond produced by HFCDV method differ from the dia Plus or Lands diamond powders. It shows similarities to the spectrums of graphite and nanotube.

IOP Conf. Series: Materials Science and Engineering **613** (2019) 012054 doi:10.1088/1757-899X/613/1/012054

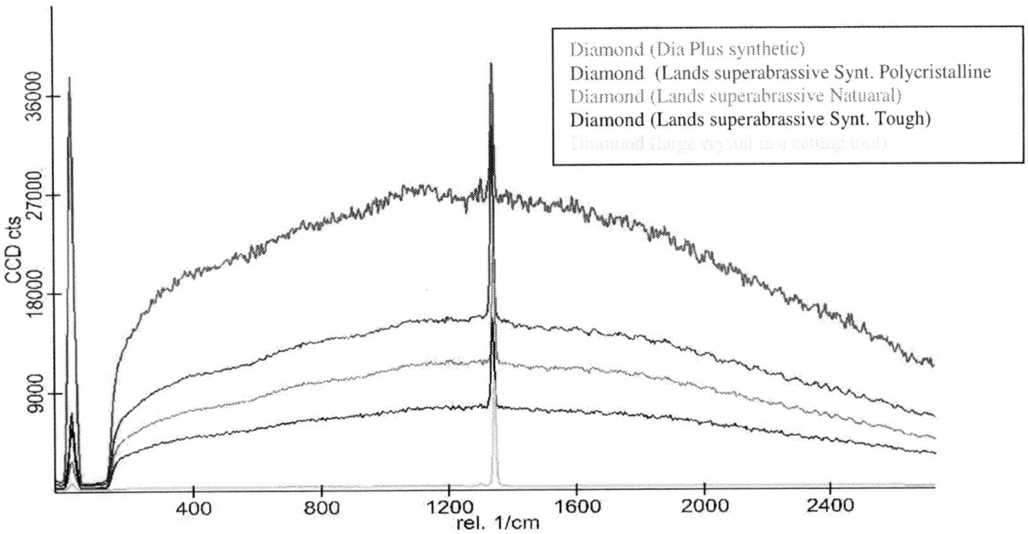

Figure 3. Raman spectrums of diamond powders and a large crystal diamond. Raman spectrometer Witec Alpha 300R. Leser wave length: 632,8 nm.

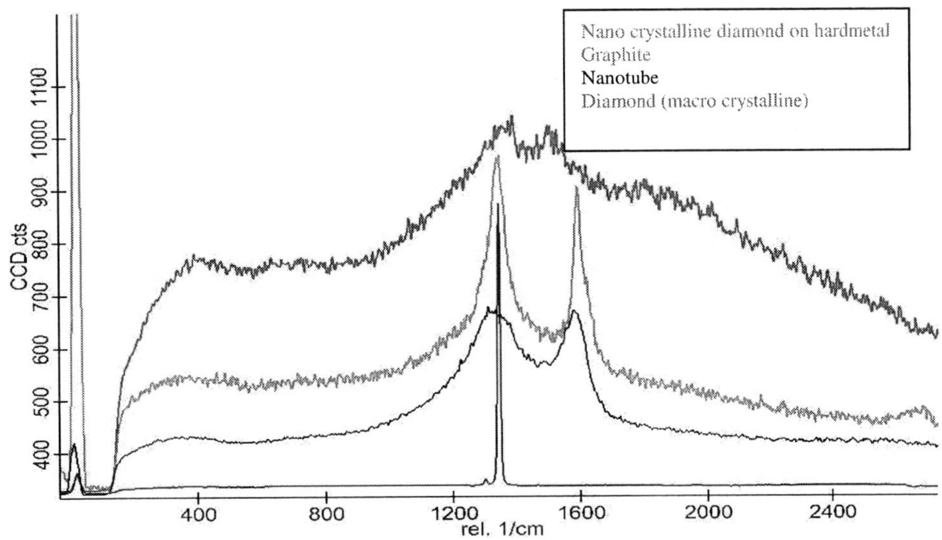

Figure 4. Raman spectrums of macro crystalline diamond, nano crystalline diamond, graphite and nanotube. Raman spectrometer Witec Alpha 300R, Leser wave length: 632,8 nm

3. Results and discussion

Oxidation of graphites in air was investigated by thermogravimetry. The mass decreasing in function of time were compared, Fig. 5. It can be seen at the Fig. 5. the curves of the artificial high purity graphite and the natural are very similar. The diamond ignition started aprox. 200 °C lower temperature than the graphite. It behaves similarly as the glassy carbon oxidation. The TG curves for air oxidation of diamonds are given on the Fig. 6. All of the diamond powders showed similar TG curves except the nano diamond deposited on the polished Mo surface by HFCVD method. The TG curve of nano diamond goes very close to the high purity graphite. The Raman spectrum also showed similarity, see Fig 4. The oxidation of carbons in solution phase were tested in 5 M sulphuric acid solutions. See Fig.7-8.

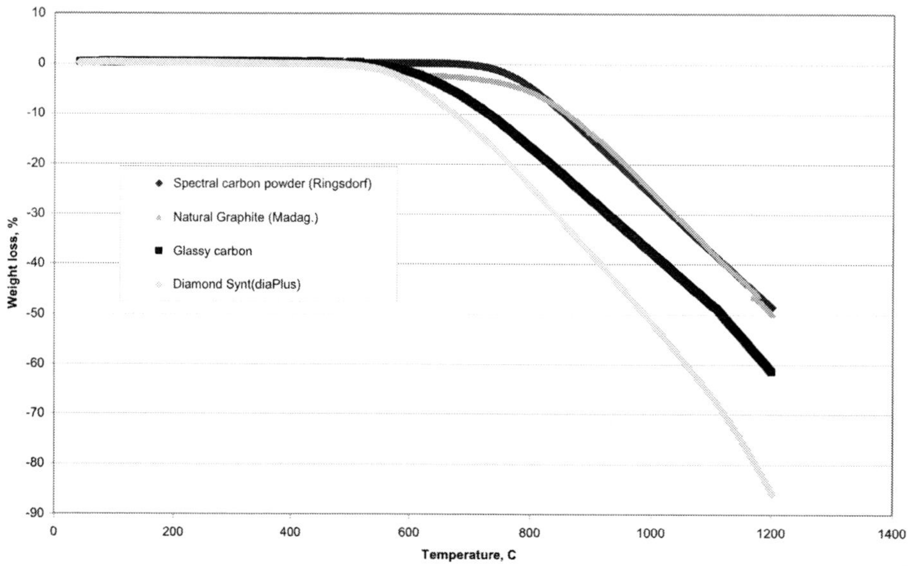

Figure 5. TG curves of air oxidation of different carbon substances.
Heating rate: 5 C/min, V air: 10 l/h

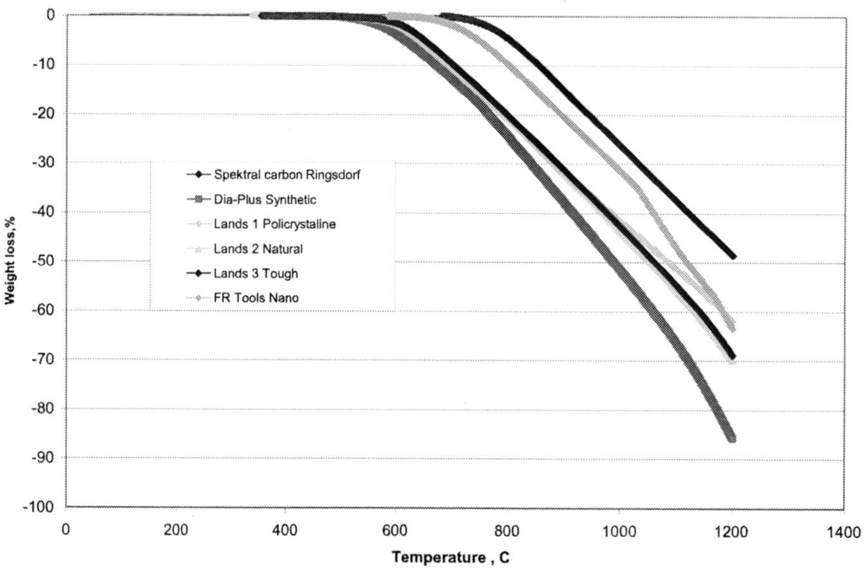

Figure 6. TG curves of air oxidation of different diamonds and graphite.
Heating rate: 5 C/min, V air: 10 l/h

On the Fig 7. the graphite the amorphous carbons coke, chars, activated carbons were compared. The less oxidizable structure was the graphite the highest the activated carbons.

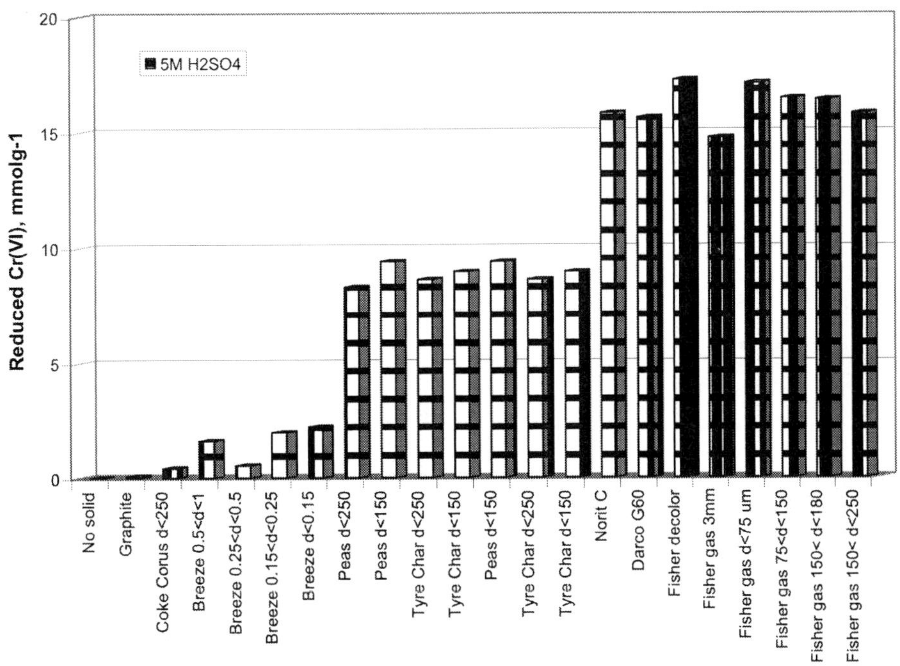

Figure 7. Cr(VI) oxidation of different type of graphite forms. (50 mM Cr(VI), 5 M H2SO4), 7 days, 1: 400 solid /liquid ratio)

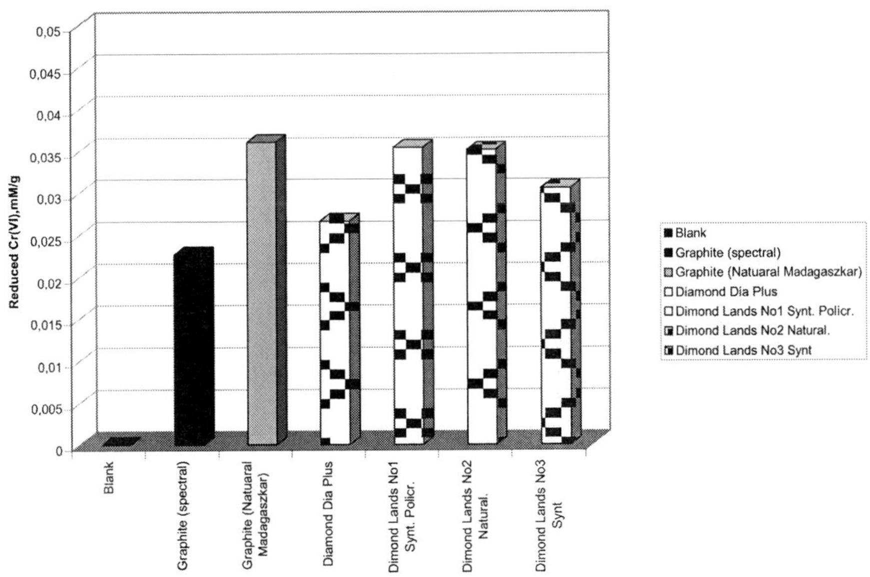

Figure 8. Cr(VI) oxidation of graphite and diamonds in 5 M H2SO4. 1 mM Cr(VI), Solid liquid ratio: 1:10

The artificial high purity graphite at room temperature less oxidizable with Cr(VI) than the natural or the diamonds, as showed on Fig 8. but the differences were not found be large. Higher rate of the oxidation can get at the boiling temperature of the solution, see Fig 9., but such high difference between the diamonds or between the graphite and diamond such was find in case of graphite and amorphous carbons could not get. Further adjustment of the conditions may improve the sensitivity to the structural differences exist between diamonds and diamonds or diamonds and graphites.

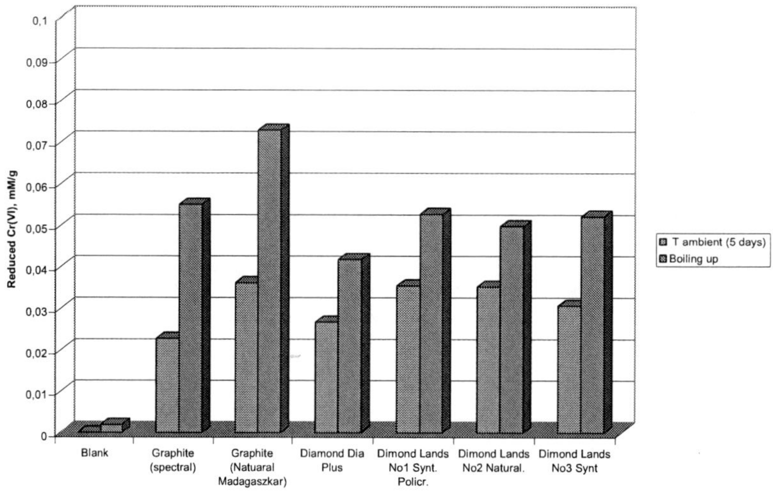

Figure 9. Cr(VI) oxidation of graphite and diamonds in 5 M H2SO4. 1 mM Cr(VI), Solid liquid ratio: 1:10.

4. Conclusions

Differences between the oxidation of carbon structures both cases in gas phase and in solution phase have been detected. Further works necessary to explain how much portion of this can attributed to the oxidation of the surface functionalities are present on the carbon materials and how much the destruction of the structure itself.

References

[1] C I Contescu, S Azad, D Miller, M J Lance, F S Baker, T D Burchell 2008 *Jurn. Nucl. Mat.* **381** 15 https://doi.org/10.1016/j.jnucmat.2008.07.020
[2] D Bom, R Andrews, D Jacques, J Anthony, B Chen, S Mark , M S Meier,. P Selegue 2002 *Nano Letters* **2** (6) 615 https://doi.org/10.1021/nl020297u
[3] J Milley J. 1988 *Fuel* **67** 143 No: 0016-2361/88/010143
[4] J Lakatos, S D. Brown, C E Snape 2002 *Fuel* **81** (5) 691 https://doi.org/6/s0016-2361(01)00159-4
[5] B Xiang , D Ling , F Gao , H Lou, H Gu , Z Guo 2016 *RSC Adv.* **6** 58354
[6] R Burgess, C Buono , P R Davies , R J Davies, T Legge, A Lai, A R Lewis ,D J Morgan, N Robinson,.D J Willock 2015 *Journal of Catalysis* **323** 10 http://dx.doi.org/10.1016/j.jcat.2014.12.021
[7] R Ljubisa, L R Radovic, C V Mora-Vilches, A J A Salgado-Casanova., A Buljan 2018 *Carbon* **130** 340 https://doi.org/10.1016/j.carbon.2017.12.112
[8] S Szunerits, A R Boukherroub 2008 *J. Solid State Electrochem.* **12** 1205 https://doi.org/10.1007/s10008-007-0473-3
[9] L Marko 2015 *Magy. Kém. Foly.* **111** (2) 46
[10] http://landssuperabrasives.com/
[11] https://www.walter-messner.de/EN/sites/EN_diamantpoliermittel.html

AUTHOR INDEX